Advances in sustainable dairy cattle nutrition

It is widely recognised that agriculture is a significant contributor to global warming and climate change. Agriculture needs to reduce its environmental impact and adapt to current climate change whilst still feeding a growing population, i.e. become more 'climate-smart'. Burleigh Dodds Science Publishing is playing its part in achieving this by bringing together key research on making the production of the world's most important crops and livestock products more sustainable.

Based on extensive research, our publications specifically target the challenge of climate-smart agriculture. In this way we are using 'smart publishing' to help achieve climate-smart agriculture.

Burleigh Dodds Science Publishing is an independent and innovative publisher delivering high quality customer-focused agricultural science content in both print and online formats for the academic and research communities. Our aim is to build a foundation of knowledge on which researchers can build to meet the challenge of climate-smart agriculture.

For more information about Burleigh Dodds Science Publishing simply call us on +44 (0) 1223 839365, email info@bdspublishing.com or alternatively please visit our website at www.bdspublishing.com.

Related titles:

Achieving sustainable production of milk Volume 2: Safety, quality and sustainability
Print (ISBN 978-1-78676-048-7); Online (ISBN 978-1-78676-050-0, 978-1-78676-051-7)

Achieving sustainable production of milk Volume 3: Dairy herd management and welfare
Print (ISBN 978-1-78676-052-4); Online (ISBN 978-1-78676-054-8, 978-1-78676-055-5)

Climate change and agriculture
Print (ISBN 978-1-78676-320-4); Online (ISBN 978-1-78676-322-8, 978-1-78676-323-5)

Reducing greenhouse gas emissions from livestock production
Print (ISBN 978-1-78676-439-3); Online (ISBN 978-1-78676-441-6, 978-1-78676-442-3)

Developing animal feed products
Print (ISBN 978-1-78676-463-8); Online (ISBN 978-1-78676-465-2, 978-1-78676-466-9)

Seaweed and microalgae as alternative sources of protein
Print (ISBN 978-1-78676-620-5); Online (ISBN 978-1-78676-622-9, 978-1-78676-623-6)

Chapters are available individually from our online bookshop:
https://shop.bdspublishing.com

climate•SMART•publishing
IN AGRICULTURAL SCIENCE

BURLEIGH DODDS SERIES IN AGRICULTURAL SCIENCE
NUMBER 133

Advances in sustainable dairy cattle nutrition

Edited by Professor Alexander N. Hristov, The Pennsylvania State University, USA

burleigh dodds
SCIENCE PUBLISHING

Published by Burleigh Dodds Science Publishing Limited
82 High Street, Sawston, Cambridge CB22 3HJ, UK
www.bdspublishing.com

Burleigh Dodds Science Publishing, 1518 Walnut Street, Suite 900, Philadelphia, PA 19102-3406, USA

First published 2023 by Burleigh Dodds Science Publishing Limited

Library of Congress Control Number: 2022949220

British Library Cataloguing in Publication Data
A catalogue record for this book is available from the British Library

ISBN 978-1-80146-205-1 (print)
ISBN 978-1-80146-208-2 (PDF)
ISBN 978-1-80146-207-5 (ePub)
ISSN 2059-6936 (print)
ISSN 2059-6944 (online)

DOI 10.19103/AS.2022.0117

Typeset by Deanta Global Publishing Services, Dublin, Ireland

Contents

Series list

Title	Series number
Achieving sustainable cultivation of maize - Vol 1 From improved varieties to local applications Edited by: Dr Dave Watson, CGIAR Maize Research Program Manager, CIMMYT, Mexico	001
Achieving sustainable cultivation of maize - Vol 2 Cultivation techniques, pest and disease control Edited by: Dr Dave Watson, CGIAR Maize Research Program Manager, CIMMYT, Mexico	002
Achieving sustainable cultivation of rice - Vol 1 Breeding for higher yield and quality Edited by: Prof. Takuji Sasaki, Tokyo University of Agriculture, Japan	003
Achieving sustainable cultivation of rice - Vol 2 Cultivation, pest and disease management Edited by: Prof. Takuji Sasaki, Tokyo University of Agriculture, Japan	004
Achieving sustainable cultivation of wheat - Vol 1 Breeding, quality traits, pests and diseases Edited by: Prof. Peter Langridge, The University of Adelaide, Australia	005
Achieving sustainable cultivation of wheat - Vol 2 Cultivation techniques Edited by: Prof. Peter Langridge, The University of Adelaide, Australia	006
Achieving sustainable cultivation of tomatoes Edited by: Dr Autar Mattoo, USDA-ARS, USA and Prof. Avtar Handa, Purdue University, USA	007
Achieving sustainable production of milk - Vol 1 Milk composition, genetics and breeding Edited by: Dr Nico van Belzen, International Dairy Federation (IDF), Belgium	008
Achieving sustainable production of milk - Vol 2 Safety, quality and sustainability Edited by: Dr Nico van Belzen, International Dairy Federation (IDF), Belgium	009
Achieving sustainable production of milk - Vol 3 Dairy herd management and welfare Edited by: Prof. John Webster, University of Bristol, UK	010
Ensuring safety and quality in the production of beef - Vol 1 Safety Edited by: Prof. Gary Acuff, Texas A&M University, USA and Prof. James Dickson, Iowa State University, USA	011
Ensuring safety and quality in the production of beef - Vol 2 Quality Edited by: Prof. Michael Dikeman, Kansas State University, USA	012
Achieving sustainable production of poultry meat - Vol 1 Safety, quality and sustainability Edited by: Prof. Steven C. Ricke, University of Arkansas, USA	013
Achieving sustainable production of poultry meat - Vol 2 Breeding and nutrition Edited by: Prof. Todd Applegate, University of Georgia, USA	014

Introduction

Global demand for milk and other dairy products continues to grow. As a result of this increased consumer demand, the sector's greenhouse gas emissions are expected to rise. It is therefore crucial that the livestock sector reacts to these developments and considers areas for improvement that could reduce the carbon footprint of milk and the sector's contribution to climate change.

This volume provides a comprehensive review of the wealth of research on recent advances in understanding and improving dairy cattle nutrition to reduce the carbon footprint of the dairy sector. The chapters are split into four parts: Part 1 chapters focus on the role of nutritional requirements in optimising gut function and overall animal health, specifically drawing attention to carbohydrates, protein and lipids. Chapters in Part 2 discuss the use of plant extracts, such as essential oils, tannins and saponins, as dietary supplements to optimise dairy cattle nutrition. Part 3 chapters review the use of direct-fed microbials and other supplements to optimise nutrition, focusing specifically on the use of probiotics, exogenous enzymes and amino acids and how feed additives can be used to mitigate enteric methane emission. Chapters in Part 4 draw attention to assessing agricultural co-products as alternative sources of fibre and protein.

Part 1 Nutritional requirements

The first chapter of the book reviews advances in understanding carbohydrate requirements and utilisation in dairy cattle. Chapter 1 begins by first describing the importance of feed analysis, focusing specifically on fibre analysis. It then moves on to discuss detergent fibres, which is then followed by an analysis of neutral detergent-soluble carbohydrates. A section on rumen carbohydrate and protein interactions is also provided, which is then followed by sections on rumen health and the importance of higher forage diets. The chapter also provides case studies that give examples of using higher forage rations in dairy cows. Current on-farm analysis tools are also described, which is then followed by a section on the use of nutrition models.

Chapter 2 draws attention to advances in understanding protein requirements and utilisation in dairy cattle. The chapter begins by first describing how to determine the supply of metabolisable protein and amino acids and how these are crucial in determining essential amino acid requirements in dairy cattle. It then moves on to discuss post-absorptive utilisation of metabolisable protein and amino acids, focusing on net protein and amino acid demand, as well as the requirements of metabolisable protein and amino acids. A case study on the estimation of protein and essential amino acid requirements is also included in the chapter.

The final chapter of Part 1 examines advances in understanding lipid requirements and utilisation in dairy cattle. Chapter 3 begins by first highlighting the major lipid constituents found in dairy cow nutrition, such as free fatty acids, triglycerides, glycolipids and phospholipids. It then goes on to discuss lipid supplements, which is followed by an examination of fat digestion and absorption in ruminant species. Sections on digestibility of dietary lipids and mammary metabolism of lipids are also provided. The chapter discusses the effects of lipids on feed intake as well as the effects of lipid supplements on milk production on composition. This is followed by an overview of how dietary lipids can be used to modify milk fat for human health, as well as the effects of lipids on reproduction. The role of dietary lipids in metabolism regulation is also reviewed, followed by an analysis of the important considerations that should be taken into account when selecting commercial inert fats. Practical recommendations for supplementing fats are also included.

Part 2 The use of dietary supplements to optimise dairy cattle nutrition: plant extracts

Part 2 opens with a chapter that examines the use of plant extracts as dietary supplements in dairy cow nutrition, specifically focusing on essential oils. Chapter 4 begins by reviewing the ruminal effects of essential oils, drawing specific attention to their mechanisms of action, how targets are selected to modulate ruminal fermentation to improve efficiency and the effects that essential oils have on fermentation both *in vitro* and *in vivo*. The chapter also discusses the post ruminal effects of essential oils, their impact on dairy cattle performance and the current problems farmers are facing in the use of essential oils in dairy cattle feeding. A case study using *in vitro* batch fermentation studies is also included.

The next chapter focuses on the use of tannins as dietary supplements in dairy cattle nutrition. Chapter 5 first identifies sources of tannins for diet supplementation, focusing specifically on the tropics and temperate climates. It then moves on to discuss the effects of tannins in dairy cows, such as the effects on palatability, digestion and performance, the use of tannins to mitigate emissions such as nitrogen and methane as well as their effects on milk quality. The chapter also draws attention to the anthelmintic properties of tannins before moving on to discuss the various factors that can influence the effects of tannins in farm animals. A section on the onset of tannin effects and long-term efficiency is also provided, which is then followed by an overview of the strategies than can be used to maximise beneficial tannin effects.

Chapter 6 looks at the use of saponins as dietary supplements in dairy cattle nutrition. It begins by identifying their sources and goes on to describe the chemistry, biosynthesis pathway and their antimicrobial properties. A section on the effects of saponins on rumen microbial populations is also included. The chapter moves on to review enteric methane production, which is followed by a discussion on nitrogen metabolism and how saponins can help reduce the production of ammonia as well as improve the synthesis of microbial proteins in the rumen. The chapter also discusses how the productivity of dairy cattle can be improved with saponins and highlights some of the commonly known toxic effects of saponins as well.

Part 3 The use of direct-fed microbials (DFM) and other supplements to optimise nutrition

The first chapter of Part 3 discusses the use of bacterial probiotics and direct-fed microbials (DFM) as dietary supplements in dairy cow nutrition. Chapter 7 first describes the commonly used DFM in dairy cattle and their modes of action in the rumen and the intestines. It moves on to discuss the effects of DFMs on the rumen and intestinal microbiomes as well as their functions, which is then followed by an overview of their effects on feed utilisation and lactation performance. A section on the effects of DFMs on the immune system and health of calves and lactating cows is also included, drawing specific attention to enteric pathogens and calf diarrhoea, the immunobiotic properties of DFMs and how their supplementation can enhance immune response to diseases such as mastitis and rumen acidosis. The chapter provides an overview of the benefits in the mitigation of methane emissions as well.

Chapter 8 reviews the use of exogenous enzymes as dietary supplements in dairy cattle nutrition. It starts by first describing exogenous fibrolytic enzymes, focusing specifically on their activities, mechanisms of action and their application and effects on dairy cow performance. This is then followed by an analysis of exogenous amylases which draws specific attention to amylolytic enzyme activity assay and the use of exogenous amylases in ruminants. A section on exogenous proteases is also included, discussing their mode of action, their use in ruminant species as well as the methods used for measuring proteolytic enzyme activity. The chapter also describes the use of fungal enzymes in dairy cattle nutrition. The role of metagenomics in understanding the effects of dietary manipulations on the rumen microbiome is also reviewed, followed by the potential use of expansin-like proteins in dairy cattle nutrition.

The subject of Chapter 9 is amino acids in dairy nutrition and how they can be used to enhance milk protein synthesis and beyond. It begins by providing some general background to amino acids and dairy nutrition, then goes on to discuss the reduction of dietary crude protein in dairy cow diets. It reviews protein feeding and amino acid balancing as well as the physiological roles of essential amino acids beyond milk protein. A section on methionine, lysine and other essential and non-essential amino acids is included.

Part 3 closes with a chapter that focuses on the use of supplements to mitigate enteric methane emission in dairy cattle. Chapter 10 begins by highlighting two dietary approaches that can decrease enteric methane (CH_4) emission, the first being the improvement of forage quality and the second the inclusion of concentrate feeds in the diet. The use of dietary lipids and lipid supplements is also discussed, followed by a review of the use of alternative electron sinks, plant-based bioactive compounds and chemical inhibitors. A section on the growing interest of macroalgae use in ruminants is also provided. The chapter also discusses the additive effect of feed supplements as well as touches on the economics of methane mitigation.

Part 4 Assessing alternative feed sources from agricultural co-products

Part 4 begins with a chapter that focuses on assessing alternative fibre sources from by-products. Chapter 11 first highlights the various plant-based by-product feeds that are currently available for use in dairy cattle and draws attention to those commonly known as non-forage fibre sources. The chapter also discusses their influence on animal health, production and efficiency and provides a case study that investigates the effect of forage physically effective neutral detergent fibre levels in diets on lactating dairy cows, focusing specifically on the use of soybean hulls or cottonseed as non-forage fibre sources.

The final chapter of the book draws attention to assessing alternative protein sources from by-products. Chapter 12 first examines the quality and digestibility of protein by drawing specific attention to rumen degradability of protein in feeds. This is then followed by an overview of ethanol production and its associated by-products via methods such as dry-milling, wet-milling, fractionation and pre- and post-fermented fractionation. A section on oilseed by-products is also provided, focusing specifically on soybean meal, canola meal, cottonseed meal, sunflower meal and safflower meal. Rendered protein meals are also discussed, such as blood meal and feather meal and the chapter reviews the environmental impact of feeding protein by-products.

Part 1

Nutritional requirements

Chapter 1

Advances in understanding carbohydrate requirements and utilization in dairy cattle

L. E. Chase, Cornell University, USA; and M. B. Hall, U.S. Dairy Forage Research Center - USDA-ARS, USA

1 Introduction

Carbohydrates comprise 65%–75% of the total dry matter (DM) in dairy cattle diets, providing the major source of energy for both the rumen microorganisms and the cow. Carbohydrates in forages and other feeds are a complex mixture of monosaccharides bound in different linkages and proportions into diverse mono-, oligo-, and polysaccharides. In terms of functions, carbohydrates are unique in that they serve both nutritional and physical needs. In providing energy, rumen fermentation of carbohydrates supports the growth of rumen microorganisms and production of short-chain fatty acids (SCFAs). The latter are a key energy source for the cow in addition to digestible starch from the

http://dx.doi.org/10.19103/AS.2022.0117.01

Table 1 Carbohydrate terms and definitions

Term	Definition	Description
ADF	Acid detergent (AD) fiber	Cellulose, lignin, AD crude protein, and AD ash
ADL	Acid detergent lignin	Lignin
ESC	Ethanol-soluble carbohydrate	Simple sugars (e.g. glucose and fructose), sucrose, short-chain carbohydrates (e.g. stachyose, raffinose, short-chain fructans, and oligosaccharides); soluble in 80% ethanol and does not include lactose
NDF	Neutral detergent (ND) fiber	ADF, hemicellulose, and ND - AD difference in crude protein and ash
aNDF		NDF analyzed with heat-stable amylase
aNDFom		aNDF - ash contamination of the aNDF residue; aNDF on an ash-free basis
iNDF	Indigestible NDF	iNDF is truly indigestible by ruminal microbes. Analytically, it is estimated as the NDF residue remaining after *in vitro* or ruminal *in situ* fermentation for 240 h or 288 h
NDFD	Neutral detergent fiber digestibility	Percentage of NDF fermented; determined at specified time points
pdNDF and pdNDFom	Potentially digestible NDF	aNDFom - uNDF at 240 h or 288 h; should be on an organic matter basis
eNDF	Effective NDF	Related to physical form of feed and fiber, associated with maintaining a desirable rumen environment and health
peNDF	Physically effective NDF	
paNDF	Physically adjusted NDF	
NDSC	Neutral detergent-soluble carbohydrates	Starch, WSC, and NDSF
NDSF	Neutral detergent-soluble fiber	Beta-glucans, pectic substances, and non-starch carbohydrates insoluble in 80% ethanol and soluble in ND
NFC	Nonfiber carbohydrates	Calculated by-difference value that includes all non-NDF carbohydrates, analytes not measured, and error
NSC	Nonstructural carbohydrates	Plant cell contents; sugars, fructans, starch; an agronomic term
ROM	Residual organic matter	100 - Ash - NDF - Starch - (fatty acids/fat factor) - (CP - 0.64* sNPNCPE)
Sugars		Monosaccharides and the disaccharides sucrose, lactose, and maltose
TTNDFD	Total tract NDF digestibility	This is an *in vitro* total tract NDF digestibility test which predicts NDF digestion
uNDF	Undigested NDF	uNDF is the analytical measure for NDF remaining undigested after a specified number of hours
WSC	Water-soluble carbohydrate	ESC plus lactose, long-chain fructans, and water-soluble polysaccharides

diet. A physical function of fiber carbohydrates is to maintain a healthy rumen environment by stimulating chewing, rumination, and rumen mat formation, which affects the utilization of the total diet. The goal in formulating diets for lactating cows is to use forages with consistently higher neutral detergent fiber (NDF) digestibility (NDFD) to support high levels of intake, milk production, and milk components while maintaining ruminal and herd health.

One challenge in working with carbohydrates in dairy diets is the proliferation of terms used to describe them. Table 1 reveals the most common terms and descriptions of these carbohydrate components. Fig. 1 is a graphical overview of carbohydrate analytical fractions.

2 Feed analysis

A major advance in feed analysis was the development of the detergent system for fiber analysis (Goering and Van Soest, 1970: Van Soest et al., 1991). This system better describes carbohydrate fractions by their nutritive value and potential digestibility. Methods for starch (Hall, 2015), neutral detergent-soluble carbohydrates (NDSC) (Hall et al., 1999), and undigested NDF (uNDF) (Raffrenato et al., 2018) have also been developed. These analyses have been useful for better describing forage and feed characteristics. The relationships of uNDF240 (% of DM), lignin, and 30-h NDFD in corn silage samples were

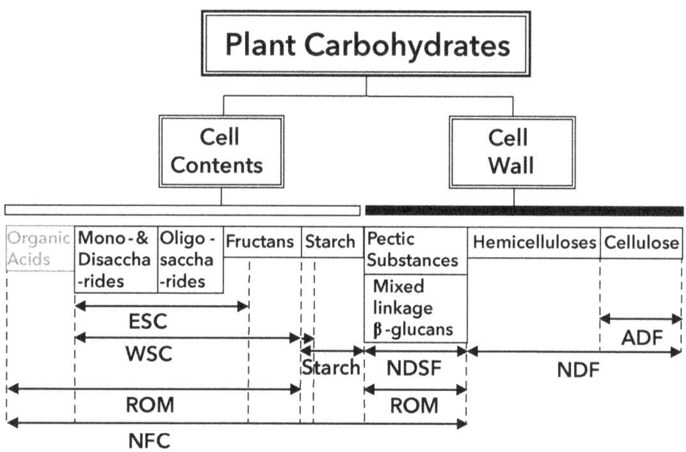

Figure 1 Carbohydrate analytical fractions. The extension of WSC into starch represents maltooligosaccharides (dextrins) that are fragments of starch soluble in water. ADF, acid detergent fiber; ESC, 80% ethanol-soluble carbohydrates which does not include lactose; NDF, neutral detergent fiber; NDSF, neutral detergent-soluble fiber (also includes other non-NDF polysaccharides of appropriate solubility); NFC, nonfiber carbohydrates by difference; ROM, residual organic matter by difference (NASEM, 2021); WSC, water-soluble carbohydrates.

reported (Hristov et al., 2020). The relationship between lignin and uNDF240 was significant ($R^2 = 0.73$; $P < 0.001$). The relationship between NDF30 and uNDF240 was also significant ($R^2 = 0.88$; $P < 0.001$). A second advance was the adoption and implementation of near-infrared reflectance spectroscopy (NIRS) by feed and forage testing laboratories. Use of NIRS requires developing calibration curves for organic feed components based on chemical analyses of the feeds; it does not detect minerals. NIRS provides rapid and lower cost analysis for use by dairy producers, nutritionists, and feed companies to better describe feeds and forages. The number of feed and forage samples analyzed each year has increased dramatically since the adoption of NIRS technology. Most forage testing laboratories offer both chemistry and NIRS analysis since there are still some analytes for which NIRS is not suitable due to lack of required calibrations.

Table 2 reveals analytical data for silage samples. Ash-corrected NDF (NDFom) values are 2-3 units lower than the NDF values. This indicates the importance of correcting NDF for ash which would otherwise be counted as carbohydrate. This is important in samples with high ash (mineral contamination) from harvesting procedures, degradation of feed in storage, or soil contamination. The magnitude of the standard deviations emphasizes the variability in the composition of these forages and the need for analysis of the feeds used in developing dairy cattle diets to meet the requirements of the cow while minimizing excretion of nutrients and gases to the environment.

In vitro fermentation of fiber and starch with mixed ruminal microbes provides estimates of ruminal degradability for use in diet formulation. Fineness

Table 2 Carbohydrate composition of silage samples[a]

Analysis	Corn silage	Small grain silage	Legume silage	Grass silage
Number of samples	6629	440	645	1470
ADF, % of DM	24.5 (3.5)	37.2 (5.1)	34.2 (3.6)	36.5 (4.9)
NDF, % of DM	41.4 (5.5)	57.4 (7.3)	43.2 (4.8)	55.9 (7.8)
NDFom, % of DM	39.1 (4.6)	53.7 (7.5)	40.4 (4.8)	53.1 (8.5)
Lignin, % of DM	2.9 (0.7)	5.2 (1.7)	7.4 (1.5)	5.4 (1.8)
30-h NDFDom, % of NDF[ss]	57.7 (6.1)	64.5 (7.6)	45.5 (9.3)	56.9 (12.9)
240-h uNDFom, % of NDF	10.2 (2.1)	15.6 (4.2)	18 (3.7)	14.8 (6.6)
ESC, % of DM	1.2 (1.0)	5.5 (4.1)	4.1 (2.2)	5.9 (3.9)
WSC, % of DM	2.6 (1.4)	7.5 (4.2	5.8 (2.3)	8.1 (4.5)
Starch, % of DM	33.2 (6.5)	6.5 (7.5)	2.2 (1.4)	2.1 (2.2)
7-h starch digestibility, % of starch	70.5 (11.1)	-	-	-

Values are expressed as mean (standard deviation).
[a]Samples analyzed at Dairy One Forage laboratory from May 1, 2019, to April 30, 2020.

of grind of samples can affect *in vitro* digestibility. A 2 mm versus 1 mm grind of forages and other samples gave lower *in vitro* NDFD when using filter bags (Damiran et al., 2008). Starch digestibility also declines with increasing particle size (Cone et al., 1989). However, a stated particle size based on screen size cannot be taken at face value without specification of the type of mill used: an abrasion mill (e.g. Cyclone and Udy) with a 2 mm screen gives a grind size equivalent to a 1 mm screen in a cutting mill (e.g. Wiley) due to the angle at which the particles hit the screen (D.R. Mertens, Mertens Innovation and Research LLC, Belleville, WI, personal communication). Current recommendations typically use 1 mm cutting mill for NDFD and a larger grind size for starch digestibility to retain some of the structure of the grain which affects starch digestibility. Typically, the values are intended to be correlated with but not necessarily identical with *in vivo* measures. Because of this, it becomes important to use the digestibility procedure specified by the diet formulation program used. More research to evaluate starch digestibility methods as they apply to different feeds and the relationship to *in vivo* responses is warranted.

3 Detergent fibers

Detergent fibers represent the bulk of material that may be recalcitrant to ruminal digestion, is indigestible by mammalian enzymes, and is related to digestibility of feeds and to feed intake. Through the actions of gut microbes, primarily those in the rumen, ruminants can convert the fibers unusable by humans into high-quality proteins and amino acids (e.g. meat and milk) for use in human diets and other useful products. Ruminants provide the opportunity to up-cycle the estimated 43.2×10^9 kg of co-products of food, fiber, and biofuel that would otherwise require disposal (White and Hall, 2017) and make use of the 168×10^6 ha of non-tillable pasture and rangeland unsuitable for growing crops (USDA, 2007). Considering that cellulose, a fiber carbohydrate, is among the most abundant carbohydrates in nature, the societal recycling service of ruminants is not inconsequential. In our analytical measures of fiber, NDF consists of cellulose, hemicellulose, and lignin and encompasses the recalcitrant fiber fraction. Acid detergent fiber (ADF) which was developed as a preparatory step for lignin determination is composed of cellulose and lignin. Hemicellulose content can be approximated as NDF – ADF and cellulose content as ADF – lignin. Both NDF and ADF also contain crude protein and ash.

In addition to its contribution to the nutrient supply, NDF and its physical form, particularly the larger sized particles associated with forage, are positively associated with chewing, rumination, and salivation which assist in maintaining rumen fermentation buffering and milk fat production. Dietary NDF is also related to rumen fill, rumen turnover, and the rate of particle size reduction which influence dry matter intake (DMI). Several approaches related to the physical form of NDF and feed have been developed to describe it and offer application in diets for maintaining ruminal health and predicting intake: percentage of total or forage NDF in the diet (Table 4), physically effective NDF (peNDF), physically adjusted NDF (paNDF), and physically effective uNDF (peuNDF).

Relationships among total and forage NDF intake, lactation performance, parity, stage of lactation, and feasible diets have been explored. Using data from several trials, Mertens (1987) reported that NDF intake of 1.2% ± 0.1% of body weight (BW) with apparently optimal diets was related to maximizing 4% fat-corrected milk (FCM) production. Mertens (2009a) developed and used an NDF intake system to examine the effect of forage NDF in diets for a 650 kg cow producing 45 kg/day of 4% FCM. Total ration NDF was held constant at 33%. Grass forage NDF ranged from 45% to 60%. Forage, as a percentage of diet DM, decreased from 52.6% to 33.4% as grass forage NDF increased from 45% to 60%. DMI, as a percentage of BW, increased from 4.15% to 4.33% as grass forage NDF increased from 45% to 60% for a 60% NDF grass forage. The decrease in forage inclusion and increased DMI were related to an increase in the amount of concentrate in the diet. Additional adjustments in Mertens' NDF intake system include changes in optimal NDF intake by parity, days in milk, filling effects of non-forage fiber sources (NFFS), and setting a minimum peNDF requirement of 21% of diet DM. Several studies have reported total diet NDF intakes of 1.3–1.8% of BW (Bargo et al., 2002; Cueva et al., 2021; Lim et al., 2015; Miller et al., 2021). These higher NDF intakes were attained by either high DMI intake of a total mixed ration (TMR) or high intakes of highly digestible pasture forages. It is important to understand that DMI has an impact of NDF expressed as a percent of BW. As an example, if a diet with 30% NDF is fed to a cow weighing 650 kg. If DMI is 23 kg, the total NDF intake is 6.9 kg which is 1.06% of BW. If DMI is 29 kg, then the total NDF intake is 8.7 kg or 1.34% of BW.

Systems specifically related to the ability of a diet to maintain rumination, rumen health, and milk production have their roots in the peNDF system developed by Mertens (1997). In this system, dried feed or TMR samples are evaluated using a vertical sieving system to determine the proportion retained on a 1.18 mm screen to give a physical effectiveness factor (pef; as percentage of NDF). Compared to the 1.18 mm screen approach, forage retained on the 4 mm screen of the Penn State Particle Separator was found to be similarly predictive of pef ($R^2 = 0.93$, slope = 0.86; Schuling et al., 2015) The pef factor is an indicator of the proportion of fiber that has the potential to provide stimulus for chewing. Recommendations are designed to maintain milk production of a given milk fat percentage. This system does not consider the digestibility of the NDF that would alter a feed particle's residence time in the rumen and thus its ability to enhance rumination. Challenges with this system include the assumption of equal distribution of NDF and particle fragility across particle sizes.

The paNDF system is focused on diet formulation for maintaining a desirable rumen environment using target pH as a proxy (White et al., 2017a,b; NASEM, 2021). Using an ensemble modeling approach with a constellation of submodels, it takes inputs of cow BW, TMR particle sizes determined on a Penn State Particle Separator on a DM basis, dietary concentrations of crude protein, starch, NDF, ADF, forage NDF, and -forage, and percentages of wet forage and cottonseed products (whole, hulls, and meal). For the particle size on a DM basis, material retained on the sieves can be analyzed for DM by using a microwave or other appropriate techniques. The DM basis gave better results than 'as is' due to variability in the DM% on each of the screens. The outputs are the recommended amounts of TMR DM on the 8 mm screen of the Penn State Particle Separator and a relative value for the minutes per day of rumination (current vs. increase or decrease with a change in the diet). Information on applying the paNDF app (MUNCH) can be found in an extension bulletin (Morris et al., 2021). The advantages of this system are that it uses a particle sizing system that is commonly used on farm and integrates a variety of readily available diet composition information. Developed from 60 publications that had 241 treatment means, the addition of more studies to expand the base of information for this approach with more different forage or feed types could expand its applicability.

The peuNDF system combines the concepts of peNDF and fiber indigestibility to approximate the impact of physical form and digestibility on DMI and ruminal residence time of fiber. peuNDF240 is calculated as pef × uNDF240, with uNDF240 as a percentage of feed DM that is the NDF undigested by rumen microbes after 240 h of *in vitro* fermentation. A lactating cow study evaluated TMR combinations of low (u) and high (U) uNDF240 and low (p) and high (P) peNDF to give peuNDF240 values of 5.4, 5.9, 5.9, and 7.1 for up, uP, Up, and UP (Grant et al., 2018). The diets contained 35% corn silage, 1.6% chopped wheat straw, and long- or short-chopped timothy hay which was used to modify the physical effectiveness of the diets. For up, uP, Up, and UP, ruminal pH (6.11, 6.17, 6.22, and 6.24), ruminal turnover rate of aNDFom (4.4%/h, 4.4%/h, 4.2%/h, and 3.9%/h), DMI (27.5, 27.3, 27.4, and 24.9 kg), and energy-corrected milk (ECM) (47.0, 45.7, 46.4, and 44.6 kg) showed differences among treatments. As peuNDF (percentage of diet DM) increased, both DMI and ECM decreased. The results of this work indicate that responses in DMI and ECM were related to peuNDF240. The responses obtained with a higher peNDF and lower uNDF240 diet were similar to the response from a diet with a higher uNDF240 and lower peNDF. These results indicate that decreasing particle size of forages with lower fiber digestibility may increase DMI and milk production. Additional studies with legume and grass forage diets are needed to refine this concept. Potential differences in response between NFFS and forage need to be explored. An observational study was conducted over 2 years on 22 commercial dairies in Italy (Serva et al., 2021). All herds were fed TMRs with grass silage, corn silage and alfalfa hay. Negative relationships for milk production (R^2 = 0.62) and feed efficiency (R^2 = 0.65) and diet peuNDF240 were reported. These relationships were significant (P = 0.001). The authors suggest that reducing the particle size of lower digestibility hay may assist in lowering ration peuNDF240 and improving milk production.

However, NDF content, source, and particle size in diets do not fully describe the impacts of NDF. A significant advance was the development and use of NDFD in formulating dairy cattle diets. Digestibility was originally and continues to be determined using ruminal *in situ* or *in vitro* fermentation techniques. The NIRS calibrations based on *in vitro* fermentation data have been developed, and most commercial forage laboratories now provide this rapid analysis for digestibility. Oba and Allen (1999) reported that a 1% unit increase in NDFD was related to an increase of 0.25 kg of 4% FCM and 0.17 kg of DMI. Mertens (2009b) expanded the database and adjusted values to a 48-h fermentation time. Results were an increase of 0.097 kg for DMI and 0.139 kg for FCM with each 1% increase in NDFD. It was also indicated that diet NDF had three times greater impact than NDFD in terms of intake and milk production (Mertens, 2009b) due to its relationship to fill and nutrient supply. Values for NDF and NDFD in Table 2 indicate large ranges in NDFD within forage type at specific fermentation time points. Table 3 reveals data on DMI and milk responses to forage NDFD. These data were calculated from 52 sets of treatment means. The responses for DMI and 4% FCM were significantly different.

Cell wall indigestibility was highly correlated with lignin content in grass and legume forages (Smith et al., 1972). Lignin times 2.4 has been used to predict indigestible NDF (iNDF). However, the approach of using a constant factor has been questioned (Van Soest et al., 2000). Several papers have reported that the 2.4 factor is not constant across feed and forage types (Raffrenato and Erasmus, 2013; Kramer et al., 2012; Palmonari et al., 2016).

Table 3 DMI and milk production responses to forage NDFD

Response	Low NDFD	High NDFD
DMI, kg/day	21.8^a	23.3^b
4% FCM, kg/day	26.8^a	28.9^b
Milk fat, %	3.35	3.43
Milk protein, %	2.93	2.90
Body weight change, kg/day	0.14	0.37
In vitro or *in situ* NDFD, %	54.5	62.9
Total tract *in vivo* NDFD, %	51.5	54.8

Source: Adapted from Oba and Allen (1999). Means within a row without common superscripts differ ($P \leq 0.02$).

An alternative to using lignin to estimate iNDF is uNDF determined by direct measurements from long-term *in vitro* fermentations or *in situ* techniques (Huhtanen et al., 2007). A standardized 240-h *in vitro* fermentation is available to determine uNDF (Raffrenato et al., 2018). Potentially digestible NDF can be calculated as aNDFom - uNDF. Rates of fermentation are only applied to pdNDF and not to total NDF. Mertens (2016) reported that uNDF is a greater improvement in determining nutritional availability beyond NDFD or lignin. Rumen fill, rumen passage rate, and DMI are influenced by uNDF. Grant (2015) summarized studies conducted at the Miner institute using corn silage-based diets that evaluated relationships among intakes and ruminal NDFom and uNDFom240. The maximum NDFom intakes were 1.26–1.47% of BW, while the maximum rumen NDFom content, an indicator of rumen fill, was about 1.28% of BW. The range in uNDFom240 intakes were 0.30–0.48% of BW. There was a ratio of 1.6 (rumen uNDF240/intake uNDFom240) across the diets. Average solids-corrected milk production was 45 kg/day and DMI averaged 27 kg/day. This initial data provides an insight into the relationships between NDFom and uNDF intakes in high-producing cows on corn silage-based diets. Additional data are needed on a larger number of diets and forages to better understand these relationships and utilize this concept in diet formulation and evaluation.

The relationship between harvest maturity and uNDF240 in alfalfa hay has been examined (Palmonari et al., 2014). Alfafa hay was harvested at prebloom (PB), first bloom, and full bloom (FB) maturity. NDFD30 decreased from 62,2 (PB) to 55.7 (FB) % of NDF. There was an increase iNDF from 15,5 (PB) to 18.3 (FB) % of NDF. A common diet was used to predict the impact of maturity on metabolizable energy (ME) and metabolizable protein (MP) milk production. A decrease in ME milk of 1.4 kg/day was predicted when FB alfalfa replaced PB alfalfa in the diet. The predicted change in MP milk was - 2.8 kg/day when FB alfalfa was substituted for PB alfalfa. These results indicate the importance of harvest maturity n NDFD and iNDF.

An unexplored factor that would influence the energy value of fiber is the SCFA yield from fermented NDF in grasses versus legumes. Some research suggests that SCFA yield per amount of NDF fermented is greater from legumes or cellulose than from grasses and xylans or hemicellulose (Strobel and Russell, 1986; Doane et al., 1997; Weimer, 2011). Ideally, SCFA data from fermentation of isolated NDF would be evaluated on a carbon basis to be most comparable as SCFA profiles change. Another component, 'soluble lignin', could partially explain a difference in SCFA yields. Soluble lignin is polyphenolic material with associated carbohydrate and protein that is soluble in rumen fluid. It was originally identified in the rumen fluid of cattle fed spear grass (Gaillard and Richards, 1975) and was recoverable in feces (Neilson and Richards, 1978). It may represent lignin not polymerized to the main mass of lignin in the cell wall which can be released by the digestion of the surrounding cell wall (Van

Soest et al., 2020). This could be more prevalent in immature grasses where lignification starts as islets which coalesce as the plant matures. Soluble lignin would represent an undigested portion of NDF that is not recovered by NDF analysis, resulting in an overestimate of NDF digestibility. The degree to which soluble lignin is produced on most dairy cattle diets is unknown.

4 Neutral detergent-soluble carbohydrates

The NDSC encompass a diverse array of carbohydrates, including sugars, oligosaccharides, fructans, pectic substances, galactans, and other carbohydrates found in water-soluble carbohydrates (WSCs), starch, and NDSF. Among these, starch, glycogen, lactose, dextrins, and trehalose (a disaccharide) can be digested by the cow in the small intestine (Kreikemeier et al., 1990) or by gut microbes; the other NDSC are digested only by gut microbes.

Starch is a polysaccharide composed of glucose units in α-1,4-glycosidic bonds with α-1,6-branches. Starch can be utilized in the rumen, the small intestine, or the hind gut of the dairy cow (Moharrery et al., 2014). Typically, total tract starch digestion is >95%, but ruminal digestion may vary from >90% to <30% (Nocek and Tamminga, 1991; Firkins et al., 2001) depending on source, processing, and other factors. In the rumen, fermented starch gives a higher proportion of propionate than do other NDSC, but this is affected by the forage or starch level of the diet: high forage (less propionate) versus high concentrate (more propionate) (Murphy et al., 1982). The propionate can be used to support gluconeogenesis, the main source of glucose for the cow.

Total dietary starch levels vary depending on the feeds used for formulation. A recent survey showed that the average dietary starch content was 26.8% of DM in dairy herds with ECM of 50 kg (Chase, 2019). The range was 20.6–33% with herds using both corn silage and corn grain. Rumen degraded starch was 5512 g/day (range 4320-6707 g/day) as calculated using the Cornell Net Carbohydrate and Protein System (CNCPS) model (Van Amburgh et al., 2015). However, digestible carbohydrate rather than starch, per se, is likely what cows require to support production, though starch is a convenient way to provide it. In a trial using a 40% forage ration and co-product feeds replacing all corn grain and soybean meal (Hall and Chase, 2014), milk production was 35.7 kg with 4.25% milk fat in late-lactation cows with a dietary starch of 11%. A trial was conducted using by-products to replace cereal grains and pulses (Ertl et al., 2015). The control diet was 5.2% starch, while the by-product diet was 3.3% starch. DMI, milk, and ECM milk were not different between the diets. A Swedish study which used grass silage diets with co-product feeds and no cereal grains achieved an ECM of 32.4–33.7 kg (Karlsson et al., 2018). The control diet was 17.3% starch, while the co-product diets were 1.4–1.7% starch. These results indicate that a wide range of diets greatly differing in dietary starch levels can support milk production. The key is that these diets still provide adequate amounts of rumen degradable carbohydrate from NDSC and degradable NDF to support microbial and cow needs. Dairy herds in the USA that produce grass-fed milk feed 100% forage diets with no grain supplementation. These diets are <5% starch.

There has been interest in using reduced starch diets for dairy cattle. High corn grain prices, improvements in milk components and cow health, and improved profitability are reasons for this interest. A trial comparing diets with 17.7-24% starch showed no differences in DMI, 3.5% FCM, milk component yields, feed efficiency, or total tract starch digestion; the average 3.5% FCM was 43 kg (Dann et al., 2014). A second trial compared diets with 21% and 26% starch (Dann et al., 2015). In that trial, one diet replaced corn grain with forage, while a second used NFFS. DMI was lower on the diet with added forage. There was no difference in 3.5% FCM, milk fat yield, or feed efficiency between diets with cows producing 52 kg of milk. Milk true protein yield was reduced for cows on the diet with added forage. Another trial compared diets with 22-28% starch (Fredin et al., 2015a). Milk production (4% FCM), DMI, milk fat, and milk true protein yields were not different between the four diets in cows producing 37 kg of ECM. A meta-analysis from 53 papers with 223 treatment means was conducted (Fredin, 2015). The starch content of the high-starch diets averaged 28.7% (range 16.9-41.9%), while the reduced starch diets were 21.9%, with a range of 5.2-34.3%. There were significant decreases in DMI of 0.1 kg/day and milk production of 0.19 kg/day for each 1% decrease in dietary starch. The decrease in DMI was 0.07 kg/day in trials that reduced starch by using NFFS and 0.12 kg/day when forage was used. Milk true protein yield decreased by 8.3 g/day and milk fat yield by 6.9 g/day for each 1% decrease in starch. Milk urea nitrogen was not different between the high- and low-starch diets. There were 24 treatment means where milk production increased when lower starch diets were fed. A recent paper used diets with 24.8% and 18.4% starch (Silvestre et al., 2022). The diet starch content was adjusted by replacing corn grain with wheat straw. This increased the diet NDF from 33.1% to 39.1%. Milk yield, ECM, and DMI were significantly lower on the reduced starch diet.

Higher starch diets have also been associated with lower NDFD (Ferraretto et al., 2013; Potts et al., 2017). Total tract NDF digestibility decreased from 53.4% to 39.6% as diet starch increased from 12.2% to 30.1% (Boerman et al., 2015). One study reported that for each 1% increase in diet starch, the NDFD would decrease by 0.59% (de Souza et al., 2018). The relationship between higher dietary starch and depression in NDFD may vary depending on the source of NDF in the diet.

Particle size, processing method, endosperm type, and conservation method can influence starch degradation rates and total tract digestibility. These factors alter the amount of starch to be included in the diet as the digestible amount rather than total starch is what matters in meeting cow requirements. Lykos and Varga (1995) compared cracked corn, chick cracked corn, finely ground corn, and steam-flaked corn. The mean particle sizes of the corn grain types were 4309, 2577, 686, and 2896 μm, respectively, and *in situ* effective rumen degradabilities of starch were 44.6%, 53.3%, 64.5%, and 75.4%. Allen and Ying, 2021a; 2021b; 2021c, compared floury versus vitreous dry ground corn and high-moisture corn (HMC) in diets for cows averaging 47.5 kg/day of 3.5% FCM. Dry ground corn had lower true ruminal digestion and apparent total tract starch digestion than HMC. Milk production was not different between the dry corn and HMC, while DMI was 1.3 kg/day higher (P = 0.02) for cows fed dry corn. Floury corn had higher ruminal and total tract starch digestion. There were no differences in DMI, milk, and 3.5% FCM between endosperm types. Fine and coarse ground corn were compared in low (18.0%) and normal (26.5%) starch diets (Fredin et al., 2015b). Corn particle size was 552 μm for the fine corn and 1270 μm for coarse ground corn. Apparent total tract starch digestion was 98% for the fine corn and 93.4% for the coarse corn. DMI was not different when diets with 52% forage were fed. Milk production and milk components were also not different. Table 4 reveals minimum NDF, FNDF, and maximum starch level recommendations (NASEM, 2021).

Corn silage is a primary forage in US dairy herds. Starch digestibility of corn silage increases with length of time in storage (Ferraretto et al., 2015a; Der Bedrosian et al., 2012). *In vitro* ruminal starch digestibility for 4 mm dried samples increased from 61.7% at day 0 to 79% after 120 days of ensiling (Ferraretto et al., 2013). Both soluble protein and ammonia-N also increased and may be good indicators of changes in starch digestibility. Brown midrib (BMR) and leafy corn silages were compared for digestibility and milk production (Ferraretto et al., 2015b). Starch, 7-h *in vitro* starch digestion, NDF, and 30-h NDFD for BMR were 30.6%, 69.9%, 38.3%, and 67.9%, respectively. Similar values for the leafy corn silage were 32.2%. 75.6%, 36%, and 57.2%. Diets fed were 41.8% corn silage (DM basis). DMI was higher ($P < 0.01$) by 1.7 kg/day for cows fed BMR. ECM was 1.5 kg/day higher for the BMR-fed cows ($P<0.07$). Total tract starch digestibility was 98% for BMR versus 93.3% for cows fed leafy corn silage

Table 4 Diet minimum fiber and maximum starch recommendations (DM basis)

Minimum FNDF, %	Minimum NDF, %	Maximum starch, %
19	25	30
18	27	28
17	29	26
16	31	24
15	33	22

Source: NASEM (2021).

($P < 0.001$). It is suggested to use FNDF rather than NDF in formulating diets. This is due to FNDF having a higher negative relationship with DMI and a higher positive relationship with rumen pH than NDF. The minimum level of FNDF is suggested to be 19-25% of diet DM (NASEM, 2021). This will vary, including DMI, forage particle size, diet starch, diet NFFS, starch degradability, and management factors such as feed bunk management and feed system (TMR vs. grain and forage fed separately).

WSCs include sugars, oligosaccharides, fructans, and other carbohydrates soluble in water. The use of WSCs rather than ethanol-soluble carbohydrates (ESCs) is recommended as WSC more completely encompasses the carbohydrates in this fraction, with ESCs excluding lactose and long-chain fructans. For measurements of WSC in feeds such as cane molasses, total sugars as invert are an acceptable value to use. Common features of the WSC characterized so far include potential for rumen microbes to use them to produce microbial glycogen, lactic acid, and a greater proportion of butyrate than typically noted with starch or fiber. The WSC content of pasture and fresh forages can be >10% of DM. During silage fermentation, WSCs are utilized by the silage microorganisms, decreasing the WSC content of silages to usually <5% of DM but can be higher in higher DM silages. Feeds such as cane or beet molasses, almond hulls, and citrus pulp may contain 30% to >60% WSC. Silage-based dairy cattle diets with no added sugar will often contain 2–4% WSC.

Sugars are utilized by rumen microorganisms and converted to gases, organic acids, glycogen, and microbial cells including protein. Weisbjerg et al. (1998) showed that the disaccharides sucrose and lactose are rapidly hydrolyzed in the rumen (>500%/h) and indicated fermentation (actually, disappearance) rates of the released monosaccharides of 300-700%/h. However, rates of sugar or WSC disappearance do not necessarily equate to carbohydrate being fermented. A portion of the WSC may be converted to the storage carbohydrate glycogen by both bacteria and protozoa. *In vitro* studies used mixed ruminal microbes and an elevated level of tryptone addition. Glucose disappeared at a rate of 65%/h, microbial N (cell growth) increased at a rate of 60%/h, organic acids at 64%/h, and glycogen at a rate of 104%/h (Hall, 2017b). For urea at the same N inclusion level, the glucose disappearance rate tended to be slower at 51%/h and the organic acid rate of production was slower at 18%/h. The rates of increasing microbial N at 39%/h and glycogen at 98%/h were not different. The rates of disappearance and products may vary with nitrogen (N) and carbohydrate source.

In vivo studies with feeds containing sucrose or pure sucrose evaluated lactating dairy cow responses to changes in nonfiber carbohydrates (NFC) source. Trial 1 used diets with 0%, 4%, 8%, or 12% added dry molasses; the base diet (0% added molasses) was 60% forage, 28% NDF, 31.5% starch, and 2.6% sugar (Broderick and Radloff, 2004; Broderick et al., 2008). As molasses inclusion increased, there was a linear increase in DMI and quadratic effects on 3.5% FCM production, milk fat percent, and milk fat yield. The calculated optimum total dietary sugar was 4.8%. Trial 2 in the same paper added 0%, 3%, 6%, or 9% liquid molasses to diets with 52% forage, 26% NDF, and 31% starch. Total diet sugar levels were 2.6%, 4.9%, 7.4%, and 10%. There were quadratic responses for DMI, milk yield, milk protein, and milk protein yield. The calculated optimum total diet sugar content in this trial was 6.3%. Several other studies have reported mixed results in milk production or milk components when supplemental sugar sources were added. In a later study, sucrose was used to replace starch in diets for lactating cows (Broderick et al., 2008). Diets on a DM basis contained 0%, 2.5%, 5%, or 7.5% sucrose and 7.5%, 5%, 2.5%, or 0% added corn starch. Diets were 60% forage and 30% NDF with total dietary sugar contents of 2.7%, 5.2%, 7.1%, and 10%. DMI, milk fat percent, and yield increased as added sugar increased. There were no differences in milk production.

An effect often seen with feeding of sugars is an increase in butterfat percentage or yield. Sucrose supplementation may decrease the *trans* 18:1 fatty acid concentration in milk through biohydrogenation of fats in the rumen by sugar-utilizing microbes (Penner and Oba, 2009), reducing the amount of this milk fat-depressing fatty acid absorbed by the cow. The sugar-mediated increase in butyrate supply could support increased *de novo* synthesis of milk fatty acids (Palmquist et al., 1969), and butyrate does play a significant role in milk triglyceride synthesis, occupying 30% of the *sn*-3 position (Jensen, 2002). Additionally, increased incorporation of butyrate into *de novo* synthesized fatty acids is more energetically efficient than synthesis from acetate alone. Using one butyrate rather than two acetates requires the formation of one less bond in the fatty acid chain, reduces the energetic demand of synthesis, and spares acetate for other uses (Baldwin, 1968). Another possibility for the milk fat synthesis effects of sugars could relate to the rate of sugar fermentation, the rate and amount of feed intake, and the options an animal has for handling 'excess' nutrients. In a study with high-producing lactating cows fed with 0%, 5.25%, or 10.5% sugarcane molasses substituted for corn grain and just focusing on the high rumen degradable protein (RDP) diets, as molasses increased, DMI declined linearly from 29.0 kg/day to 26.7 kg/day, milk declined linearly from 47.6 kg/day to 42.3 kg/day, and protein and lactose yields declined linearly; milk fat yield did not change and fat percentage increased from 3.39% to 3.93%, and ECM feed efficiency remained unchanged at 1.65 (Hall and Zanton, 2022). In a separate study, cows were recorded consuming 30% of the daily TMR ration in the first 3 h post-feeding. The hypothesis developed but not yet tested from this work was: If cows rapidly consume a large amount of a rapidly fermentable feed, it will quickly produce a substantial mass of SCFA. If the SCFA mass passes a threshold for what the animal requires for energy/oxidation, will acetate and butyrate be shunted to produce butterfat or adipose, depending on the physiologic state? This would mean focusing on a shorter time step than 'day'. And could the propionate made in the same time frame depress intake? A study using ruminal infusion of propionate suggests that this could be the case. Ruminal infusions of 1 kg of propionate adjusted with salts to be isotonic at rates of 2.25 mol/h or 0.75 mol/h versus its isotonic control reduced DMI in dairy cows, with the high rate causing greater DMI depression (Boudon et al., 2009). This concept needs to be further evaluated. Oba et al. (2015) compared the impact of sucrose, lactose, and corn starch on rumen fermentation. At 30 min after dosing, total SCFA was 65, 50, and 40 mM for sucrose, lactose, and corn starch. This indicates that the SCFA response to added carbohydrates in dairy diets depends on the source added.

Dietary sugar can also have effects on rumen ammonia, microbial protein synthesis, rumen pH, and fiber digestion (Hall, 2017a; de Ondarza et al., 2020). Rumen pH is not negatively affected in most studies. Nonlinear effects on fiber digestion have been observed (Broderick and Radloff, 2004,). Microbial protein synthesis has increased in some studies but not in others. Ruminal microbes utilizing different carbohydrates rely on different N sources. When dextrose or corn starch was dosed into the rumens of lactating dairy cows in which ammonia was labeled with ^{15}N, the duodenal flow of microbial N formed from ammonia was less for dextrose than for starch (Hristov et al., 2005). Rumen ammonia levels are often reduced when sugar is added, but it is unknown whether this is due to increased ruminal use of ammonia, increased use of amino acids and peptides so that less ammonia is generated ruminally, or some combination. The interaction between sugar, starch, fiber, and RDP has major impacts on results obtained with added sugar.

A meta-analysis of the effects of added sugar in dairy diets has been reported (de Ondarza et al., 2017). Cows producing >33 kg/day of milk had an increase in 3.5% FCM of 2.14 kg/day compared with 0.77 kg/day in cows producing <33 kg/day of milk when 5-7% sugar was added. A total diet sugar content of 6.75% was calculated to be optimum in this paper. The authors suggest that diets with 6-8% soluble fiber, 10-11% RDP, and 22-27% starch may be needed to optimize milk yield when adding sugar to diets. An additional consideration, the impact of which has not been well explored, is that sugar present in plant feedstuffs may ferment more slowly than feeds comprising added sugars not held in plant cells (Hoover et al., 2006). The type of sugar supplemented and its source bear further investigation.

It has been suggested that the by-difference NFC fraction be separated into two components to be used for calculating dietary energy values (Tebbe et al., 2017). These components are starch and residual organic matter (ROM). ROM is calculated by difference (Table 1) and includes WSC, NDSF, and all organic compounds not found in the crude protein, fatty acid, NDF, and starch fractions, as well as analytical error. The rationale is that starch is variable in digestibility, while ROM in the diets evaluated was nutritionally uniform with a digestibility value of 0.96. Starch digestibility of corn grain varies between 0.77 and 0.96 depending on the particle size, moisture content, or processing (NASEM, 2021). The summative equation to predict feed energy values in the new NASEM (2021) was modified to use both starch and ROM. The revised equation no longer uses total digestible nutrients and instead uses gross energy and digestible energy.

5 Rumen carbohydrate and protein interactions

Rumen microbes require N sources to make amino acids and protein. Ruminal ammonia levels at 50 mg NH_3-N/L or greater have been suggested to be adequate to support microbial needs for nonprotein N and maintain fiber fermentation (Satter and Slyter, 1974). This was an *in vitro* study, and the relationship to *in vivo* responses is unclear. The impact of RDP and NDSC supplementation on fiber fermentation was illustrated when sugar and starch supplementation depressed NDF digestion at a low level of dietary RDP, but greater RDP supplementation improved NDF digestion (Heldt et al., 1999). Microbial protein needs must be considered in light of rate and mass of carbohydrate fermentation which establishes demand relative to the same factors for protein supply (Van Soest, 1994).

Dietary protein supply and type also affect what products are formed from carbohydrates by microbial fermentation. More lactic acid is produced (Counotte and Prins, 1981), more microbes are produced (Argyle and Baldwin, 1989), and less microbial glycogen is produced (McAllan and Smith, 1974) when relatively more RDP/amino acids/soluble protein are provided with rapidly fermentable carbohydrates such as sugar. Similarly, supplementation with amino acids increased growth efficiency and gave a reduction in the specific rate of heat production (energy spilling) with *Streptococcus bovis* (Russell, 1993). One could consider both glycogen accumulation and energy spilling as differing methods by which microbes process carbohydrate in excess of immediate growth needs. It has been suggested and incorporated into the CNCPS that availability of peptides can increase the yield of nonstructural carbohydrate-utilizing bacteria by 18.7% (Russell et al., 1992). With glucose fermentation, the increase in yield of microbial N and a decrease in glycogen was more marked with peptides than with urea, and rates of microbial N and organic acid production were less with urea, as described earlier (Hall, 2017b). The increase in lactic acid production is interpreted as reflecting an increase in the rate/flux of hexose through glycolysis. Related to the energetic needs of the rumen microbes, the reduced amount of adenosine triphosphate (ATP) derived from a fermentation producing lactate may be counterbalanced with a reduction in ATP required for glycogen formation, and an increase in ATP produced per unit of time to support microbial growth. Determining how to best supplement N to the rumen to carbohydrate use remains a challenge. Estimating the supply of ruminally available peptides as they are simultaneously produced, degraded, and utilized and sometimes accumulated (Wallace and Cotta, 1988) is uncertain, as is predicting supply and utilization of feed-derived hydrophobic or hydrophilic peptides (Chen et al., 1987).

Optimizing rumen function and metabolism requires the integration of diet energy and protein (Nocek and Russell, 1988; Firkins et al., 2006). The importance of matching rumen fermentation rates was described by Johnson (1976). Russell and Hespell (1981) stated that the balance of rumen fermentation products and the rumen microorganisms are key determinants of nutrient use in ruminants. Hoover and Stokes (1991) used a continuous culture system to examine the roles of proteins and carbohydrates on microbial production in the rumen. They used NSC and degradable intake protein (DIP) in this analysis. Carbohydrate digestion and microbial efficiency were positively and linearly related to the diet DIP level. Diets with a ratio of NSC:DIP <3 and >37% NSC maximized rumen fermentation (Stokes et al., 1991b). A study using lactating cows reported that diets >9% DIP and >24% NSC increased microbial flow from the rumen (Stokes et al., 1991a). Microbial N production increased as both diet urea and RDP increased in a continuous culture system (Griswold et al., 2003). The authors attributed this response to the protein treatments to the increased digestibility of NDF, hemicellulose, and NSC.

Studies using combinations of barley, milo, brewer's grain, and cottonseed meal examining the effects on microbial protein synthesis and milk production were conducted (Herrera-Saldana et al., 1990). Microbial N production in g/kg of truly fermented organic matter was higher on the barley diets. Milk production was highest for cows fed the barley-cottonseed meal diets (Herrera-Saldana and Huber, 1989). The barley and cottonseed meal diet had a better balance of rumen carbohydrate and protein availability. Oba and Allen (2003) reported that the rate of starch passage and the rate of starch availability were important in determining the *in vivo* efficiency of microbial protein synthesis. A meta-analysis found a linear relationship between organic matter intake and the flow of microbial N to the omasum (Broderick et al., 2010). Higher rumen availabilities of protein and NSC were associated with increased flow of bacterial N to the duodenum (Aldrich et al., 1993). Dietary starch has also been associated with improvements in providing glucogenic nutrients to increase the efficiency of transfer of amino acids into milk protein and of milk protein production (Cantalapiedra-Hijar et al., 2014; LaPierre et al., 2020).

6 Rumen health

An imbalance of rumen-available carbohydrates and physically effective fiber can result in clinical or subacute ruminal acidosis (SARA) in dairy cows. This disorder is estimated to cost US$400+ per case (Stone, 1999). SARA is defined as rumen pH being at or below 5.5 for several hours per day. This topic has been the subject of review papers (Krause and Oetzel, 2006; Prasanth and Ajithkumar, 2016; Stone, 2004; Lean and Golder, 2020).

Animal indications of SARA include decreased DMI, low milk fat, higher risk of foot problems or laminitis, decreased feed efficiency, lower microbial protein synthesis, and increased involuntary culling. A primary cause is an excess of rumen-available carbohydrates. Rapid diet changes and low physically effective fiber are also involved. TMR versus component feeding, frequency of feeding, frequency of feed pushups, feed sorting, slug feeding, overcrowding, and empty bunk syndrome can be significant contributing management factors. Table 4 reveals guidelines for diet NDF and starch levels to minimize the risk of SARA. Donovan et al. (2004) reported that the shift from a low-energy pre-calving diet to a high-energy post-calving diet can increase the risk of subclinical laminitis.

Feed sorting can be a key factor that can increase the risk of SARA. Sorting alters NDF and starch content of the diet consumed since cows tend to sort against coarse and long fiber particles (Miller-Cushion and DeVries, 2017). The result is that the intake of fermentable carbohydrates increases while the fiber intake decreases. Higher forage and drier diets are also more prone to sorting. Primiparous cows exhibited more sorting, had more time below pH 5.8, and had a higher SARA index than multiparous cows (Stauder et al., 2020). Cows at elevated risk of acidosis altered sorting behavior by sorting for longer particles which may have been an attempt to counteract low rumen pH (Devries et al., 2008). Adding water to drier diets is one approach that may decrease sorting (Shaver, 2002; Leonardi et al., 2005). The use of added liquid molasses has also had benefit in terms of reduced sorting (Broderick and Radloff, 2004; DeVries and Gill, 2012). A series of papers examined sorting in high-straw dry cow diets. Adding water or liquid molasses can improve the feed intake and decrease sorting against long particles (Havekes et al., 2020a,b). Sorting was reduced which improved feed intake in high-straw dry cow diets with shorter straw particle size (Havekes et al., 2020c).

Rumen pH can be a factor that shifts the biohydrogenation pathway in the milk fat depression syndrome (Jenkins, 2019; Harvatine, 2018; Souza et al., 2022). The shift in biohydrogenation results in the production of *trans*-10, *cis*-12 conjugated linoleic acid. Small amounts of these compounds (1–2 g/day) can decrease the milk fat percentage. One mechanism for this effect of rumen pH could be a shift in the bacterial population. Diet carbohydrate factors that are risk factors for milk fat depression include the amount and types of carbohydrates fed, extent and rate of fermentation and diet effective fiber. The net effect is that diet carbohydrates and feeding management can have a significant impact on rumen health and milk fat content.

7 Higher forage diets

Forages are the foundation of dairy cattle diets and have impacts on both productivity and animal health. The level of forage used in dairy cattle diets has increased, while milk production per cow has also increased over the last 15-20 years. This increase is only possible with a concomitant increase in forage quality. A survey of six high-producing Holstein herds was conducted in Wisconsin (Shaver and Kaiser, 2004). These herds averaged 44.4 kg of milk/cow/day with milk fat of 3.75% and milk true protein of 3%. The average forage inclusion in the diet was 48% on a DM basis with a range of 45-53%. Total diet NDF was 29.5% with a range of 26.2-32.3%. Forage NDF averaged 68% of total NDF.

A survey was conducted of Pennsylvania dairy herds in 2013 (Buza and Holden, 2016). Herds feeding 60% forage were 58.5% of the respondents, with 17.1% of the herds feeding 70% forage diets. A survey of nutritionists was done by Cornell researchers in 2018. Ninety percent of the responses indicated that diet forage levels had increased in the last 10-15 years. Herds feeding 60-70% forage were 27% of the responses, while 10% indicated herds were feeding >70% forage. A survey of high-producing dairy herds was recently reported (Chase, 2019). Seventy-two herds had an average daily milk production of 50 kg/cow. Forage inclusion rate in the diet averaged 54.7% of total ration DM with a range of 33-72%. Diet NDF averaged 30%, with 74% of the NDF from forage. Table 5 reveals high-forage diets for five herds in this survey. These herds are in the Eastern and Midwest areas of the USA where corn and alfalfa silages are typically fed. Diet NDF and starch levels are within the diet nutrient content ranges suggested by NASEM (2021). The key to higher forage diets in these herds is a consistent supply of higher NDFD forages.

Two trials were conducted using diets containing 50% or 60% forage (Martinez et al., 2009). Trial 1 used diets with an alfalfa silage to corn silage ratio of 54:46. Diets were 31.4-31.7% NDF. DMI was 2.3 kg/day lower ($P < 0.01$) on the 60% forage diet, while 3.5% FCM was not different (46.5 vs. 46.8 kg/day). Income over feed cost was numerically higher for cows on the 60% forage diet (US$8.88 vs. 8.53/day). Trial 2 used diets with a corn silage to alfalfa silage a ratio of 69:32. Diets were 35% and 34% NDF and 38.9% and 39.6% NFC. DMI was 1.4 kg lower ($P < 0.01$) on the higher forage diet, while 3.5% FCM was not different (41.9 vs. 42.8 kg/day). Income over feed cost was again higher

Table 5 Example of herds feeding higher forage diets

Item	Herd A	Herd B	Herd C	Herd D	Herd E
Milk, kg/day	50	53.5	55	50	50
Forage, % of DM	68	69	66	64	59
NDF, % of DM	29.8	30.7	26.8	32.4	25
FNDF[a], % of NDF	26	24.7	21.7	25	19.3
FNDF, % of NDF	87	80.4	80.9	77.2	77
FNDF, % of BW	1.0	1.2	0.8	1.1	0.8
Starch, % of DM	26	27.4	30	21.1	24.7
Fat, % of DM	4.2	5.3	5.8	5.3	4.3
BMR[b] corn silage, % of DM	41	46	35		
Non-BMR corn silage, % of DM				16.6	23.8
Alfalfa silage, % of DM		23	31	44.6	35.2
Mixed legume-grass silage, % of DM	27			2.8	

Source: Adapted from Chase (2019).
[a]Forage NDF.
[b]Brown midrib.

for cows on the 60% forage diet (US$7.07 vs. 6.65/day). The results of this trial indicate that higher forage diets can support milk production similar to lower forage diets while improving profitability by lowering purchased feed costs. An important consideration when implementing higher forage diets is a higher level of forage inventory since daily forage needs increase. Frequent forage testing, including NDFD and uNDF240, and TMR mixer wagon capacity also need to be considered (Chase and Grant, 2013). Higher forage diets are bulkier and fewer cows can be fed from the same size TMR mixer. Feed bunk management is also important since it will take cows more time to consume the bulkier high forage diets.

8 Case studies

As discussed in the previous section, the trend in the Midwest and Northeast areas of the USA is to feed higher forage rations to dairy cows. A herd in New York increased from 50% to 65% of the total ration DM as forage over a 4-month period. Corn silage NDF decreased from 43% to 36%, while hay crop silage NDF went from 52% to 40%. The changes in NDF indicate that forage quality improved during the time of the study, and the herd nutritionist used this information to adjust diet formulation. The amount of forage fed increased from 11.5 kg/day to 15.4 kg/day. Forage NDF intake increased from 0.84% to 0.95% of BW. Milk production increased by 1.8 kg/cow/day, while income over feed cost increased by US$130 per day in this 100-cow herd. The change in milk production was associated with lower days in milk, a lower NDF haylage with higher NDFD, and a higher starch corn silage. A second New York herd increased the diet forage from 50% to 60% after shifting from conventional to BMR corn silage. Milk production increased by 1360 kg per lactation and herd health improved. The herd nutritionist removed 1.8 kg of forage fiber extenders from the diet, which saved US$21 000 per year in the purchased feed cost in this 1000-cow herd.

The impact of corn grain starch digestibility on milk production was observed in a Wisconsin herd. Herd milk production dropped from 45 kg/cow/day to 42 kg/cow/day when a new delivery of corn grain was fed. The herd was fed 2.7 kg of corn grain. The new delivery of corn grain was 15 points lower in 7-h starch digestibility. The herd replaced corn grain with the same amount of wheat, with much greater starch digestibility than corn grain, and milk production returned to 45 kg/day within a week.

An example of the importance of NDFD on milk production and DMI was evaluated using BMR corn silage in transition cow diets (Stone et al., 2012). BMR corn silage replaced conventional corn silage 3 weeks pre-calving and the first 3 weeks after calving. At weeks 4–15 of lactation, all cows were fed conventional corn silage. The conventional corn silage was 44% NDF and 26% starch and had a 30-h NDFD of 57%. The BMR corn silage was 45.5% NDF and 23.5% starch and had an NDFD of 74%. Prepartum DMI was 1.1 kg/day higher for cows on the BMR diet. DMI in the first 3 weeks of lactation was 2 kg/day higher for cows on the BMR diet, while milk production was 3.2 kg/day higher ($P < 0.03$). Milk production was 2 kg/day higher for the BMR cows during weeks 4–15 of lactation. These results emphasize the importance of feeding a highly digestible forage to early lactation cows. Even though these cows were switched to conventional corn silage at 4 weeks of lactation, there was still a positive carryover effect on milk production.

A 500-cow New York herd is an example of the impact of SARA (Stone, 1999). Herd milk production was 30 kg/cow/day with milk fat of 3.4% and milk true protein of 2.8%. Milk production was 2.7 kg/day less than expected and the herd had signs of laminitis. The diet fed was 50% forage with 34% NDF, 25% effective NDF 27% starch, and 77% fermentable starch. The diet energy sources were high-moisture ear corn (3.2 kg/day) and corn meal (1.8 kg/day). Feed refusals were similar to the TMR fed in terms of particle size distribution. Rumen samples for pH were taken by ruminocentesis. Greater than 50% of the cows sampled had a rumen pH < 5.8 and about 33% had pH < 5.6. Diets were adjusted by replacing the high-moisture ear corn with corn meal. Diet NDF was 33%, starch was 28%, and fermentable starch was 73% in the adjusted diet. This change decreased the amount of starch being digested in the rumen. Milk production increased to 34 kg in about 3 weeks, with milk fat of 3.7% and milk true protein of 2.9%. At that point, only one cow had a rumen pH < 5.8. Financial impact was estimated using two milk pricing systems. One system used a milk blend price using only milk fat. The second used multiple component pricing, including fat, protein, and solids not fat. The cost of SARA was estimated at US$400–475 per cow.

9 On-farm analysis tools

There are several tools available that can be used to assess aspects of carbohydrate nutrition at the farm level. These include:

1. Forage and TMR particle size - using the Penn State Particle Separator. Wet samples are sieved on a set of three screens (19 mm, 8 mm, and 4 mm) and a pan. Details on use are available (Heinrichs, 2013). Table 6 reveals the recommended particle size distribution for forages and TMRs.
2. Manure evaluation (Hall, 2002) - This is done by walking through the animal housing facility and observing and scoring fresh cow manure. The consistency, shape, and presence of fiber or corn particles are the primary things to observe. A manure scorecard is available (https://nagonline.net/wp-content/uploads/2018/04/Manure-score-card.pdf) as is a PowerPoint presentation on the topic (https://www.ars.usda.gov/ARSUserFiles/50901500/px-based_v3.2/educ-matrls/pdfs/PP_manure-evaluation_hall.pdf?_sm_au_=iHVQDNDkSPk0275l).
3. Fecal starch (Fredin et al., 2014) - This tool provides a way to assess total tract starch digestibility. Samples of fresh manure are obtained from 8 to 12 cows per group. Samples are sent to a laboratory for analysis. Total tract starch digestibility is calculated as: 100 - 1.25 × fecal starch. A value <3% fecal starch is considered good and should represent >96% total tract starch digestibility. Ferguson (2003) estimated that for every 1% increase above 3% fecal starch, milk production would decrease by 0.32 kg/day. A summary of 596 fecal starch samples from Dairyland Laboratories (www.dairylandlabs.com, 2013) reported that 34% of the samples were <3% fecal starch, while 45% were greater than 5.5%. Fecal samples were obtained from 19 Holstein herds in Illinois (Lidy et al., 2009). The average fecal starch was 6% with a range of 3.9-9.9%. The calculated total tract starch digestibility average was 84.6% with a range of 70-96%.

Table 6 Particle size recommendations for lactating cows, % on screen

Screen	Screen opening, mm	Corn silage	Haylage	TMR
Upper	**19**	3-8	5-15	<5
Middle	**8**	50-65	50-75	>50
Lower	**4**	30-40	25-30	10-20
Pan	-	<5	<5	25-30

Source: Adapted from Grant (2019).

4 Corn silage processing score (Ferreira and Mertens, 2005) - Samples of corn forage are taken at harvest and sent to a forage analysis laboratory. Samples are dried and sieved using a Ro-Tap shaker. The material passing through a 4.75 mm sieve is collected and analyzed for starch. The percent of the total starch passing through this screen is the corn silage processing score. A value >70% is considered as optimal processing. The disadvantage of this analysis is that it cannot be used to make decisions for adjusting forage harvester settings during harvest to optimize kernel processing.
5 Harvest monitoring of kernel processing (Luck et al., 2019) - This can be used to make rapid changes in adjusting harvester setting for kernel processing. One system counts the number of kernels in 0.95 L. If >2 whole or half corn kernels are present, processing is not optimal and harvester adjustments are necessary. A second approach floats a fresh corn forage sample in a water containing pan. The coarse material floats to the top and is discarded. The grain that sinks to the bottom can be evaluated for degree of processing.

10 Nutrition models

Nutrition models provide the mechanism to formulate diets for carbohydrates while integrating the many factors affecting their digestibility, interactions with other dietary components, and impacts on animal performance. The models are useful to evaluate and formulate dairy cattle diets for on-farm application, can be useful for research to identify areas in need of better characterization, and can be useful for teaching nutrition in terms of an integrated animal-microbial feed system. Models vary in depth of detail, capability, and ease of use. One goal is to aid in improving the efficiency of animal performance while lowering environmental impact and improving profitability. Review papers on models and model concepts are available (Tedeschi, 2006, 2019; Ellis, 2021). A recent paper reported that nine diet balancing programs were used by dairy cattle nutritionists in the USA (Prestegaard-Wilson et al., 2021). This number did not include programs developed and used by commercial feed companies. The programs listed ranged from an Excel spreadsheet to models incorporating the CNCPS model (Van Amburgh et al., 2015). Several models are available in both the USA and Europe that have the capability to integrate animal and feed characteristics, digestion, and passage rates to enhance the efficiency of animal performance and lower the environmental impact of milk production. Several nutritional models available from universities or research institutions are listed in Table 7. There are other models available and used for teaching, research, and diet formulation.

Table 7 Nutritional models available from universities or research institutions

Model	Country	Website	References
CNCPS[a]	USA	www.cncps.cornel.edu	Tylutki et al. (2008); Van Amburgh et al. (2015)
Molly	USA	www.vmtrc.ucdavis.edu/lboratories/molly	Baldwin (1995); Johnson and Baldwin (2008)
NASEM	USA	www.nap.edu/catalog/35806	NASEM (2021)
NorFor	Norway, Sweden		Volden et al. (2011)
Karoline	Sweden		Danfaer et al. (2006a); Danfaer et al. (2006b)
InRation	France	www.inration-rminsl-fr/en/inration-V5-ruminal	INRA (2018)

The evaluation of model predictions compared to *in vivo* research results is invaluable to providing the data needed for further improvement of our understanding of ruminant nutrition, and, thus, of the models. Sufficiently complete data sets including descriptions of feeds, animals, and responses are essential for data sets to be used. The *in vivo* responses can be compared to specific predictions from models. Table 8 is an example of CNCPS model outputs when the same diet is fed at three levels of DMI. The diet was 68% forage and 32% concentrate. Diet specifications were 17% crude protein (CP), 29.8% NDF, 26% starch, 4.2% fat, and 4% sugar. The accuracy and directionality of predictions of ECM, microbial protein supply, and methane production are all testable with animal studies. This information can be used to help better understand the impacts of diet formulation, animal definition, environmental condition, or DMI on predicted animal performance. The data in Table 8 show the impact of changes in DMI on predicted milk production and the partition of carbohydrate fraction utilized in the rumen when the same diet is fed. A 3 kg/day change in DMI is predicted to shift milk production by about 7 kg/day. This emphasizes the importance of DMI in formulating or evaluating diets. Predictions of methane (CH_4) emissions from a diet also vary depending on DMI.

11 Methane

Methane (CH_4) is a natural co-product of carbohydrate fermentation in the rumen. Carbon dioxide and hydrogen ions are the building blocks for CH4 synthesis. Hydrogen ion sources are acetic and butyric acids while propionic acid and fatty acids are hydrogen sinks (Bannink et al., 2011). It is estimated that the energy associated with CH_4 formation represents 2–12% of diet gross energy (Broucek, 2014). The types and quality forages fed, the type and availability of diet carbohydrate sources and the type and quantity of fatty acids in the diet influence the volatile fatty acids produced and the hydrogen ion balance.

Key drivers of the amount of CH_4 produced are dry matter intake, milk production and the quantity of carbohydrates digested in the rumen. A correlation of 0.937 was reported between dry matter intake and the grams of CH_4 produced (Kriss, 1930). This was derived from 131 cattle trials. A second study reported a correlation of 0.94 between the grams of carbohydrate digested and the grams CH_4 produced (Bratzler and Forbes, 1940). This relationship was based on 130 measurements in beef and dairy animals.

There are several nutritional approaches that can be used to mitigate CH_4 emissions in dairy cows. Two recent review papers provide an overview of these approaches (Beauchemin et al., 2022; Hristov et al., 2022).

12 Conclusion and future trends in research

There have been significant advances in defining, describing, and utilizing carbohydrates in dairy cattle diets. The availability of analytical procedures to better describe feed fractions and provide multiple time points was key in developing digestion rates using kinetic models. Establishment of commercial feed analysis laboratories was a major advance in providing data for use in

Table 8 Example outputs from CNCPS 6.5.5 model

Item	Base DMI	Decrease DMI	Increase DMI
DMI, kg/day	26	23	29
ME allowable milk, kg	50.4	43.2	57.4
MP allowable milk, kg	50.3	43.5	57.2
ME, Mcal/day	71.8	63.6	79.9
MP, g/day	3097	2704	3496
Microbial protein, g/day	1771	1892	1951
Rumen degradable sugar, g/day	294	274	312
Rumen degradable starch, g/day	5406	4877	5915
Rumen degradable fiber, g/day	4671	4147	5193
Rumen CHO degraded, g/day	13 047	11 688	14 378
Rumen CHO escape, g/day	5643	4846	6469
CH_4, g/day	495	457	527
CH_4, g/kg DMI	19.0	19.9	18.2
CH_4, g/kg ME milk	9.8	10.6	9.2
Forage NDF intake, % of BW	1.2	0.9	1.14

ME, metabolizable energy; MP, metabolizable protein; CHO, carbohydrate.

utilizing these concepts by feed companies and nutritionists in formulating grain mixes and dairy cattle diets. This was accomplished by the development and application of computer models that could integrate nutritional concepts with feed and forage analytical information. Improving the understanding of starch utilization by accounting for factors including endosperm texture, particle size, moisture content, processing method, and conservation system was key to enhancing milk and milk component production. Development of summative equations to predict feed energy values, using uNDF to estimate indigestibility, and the role of NDF in stimulating chewing, rumination, and maintaining the rumen environment were pivotal in improving animal productivity and health. During this same time, there were significant improvements in plant and forage genetics which improved NDF digestibility. These improvements provided the base to utilize higher levels of forage in the diets of high-producing dairy cows.

There are still many areas in which additional research is needed. Some of these require innovative approaches and insights. In other cases, additional research is needed to validate or expand preliminary data on recently proposed concepts. Examples of needed research include:

- Equivalency of yield of SCFA from different NDF sources;
- Why sugars tend to support milk fat production compared to starch that tends to enhance milk volume?
- Integration of diet composition and physical form relative to intake, passage, and milk production;
- Factors that influence passage and impact the nutrient supply from the rumen;
- Modeling the interaction of protein and carbohydrates in the rumen using new dietary approaches;
- Determine the impact of differing rates of ruminal fermentation of carbohydrates on nutrient supply using a time step shorter than 'day';
- Additional animal trials, both intensive and long term, with a wider variety of forages to better define the guidelines for using uNDF and peuNDF240 in dairy diets;
- Ruminal kinetics: Development of multiple time point analyses for starch to be used as a base for determining digestion rates that are correlated to in vivo responses. Do nutrient fractions fit more than one potentially digestible pool? Develop methods to assess liquid and solid passage rates that are reliable and agree with animal responses/flow from the rumen. Develop methods for quantifying microbial product formation in vivo;
- Is ROM acceptable for prediction of energy values or is more refinement into specific carbohydrate fractions warranted?

- Relationships between forage fragility, NDFD, and iNDF in dairy cattle diets as they impact DMI, ECM, and rumen turnover;
- Methods for describing forage fragility to integrate with other NDF descriptors; and
- Does ruminal-soluble lignin matter in dairy cattle diets?

13 Where to look for further information

13.1 Books

- Dairy nutrition 2014. Veterinary Clinics of North America: Food Animal Practice, Vol. 30, Issue 3. Elsevier, Inc. ISBN: 978-0-323-32688-9.
- Forages, 7th edition, Vol. I, An introduction to grassland agriculture. 2017. West Sussex: John Wiley & Sons, Ltd..
- Forages, 7th edition, Vol. II, The Science of Grassland Agriculture. 2020. West Sussex: John Wiley & Sons, Ltd..
- NASEM (National Academies of Sciences, Engineering, and Medicine). 2021. Nutrient Requirements of Dairy Cattle. 8th rev. ed. Washington, DC: Natl. Acad. Sci.. https://nap.nationalacademies.org/catalog/25806.
- Van Soest, P. J. 1994. Nutritional Ecology of the Ruminant. 2nd ed. Ithaca, NY: Cornell University Press.
- Large Dairy Herd Management. 2017. 3rd ed. Beede, D. K. (ed.), American Dairy Science Association. http://ldhm/adsa.org.
- Rumen Microbiology and Its Role in Ruminant Nutrition. 2002. Russell, J. B. https://www.ars.usds.gov/research/software/download/?softwareid-409.

13.2 Nutrition conferenced proceedings

Searchable proceedings of animal conferences (SPAC) – This database currently includes 63 conferences and 676 conference proceedings. There is a fee to access SPAC: https://spac.adsa.org.

13.3 Review papers

Beauchemin, K. A. 2018. Invited review: current perspectives on eating and rumination activity in dairy cows. *J. Dairy Sci.* 101(6):4762–4784. doi:10.3168/jds.2017-13706.

Beauchemin, K. A., Ungerfeld, E. M., Eckard, R. J. and Wang, M. 2020. Review: fifty years of research on rumen methanogenesis: lessons learned and future challenges for mitigation. *Animal* 14 (Suppl.1):s2–s16. doi:10.1017/S1751731119003100.

Firkins, J. L. 2021. Invited review: advances in rumen efficiency. *Appl. Anim. Sci.* 37:399-403. doi:10.15232/aas.2021-02163.

Guiberti, G., Gallo, A., Masoero, F., Ferraretto, L. F., Hoffman, P. C. and Shaver, R. D. 2014. Factors affecting starch utilization in large animal food production system: a review. *Starch* 66(1-2):72-90. doi:10.1002/star.201300177.

Hackman, T. J. and Firkins, J. L. 2015. Maximizing efficiency of rumen microbial protein production. *Front. Microbiol.* 6:465. doi:10.3389/fmicb.2015.00465.

Hall, M. B. and Mertens, D. R. 2017. A 100-year Review: carbohydrates - characterization, digestion, and utilization. *J. Dairy Sci.* 100(12):10078-10093. doi:10.3168/jds.2017-13311.

Jenkins, T. C. and Harvatine, K. J. 2014. Lipid feeding and milk fat depression. *Vet. Clin. North Am. Food Anim. Pract.* 30(3):623-642. doi:10.1016/j.cvfa.2014.07.006.

Klevenhusen, F. and Zebelli, Q. 2021. A review on the potentials of using feeds rich in water-soluble carbohydrates to enhance rumen health and sustainability of dairy cattle production. *J. Sci. Food Agric.* 101(14):5737-5746. doi:10.1002/jsfa.11358.

Krause, K. M. and Oetzel, G. R. 2006. Understanding and preventing subacute ruminal acidosis in dairy herds: a review. *Anim. Feed Sci. Technol.* 126(3-4):215-236. doi:10.1016/j.anifeedsci.2005.08.004.

Moharrey, A., Larsen, M. and Weisberg, M. R. 2014. Starch digestion in the rumen, small intestine, and hind gut of dairy cows - A meta-analysis. *Anim. Feed Sci. Technol.* 192:1-14. doi:10.1016/j.anifeedsci.2014.03.001.

Patra, A., Park, T., Kim, K. and Yu, Z. 2017. Rumen methanogens and mitigation of methane emission by anti-methanogenic compounds and substances. *J. Anim. Sci. Biotechnol.* 8:1-18. doi:10.1186/s40104-017-0145-9.

Reynolds, C. K. 2006. Production and metabolic effects of site of starch digestion in dairy cattle. *Anim. Feed Sci. Technol.* 130(1-2):78-94. doi:10.1016/j.anifeedsci.2006.01.019.

Trotta, R. J., Harmon, D. L., Matthews, J. C. and Swanson, K. C. J. 2022. Nutritional and physiological constraints contributing to limitations in small intestinal starch digestion and glucose absorption in ruminants. *Ruminants* 2(1):1-26. doi:10.3390/ruminants2010001.

Zebeli, Q., Aschenbach, J.R., Tafaj, M., Boguhn, J., Ametaj, B.N., and Drochner, W. 2012.Invited review: Role of ohysically effective fiber and estimation of dietary fiber adequacy in high-producing dairy cattle. J. Daiy Sci. 85:1041-1056. http://dx.doi.org/10.3168/jd011-4421.

14 References

Aldrich, J. M., Muller, L. D., Varga, G. A. and Griel, L. C. 1993. Nonstructural carbohydrate and protein effects on rumen fermentation, nutrient flow, and performance of dairy cows. *J. Dairy Sci.* 76(4):1091–1105.

Allen, M. S., Longuski, R. A. and Ying, Y. 2021a. Effects of corn endosperm type and fineness of grind on feed intake, feeding behavior, and productive performance of lactating dairy cows. *J. Dairy Sci.* 104(7):7630–7640. https://doi.org/10.3168/jds.2020-18991.

Allen, M. S., Longuski, R. A. and Ying, Y. 2021b. Effects of corn grain endosperm type and fineness of grind on site of digestion, ruminal digestion kinetics, and flow of nitrogen fractions to the duodenum in lactating dairy cows. *J. Dairy Sci.* 104(7):7641–7652. https://doi.org/10.3168/jds.2020-18992.

Allen, M. S. and Ying, Y. 2021c. Effects of corn grain endosperm type and conservation method on site of digestion, ruminal digestion kinetics, and flow of nitrogen fractions to the duodenum in lactating dairy cows. *J. Dairy Sci.* 104(7):7617–7629. https://doi.org/10.3168/jds.2020-18882.

Argyle, J. L. and Baldwin, R. L. 1989. Effects of amino acids and peptides on rumen microbial growth yields. *J. Dairy Sci.* 72(8):2017–2027.

Baldwin, R. L. 1968. Estimation of theoretical calorific relationships as a teaching technique: A review. *J. Dairy Sci.* 51(1):104–111.

Baldwin, R. L. 1995. Modeling ruminant digestion and metabolism. https://link.springer.com/book/9780412591600.

Bannink, A., van Schjindel, M. W. and Dijkstra, J. 2011. A model of enteric fermentation in dairy cows to estimate methane emission for the Dutch National Inventory Report using the IPCC Tier 3 approach. *Anim. Feed Sci. Tech.* 166–167:603–618. https://doi.org/10.1016/anifeedsci.2011.04.043.

Bargo, F., Muller, L. D., Delahoy, J. E. and Cassidy, T. W. 2002. Milk response to concentrate supplementation of high producing dairy cows grazing at two pasture allowances. *J. Dairy Sci.* 85(7):1777–1792.

Beauchemin, K. A., Ungerfeld, E. M., Abdalla, A. L., Alvarez, C., Arndt, C., Becquet, P., Benchaar, C., Berndt, A., Mauricio, R. M., McAllister, T. A., Oyhantcabal, W., Salami, S. A., Shalloo, L., Sun, Y., Tricarico, J., Uwizeye, A., De Camilia, C., Bernoux, M., Robinson, T. and Kebreab, E. 2022. Invited review: Current enteric methane mitigation options. *J. Dairy Sci.* 105(12):9297–9326. https://doi.org/10.3168/jds.2022-22091.

Boerman, J. P., Potts, S. B., VanderHaar, M. J., Allen, M. S. and Lock, A. L. 2015. Milk production responses to a change in dietary starch concentration vary by production level in dairy cattle. *J. Dairy Sci.* 98(7):4698–4706. https://doi.org/10.3168/jds.2014-8999.

Boudon, A., Juton, J., Delagarde, R., Faverdin, P. and Peyraud, J. L. 2009. Rate of propionate absorption influences intake in dairy cows fed ryegrass. Ruminant physiology. Digestion, metabolism, and effects of nutrition on reproduction and welfare. In: Chilliard, Y., Glasser, F., Faulconnier, Y., Bocquier, F., Veissier, I. and Doreau, M. (Eds.), *Proceedings of the XIth Int. Symp. Ruminant Phys.* The Netherlands: Wageningen Academic Publishers, pp. 70–71.

Bratzler, J. W. and Forbes, E. B. 1940. The estimation of methane production by cattle. *J. Nutr.* 19:611-613.

Broderick, G. A., Huhtanen, P., Ahvenjarvi, S., Reynal, S. M. and Shingfield, K. J. 2010. Quantifying ruminal nitrogen metabolism using omasal sampling technique in cattle - A meta-analysis. *J. Dairy Sci.* 93(7):3216-3230. https://doi.org/10.3168/jds.2009-2989.

Broderick, G. A., Luchini, N. D., Reynal, S. M., Varga, G. A. and Ishler, V. A. 2008. Effect on production of replacing starch with sucrose in lactating dairy cows. *J. Dairy Sci.* 91(12):4801-4810. https://doi.org/10.3168/jds.2008-1480.

Broderick, G. A. and Radloff, W. J. 2004. Effect of molasses supplementation on the performance of lactating dairy cows fed diets based on alfalfa and corn silages. *J. Dairy Sci.* 87(9):2997-3009.

Broucek, J. 2014. Production of methane emissions from ruminant husbandry: A review. *J. Environ. Prot.* 5:1482-1483. http://doi.org/10.4236/jep.2014.515141.

Buza, M. H. and Holden, L. A. 2016. A survey of feeding management practices and by-product feed usage on Pennsylvania dairy farms. *Prof. Anim. Sci.* 32(2):248-252. http://doi.org/10.15232/pas.2015-01481.

Cantalapiedra-Hijar, G., Lemosquet, S., Rodriguez-Lopez, J. M., Messad, F. and Ortigues-Marty, I. 2014. Diets rich in starch increase the posthepatic availability of amino acids in dairy cows fed diets at low and normal protein levels. *J. Dairy Sci.* 97(8):5151-5166. http://doi.org/10.3168/jds.2014-8019.

Chase, L. E. 2019. What do high producing herds really feed? *Proceedings of the Cornell Nutr. Conf.* https://hdl.handle.net/1813/69483, pp. 225-230.

Chase, L. E. and Grant, R. J. 2013. High forage rations - What do we know? *Proceedings of the Cornell Nutr. Conf.* https://ecommons.cornell.edu/handle/1813/36474.

Chen, G., Stobel, H. J., Russell, J. B. and Sniffen, C. J. 1987. Effect of hydrophobicity on utilization of peptides by rumen bacteria in vitro. *Appl. Environ. Microbiol.* 53(9):2021-2025.

Cone, J. W., Cliné-Theil, W., Malestein, A. and Van't Klooster, Th. A. 1989. Degradation of starch by incubation with rumen fluid. A comparison of different starch sources. *J. Sci. Food Agric.* 49:173183.

Counotte, G. H. M. and Prins, R. A. 1981. Regulation of lactate metabolism in the rumen. *Vet. Res. Commun.* 5(2):101-115.

Cueva, S. F., Stefani, S., Melgar, A., Raisanen, S. E., Lage, C. F. A., Wasson, D. E., Fetter, M. E., Pelaez, A. M., Roth, G. E. and Hristov, A. N. 2021. Lactational performance, rumen fermentation, snd enteric methane emissions of dairy cows fed an amylase-enabled corn silage. *J. Dairy Sci.* 104:9827-9841. https://doi.org/10.3168/jds.2021-20251.

Damiran, D., DelCurto, T., Bohnert, D. W. and Findholt, S. L. 2008. Comparison of technique and grinding size to estimate digestibility of forage based ruminant diets. *Anim. Feed Sci. Technol.* 141(1-2):15-35. https://doi.org/10.1016/j.anifeedsci.2007.04.007.

Danfaer, A., Huhtanen,P., Uden, P., Sveinbjortnsson, J. and Volden, H. 2006a. The Nordic dairy cow model, Karoline-description. In: Kebreab, E., Dikstra, J., Bannink, A., Gerrits, W. J. J. and France, J. (Eds.), *Nutrient Digestion and Utilization in Farm Animals Modeling Approaches*. CABI Publishing, pp. 383-406.

Danfaer, A., Huhtanen, P., Uden, P., Sveinbjornsson, J. and Volden, H. 2006b. The Nordic dairy cow model, Karoline-evaluation. In: Kebreab, E., Dikstra, J., Bannink, A., Gerrtis,

W. J. J. and France, J. (Eds). *Nutrient Digestion and Utilization in Farm Animals Modeling Approaches*. CABI Publishing, pp. 407–415.

Dann, H. M., Fredin, S. M., Contanch, K. W., Grant, R. J., Kokko, C., Ji, P. and Fujita, K. 2015. Effects of corn-based reduced-starch diets using alternative carbohydrate sources on performance of lactating Holstein cows. *J. Dairy Sci.* 98(6):4041–4054. http://doi.org/10.3168/jds.2014-9078.

Dann, H. M., Tucker, H. A., Cotanch, K. W., Krawczel, P. D., Mooney, C. S., Grant, R. J. and Eguchi, T. 2014. Evaluation of lower starch diets for lactating Holstein cows. *J. Dairy Sci.* 97(11):7151–7161. http://doi.org/10.3168/jds.2014-8341.

de Ondarza, M. B., Emanuele, S. M. and Sniffen, C. J. 2017. Effect of increased dietary sugar on dairy cow performance as influenced by diet nutrient components and level of milk production. *Prof. Anim. Sci.* 33(6):700–707. https://doi.org/10.15232/pas.2017-01648.

de Ondarza, M. B., Emanuele, S. M. and Sniffen, C. J. 2020. Dietary sugars for optimizing rumen function and dairy cow performance. *Proceedings of the Cornell Nutr. Conf. 11 pgs*. https://hdl.handle.net/1814/72888.

de Souza, R. A., Templeman, R. J., Allen, M. S., Weiss, W. P., Bernard, J. K. and VandeHaar, M. J. 2018. Predicting nutrient digestibility in high producing dairy cows. *J. Dairy Sci.* 101(2):1123–1135. https://doi.org/10.3168/jds.2017-13344

Der Bedrosian, M. C., Nestor, K. E. Jr. and Kung, L. Jr. 2012. The effects of hybrid, maturity, and length of storage on the composition and nutritive value of corn silage. *J. Dairy Sci.* 95(9):5115–5126. http://doi.org/10.3168/jds.2011-4833.

Devries, T. J., Dohme, F. and Beauchemin, K. A. 2008. Repeated ruminal acidosis challenges in lactating dairy cows at high and low risk for developing acidosu: Feed sorting. *J. Dairy Sci.* 91(10):3958–3967. https://doi.org/10.3168/jds.2008-1347.

Devries, T. J. and Gill, R. M. 2012. Adding liquid feed to a total mixed ration reduces feed sorting behavior and improves productivity of lactating dairy cows. *J. Dairy Sci.* 95(5):2648–2655. http://doi.org/10.3168/jds.2011-4965.

Doane, P. H., Schofield, P. and Pell, A. N. 1997. Neutral detergent fiber disappearance and gas and volatile fatty acid production during the in vitro fermentation of six forages. *J. Anim. Sci.* 75(12):3342–3352. https://doi.org/10.2527/1997.75123342x.

Donovan, G. A., Risco, C. A., DeChant Temple, G. M., Tran, T. Q. and van Horn, H. H. 2004. Influence of transition diets on occurrence of subclinical laminitis in Holstein dairy cows. *J. Dairy Sci.* 87(1):73–84.

Ellis, J. L. 2021. The role of mechanistic models in precision dairy nutrition and on-farm decision support. In: *Proceedings of the Anim. Nutr. Conf.*, Canada.

Ertl, P., Zebeli, Q., Zollitsch, W. and Knaus, W. 2015. Feeding of by-products completely replaced cereal grains and pulses in dairy cows and enhanced edible feed conversion ratio. *J. Dairy Sci.* 98(2):1225–1233. http://doi.org/10.3168/jds.2014-8810.

Ferguson, J. D. 2003. Monitoring feeding programs on dairy farms. Proceedings of Nutrition and Management of Dairy Cattle. Consorzio Ricerca Fileria Lattiero-Casearia Regione Siciliana.

Ferraretto, L. F., Crump, P. M. and Shaver, R. D. 2013. Effect of cereal grain type and corn grain harvesting and processing methods on intake, digestion, and milk production by dairy cows through a meta-analysis. *J. Dairy Sci.* 96(1):533–550. http://dx.oi.org/10.3168/jds.2012-5932.

Ferraetto, L. F., Fonesca, A. C., Sniffen, C. J., Formigoni, A. and Shaver, R. D. 2015a. Effect of corn silage hybrids differing in starch and neutral detergent fiber digestibility on lactation performance and total-tract digestibility by dairy cows. *J. Dairy Sci.* 98(1):395–405. http://doi.org/10.3168/jds.2014-8232.

Ferraretto, L. F., Shaver, R. D., Massie, S., Singo, R., Taysom, D. M. and Brouillette, J. P. 2015b. Effect of ensiling time and hybrid type on fermentation profile, nitrogen fractions, and ruminal in vitro starch and neutral detergent fiber digestibility in whole-plant corn silage. *Prof. Anim. Sci.* 31(2):146–152. http://doi.org/10.15232/pas.2014-01371.

Ferreira, G. and Mertens, D. R. 2005. Chemical and physical characteristics of corn silages and their effects on in vitro disappearance. *J. Dairy Sci.* 88(12):4414–4425.

Firkins, J. L., Eastridge, M. L., St-Pierre, N. R. and Noftsger, S. M. 2001. Effects of grain variability and processing on starch utilization by lactating dairy cattle. *J. Anim. Sci.* 79(E-Suppl):E218–E238.

Firkins, J. L., Hristov, A. N., Hall, M. B., Varga, G. A. and St-Pierre, N. R. 2006. Integration of ruminal metabolism in dairy cattle. *J. Dairy Sci.* 89(E) (Suppl. 1):E31–E51.

Fredin, S. M. 2015. How to get more out of dietary starch and low starch diets. *Proceedings of the Cornell Nutr. Conf.* https://hdl.handle.net/1813/41225, pp. 91–101.

Fredin, S. M., Akins, M. S., Ferraretto, L. F. and Shaver, R. D. 2015a. Effects of corn-based diet starch content and neutral detergent fiber source on lactation performance, digestibility, and bacterial protein flow in dairy cows. *J. Dairy Sci.* 98(1):554–565. http://fx.doi.org/10.3168/jds.2014-8592.

Fredin, S. M., Ferraretto, L. F., Akins, M. S., Bertics, S. J. and Shaver, R. D. 2015b. Effects of corn-based diet starch content and corn particle size on lactation performance, digestibility, and bacterial protein flow in dairy cows. *J. Dairy Sci.* 98(1):541–553. http://doi.org/10.3168/jds.2014-8502.

Fredin, S. M., Ferraretto, L. F., Akins, M. S., Hoffman, P. C. and Shaver, R. D. 2014. Fecal starch as an indicator of total-tract starch digestibility by lactating dairy cows. *J. Dairy Sci.* 97(3):1862–1871. http://doi.org/10.3168/jds.2013-7395.

Gaillard, B. D. E. and Richards, G. N. 1975. Presence of soluble lignin – Carbohydrate complexes in the bovine rumen. *Carbohydr. Res.* 42(1):135–145.

Goering, H. K. and Van Soest, P. J. 1970. *Forage Fiber Analysis (Apparatus, Reagents, Procedures, and Some Applications).* Washington, DC: USDA Agricultural Research Service, Handbook No. 379.

Grant, R. 2015. Making milk with forage: Understanding rumen fiber dynamics. *Proceedings of the Four-State Dairy Nutr. and Management Conf.* https://www.agribusiness.org/fourstatedairy/2015/2015-4state.proceedings, pp. 63–69.

Grant, R. 2019. Rethinking forage particle size. *Hay Forage Grower*, November 30.

Grant, R. J., Smith, W. A., Miller, M. D., Ishida, K. and Obata, A. 2018. Relationships between undigested and physically effective fiber in lactating dairy cows. *Proceedings of the Cornell Nutr. Conf. for Feed Manufacturers*, East Syracuse, NY. https://hdl.handle.net/1813/59839, October 16–18, pp. 35–45.

Griswold, K. E., Apgar, G. A., Bouton, J. and Firkins, J. L. 2003. Effects of urea infusion and ruminal degradable protein concentration on microbial growth, digestibility, and fermentation in continuous culture. *J. Anim. Sci.* 81(1):329–336.

Hall, M. B. 2002. Working with non-NDF carbohydrates: With manure evaluation and environmental considerations. *Mid-South Nutr. Conference*, Arlington, TX, pp. 37–48.

Hall, M. B. 2015. Determination of dietary starch in animal feeds and pet food by an enzymatic-colorimetric method: collaborative study. *J. Assoc. Off. Anal. Chem.* 98:397–409.

Hall, M. B. 2017a. Nitrogen source and concentration affect utilization of glucose by mixed ruminal microbes in vitro. *J. Dairy Sci.* 100(4):2739–2750. https://doi.org/10.3168/jds.2016-12091.

Hall, M. B. 2017b. Sugars in dairy cattle rations. *Proceedings of the Tri-State Dairy Nutrition Conference*, pp. 135–146.

Hall, M. B. and Chase, L. E. 2014. Responses of late-lactation cows to forage substitutes in low-forage diets supplemented with by-products. *J. Dairy Sci.* 97(5):3042-3052. http://doi.org/10.3168/jds.2013-7539.

Hall, M. B., Hoover, W. H., Jennings, J. P. and Miller Webster, T. K. M. 1999. A method for partitioning neutral detergent-soluble carbohydrates. *J. Sci. Food Agric.* 79(15):2079–2086.

Hall, M. B. and Mertens, D. R. 2017. A 100-year review: Carbohydrates-characterization, digestion, and utilization. *J. Dairy Sci.* 100(12):10078–10093. https://doi.org/10.3168/jds.2017-13311.

Hall, M. B. and Zanton, G. I. 2022. Substitution of cane molasses for corn grain at two levels of degradable protein I. Lactating cow performance, nutrition model predictions, and potential basis for butterfat and intake responses. *J. Dairy Sci.* 105(5):3939-3953. https://doi.org/10.3168/jds.2021-21241.

Harvatine, K. J. 2018. Understanding milk fat depression and rumen health through milk fatty acid profile. *Proceedings of the Minn. Nutr. Conf.*, pp. 91–99.

Havekes, C. D., Duffield, T. F., Carpenter, A. J. and DeVries, T. J. 2020a. Effects of wheat straw chop length in high-straw dry cow diets on intake, health, and performance of dairy cows across the transition period. *J. Dairy Sci.* 103(1):254–271. https://doi.org/10.3168/jds.2019-17033.

Havekes, C. D., Duffield, T. F., Carpenter, A. J. and DeVries, T. J. 2020b. Moisture content of high-straw dry cow diets affects intake, health, and performance of transition dairy cows. *J. Dairy Sci.* 103(2):1500–1515. https://doi.org/10.3168/jds.2019-17557.

Havekes, C. D., Duffield, T. F., Carpenter, A. J. and DeVries, T. J. 2020c. Effects of molasses-based liquid feed supplementation to a high-straw dry cow diet on feed intake, health, and performance of dairy cows across the transition period. *J. Dairy Sci.* 103(6):5070–5089. https://doi.org/10.3168/jds.2019-18085.

Heinrichs, J. 2013. The Penn State particle separator. DSE 2013-186. Department of Animal Science, The Pennsylvania State University.

Heldt, J. S., Cochran, R. C., Stokka, G. L., Farmer, C. G., Mathis, C. P., Titgenmeter, E. C. and Nagaraja, T. G. 1999. Effects of different supplemental sugars and starch fed in combination with degradable intake protein on low-quality forage use by beef steers. *J. Anim. Sci.* 77(10):2793–2802.

Herrera-Saldana, R., Gomez-Alarcon, R., Torabi, M. and Huber, J. T. 1990. Influence of synchronizing protein and starch degradation in the rumen on nutrient utilization and microbial protein synthesis. *J. Dairy Sci.* 73(1):142–148.

Herrera-Saldana, R. and Huber, J. T. 1989. Influence of varying protein and starch degradabilities on performance of lactating cows. *J. Dairy Sci.* 72(6):1477–1483.

Hoover, W. H. and Stokes, S. R. 1991. Balancing carbohydrates and proteins for optimum rumen microbial yield. *J. Dairy Sci.* 74(10):3630–3644.

Hoover, W. H., Tucker, C., Harris, J., Sniffen, C. J. and de Ondarza, M. B. 2006. Effects of nonstructural carbohydrate level and starch:sugar ratio on microbial metabolism in continuous culture of rumen contents. *Anim. Feed Sci. Technol.* 128(3-4):307–319. https://doi.org/10.1016/j.anifeedsci.2005.11.011.

Hristov, A. N., Harper, M. T., Roth, G., Canale, C., Huhtanen, P., Richard, T. L. and Dimarco, K. 2020. Effects of ensiling time on corn silage neutral detergent fiber degradability and relationship laboratory fiber analysis and in vivo digestibility. *J. Dairy Sci.* 103(3):2333–2346. https://doi.org/10.3168/jds.2019-16917.

Hristov, A. N., Melgar, A., Wasson, D. and Arndt, C. 2022. *Symposium* review: Effective nutritional strategies to mitigate enteric methane in dairy cattle. *J. Dairy Sci.* 105(10):8543–8557. https://doi.org/10.3168/jds.2021-21398.

Hristov, A. N., Ropp, J. K.,Grandeen, K. L., Abedi, S., Etter, R. P., Melgar, A. and Foley, A. E. 2005. Effect of carbohydrate source on ammonia utilization in lactating dairy cows. *J. Anim. Sci.* 83(2):408–421.

Huhtanen, P., Asikainen, U., Arkkila, M. and Jaakkola, S. 2007. Cell wall digestion and passage kinetics estimated by marker and in situ methods or by rumen evacuations in cattle fed hay 2 or 18 times daily. *Anim. Feed Sci. Technol.* 133(3-4):206–227. https://doi.org/10.1016/j.anifeedsci.2006.05.004

INRA. 2018. *INRA Feeding System for Ruminants*. Wageningen, the Netherlands: Wageningen Academic Publishers, p. 640. www.wageningenacademic.com/bookshop.

Jenkins, T. 2019. Milk fat depression: How important is rumen pH? www.fattyavidforum.com/wp-content/uploads/2019/04/Jenkins-MFD-Rumen-pH.pdf.

Jensen, R. G. 2002. The composition of bovine milk lipids: January 1995 to December 2000. *J. Dairy Sci.* 85(2):295–350.

Johnson, H. A. and Baldwin, R. L. 2008. Evaluating model predictions of partitioning nitrogen excretion using the dairy cow model, Molly. *Anim. Feed Sci. Technol.* 143(1-4):104–126. https://doi.org/10.1016/j.anifeedsci.2007.05.007.

Johnson, R. R. 1976. Influence of carbohydrate solubility on non-protein nitrogen utilization in the ruminant. *J. Anim. Sci.* 43(1):184–191.

Karlsson, J., Sporndly, R.,Lindberg, M. and Holtenius, K. 2018. Replacing human-edible feed ingredients with by-products increases net food production efficiency in dairy cows. *J. Dairy Sci.* 101(8):7146–7155. https://doi.org/10.3168/jds.2017-14209.

Kramer, K., Weisbjerg, M. R., Lund, P., Jensen, C. S. and Pedersen, M. G. 2012. Estimation of indifestible NDF in forages and concentrates from cell wall composition. *Anim. Feed Sci. Technol.* 177:40–51.

Krause, K. M. and Oetzel, G. R. 2006. Understanding and preventing subacute ruminal acidosis in dairy herds: A review. *Anim. Feed Sci. Technol.* 126(3-4):215–236. https://doi.org/10.1016/j.anifeedsci.2005.08.004.

Kreikemeier, K. K., Harmon, D. L., Peters, J. P., Gross, K. L., Armendariz, C. K. and Krehbiel, C. R. 1990. Influence of dietary forage and feed intake on carbohydrate activities and small intestinal morphology of calves. *J. Anim. Sci.* 68(9):2916–2929.

Kriss, M. 1930. Quantitative relationships of the dry matter of the food consumed, the heat production, the gaseous outgo, and the insensible loss in body weight of cattle. *J. Agric. Res.* 40:283–295.

LaPierre, P., Fredin, S., Ross, D. and Van Amburgh, M. E. 2020. Impact of starch and energy on amino acid requirements of lactating cows. *Proceedings of the Cornell Nutr. Conf.* https://hdl.handle.net/1813/72897.

Lean, K. I. J. and Golder, H. M. 2020. Ruminal acidosis: beyond pH snd rumen. *Proceedings of the Cornell Nutr. Conf.* https://hdl.handle.net/1813/67040.
Leonardi, C., Giannico, F. and Armentano, L. E. 2005. Effect of water addition on selective consumption (sorting) of dry diets by dairy cattle. *J. Dairy Sci.* 88(3):1043-1049.
Lidy, D., Osorio, J. S., Hutjens, M. F. and Meyer, D. W. 2009. Evaluating total tract starch digestibility. Illinois Dairy Report. Department of Animal Sciences, University of Illinois, pp. 27-29.
Lim, J. M., Nestor, K. R. E., Jr. and Kung, L., Jr. 2015. The effect of hybrid type snd dietary proportions of corn silage on the lactation performance of high producing dairy cows. *J. Dairy Sci.* 98:1195-1203. http://dx.doi.org/10.3168/jds.2014.
Luck, B., Willett, R., Drewry, J. and Ferraretto, L. 2019. Monitoring kernel processing during harvest. A4174. University of Wisconsin, Extension-Madison. https://learningstore.extension.wisc.edu/products/krnl-monitoring-processing-during-harvest.
Lykos, T. and Varga, G. A. 1995. Effects of processing method on degradation characteristics of protein and carbohydrate sources in vitro. *J. Dairy Sci.* 78:1789-1801.
Martinez, C. M., Chung, Y. H., Ishler, V. A., Bailey, K. W. and Varga, G. A. 2009. Effects of dietary forage level and monensin on lactation performance, digestibility and fecal excretion of nutrients, and efficiency of feed nitrogen utilization of Holstein dairy cows. *J. Dairy Sci.* 92(7):3211-3221. https://doi.org/10.3168/jds.2008-1306.
McAllan, A. B. and Smith, R. H. 1974. Carbohydrate metabolism in the ruminant: Bacterial carbohydrates formed in the rumen and their contributions to digesta entering the duodenum. *Br. J. Nutr.* 31(1):77-88.
Mertens, D. R. 1987. Predicting intake and digestibility using mathematical models of rumen function. *J. Anim. Sci.* 64(5):1548-1558.
Mertens, D. R. 1997. Creating a system for meeting the fiber requirements of dairy cows. *J. Dairy Sci.* 80(7):1463-1481.
Mertens, D. R. 2005. Particle size, fragmentation index, and effective fiber: tools for evaluating physical attributes of corn silages. *Proceedings of the Four State Dairy Nutrition and Management Conference*, pp. 211-220.
Mertens, D. R. 2009a. Maximizing forage use by dairy cows. *West. Can. Dairy Seminar Advances in Dairy Technology*, Vol. 21, pp. 303-319.
Mertens, D. R. 2009b. Impact of NDF content and digestibility on dairy cow performance. *West. Can. Dairy Seminar Advances in Dairy Technology*, Vol. 21, pp. 191-201.
Mertens, D. R. 2016. Using uNDF to predict dairy cow performance and design rations. *Proceedings of the Four-State Dairy Nutrition and Management Conference*, June 15 and 16, Dubuque, I. A., pp. 12-19. https://wiagribuiness.org/fourstatedairy/2016/2026-4state.proceedings.
Miller, M. D., Kokko, C., Ballard, C. S., Dann, H. M., Fustini, M., Pamonari, A., Formigoni, A., Cotanch, K. W., and Grant, R. J. 2021. Influence of fiber degradability of corn silage in diets with lower and higher fiber content on lactational performance, nutrient digestibility, and ruminal characteristics in lactating Holstein cows. *J. Dairy Sci.* 104:1728-1743. https://doi.org/10.3168/jds.2020-19088.
Miller-Cushion, E. K. and Devries, T. J. 2017. Feed sorting in dairy cattle: Causes, conseqeuences, and management. *J. Dairy Sci.* 100(5):417-428. https://doi.orgh/10.3168/jds.2016-11983.

Moharrery, A., Larsen, M. and Weisbjerg, M. R. 2014. Starch digestion in the rumen, small intestine, and hind gut of dairy cows – A meta-analysis. *Anim. Feed Sci. Technol.* 192:1–14. http://doi.org/10.1016/j.anifeedsci.2014.03.001.

Morris, D. L., Clark, K., Kononoff, P. J., White, R. R., Hall, M. B. and Firkins, J. L. 2021. *MUNCH: A Smartphone Application for Effective Fiber for Dairy Cows. NebGuide G2316. Index: Animal Agriculture/Dairy.* Lincoln: University of Nebraska. https://dairy.unl.edu/munch-effectivde-fiber-app.

Murphy, M. R., Baldwin, R. L. and Koong, L. J. 1982. Estimation of stoichiometric parameter for rumen fermentation of roughage and concentrate diets. *J. Anim. Sci.* 55(2):411–421.

NASEM (National Academies of Sciences, Engineering, and Medicine). 2021. *Nutrient Requirements of Dairy Cattle.* 8th Revised Edition. Washington, DC: National Academy of Sciences. https://nap.nationalacademies.org/catalog/25806.

Neilson, M. J. and Richards, G. N. 1978. The fate of the soluble lignin-carbohydrate complex produced in the bovine rumen. *J. Sci. Food Agric.* 29(6):513–519.

Nocek, J. E. and Russell, J. B. 1988. Protein and energy as an integrated system: Relationship of ruminal protein and carbohydrate availability to microbial protein synthesis and milk production. *J. Dairy Sci.* 71(8):2070–2107.

Nocek, J. E. and Tamminga, S. 1991. Site of digestion of starch in the gastrointestinal tract of dairy cows and its effect on milk composition. *J. Dairy Sci.* 74(10):3598–3629.

Oba, M. and Allen, M. S. 1999. Evaluation of the importance of digestibility of neutral detergent fiber from forage: Effects on dry matter intake and milk yield of dairy cows. *J. Dairy Sci.* 82(3):589–596.

Oba, M. and Allen, M. S. 2003. Effects of diet fermentability on efficiency of microbial nitrogen production in lactating dairy cows. *J. Dairy Sci.* 86(1):195–207.

Oba, M., Merwin, J. L. and Zhining, Z. 2015. Effects of rumen doses of sucrose, lactose, and corn starch on ruminal fermentation and expression of genes in rumen epithelial cells. *J. Dairy Sci.* 98:586–594. https://doi.org/10.3168/jds.2014-8697.

Palmonari, A., Fustini, M., Canestrari, G., Grilli, E. and Formigoni, A. 2014. Influence of maturity on alfalfa hay nutritional fractions and indigestible fiber content. *J. Dairy Sci.* 97(12):7729–7734. http://doi.org/10.3168/jds.2014-8123.

Palmonari, A., Gallo, A., Fustini,M., Canestrari, G., Masoero, F., Sniffen, C. J. and Formigoni, A. 2016. Estimation of the indigestible fiber in different forage types. *J. Anim. Sci.* 94(1):248–254.

Palmquist, D. L., Davis, C. L., Brown, R. E. and Sachan, D. S. 1969. Availability and metabolism of various substrates in ruminants. V. Entry rate into the body and incorporation into milk fat of D (-)ß-hydroxybutyrate. *J. Dairy Sci.* 52(5):633–638.

Penner, G. B. and Oba, M. 2009. Increasing dietary sugar concentration may improve dry matter intake, ruminal fermentation, and productivity of dairy cows in the postpartum phase of the transition period. *J. Dairy Sci.* 92(7):3341–3353. https://doi.org/10.3168/jds.2008-1977.

Potts, S. B., Boerman, J. P., Lock, A. L., Allen, M. S. and VandeHaar, M. J. 2017. Relationship between residual feed intake and digestibility for lactating Holstein cows fed high and low starch diets. *J. Dairy Sci.* 100(1):265–278. https://doi.org/10.3168/jds.2016-11079.

Prasanth, C. R. and Ajithkumar, S. 2016. Effect of sub-acute ruminal acidosis (SARA) on milk quality and production performances in commercial dairy farms: A review. *Int. J. Sci. Environ. Technol.* 5:3731–3741.

Prestegaard-Wilson, J. M., Daley, V. L., Drape, T. A. and Hanigan, M. D. 2021. A survey of United States dairy cattle nutritionists' practices and perceptions of reducing crude protein in dairy cow diets. *Appl. Anim. Sci.* 37:697–709. https://doi.org/10.15232/aas.2021-02179.

Raffrenato, E. and Erasmus, L. J. 2013. Variability of indigestible NDF in C_3 and C_4 forages and implications on the resulting feed energy values and potential microbial protein synthesis in dairy cattle. *South Afr. J. Anim. Sci.* 43(5):593–597.

Raffrenato, E., Ross, D. A. and Van Amburgh, M. E. 2018. Development of an in vitro method to determine rumen undigested aNDFom for use in feed evaluation. *J. Dairy Sci.* 101(11):9888–9900 https://doi.org/10.3168/jds.2018-15101.

Russell, J. B. 1993. Effect of amino acids on the heat production and growth efficiency of *Streptococcus bovis*: balance of anabolic and catabolic rates. *Appl. Environ. Microbiol.* 59(6):1747–1751.

Russell, J. B. and Hespell, R. B. 1981. Microbial rumen fermentation. *J. Dairy Sci.* 64(6):1153–1169.

Russell, J. B., O'Connor, J. D., Fox, D. G., Van Soest, P. J. and Sniffen, C. J. 1992. A net carbohydrate and protein system for evaluating cattle diets: I. Ruminal fermentation. *J. Anim. Sci.* 70(11):3551–3561.

Satter, L. F. and Slyter, L. L. 1974. Effect of ammonia concentration on rumen microbial protein production in vitro. *Br. J. Nutr.* 3:199–208.

Schuling, S. E., Staudinger, E. J., Rortvedt, J. A., Windschitl, P. M., Golombeski, G. L. and Cotanch, K. W. 2015. Evaluation of an on-farm tool to estimate physically effective neutral detergent fiber of forages and total mixed rations. *J. Dairy Sci.* 98 (Suppl. 1):750 (Abstr.).

Serva, L., Magrin, L., Andrighetto, I. and Marchesini, G. 2021. Short communication: The relationship between dietary particle size and undegraded neutral detergent fibre in lactating dairy cows: A prospective cohort observational study. *Anim. Feed Sci. Technol.* 279:115017. https://doi.org/10.1016/j.anifeedsci.2021.115017.

Shaver, R. and Kaiser, R. 2004. Feeding programs in high producing dairy herds. *Proceedings of the Tri-State Dairy Nutr. Conf.*, pp. 143–170.

Shaver, R. D. 2002. Rumen acidosis in dairy cattle: Bunk management considerations. *Adv. Dairy Technol.* 14:241–249.

Silvestre, T., Fetter, M., Raisanen, S. E., Lage, C. F. A., Stefenoni, H., Melgar, A., Cueva, S. F., Wasson, D. E., Martins, L. F., Karnezos, T. P. and Hristov, A. N. 2022. Performance of dairy cows fed normal or reduced-starch diets supplemented with an exogenous enzyme preparation. *J. Dairy Sci.* 105(3):2288–2300. https://doi.org/10.3168/jds.2021-21264.

Smith, L. W., Goering, H. K. and Gordon, C. H. 1972. Relationship of forage compositions with rates of cell wall digestion and indigestibility of cell walls. *J. Dairy Sci.* 55(8):1140–1147.

Souza, J. G., Riberio, C. V. D. M. and Harvatine, K. J. 2022. Meta-analysis of rumination behavior and its relationship with milk and milk fat production, rumen pH, and total-tract digestibility in lactating dairy cows. *J. Dairy Sci.* 105(1):188–200. https://doi.org/10.3168/jds.2021-20535.

Stauder, A., Humer, E., Neubauer, V., Reisinger, N., Kaltenegger, A. and Zebeli, Q. 2020. Distinct responses in feed sorting, chewing behavior, and ruminal acidosis between primiparous and multiparous Simmental cows fed diets differing in forage and starch levels. *J. Dairy Sci.* 103(9):8467–8481. https://doi.org/10.3168/jds.2019-17760.

Stokes, S. R., Hoover, W. H., Miller, T. K. and Blauweikel, R. 1991a. Ruminal digestion and microbial utilization of diets varying in type of carbohydrate and protein. *J. Dairy Sci.* 74(3):871–881.

Stokes, S. R., Hoover, W. H., Miller, T. K. and Manski, R. P. 1991b. Impact of carbohydrate and protein levels on bacterial metabolism in continuous culture. *J. Dairy Sci.* 74(3):860–870.

Stone, W. C. 1999. The effect of subclinical rumen acidosis on milk components. *Proceedings of the Cornell Nutr. Conf.*

Stone, W. C. 2004. Nutritional approaches to minimize subacute ruminal acidosis and laminitis in dairy cattle. *J. Dairy Sci.* 87 (Suppl.):E13–E26.

Stone, W. C., Chase, L. E., Overton, T. R. and Nestor, K. E. 2012. Brown midrib corn silage fed during the peripartal period increased intake and resulted in a persistent increase in milk solids yield of Holstein cows. *J. Dairy Sci.* 95(11):6665–6676. http://doi.org/10.3168/jds.2012-5531.

Strobel, H. J. and Russell, J. B. 1986. Effect of pH and energy spilling on bacterial protein synthesis by carbohydrate-limited cultures of mixed rumen bacteria. *J. Dairy Sci.* 69(11):2941–2947.

Tebbe, A. W., Faulkner, M. J. and Weiss, W. P. 2017. Effect of partitioning the nonfiber carbohydrate fraction and neutral detergent fiber method on digestibility of carbohydrate by dairy cows. *J. Dairy Sci.* 100(8):6218–6228. https://dpi.org/10.3168/jds.2017-12719.

Tedeschi, L. O. 2006. Assessment of the adequacy of mathematical models. *Agric. Syst.* 89(2–3):225–247. https://doi.org/10.1016/j.agsy.2005.11.004.

Tedeschi, L. O. 2019. Mathematical modeling in ruminant nutrition approaches and paradigms, extant models, and thoughts for upcoming predictive analytics. *J. Anim. Sci.* 97:1921–1944.

Tylutki, T. P., Fox, D. G., Durbal, V. M., Tedeschi, L. O., Russell, J. B., Van Amburgh, M. E., Overton, T. R., Chase, L. E. and Pell, A. N. 2008. Cornell net carbohydrate and protein system: A model for precision feeding of dairy cattle. *Anim. Feed Sci. Technol.* 143(1–4):174–202. https://doi.org/10.1016/j.anifeedsci.2007.05.010.

US Department of Agriculture/National Agricultural Statistics Service. 2007. *Census of Agriculture: Farm and Ranch Irrigation Survey.* Washington, DC: United States Department of Agriculture/North American Spine Society.

Van Amburgh, M. E., Collao-Saenz, E. A., Higgs, R. J., Ross, D. A., Reckentenwald, E. B., Raffrenato, E., Chase, L. E., Overton, T. R., Mills, J. K. and Foskolos, A. 2015. The Cornell Net carbohydrate and protein system: Updates to the model and evaluation of version 6.5. *J. Dairy Sci.* 98(9):6361–6380. https://doi.org/10.3168/jds.2015-9378.

Van Soest, P. J. 1994. *Nutritional Ecology of the Ruminant*. 2nd Edition. Ithaca, NY: Cornell University Press.

Van Soest, P. J., Robertson, J. B., Hall, M. B. and Barry, M. C. 2020. Klason lignin is a nutritionallheterogeneous fraction unsuitable for the prediction of forage neutral-detergent fibre digestibility in ruminants. *Br. J. Nutr.* 124(7):693–700. https://doi.org/10.1017/S0007114520001713.

Van Soest, P. J., Robertson, J. B. and Lewis, B. A. 1991. Methods for dietary fiber, neutral detergent fiber, and non-starch polysaccharides in relation to animal nutrition. *J. Dairy Sci.* 74(10):3583–3597.

Van Soest, P. J., Van Amburgh, M. E. and Tedeschi, L. O. 2000. Rumen balance and rates of fiber digestion. *Proceedings of the Cornell Nutr. Conference*.

Volden, H., Nielsen, N. I., Akerlind, M. and Rygh, A. J. 2011. System evaluation. In: Volden, H. (Ed.), *EAAP Publ. NorFor- The Nordic Feed Evaluation System*. The Netherlands: Wageningen Academic Publishers, Vol. 130, pp. 141–158.

Wallace, R. J. and Cotta, M. A. 1988. Metabolism of nitrogen-containing compounds. In Hobson, P. N. (Ed.), *The Rumen Microbial Ecosystem*. London and New York: Elsevier Applied Science, pp. 217–249.

Weimer, P. J. 2011. End product yields from the extraruminal fermentation of various polysaccharide, protein, and nucleic acid components of biofuels feedstocks. *Bioresour. Technol.* 102(3):3254–3259. https://doi.org/10.1016/j.biortech.2010.11.050.

Weisbjerg, M. R., Hvelplund, T. and Bibby, B. M. 1998. Hydrolysis and fermentation rate of glucose, sucrose, and lactose in the rumen. *Acta Agric. Scand. A* 48(1):12–18.

White, R. R. and Hall, M. B. 2017. Nutritional and greenhouse gas impacts of removing animals from US agriculture. *Proc. Natl. Acad. Sci. USA* 114(48):E10301–E10308. https://doi.org/10.1073/pnas.1707322114; www.pnas.org/cgl/doi/10.1073/pnas.1707322114.

White, R. R., Hall, M. B., Firkins, J. L. and Kononoff, P. J. 2017a. Physically adjusted neutral fiber system for lactating dairy cow rations. I: Deriving equations that identify factors that influence effectiveness of fiber. *J. Dairy Sci.* 100(12):9551–9568. https://doi.org/10.3168/jds.2017-12765.

White, R.R.R. Hall., Fir kins, J. L. and Kononoff, P. J. 2017b. Physically adjusted neutral detergent fiber system for lactating dairy cow rations. II. Development of feeding recommendations. *J. Dairy Sci.* 100(12):9569–9584. https://doi.org/10.3168/jds.2017-12766.

Chapter 2

Advances in understanding protein requirements and utilization in dairy cattle

Jeffrey L. Firkins, The Ohio State University, USA; and Hélène Lapierre, Sherbrooke Research and Development Centre, Agriculture and Agri-Food Canada, Canada

1 Introduction

Dairy cattle requirements of protein and amino acids (AA) have made major advances based on landmark research using quantitative approaches, concepts, assumptions, and limitations (Schwab and Broderick, 2017; Hristov et al., 2019). Many approaches for metabolizable protein (MP) root back to a conceptual approach (NRC, 1985) that has since been adapted to quantitative models resembling that shown in Fig. 1. Some of the original coefficients are still used partly because they have lasted the test of time and partly because of lack of institutional research to fill critical knowledge gaps.

For ruminants, the critical juncture of ruminal metabolism has consistently complicated research efforts. For example, splitting dietary protein into rumen-degraded protein (RDP) and rumen-undegraded protein (RUP) was recognized several decades ago, yet we will describe persistent problems in measurement among individual feeds. More mechanistic approaches have limitations in scale and cannot be replicated for every experiment. Therefore, requirement models invariably rely on a feed library that has nutrients and predicted coefficients used in various equations needed to derive MP.

http://dx.doi.org/10.19103/AS.2022.0117.02

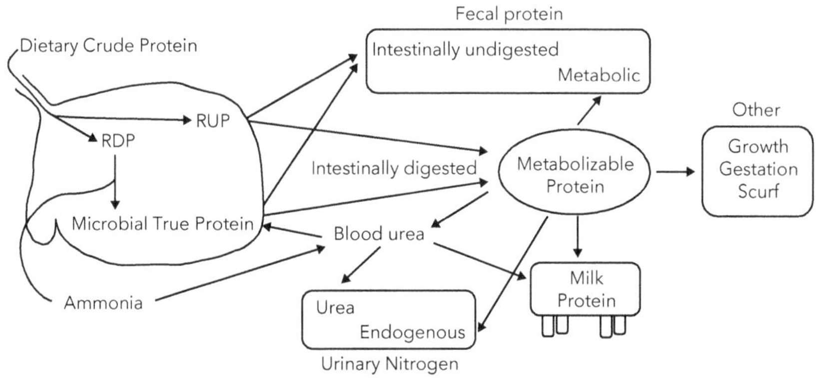

Figure 1 General schematic of metabolizable protein (MP) derivation and usage in lactating dairy cattle. Dietary crude protein (nitrogen x 6.25) is fractionated into rumen-degraded (RDP) and rumen-undegraded (RUP) proteins. Quantitative amino acid (AA) recovery converts microbial crude protein (MCP) into microbial true protein flowing from the rumen. The AA profiles of RUP and intestinally dRUP are typically factored from feed libraries. Microbial AA and digestibility are factored from constants. Energy availability in the rumen and total tract influence microbial protein supply and efficiency of metabolizable AA conversion into milk protein. Factorial requirements models require correction for secretions or losses of metabolic fecal, endogenous urinary, and scurf protein equivalent.

Increasingly, the feed library is used to derive coefficients or response variables that are used as explanatory variables in subsequent equations and models, particularly when extending from MP to metabolizable AA. Expanded computational power allows models that are more reliant on values measured from feeds and thus adaptable to feeds used on commercial dairy farms. However, usage of model outputs from centralized nutrients in feed libraries as model inputs also requires further assessment when feed nutrients in field situations deviate beyond ranges associated with those library means.

Whether for derivation or evaluation, equations on MP supply must harness a robust database from studies reporting total N and microbial N flowing out from the rumen. The RUP in a total diet can be estimated as non-ammonia N subtracting microbial N and a predicted value for endogenous N. The crude protein (N x 6.25) and AA profile are typically factored because a relatively few studies have measured AA composition, and even fewer have quantitatively recovered all of the AA comprising true protein. Landmark reviews (Clark et al., 1992; Stern et al., 1994; Santos et al., 1998) started the process of compiling source studies most of which are still used. Although comparing studies with test protein sources versus a soybean meal control helps to control for among-study variation, many of these papers fed a high amount of test protein and limited the RDP fraction, which limited microbial protein outflow (Ipharraguerre and Clark, 2005). Hence, evaluation of MP requirements typically now relies on a broader inclusion of studies that did not necessarily have hypotheses related to partitioning N fractions in the post-ruminal digesta. These approaches rely on the statistical inclusion of the random effect of study and the assumption that study does not interact with other variables (St-Pierre, 2001). The degree to which the cumulative of these studies has bias against true (but unknown) values is therefore embedded in the framework of the requirement model.

Over the decades, emphasis has been put on the determination of the supply of MP and metabolizable AA from dairy rations. This was needed as a starting point to determine with which efficiency these supplies are being used to support protein functions. Although these predictions still need to be improved, we can begin to tackle how efficiently they are being used. This information will help to establish recommendations and pinpoint which AA might be in short supply for a given diet and expected milk protein yield (MPY).

Our objectives are to highlight data limitations, model assumptions, and requirements considerations that arose during our experience working on the protein model of NASEM (2021). We will comment on research needs to fill knowledge or data gaps. We will refer to other requirements or evaluation models when the comparison emphasizes these gaps and support future research efforts to compare performance of these models as new data emerge.

2 Supply of metabolizable protein and amino acids

Determining the supplies of MP and its essential AA (EAA) are critical components for the estimation of EAA requirements for ruminants, including dairy cattle. We note that 'digested' and 'absorbed' have been used in place of 'metabolizable'. However, NASEM (2021) chose the latter because digestible infers delivery and disappearance from the small intestine, yet absorbed infers accounting for intestinal metabolism. No studies have ever actually measured MP or metabolizable AA supply directly because that would require accounting for various fluxes including flows of microbial crude protein (MCP) and RUP from the rumen, disappearance of AA from the small intestine using ileal cannulation, accounting for endogenous fractions, and appearance of AA using blood catheterization. Difficult procedures and limited results restrict such studies as that of Berthiaume et al. (2001) to establishing 'proof in principle' for MP systems rather than evaluation of those systems.

All current MP systems rely on some way to partition dietary protein into RDP and RUP using a standardized approach to account for normal variability among feedstuffs. Various *in vitro* and *in situ* approaches have been tested (Stern et al., 1997; Schwab and Broderick, 2017), but all have limitations. For example, exogenous enzymes from microbial preparations vary in activity and might not fully represent the major rumen microbial populations. Ruminal inocula can be obtained from donor animals fed under standardized conditions to normalize results among *in vitro* runs or laboratories. However, these standardized conditions might not represent all dietary situations (such as feed particle size) or even differences among animals. More research is needed to understand relative differences among feeds both for means (accuracy) and repeatability (precision) of RDP and intestinally digestible RUP (dRUP) measurements. Examples of relative differences among feeds include wilting (Hartinger et al., 2018), usage of inoculants (Muck et al., 2018), and overheating of protein sources that could increase RUP at the expense of intestinal availability (Hartinger et al., 2018; Schwab and Broderick, 2017).

To assess accuracy and to calibrate requirements systems, all approaches have in common the critical need to be standardized against *in vivo* measurements. Most but not all studies measuring microbial N also measured non-ammonia non-microbial N (NANMN) flows (Roman-Garcia et al., 2016). Endogenous N is virtually never directly measured in these studies; a prediction was made using limited studies as described in NASEM (2021). Correction of NANMN by subtracting a predicted endogenous N (multiplied x 6.25) to estimate RUP is useful but has limitations. For example, any overestimation of MCP will underestimate RUP derived this way, and vice versa; these corrected NANMN approaches should not be directly used to assess the effect of RDP on microbial N because their error is correlated through common calculations.

A useful case study is when Broderick et al. (2010) compared RUP using omasal flow studies (using corrected NANMN) to RUP predicted by NRC (2001). First, omasal flows in the evaluation data were compared with NRC (2001) flows nearly all of which were derived using duodenal sampling. However, Roman-Garcia et al. (2016) had to include an intercept shift for site of sampling to remove interactions with the random effect of study. Microbial N from omasal sampling is uniformly higher than that derived using duodenal sampling. Almost no studies have measured flows using both approaches, and some criticism of using a single marker such as chromic oxide for duodenal sampling can be countered with proper attention to dosing and sampling (Firkins et al., 1998). Second, the comparisons are almost always made against library-derived values for RUP that are normalized to a mean and need not actually represent well the RUP from those source evaluation studies. The error associated with using library values would need to be random across all studies, which seems likely (although not checked), whereas the NRC (2001) library for RUP has been updated in NASEM (2021) with more accurate values including sometimes to correct for misclassification of feeds (Tran et al., 2020).

Although a discussion of flow markers is beyond our scope, a triple marker system might be required because of the technical challenges in getting a representative sample from omasal sampling (Firkins et al., 1998). No marker has ideal properties (Owens and Hanson, 1992). Although this triple marker method is sound in theory, limitations of markers and their measurements are still problematic (Reveneau et al., 2012). Error from flow markers is amplified when error from microbial markers is included for microbial N flow (Broderick and Merchen, 1992; Hristov et al., 2019). Without an absolute standard, the NASEM (2021) committee merged omasal and duodenal sampling approaches to predict MCP and to calibrate RUP, as described subsequently. This limitation needs further controlled research to sort out differences among sampling and marker approaches.

2.1 Rumen-undegraded protein derivation and intestinal digestibility

In many ruminant requirements systems, including NASEM (2021), the protein in feeds is typically partitioned into the proportion that solubilizes (A) and the proportion that resists degradation with unlimited incubation (C). The potentially degradable (B) fraction is derived by difference of total CP minus that in A and C (Fig. 2). Most approaches assume a first-order compartment model that requires derivation of fractional rates of degradation (kd) and passage (kp) to estimate the degraded B as kd/(kd + kp). Some systems separate protein fractions by solubility in different buffers, but they will not be discussed here because they tend not to be part of requirement models affiliated with national projects. Moreover, the dRUP must be estimated because it is rarely measured.

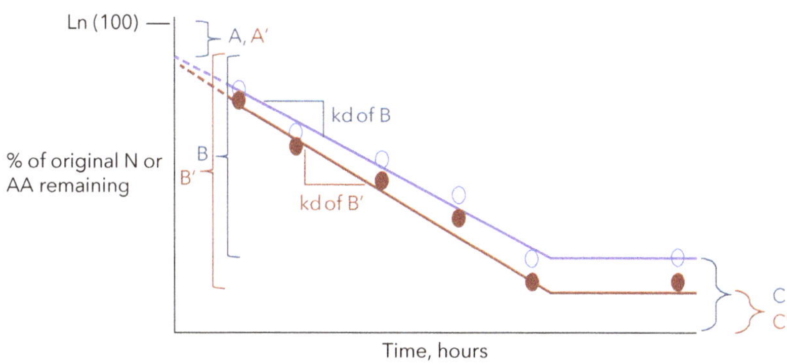

Figure 2 Simulation of first-order (linearized by natural logarithm transformation here), single compartment loss of feed N or AA originally weighed into Dacron bags incubated in the rumen *in situ* that have not (blue) or, rarely, have been corrected (red, constants with a prime) for microbial contamination. A progressive increase in microbial contamination relative to remaining feed protein increases C and thereby decreases the potentially degradable B fraction (total - A - C) and often its degradation rate (kd). The degradability of B is a function of the kd and a predicted passage rate (kp). A portion of the A fraction is now recognized to pass from the rumen.

2.1.1 Outflow of soluble protein in the A fraction

Outflow of soluble protein and AA and limitations of *in situ* techniques have been discussed (Broderick et al., 2010). Some protein is solubilized, and some insoluble protein could be of small enough size to pass through the pores of Dacron bags. Moreover, reliance on prediction of kd (and its expense for routine usage) and especially first-order passage rates (kp) justify the usage of computational solutions avoiding those rates (White et al., 2017). However, these alternatives removing kd and kp also have their own limitations, particularly for characterizing differences in feeds intentionally processed to increase RUP, which is why NASEM (2021) retained the *in situ* first-order modeling approach of NRC (2001) with reservation.

Current researchers (even if they are not aware) rely on foundational principles of (1) what makes proteins resistant to degradation and by which microbes (Wallace, 1996; Walker et al., 2005), (2) the importance of research evaluating RUP with dairy cattle (Stern et al., 1994), and (3) the ways to estimate RUP and its dRUP (Stern et al., 1997). Van Amburgh et al. (2015) have revised the Cornell Net Carbohydrate and Protein System approach to assessing RUP with respect to predicted kp of particles and fluid exiting the rumen. Similar approaches have been used in other models such as the Dutch model (van Duinkerken et al., 2011). However, computational solutions rather than models based on measured kp (Hanigan et al., 2021) were chosen by NASEM (2021) based on issues described subsequently.

A constant 6.4% of the A fraction passing from the rumen was used in NASEM (2021) as justified by Hanigan et al. (2021). Although values higher than 6.4% might be realistic (Hristov et al., 2019), some soluble protein actually is recycled microbial protein (therefore not technically in the A fraction). This recycled fraction is challenging to quantify but might exceed half of the soluble protein passing from the rumen (Choi et al., 2002). Predicting the proportion of A outflow based on measured NANMN flows should theoretically account for the soluble protein that is of microbial origin only if the microbial marker also represents this soluble fraction in the same proportion as it represents intact cells. In contrast, values for A passage might be overestimated based on how samples were collected and stored (Reynal et al., 2007). Deriving a liquid kp to estimate passage of A would complicate prediction because of relatively limited predictability with liquid-phase markers (Seo et al., 2006a). Given the importance of estimating RUP, future controlled research should address methodological issues related to measurement of this fraction (as distinguished from microbial sources) flowing from the rumen and be associated with measurement of washout from bags, particularly for proteins such as from blood meal that resist degradation but have a high A fraction.

2.1.2 Estimation of degradation kinetics of the B protein fraction

The potentially degradable B fraction is measured by difference and therefore can be biased progressively with increasing bias of A and C estimation. The most important contributors to this problem are related to Dacron bag efflux or influx of particulates to or from the rumen fluid and microbial contamination of residual feed. The former problem can be partially corrected with blank bags so long as the bags and their handling are controlled and standardized as reported in NRC (2001). Although we have noted improvements in data integrity over the years, requirements systems might have to rely on older data or extrapolation from similar feeds to fill in database gaps.

Contamination of residual feed protein from adherent microbes after incubation *in situ* or *in vitro* has been noted for decades (NRC, 2001). Based on limited data, however, contamination is often ignored. Problems associated with microbial protein contamination are minor for feeds commonly supplemented as protein sources but can become significant for feeds that are low in protein and high in NDF providing both substrate for growth and a vehicle for adherence by microbes. A similar limitation in source studies with microbial contamination was extended to NASEM (2021). The C fraction is thus overestimated particularly for forages and fibrous byproducts, which results in mathematical underestimation of the potentially degradable B fraction (Fig. 2). Prior to the endpoint, an increasing percentage of adherent microbial protein with increasing incubation time *in situ* or in *vitro* leads to underestimation of the kd. Because parameter estimates are highly correlated during fitting (Woods et al., 2003), the RDP can be underestimated by up to 5% for concentrate feeds and even more for forages (Alexandrov, 1998; Krawielitzki et al., 2006). Although ^{15}N is becoming the microbial marker of choice because of problems with other markers (Hristov et al., 2019), even ^{15}N would be biased if microbial samples are not taken to represent the diurnality and spatial gradient of their ^{15}N enrichment relative to enrichment in bags.

Earlier attempts to account for microbial contamination were based on microbial markers such as cytosine (Cozzi et al., 1993), which suggested a peaked contamination earlier in the incubation *in situ* (Cozzi et al., 1995). However, a small but potentially significant escape of purines (probably similar to pyrimidines such as cytosine) was noted in continuous culture (Calsamiglia et al., 1996) and would likely be higher with earlier incubation. Because of the cumulative errors, this assay is not recommended for assessing microbial contamination *in situ*. Diaminopimelic acid has been used (Alexandrov, 1998). However, because of its potential biases (Broderick and Merchen, 1992), it is not recommended unless perhaps if one is already doing AA analyses and accounts for potential measurement issues.

Two assays most likely assessing microbial contamination *in situ* are ^{15}N infused into the rumen and quantitative polymerase chain reaction (qPCR). Researchers have labeled forages grown in ^{15}N-labeled fertilizer, but this approach is potentially problematic because it assumes bacteria are not labeled with the ^{15}N from the degraded protein (Kamoun et al., 2014; Krawielitzki et al., 2006) and is impractical for other feeds such as byproducts. In addition to its relative ease of measurement, the infusion of $^{15}NH_4$ salts allows multiple analyses of feeds per animal and is useful so long as researchers limit contaminating feed N in bacterial samples. A qPCR assay was used to assess bacterial contamination of byproducts (Paz et al., 2014). However, DNA would need to be extracted from all bacteria and for derivation of its own standard. This method offers considerable potential when researchers are simultaneously evaluating DNA-based changes in community development of bacteria colonizing differing forages over time (Gruninger et al., 2019).

Considerably more research is needed to describe the RUP of forages and byproducts because their high inclusion rate in diets can provide significant RUP and because of the large as-of-yet unexplained variability associated with natural differences among forage, harvesting, and processing. Bacteria (and perhaps fungi) in the ruminal particulates also are the most likely to adhere and contaminate feeds and should thus be used as standards for marker assays. Such a sample could be collected at each incubation time and used as a standard for all feeds incubated *in situ* per animal.

2.1.3 Estimation of the fractional rate of passage of the B fraction

Estimation of the kp of the B fraction is an elusive problem for requirement models relying on kd. Diligent and thoughtful effort girded the prediction of the kp of wet forages, dry forages, and concentrates (NRC, 2001). Seo et al. (2006b) corrected one of the equations and further evaluated those equations to have minimal bias compared with the measured kp. However, measured kp of marked feeds does not necessarily represent the kp of the B fraction. Hanigan et al. (2021) derived static kp for forage (4.87%/h) and for concentrate (5.28%/h), which were used in NASEM (2021). Although expected based on both mechanism and precedent, inclusion of dry matter intake (DMI) did not improve model fit. These static kp values were generally consistent with mean values from the database from which new equations were derived for forage, concentrate, and liquid (Seo et al., 2006a). Mechanistic models to predict kp (Seo et al., 2009; Gregorini et al., 2015) offer some promise to provide flexibility among dietary conditions but still maintain the same drawback that the source data are not specifically measuring the kp of the B protein fraction. Further complications with deriving and using kp of differing fractions in nutrition models are detailed by van Duinkerken et al. (2011).

Although flow markers also have error, critical problems in passage rate markers have received surprisingly little attention since their problems were identified (Firkins et al., 1998; Bernard and Doreau, 2000). Rumen evacuation approaches based on undegradable fiber typically yield lower values for kp (Krizsan et al., 2010), yet Hanigan et al. (2021) argued that protein from feeds leaches out and is likely to pass faster than the undegradable fiber. This research area to measure kp of the B fraction needs further investigation, especially for higher producing dairy cows with high DMI (and presumably faster kp).

2.2 Microbial protein outflow from the rumen and intestinal digestibility

2.2.1 Prediction of microbial protein outflow

Previous requirements systems have typically predicted MCP contribution to MP supply by one of two ways. The first is per some unit related to that system's energy measurement such as intake of total digestible nutrients (TDN) or net energy of lactation. The second approach uses mechanistic elements to predict an efficiency of microbial protein synthesis (EMPS). The product of this predicted EMPS and its denominator (predicted amount of rumen-degraded organic matter (OM) or carbohydrate) is MCP supply. The 130 g of microbial protein/kg of adjusted TDN intake (NRC, 2001) is, in essence, a constant EMPS. The beef NASEM (2016) had two similar empirical equations depending on dietary fat percentage (Galyean and Tedeschi, 2014) because supplemental fat is not used as an energy source by anaerobes in the rumen. However, a continuous response to supplemental fat would be limited by data issues (Roman-Garcia et al., 2016) and was not used in NASEM (2021). Moreover, supplemental fatty acids might improve EMPS by decreasing protozoal predation and recycling of bacterial protein or by direct insertion into membranes, sparing carbon that can be used to produce VFA for ATP. More rumen-degradable starch lowers EMPS in typical dairy diets, particularly with a shortage of RDP, and high starch also can interfere with biohydrogenation of supplemental fat (Hackmann and Firkins, 2015). Mechanistic models generally increase the number of variables needed to predict EMPS and rumen-degraded fractions (van Duinkerken et al., 2011), with some models containing several dozen variables (Fox et al., 2014). Greater complexity can better reflect biology but increases risks for poor fit in certain conditions resulting from over-parameterization.

Nutrients digested post-ruminally obviously cannot be used for ruminal synthesis of MCP, and site of digestion is not necessarily constant among dietary situations. The prediction of rumen-degraded carbohydrate or OM is more appropriate mechanistically, but the question is if it should be estimated *in situ* or based on *in vivo* predictions and resulting equations. Processing forages and grains for inclusion in Dacron bags does not necessarily represent the cow's mastication or normal comminution processes, and increasing surface area can increase fiber degradability, particularly for forages (Bowman and Firkins, 1993). A similar limitation in measuring the kp of the B fraction of protein would limit estimation or the kp of the B fraction to estimate rumen-degraded starch and NDF *in situ*. Even standardization might require using a standardized sample to correct for degradation of fiber *in situ* across laboratories (Krizsan et al., 2015). Ruminal starch digestibility estimated *in situ* was adjusted for statistical bias against *in vivo* measurements (Offner and Sauvant, 2004). Estimation of NDF degradation *in situ* also is complicated because the kp of small particle byproducts depends on consistency of the ruminal mat (Grant, 1997). Presumably similar interactions occur for starch in ground grains. Both methods (*in situ* or *in vivo* prediction) are biased to the degree to which microbes, especially protozoa, contribute reserve carbohydrate that would be measured as starch.

Both the digestibility of potentially degradable fiber and EMPS likely are decreased with increased inclusion of starch to dairy rations, particularly with an inadequacy of RDP (Firkins, 2021). Although demonstrated many times in laboratory studies, the concept of synchronizing carbohydrate with RDP is not well supported *in vivo* because feeding frequency helps to spread RDP over meals (Firkins et al., 2006). Those authors explained how blood urea-N (BUN) transfer seems to help mediate deficiencies in these cases. However, with further attempts to limit dietary protein to decrease urinary N excretion, further consideration for AA in the RDP and the potential substitution for branched chain short-chain fatty acids is needed to increase EMPS (Hackmann and Firkins, 2015).

The assumption that MPS is not limited by RDP until predicted energy-supported MPS (from TDN) would exceed RDP intake (NASEM, 2016) was not extended to dairy cattle (NASEM, 2021). First, blood urea transfer to the rumen bacteria based on ^{15}N-urea kinetics (Batista et al., 2017) is likely overestimated because any ^{15}N not recovered in excreted feces and urine is added to this mass balance for the anabolic N calculation assumed to represent incorporation of urea into MCP. In contrast, N balance is well known to be overestimated in dairy cows because of under-recovery of fecal and urinary N (Spanghero and Kowalski, 2021; Hristov et al., 2019). Moreover, this anabolic N calculation is overestimated the degree to which the ^{15}N in urea randomizes through normal biological reactions.

A second issue is representation of data. Reynolds and Kristensen (2008) related recycling of BUN to the gut versus the percentage of dietary protein.

In contrast, microbial N being linearly associated with N intake (Batista et al., 2017) ignores the common association with feed intake in both calculations. For example, if intake increased more than CP decreased, a greater N intake could correspond with greater intake of rumen-degradable carbohydrate and more transfer of BUN into microbial N on an amount basis. Moreover, the error is correlated by computation in both dependent and independent variables. If N intake is overestimated, both it and also anabolic N are increased; if N intake is underestimated, both it and anabolic N are underestimated. The current author (JLF) recommends multiple regression to separate DMI from dietary crude protein percentage as separate variables; collinearity associated with DMI might have been reduced, and the conversion of BUN to microbial N would likely have supported increased reliance on BUN transfer with declining CP percentage, as reported by Reynolds and Kristensen (2008), who used much of the same data as did Batista et al. (2017).

Although seemingly logical to feed RDP only when there is negative ruminal N balance, collinearity again should be considered more carefully. For example, if CP balance is computed as (N intake x 6.25) - (RUP + MCP), statistical assumptions would require both RUP and MCP to be independent. If RUP is derived as (non-ammonia-non-microbial N - endogenous N) x 6.25 and RDP = CP - RUP based on previous discussion, then equations deriving N balance from *in vivo* data would again likely have non-random error. For example, an overprediction of MCP would result in an underprediction of RUP such that N balance is not changed. Moreover, if a rumen N balance approaches 0 and DMI decreases, as might be expected if RDP limits NDF digestibility and DMI through fill restriction (Allen, 2000), then rumen N balance need not be as little as zero before a model would predict a need for more RDP.

Limiting RDP to zero rumen N balance clearly would decrease N excretion, which is an important consideration for animal feeding operations, but this objective should be evaluated separately from the role of RDP to influence EMPS. An example study to be considered by future researchers is outlined by Kand et al. (2021). Limiting rumen N balance decreased NDF digestibility and minutes at the feed trough per cow per day for cows fed two diets. However, limited rumen N balance decreased DMI significantly (by 0.9 kg/day) in only one of two dietary treatments. Cows fed this diet had a lower EMPS (Kand and Dickhoefer, 2021). Moreover, variability in model inputs and dietary nutrient composition is greater under field conditions.

The MCP system in NASEM (2021) is sensitive to RDP while simultaneously relating to rumen-degraded carbohydrate in a semi-mechanistic manner. Although some inputs are derived empirically, the model form is an expression of EMPS. Because no attempt would yield an asymptotic relationship, the increased MCP prediction from RDP was truncated at 12%. The need to provide adequate RDP not just for EMPS but for NDF digestibility and DMI has been consistently recognized (Schwab and Broderick, 2017). The dietary CP also should be kept to under about 21.4% (mean + 2 standard deviations) to prevent over-sensitivity in the denominator of the MCP equation. Hopefully, future researchers will be able to improve this relationship, particularly if RDP measurement is improved in feeds such as forages that are not well represented in the feed library.

2.2.2 Prediction of microbial AA profile

MCP contains a lot of non-protein N forms such as glucosamine and especially nucleic acids that will not contribute to MP (NASEM, 2021). Faster microbial growth with increasing DMI by high-producing cows should increase the nucleic acid proportion of CP while simultaneously decreasing the proportion of true protein (Bach et al., 2005). Sok et al. (2017) surveyed the literature, screened out studies that used formalin (which can bind AA), and estimated particulate-associated bacteria, fluid-associated bacteria, and protozoa as 33.4, 50.1, and 16.5%, respectively, of the true protein, which was 82.4% of CP. Archaea were grouped with bacteria, but fungi were ignored because of limited quantification in the literature. Compared with older models that ignored protozoa, more recent studies with dairy cattle suggest that the proportion of microbial AA as protozoal AA might even be higher than 16.5% in high-producing dairy cattle (Ahvenjärvi et al., 2018; Dineen et al., 2021).

The approach of Sok et al. (2017) compares favorably with similar approaches (Fessenden et al., 2017; Fessenden et al., 2019). In particular, protozoa increase Lys composition compared with bacteria. Most EAA in microbial true protein are in a relatively good balance relative to the needs for milk protein with the exception of histidine (His), which becomes limiting with lower CP diets (Schwab and Broderick, 2017). Histidine also has unique properties for tissue mobilization (Hristov et al., 2019) and excretion in endogenous fractions (NASEM, 2021), so ramifications for such a low His profile in microbial protein should be investigated with respect to tissue metabolism, not just for milk protein, in future studies.

2.2.3 Prediction of microbial AA digestibility

The true digestibility of true microbial true protein has been considered to be a constant 80% (NRC, 2001) and retained for beef cattle (NASEM, 2016) and the French system (Sauvant and Nozière, 2016). Although digestibility is assumed to be higher (85%), the true protein in microbial true protein is assumed lower (75% vs. 80%) in the Dutch system (van Duinkerken et al., 2011), yielding a similar net digestibility of true protein. Hristov et al. (2019) reviewed the studies that assessed microbial AA digestibility in the small intestine, ranging from *in vitro* to *in vivo* assessments; many studies reported values higher than 80%. Most of these studies ignore potential sources of variation that could influence digestibility beyond the experimental conditions. With considerations discussed subsequently and a void for *in vivo* true digestibility values in dairy cattle, the 80% digestibility of microbial true protein remained unchanged (NASEM, 2021).

Bacterial AA digestibility was estimated to be higher than 80% when incubated with digestive enzymes *in vitro* (Fessenden et al., 2017). Protein disappearance exceeded 90% from mixed bacterial cells incubated with pepsin and pancreatin (Wallace, 1983). In contrast, some pure cultures in the latter study had markedly low values (19% for *Streptococcus bovis*); clumping during collection and processing was among the potential reasons. A likely complicating factor is that, if bacterial cell wall 'ghosts' are escaping digestion (Broderick and Merchen, 1992), bacterial protein digestibility would be overestimated to the degree to which these solubilized cell wall fragments pass through the filters in these assays.

Other considerations deserve further research attention. Intra-gastric infusion could lessen expression or secretion of enzymes (Larsen et al., 2001). Those authors also explained the challenge of accounting for endogenous secretions. On the other hand, infusion studies still separate bacterial fractions from the surrounding undigested feed matrix. Bacteria flowing with the fluid phase of the rumen were more digestible than the particulate-phase bacteria in cecectomized roosters (Fonseca et al., 2014). In addition to differences in cell wall composition, there is limited evidence regarding how well pancreatic proteases mix within the intestinal chyme to digest bacteria embedded within particulate matter. The digestibility of the particulate-phase bacteria could be even lower when they are not pre-extracted.

In non-lactating cattle, the apparent digestibility of ^{15}N-labeled non-ammonia nitrogen between the duodenum or omasum and ileum (< 80%) was similar to that of total non-ammonia nitrogen (Firkins et al., 1987; Mariz et al., 2018a). Ileal bacteria that could be partially enriched with ^{15}N from blood urea transfer, potentially leading to an overestimation of true digestibility (86%) of microbial N (Mariz et al., 2018b). By attempting to account for endogenous AA reaching the ileum, Larsen et al. (2001) estimated true digestibilities of bacterial AA between the duodenum and terminal ileum of dairy cattle to be slightly lower than 80%. The authors explained likely compensating errors in these approaches. Studies evaluating intestinal disappearance using ileal cannulas in lactating dairy cattle are relatively few, and the dairy cattle (Larsen et al., 2001) can be estimated to have consumed DM only at slightly above 2% of body weight.

The ruminant small intestine contains the array of pancreatic and intestinal proteases needed to digest a relatively more constant supply of protein from microbial outflow (Harmon and Swanson, 2020). Increasing DMI from 1.4% to 2.1% (Firkins et al., 1987) or 1.2% to 2.1% (Mariz et al., 2018a) of body weight in beef cattle did not affect intestinal digestibility of microbial N. However, lactating dairy cattle could easily double this relative intake, which could further shorten the transit time and availability for digestion within the small intestine. There is a surprising lack of experiments for how level of feed intake affects kp and digestibility in the rumen and especially the intestines of dairy cattle. However, the presumed passage rate in the lower tract increased by about 80% for lactating cows consuming about 3.4% of body weight versus dry cows (Colucci et al., 1989).

Few studies have assessed the digestibility of other microbes besides bacteria (and co-collected archaea). Fungal and especially protozoal AA likely have higher digestibilities than bacteria (Jouany, 1996). The digestibility of diaminopimelic acid found only in bacterial cell walls (or within protozoa consuming bacteria) is much lower than that of other AA (Fonseca et al., 2014; Larsen et al., 2001), presumably because the cell wall is less digestible than the rest of the cell (Broderick and Merchen, 1992). Protozoa convert the diaminopimelic acid in the poorly digestible peptidoglycan to lysine (Lys) (Onodera, 1993). Moreover, some bacteria have significant proportions of alanine and glutamate (and somewhat of aspartate) in D-configurations (Schieber et al., 1999), which could lessen absorption of those non-EAA.

Although microbial samples (unless treated with formalin) might not be expected to have marked differences in intestinal digestibility among the individual AA, few studies are available to support these assertions. Of the EAA, phenylalanine (Phe) and valine (Val) sometimes but not consistently had lower intestinal digestibility than the mean EAA in microbial protein (Larsen et al., 2001; Mariz et al., 2018a). Perhaps, a better evaluation of particle- versus fluid-associated bacteria can help rectify this variation (Fonseca et al., 2014). Digestibility of AA will be continually factored across constant compositions of microbial AA and constant digestibilities until further research is done.

2.3 Amino acid profile of metabolizable protein of dietary origin

Measurement approaches for dRUP are needed both for derivation of MP models but also for using them with local feeds. Within available studies, typically only one approach was used to estimate dRUP and without true calibration against *in vivo*. Hence, the method used to estimate RDP and dRUP interacted with feed category (Liebe et al., 2018) and hampered efforts of NASEM (2021). If there were no interactions but one approach just increased or decreased predictions, then an intercept shift could be applied. In addition, the interactions beg the question whether or not different feeds are better assayed with differing methods. For example, mobile bags represent the true intestinal process but might not represent all feeds, depending on factors such as CP concentration (Liebe et al., 2018). Another issue is that an assay should be readily assessable and repeatable across laboratories. For example, incubation with digestive enzymes can be standardized. Both approaches should have pre-exposure to rumen inoculum to better represent intestinal digestibility.

Liebe et al. (2018) documented the strong relationship between the RUP and the total AA (TAA) escape from the rumen plus the relationship between dRUP and digestibility of TAA. Thus, a narrower group of comprehensive studies could be done to assess disappearance of individual AA normalized to that of TAA; subsequent studies would not need to measure AA but could substitute the TAA response for dRUP to have a broader estimate of degradation or intestinal digestibility among individual AA. Those authors also noted that the large proportion of variation explained by the random effect of study could be removed when using ruminal TAA degradation or intestinal digestibility of TAA.

Without a system to robustly predict the ruminal degradability and the intestinal digestibility of individual AA in the RUP, ration evaluation systems must default to factoring AA from feed libraries. Such approaches condense true error while missing biological variation that otherwise could be explained with a more robust equation. We note that an improved library such as that used by NASEM (2021), particularly for sulfur AA and tryptophan (Trp), reduces but cannot eliminate these issues. In addition, these approaches miss the opportunity to assess true differences in metabolizability of individual AA among different feed sources. For example, heat-damaged protein sources should yield less absorbed Lys compared with properly heated sources (Boucher et al., 2009). Finally, Liebe et al. (2018) noted how pore size of bags or nutrient composition besides AA could be used to minimize variability. Unlike with measuring microbial N or NANMN, we have no robust *in vivo* standard for assessing intestinal digestibility. As stated previously, ileal cannulated dairy cows are impractical and still need correction for endogenous AA discharged in ileal digesta.

Tedeschi (2019) provided an example of how overreliance on predicted values potentially collapses actual variability needed to prevent destabilization of nutrition models. Retained energy (RE) and retained protein (RP) are important for models of beef cattle growth. When predicted values for RE were used to predict RP, the correlation was very high ($r = 0.94$), but when observed values for RE were used to predict RP, the correlation decreased considerably ($r = 0.68$) even though observed and predicted RE were highly correlated ($r = 0.93$). Thus, models will work well when predicting outputs for normal conditions (near the means) but potentially not for conditions outside of those conditions. More observed values are critically needed for RUP and dRUP combined with individual AA plus TAA disappearances for numerous feeds, most notably for forages.

3 Post-absorptive utilization of metabolizable protein and amino acids

Requirements of MP and EAA are based on their net utilization rather than their total utilization. Whole-body protein synthesis cannot be used directly to estimate MP requirements: due to extensive protein degradation, whole-body protein synthesis exceeds MP supply. In lactating dairy cows, whole-body protein synthesis exceeded MP supply by almost 50%, averaging 2.55 and 1.74 kg/day, respectively (Blouin et al., 2002; Lapierre et al., 2002; Raggio et al., 2006a). Specifically, the total usage of EAA is much higher than the metabolizable EAA supply. For example, on a daily basis, in lactating dairy cows fed 2 levels of MP, the net supply of leucine (Leu), measured as total Leu absorbed in the portal vein, averaged only one-third of its whole-body rate of disappearance or usage (Lapierre et al., 2002). On average, in lactating dairy cows, protein degradation estimated using Leu kinetics averaged slightly more than half of whole-body protein synthesis, whereas the sum of Leu in exported proteins, as detailed below, plus Leu oxidation averaged 60% of the whole-body usage of Leu (Lapierre et al., 2002; Thivierge et al., 2002; Lapierre et al., 2005; Raggio et al., 2006b). This means that approximately 40% of the Leu usage at the whole-body level was not using a net supply but was using Leu 'recycled' through protein degradation between the pool of EAA in proteins and the pool of free EAA.

Altogether, these data indicate that the primary driver of the net demand of EAA is not the whole protein synthesis, but the synthesis of proteins that are 'irreversibly' pulling the EAA out of the pool of free EAA. Simplistically, these are the proteins secreted out of the body of the cow plus proteins contributing to protein accretion through growth or gestation, albeit these latter accretions are dynamic. In addition, because the absorbed AA are not used at a 100% efficiency to support protein synthesis, we need to add a catabolic usage to this net anabolic demand. These two types of demand, net anabolic and catabolic, are described in the following sections.

3.1 Net protein and amino acid demand

3.1.1 Protein

The different feed evaluation models used to balance dairy rations present a simplification of the post-absorptive metabolism of protein and EAA, with one of the primary goals being to identify the net protein demand. The main anabolic functions examined are scurf, metabolic fecal protein, milk protein, growth, gestation, and pre-duodenal endogenous secretion (used in one model) to which is added a basal endogenous urinary loss. A summary of the propositions of the recent NASEM (2021) will be presented next, plus references to the concepts and detailed equations from five other feed evaluation models reported previously (Lapierre et al., 2018). These five models are the Cornell Net Carbohydrate and Protein System, v6.5.5, from the United States (Fox et al., 2004; Van Amburgh et al., 2015); the DVE/OEB system in the Netherlands, where DVE [DarmVerteerbaar Eiwit is ileal digestible protein and OEB (Onbestendig Eiwit Balans is rumen-degraded protein balance (Tamminga et al., 1994; Van Duinkerken et al., 2011)]; the Institut National de la Recherche Agronomique model from France (INRA, 2018); the National Research Council from United States (NRC, 2001); and NorFor, the Nordic feed evaluation model (2011). For the sake of simplicity and because there is very limited reference to the growth or gestation status of the cows in published papers used to develop the models, only protein secretion estimations are presented next.

First, the endogenous urinary loss is estimated by most of the models, namely NRC, CNCPS, Norfor, and DVE/OEB, using an equation developed by Swanson (1977, in g/day: $2.75 \times BW^{0.5}$). However, using a different approach evaluating N loss when apparent N digestibility is 0 rather than the minimum unavoidable N loss with minimal N supply, INRA estimates the endogenous urinary N loss, in g/day, at $0.3125 \times BW$. In an attempt to better define the AA composition of the endogenous urinary loss through a literature review, NASEM (2021) proposed an estimate very close to the INRA model at $0.3313 \times BW$. In the same paper, Swanson (1977) also developed an equation to estimate the secretion of proteins as scurf (in g/day: $0.2 \times BW^{0.6}$). These secretions, being minimal, received little attention, and this equation has been adopted by all the models, with inclusion of the ratio of true protein:CP in NASEM (2021). A requirement for the endogenous proteins flowing at the entrance of the small intestine is only present in NRC (2001).

The metabolic fecal protein has the most divergent estimation across the different protein secretions among the six models. This large difference is due to the definition of what should be included in this fraction and the challenge of measuring it. Its estimation is based (for detailed equations, see Table 1 in Lapierre et al., 2018) either on undigested dry matter (CNCPS), on DMI and undigested OM (INRA), on the flow of OM across different gut segments (NorFor), or on DMI and undigested bacteria (NRC). The recent NASEM (2021) model stated that this secretion should represent the endogenous proteins secreted along the digestive tract and flowing undigested at the end of the ileum plus hindgut secretions. The accounted secretions were only those directly using EAA and do not include undigested bacteria synthesized from

Table 1 Combined efficiency and optimum supply in CNCPS[a] and target efficiency in NASEM[b] for EAA

Amino acid	Combined efficiency	Optimum supply (g EAA/ Mcal ME)[c]	Target efficiency
	CNCPS v6.5	CNCPS v7	NASEM
Arg	0.58	2.04	ND
His	0.76	0.91	0.75
Ile	0.67	2.16	0.71
Leu	0.61	3.42	0.73
Lys	0.69	3.03	0.72
Met	0.66	1.14	0.73
Phe	0.57	2.15	0.60
Thr	0.66	2.14	0.64
Trp	0.65	0.49	0.86
Val	0.66	2.48	0.74

[a]CNCPS: Cornell Net Carbohydrate and Protein System, version 6.5 (Van Amburgh et al., 2015) and version 7 (Van Amburgh et al., 2018).
[b]NASEM: National Academies of Sciences, Engineering, and Medicine (2021).
[c]EAA: essential amino acid; ME: metabolizable energy.

recycled urea. In this context, this estimation was based on studies directly focused on the measurement of endogenous secretions in dairy cows (Ouellet et al., 2002; Ouellet et al., 2010) and adjusted with studies conducted in cattle (Marini et al., 2008) and sheep (Sandek et al., 2001). The final equation is based on DMI and NDF percentage of the ratio (in g of true protein/day: $[8.5 + 0.1 \times \text{NDF}] \times \text{DMI}$). However, in DVE/OEB, where its estimation is based on DMI, the metabolic fecal output is not included in the requirement and is excluded from the supply because it is considered to be a consequence of the diet and not the animal. Milk protein output is certainly the simplest secretion protein to determine because it is directly measured as true protein.

3.1.2 Essential amino acids

Currently, DVE/OEB, INRA, NorFor, and NRC models are using a proportional approach to estimate EAA requirements, i.e. that their recommendations are based on a percentage of each EAA relative to MP supply. As such, they had no need to determine a net demand of EAA.

The CNCPS and NASEM models have, however, developed recommendations of EAA using a factorial approach and, therefore, estimate a net demand of EAA. The CNCPS model estimated the net demand of EAA from the net demand of protein for the non-productive functions multiplied by tissue AA composition (Fox et al., 2004) with a revisited body composition (Van

Amburgh et al., 2015) plus MPY times the AA composition of milk protein (Fox et al., 2004). For NASEM (2021), an AA composition of each protein function was determined from literature review and has already been reported based on net true protein (Lapierre et al., 2020). The AA composition proposed was corrected to account for incomplete recovery of AA with the classical 24-h protein hydrolysis (Lapierre et al., 2019). In addition, after examination of the synthesis process of the N-compounds of the endogenous urinary loss, only the endogenous urea and 3-methyl-His excretions create a direct demand on EAA utilization (considering arginine, Arg, as a conditional EAA). Therefore, endogenous urinary urea excretion (in g of protein equivalent/day: $0.0625 \times BW$) is used to estimate the EAA endogenous urinary loss, assuming that this loss has the AA composition of the whole empty body. For His, to account for 3-methyl-His excretion [in g/day: $(7.82 + 0.55 \times BW)/1000$] is added to His loss calculated from endogenous urea excretion.

These previously described equations represent the classical net demands of MP and EAA quantified after years of research. It is acknowledged that AA are used for functions other than directly supporting protein synthesis. Methionine (Met) is a clear example of an EAA involved in more than a hundred metabolic pathways, mainly through production of the universal methyl donor, S-adenosylmethionine. There are increasing reports of the positive effect that Met supplementation would have, especially in the peripartum period, in improving the inflammatory status and the immune cell function, reducing oxidative stress, and enhancing liver function (Lopreiato et al., 2020; McFadden et al., 2020). However, despite this acknowledged positive role of Met in the transition cow, there is yet no quantification of the needed total supply during this period when DMI has not yet reached its peak. Similarly, non-EAA also have critical roles in many metabolic pathways (Wu, 2017): however, research is very limited to quantify the *de novo* rate of synthesis, making it almost impossible for now to determine recommendations. Therefore, not denying the importance of certain EAA on specific functions not directly related to protein synthesis and the role of non-EAA, the net demands detailed in this paper are limited to the functions described above and to EAA.

3.2 Requirements of metabolizable protein and amino acids

3.2.1 Metabolizable protein

To determine the requirements, or more correctly the recommendations, of MP, each model determines the efficiencies with which MP supply is being used for the different accretions or secretions they have recognized. In fact, unity minus the efficiency plus the endogenous urinary loss quantifies the MP or AA that are estimated to be diverted to the catabolic route. These efficiencies, fixed constants or the equations used to derive them, are detailed in Table 3 of Lapierre et al. (2018). The oldest models, NRC and CNCPS, are using a fixed efficiency of utilization of MP at 0.67 for all the secretions except the metabolic fecal protein to which there is no efficiency assigned, which means that an efficiency of 1 is effectively being used.

The DVE/OEB and NorFor models are using a fixed efficiency of 0.67 for the non-productive functions (except an efficiency of 1 for the metabolic fecal protein originating from the small intestine in NorFor) and a variable efficiency for lactation, based on the ratio of MP/net energy of lactation supplies. The INRA model uses variable efficiency for all the functions except for endogenous urinary loss, which has an efficiency of 1. The efficiency is calculated as the ratio of the sum of protein secretion and accretion divided by (supply minus endogenous urinary loss). A proxy for the estimation of this efficiency was based on the ratio of MP supply/DMI (Sauvant et al., 2015) when MPY needs to be estimated. Although the strong relationship between the efficiency of utilization of MP and the ratio of MP/metabolizable energy (ME) supplies has been recognized (Lapierre et al., 2020), as a first step, NASEM estimates MP requirement assuming a target efficiency of 0.69 for all functions except the endogenous urinary loss at an efficiency of 1 and assuming that digestible energy requirement is adequately met.

3.2.2 Essential amino acids

As mentioned previously, most of the models are using a proportional approach to estimate EAA requirements, i.e. that their recommendations are based on a percentage of each EAA relative to MP supply. The initial recommendations of NRC were only for Lys and Met at 7.2% and 2.4% of MP, respectively, and have been revisited with an expanded database and updated to 6.89% and 2.23% of MP to optimize MPY and 6.95% and 2.38% of MP to optimize milk protein concentration (Whitehouse et al., 2010a; Whitehouse et al., 2010b). These authors also reported that these proportions differed between three feed evaluation models and should therefore be evaluated within each feed evaluation model to be accurate. Later, in NorFor, recommendations for His were added; recommendations for His, Lys, and Met are, respectively, 2.2%, 6.6% and 2.2% of MP. The INRA model extended their recommendation to all EAA, except Trp, and proposed the following ideal profile (% of MP): Arg (3.4), His (2.4), isoleucine (⩾5.0), Leu (8.5), Lys (7.0), Met (2.4), Phe (4.4-5.0), threonine (⩾ 3.8), and Val (4.9-5.7).

The two models that have developed recommendations of EAA using a factorial approach, CNCPS and NASEM, used a completely different perspective. Initially, CNCPS based its factorial estimation as being the sum of each protein function times its AA composition divided by an efficiency, which was different for non-productive functions (but unity for metabolic fecal protein) and lactation (Fox et al., 2004). Later, the model opted for a combined efficiency for all functions, except metabolic fecal protein (Table 1, v 6.5; Van Amburgh et al., 2015). More recently, CNCPS proposed to estimate EAA requirements using a ratio of g of EAA relative to Mcal of ME (Table 1, v 7; Van Amburgh, 2018). For NASEM (2021), although it is recognized that the efficiency of utilization of EAA, as for MP, is variable (Lapierre et al., 2020), as a first step, a target efficiency was proposed for each EAA (Table 1), which is estimated assuming an adequate supply of digestible energy (Table 1). These target efficiencies are used for all protein functions except endogenous urinary loss. The use of a combined efficiency for EAA has been chosen to reflect observations of EAA metabolism. Indeed, in dairy cows, for AA belonging to Group 1 (i.e. His, Met, Phe, Trp), all measurable catabolism occurred across the liver (Larsen et al., 2015; Doepel et al., 2007; Lapierre et al., 2021) and not at the site of the synthesis of the proteins being exported (i.e., mammary gland for milk and gut for the metabolic fecal protein). However, in NASEM (2021), these efficiencies are not used to predict MPY, but a multivariate equation was derived based on the supply of His, Ile, Leu, Lys, Met, the sum of all other AA, digestible energy, NDF, and BW. In this context, these target efficiencies are also proposed as a threshold to consider, above which it might not be possible to reach the MPY predicted by the multivariate equation.

3.3 A case study on the estimation of protein and essential amino acid requirements

3.3.1 Protein

Although the previous discussion provides evidence that the driving force of MPY is not MP per se, but individual AA, we will have a glance at the impact of the concepts beneath the calculation of MP requirements in the lactating dairy cow. Very simplistically, to compare the implication of the differing concepts underlying the estimations of the different protein functions used by the different models, the protein requirements of an average cow corresponding to the treatment mean of the diet at 16.5 %CP in Olmos Colmenero and Broderick (2006) were used. This treatment was the middle treatment of five incremental protein supplies. The NASEM requirements have been added to the comparison reported previously (Lapierre et al., 2018) and are detailed in Table 2.

First, even though both INRA and NASEM assign an efficiency of 1 to the endogenous urinary protein-equivalent loss, the final MP requirement was twice as large for these two models compared to the others, averaging 190 versus

Table 2 Estimations of the MP requirements and supply of one example cow[a] and predicted MPY using six feed evaluation models

Feed evaluation	Efficiency[c]		Requirements[d,e]							Predicted
model[b]	Non-prod.[g]	Lactation	EndoUri	Scurf	MFP	DuoEndo	Milk	Total	Supply[d]	MPY[d,f]
CNCPS	0.67[h]	0.67	100	14	621		1766	2501	2573	1278
DVE/OEB	0.67	0.73	100	14	386		1619	2119	2273	1293
INRA	0.73[i]	0.73	184	13	411		1621	2229	2290	1141
NorFor	0.67	0.69	100	14	337		1722	2172	2085	1246
NRC	0.67[h]	0.67	100	14	583	176	1766	2638	2427	1082
NASEM	0.69[i]	0.69	195	13	356		1715	2279	2120	1124

[a]Average cow of the 16.5% CP diet treatment in Olmos Colmenero and Broderick (2006): 38.3 kg/day of milk at 3.09% true protein, 3.27% fat, eating 23 kg/day of dry matter at 120 days in milk, not changing BW (589 kg), and not gestating; adapted from Lapierre et al. (2018).

[b]CNCPS: Cornell Net Carbohydrate and Protein System, v 6.5 (van Amburgh et al., 2015); DVE/OEB: Dutch protein evaluation system (Tamminga et al., 1994 and VanDuinkerken et al., 2011); INRA: Institut National de la Recherche Agronomique (2018); NRC (National Research Council, 2001); NorFor (2011); NASEM: National Academies of Sciences, Engineering, and Medicine (2021).

[c]g of protein (secretion + accretion)/g of protein supply.

[d]In g/d.

[e]EndoUri: endogenous urinary loss; MFP: metabolic fecal protein; DuoEndo: endogenous duodenal flow. In DVE/OEB model, the metabolic fecal protein loss is not included in the requirement and is excluded from the supply: it is considered to be due to the diet and not the animal; for comparison purpose with the other feed evaluation models, this fraction has been added to the requirements.

[f]MPY: milk protein yield.

[g]Non-productive functions include endogenous urinary, scurf, metabolic fecal and duodenal endogenous proteins.

[h]Metabolic fecal protein has an efficiency of 1 in this model.

[i]Endogenous urinary loss has an efficiency of 1 in this model.

100 g/day. Estimation of requirements for metabolic fecal protein were the highest for CNCPS and NRC, and in a fairly similar range for the other models, albeit slightly lower for Norfor, averaging 602 versus 366 g/day. These different estimations of endogenous loss in urine and feces not only had an impact on the estimated amount of N secreted to support these non-productive functions, but the partition of endogenous losses between urine and feces also differed greatly: the ratio of loss in urine:loss in feces was the lowest in CNCPS at 0.16 versus the highest ratio in NASEM at 0.55. The NRC model is the only model including a requirement to cover the flow of endogenous proteins flowing at the duodenum and also includes a supply from the digestion of these proteins. However, if we want to follow the principle that the net requirement only includes the proteins creating a net demand on AA, the proportion of these secretions digested in the small intestine and returned to blood circulation does not need to be accounted for, and the proportion that is not digested is already included in the definition and estimation of the metabolic fecal protein. Finally, the MP requirement for milk protein, calculated as MPY divided by the efficiency of lactation, varied according to the efficiency used in each model, which varied between 0.67 and 0.73.

Overall, estimates of total MP requirements averaged 2323 g/day with a CV of 8.7% across the models. Nevertheless, estimates of requirements need to be aligned with estimates of supply within each model (Table 2). The NorFor, NRC, and NASEM models estimated that the ration did not supply sufficient MP to support requirements, whereas CNCPS, DVE/OEB, and INRA predicted that the supply was larger than the requirement. For this treatment mean, MPY averaged 1180 g/day with a standard error of the least squares means of 30 g/day (Olmos Colmenero and Broderick, 2006). The models with the closest relationship between predicted supply and requirements and accounting for energy supply in their estimation (i.e. INRA and NorFor) yielded the closest prediction of MPY (Table 2). Obviously, this exercise needs to be performed with a large number of studies representing various conditions to generalize a conclusion, but the intention was to demonstrate the impact of the definition of the concepts underlying the definition of the protein functions needing to be accounted for in the estimation of MP requirements.

3.3.2 Essential amino acids

In addition to the estimation of MP requirements detailed above, EAA requirements, or recommendations, have also been calculated for the average cow fed the 16.5% CP diet in Olmos Colmenero and Broderick (2006). The efficiencies of utilization of EAA used for CNCPS and NRC are detailed in Table 1, and the supplies and requirements for His, Lys, and Met for the six models are presented in Table 3.

Table 3 Estimations of AA supply and requirements (g/day) of one example cow[a] using six feed evaluation models

	Model[a]					
	CNCPS	DVE/OEB[b]	INRA	NorFor	NRC	NASEM
Supply						
MP[c], g/day	2573	2273	2290	2085	2427	2120
Metabolizable[d] His, g/day	66	NA	51	53	52	51
Metabolizable Lys, g/day	174	123	152	139	157	162
Metabolizable Met, g/day	57	42	42	45	45	47
His, %MP	2.58	NA	2.24	2.55	2.14	2.41
Lys, %MP	6.75	6.52	6.63	6.67	6.47	7.64
Met, %MP	2.20	2.20	1.84	2.18	1.85	2.22
Requirement[e]						
His	62	NA	2.4	2.2	NA	59
Lys	183	NA	7.0	6.6	6.89[f]	174
Met	64	NA	2.4	2.2	2.23[f]	56
Ratio of supply/requirement						
His	1.06	NA	0.93	1.16	NA	0.86
Lys	0.95	NA	0.95	1.01	0.94	0.93
Met	0.88	NA	0.77	0.99	0.83	0.84

[a]CNCPS: Cornell Net Carbohydrate and Protein System, v 6.5 (van Amburgh et al., 2015); DVE/OEB: Dutch protein evaluation system (Tamminga et al., 1994 and VanDuinkerken et al., 2011); INRA: Institut National de la Recherche Agronomique (2018); NRC (National Research Council, 2001); NorFor (2011); NASEM: National Academies of Sciences, Engineering, and Medicine (2021).

[b]In the DVE/OEB model, metabolic fecal protein (MFP) loss is not included in the requirement and is excluded from the supply: it is considered to be due to the diet and not the animal; for comparison purpose with the other feed evaluation models, this fraction has been added to the supply of MP. However, the ratio of metabolizable AA:MP supply has been calculated using the MP supply excluding MFP, which is equal to 1887 g/day.

[c]Metabolizable protein.

[d]Equivalent to the flow of digestible AA.

[e]CNCPS and NASEM in g/day and NRC, NorFor, and INR in % of MP.

[f]Updated requirements to optimize milk protein yield, based on Whitehouse et al. (2010a and b).

The supplies of His, Lys, and Met were the largest for the CNCPS model. The supply of His was not reported in the DVE/OEB model, but it estimated the lowest supplies of Lys and Met in terms of absolute amount. To remove the impact of the MP supply per se, if we look at the proportion of the digestible flows of these three EAA relative to MP, the CNCPS model is also yielding the highest proportions of His, Lys, and Met relative to MP, except for NASEM yielding a highest proportion of Lys. The DVE/OEB and NorFor models were very close to the CNCPS, whereas the INRA and NRC yielded lower proportions.

The major point of this exercise is that models differ substantially in the ratio between the estimated supply and requirement, as detailed in Table 3. Histidine presented the largest variation, whereas for Met, the ratio of supply:requirement was below 1 for all models but was close to 0.99 for NorFor. For these three EAA, the NorFor model estimated that the supply was sufficient to cover the requirements, whereas the other models would suggest that Lys and mainly Met might be in short supply. Although significant progress has been made on the estimations of the requirements of EAA, this exercise indicates that more research is still needed to improve our understanding of EAA biology and its mathematical representation in feed evaluation models for dairy cows.

4 Conclusion

Numerous models and data-driven approaches are improving the integration between biological principles and predictability of outputs. However, critical gaps remain. For example, the large diversity among feed sources, especially forages, is better represented but still needs improved characterization for improved prediction of RDP, dRUP, and especially the individual AA within those measurements. Some improvements, such as the AA profile of microbial protein, are recognized in this chapter. Although efficiency of utilization of MP and metabolizable AA to support the various protein functions, including milk protein is widely recognized to vary among dietary conditions, progress is still to be made to quantify factors affecting it, including the type of energy. Further advances are needed to improve recommendations of the supply of AA and energy needed to sustainably produce milk protein and cow health while maximizing production efficiency and minimizing environmental impact of dairy cattle enterprises.

5 Future trends in research

Passage rates have been predicted with relatively poor precision. Accuracy of measurement of the kp of respective rumen-degradable fractions of protein, starch, and fiber are assumed to be represented by markers that are not specific to those fractions. Improved techniques would thus improve models that rely on predicted kp. An approach is described that would allow ruminal degradability and intestinal digestibility of individual AA to be normalized relative to total AA, which then could then be substituted for CP in more routine analyses and could potentially be scaled for differences among laboratory methods. Many more studies were done to assess supplemental protein sources, yet parameterization into models often relies also on library values for commodities and forages for which the range in the parameters is not captured. Predictions for nutrient composition of byproducts and forages is likely missing the true variability that could be captured; moreover, a continued overreliance on library values for forages will continue to ignore how plant breeding has changed those values (e.g., today's corn silage probably has higher starch and digestibility of starch and fiber). The role of processing of forages and grains is not developed enough for robust prediction.

Older predictions of microbial protein supply were based on or evaluated against MCP flow to the duodenum. As models progress, though, we need to address the greater values for omasal sampling, which has largely predominated over duodenal sampling in recent studies. Reasons for higher MCP flows from omasal sampling versus duodenal sampling are not well described using comparative data within only a few studies. Starch is now being routinely measured for various grains and forages, but the soluble fiber fractions in feed and digesta should be measured more to understand their role to support microbial protein synthesis. Meta-analyses rely on predicting MCP largely using studies that intentionally overfed or underfed RDP for all treatments within the study. This imbalance in datasets could be rectified with more studies intentionally designed to evaluate a nonlinear role for RDP to optimize microbial protein supply. Similarly, more attention to AA in microbes, including protozoa and possibly fungi would allow a better assessment of whether AA profile changes under differing dietary conditions.

More definitive research effort needs to be devoted to the quantification of the efficiency of utilization of the predicted supply of MP and metabolizable AA. Data are limited in terms of the high-producing cow's response to gradual incremental supply of a single AA, the response to one AA when the supply of the others is estimated to be short, and certainly to the interaction between AA and energy supply, specifically looking at the source of energy. Another gap is to establish roles of EAA in functions other than directly supporting protein synthesis and to quantify the extra supply required to support these functions. The challenge is to be able to represent all these various and complex metabolic pathways in a simplified mathematical model that can represent the enormous flexibility of the cow and its mammary gland.

6 Where to look for further information

Through the National Academies of Sciences, Engineering, and Medicine, there are several resources such as the nutrient requirements of beef, small ruminants, and dairy available through their website: https://www.nap.edu/topic/276/agriculture. In addition to feed tables and nutrient requirements, seminal components supporting those requirements are embedded therein. The Australian (CSIRO; ISBN 9780643092624), Dutch (DVE/OEB; doi:10.1017/S0021859610000912), French (INRA; doi.org/10.3920/978-90-8686-292-4), and Nordic (doi.org/10.3920/978-90-8686-718-9) systems are among those with similar resources. Otherwise, more information on the DVE/EOB and CNCPS models can be found in the literature quoted in the current review. The Encyclopedia of Dairy Sciences (ISBN 978-0-12-374407-4) has a vast array of divergent chapters. Many proceedings from regional nutrition conferences across the United States and Canada are categorized by The Searchable Proceedings of Animal Conferences (https://spac.adsa.org/index.asp).

7 References

Ahvenjärvi, S., Vaga, M., Vanhatalo, A. and Huhtanen, P. 2018. Ruminal metabolism of grass silage soluble nitrogen fractions. Journal of Dairy Science 101(1), 279-294. doi: 10.3168/jds.2016-12316.

Alexandrov, A. N. 1998. Effect of ruminal exposure and subsequent microbial contamination on dry matter and protein degradability of various feedstuffs. Animal Feed Science and Technology 71(1-2), 99–107. doi: 10.1016/S0377-8401(97)00129-6.

Allen, M. S. 2000. Effects of diet on short-term regulation of feed intake by lactating dairy cattle. Journal of Dairy Science 83(7), 1598–1624. doi: 10.3168/jds.S0022-0302(00)75030-2.

Bach, A., Calsamiglia, S. and Stern, M. D. 2005. Nitrogen metabolism in the rumen. Journal of Dairy Science 88(E), E9–E21. doi: 10.3168/jds.S0022-0302(05)73133-7.

Batista, E. D., Detmann, E., Valadares Filho, S. C., Titgemeyer, E. C. and Valadares, R. F. D. 2017. The effect of CP concentration in the diet on urea kinetics and microbial usage of recycled urea in cattle: A meta-analysis. Animal 11(8), 1303–1311. doi: 10.1017/S1751731116002822.

Bernard, L. and Doreau, M. 2000. Use of rare earth elements as external markers for mean retention time measurements in ruminants. Reproduction, Nutrition, Development 40(2), 89–101. doi: 10.1051/rnd:2000122.

Berthiaume, R., Dubreuil, P., Stevenson, M., McBride, B. W. and Lapierre, H. 2001. Intestinal disappearance and mesenteric and portal appearance of amino acids in dairy cows fed ruminally protected methionine. Journal of Dairy Science 84(1), 194–203. doi: 10.3168/jds.S0022-0302(01)74469-4.

Blouin, J. P., Bernier, J. F., Reynolds, C. K., Lobley, G. E., Dubreuil, P. and Lapierre, H. 2002. Effect of supply of metabolizable protein on splanchnic fluxes of nutrients and

hormones in lactating dairy cows. Journal of Dairy Science 85(10), 2618–2630. doi: 10.3168/jds.S0022-0302(02)74347-6.
Boucher, S. E., Calsamiglia, S., Parsons, C. M., Stein, H. H., Stern, M. D., Erickson, P. S., Utterback, P. L. and Schwab, C. G. 2009. Intestinal digestibility of amino acids in rumen undegradable protein estimated using a precision-fed cecectomized rooster bioassay: I. Soybean meal and SoyPlus. Journal of Dairy Science 92(9), 4489–4498. doi: 10.3168/jds.2008-1884.
Bowman, J. G. P. and Firkins, J. L. 1993. Effects of forage species and particle size on bacterial cellulolytic activity and colonization in situ. Journal of Animal Science 71(6), 1623–1633. doi: 10.2527/1993.7161623x.
Broderick, G. A., Huhtanen, P., Ahvenjärvi, S., Reynal, S. M. and Shingfield, K. J. 2010. Quantifying ruminal nitrogen metabolism using the omasal sampling technique in cattle–A meta-analysis. Journal of Dairy Science 93(7), 3216–3230. doi: 10.3168/jds.2009-2989.
Broderick, G. A. and Merchen, N. R. 1992. Markers for quantifying microbial protein synthesis in the rumen. Journal of Dairy Science 75(9), 2618–2632. doi: 10.3168/jds.S0022-0302(92)78024-2.
Calsamiglia, S., Stern, M. D. and Firkins, J. L. 1996. Comparison of nitrogen-15 and purines as microbial markers in continuous culture. Journal of Animal Science 74(6), 1375–1381. doi: 10.2527/1996.7461375x.
Choi, C. W., Ahvenjärvi, S., Vanhatalo, A., Toivonen, V. and Huhtanen, P. 2002. Quantitation of the flow of soluble non-ammonia nitrogen entering the omasal canal of dairy cows fed grass silage based diets. Animal Feed Science and Technology 96(3–4), 203–220. doi: 10.1016/S0377-8401(01)00348-0.
Clark, J. H., Klusmeyer, T. H. and Cameron, M. R. 1992. Microbial protein synthesis and flows of nitrogen fractions to the duodenum of dairy cows. Journal of Dairy Science 75(8), 2304–2323. doi: 10.3168/jds.S0022-0302(92)77992-2.
Colucci, P. E., Macleod, G. K., Grovum, W. L., Cahill, L. W. and McMillan, I. 1989. Comparative digestion in sheep and cattle fed different forage to concentrate ratios at high and low intakes. Journal of Dairy Science 72(7), 1774–1785. doi: 10.3168/jds.S0022-0302(89)79294-8.
Cozzi, G., Andrighetto, I., Berzaghi, P. and Polan, C. E. 1995. In situ ruminal disappearance of essential amino acids in protein feedstuffs. Journal of Dairy Science 78(1), 161–171. doi: 10.3168/jds.S0022-0302(95)76626-7.
Cozzi, G., Bittante, G. and Polan, C. E. 1993. Comparison of fibrous materials as modifiers of in situ ruminal degradation of corn gluten meal. Journal of Dairy Science 76(4), 1106–1113. doi: 10.3168/jds.S0022-0302(93)77439-1.
Dineen, M., McCarthy, B., Dillon, P., Matthews, C., Ross, D. and Van Amburgh, M. E. 2021. Microbial composition and omasal flows of bacterial, protozoal, and nonmicrobial amino acids in lactating dairy cows fed fresh perennial ryegrass (Lolium perenne L.) not supplemented or supplemented with rolled barley. Journal of Dairy Science 104(4), 4192–4205. doi: 10.3168/jds.2020-19203.
Doepel, L., Lobley, G. E., Bernier, J. F., Dubreuil, P. and Lapierre, H. 2007. Effect of glutamine supplementation on splanchnic metabolism in lactating dairy cows. Journal of Dairy Science 90(9), 4325–4333. doi: 10.3168/jds.2007-0124.
Fessenden, S. W., Hackmann, T. J., Ross, D. A., Block, E., Foskolos, A. and Van Amburgh, M. E. 2019. Rumen digestion kinetics, microbial yield, and omasal flows of nonmicrobial,

bacterial, and protozoal amino acids in lactating dairy cattle fed fermentation by-products or urea as a soluble nitrogen source. Journal of Dairy Science 102(4), 3036–3052. doi: 10.3168/jds.2018-15448.

Fessenden, S. W., Hackmann, T. J., Ross, D. A., Foskolos, A. and Van Amburgh, M. E. 2017. Ruminal bacteria and protozoa composition, digestibility, and amino acid profile determined by multiple hydrolysis times. Journal of Dairy Science 100(9), 7211–7226. doi: 10.3168/jds.2016-12531.

Firkins, J. L. 2021. Invited review: Advances in rumen efficiency. Applied Animal Science 37(4), 388–403. doi: 10.15232/aas.2021-02163.

Firkins, J. L., Allen, M. S., Oldick, B. S. and St-Pierre, N. R. 1998. Modeling ruminal digestibility of carbohydrates and microbial protein flow to the duodenum. Journal of Dairy Science 81(12), 3350–3369. doi: 10.3168/jds.S0022-0302(98)75901-6.

Firkins, J. L., Hristov, A. N., Hall, M. B., Varga, G. A. and St-Pierre, N. R. 2006. Integration of ruminal metabolism in dairy cattle. Journal of Dairy Science 89(E) (Suppl. 1), E31–E51. doi: 10.3168/jds.S0022-0302(06)72362-1.

Firkins, J. L., Lewis, S. M., Montgomery, L., Berger, L. L., Merchen, N. R. and Fahey, G. C. 1987. Effects of feed intake and dietary urea concentration on ruminal dilution rate and efficiency of bacterial growth in steers. Journal of Dairy Science 70(11), 2312–2321. doi: 10.3168/jds.S0022-0302(87)80291-6.

Fonseca, A. C., Fredin, S. M., Ferraretto, L. F., Parsons, C. M., Utterback, P. L. and Shaver, R. D. 2014. Short communication: Intestinal digestibility of amino acids in fluid- and particle-associated rumen bacteria determined using a precision-fed cecectomized rooster bioassay. Journal of Dairy Science 97(6), 3855–3859. doi: 10.3168/jds.2013-7880.

Fox, D. G., Tedeschi, L. O., Tylutki, T. P., Russell, J. B., Van Amburgh, M. E., Chase, L. E., Pell, A. N. and Overton, T. R. 2004. The Cornell Net carbohydrate and Protein System model for evaluating herd nutrition and nutrient excretion. Animal Feed Science and Technology 112(1–4), 29–78. doi: 10.1016/J.ANIFEEDSCI.2003.10.006.

Galyean, M. L. and Tedeschi, L. O. 2014. Predicting microbial protein synthesis in beef cattle: Relationship to intakes of total digestible nutrients and crude protein. Journal of Animal Science 92(11), 5099–5111. doi: 10.2527/jas.2014-8098.

Grant, R. J. 1997. Interactions among forages and nonforage fiber sources. Journal of Dairy Science 80(7), 1438–1446. doi: 10.3168/jds.S0022-0302(97)76073-9.

Gregorini, P., Beukes, P., Waghorn, G., Pacheco, D. and Hanigan, M. 2015. Development of an improved representation of rumen digesta outflow in a mechanistic and dynamic model of a dairy cow, Molly. Ecological Modelling 313(10), 293–306. doi: 10.1016/j.ecolmodel.2015.06.042.

Gruninger, R. J., Ribeiro, G. O., Cameron, A. and McAllister, T. A. 2019. Invited review: Application of meta-omics to understand the dynamic nature of the rumen microbiome and how it responds to diet in ruminants. Animal 13(9), 1843–1854. doi: 10.1017/S1751731119000752.

Hackmann, T. J. and Firkins, J. L. 2015. Maximizing efficiency of rumen microbial protein production. Frontiers in Microbiology 6, 465. doi: 10.3389/fmicb.2015.00465.

Hanigan, M. D., Souza, V. C., Martineau, R., Daley, V. L. and Kononoff, P. 2021. Predicting ruminally undegraded and microbial protein flows from the rumen. Journal of Dairy Science 104(8), 8685–8707. doi: 10.3168/jds.2020-19672.

Harmon, D. L. and Swanson, K. C. 2020. Review: Nutritional regulation of intestinal starch and protein assimilation in ruminants. Animal 14(S1), s17–s28. doi: 10.1017/S1751731119003136.

Hartinger, T., Gresner, N. and Südekum, K. H. 2018. Does intra-ruminal nitrogen recycling waste valuable resources? A review of major players and their manipulation. Journal of Animal Science and Biotechnology 9(1), 33. doi: 10.1186/s40104-018-0249-x.

Hristov, A. N., Bannink, A., Crompton, L. A., Huhtanen, P., Kreuzer, M., McGee, M., Nozière, P., Reynolds, C. K., Bayat, A. R., Yáñez-Ruiz, D. R., Dijkstra, J., Kebreab, E., Schwarm, A., Shingfield, K. J. and Yu, Z. 2019. Invited review: Nitrogen in ruminant nutrition: A review of measurement techniques. Journal of Dairy Science 102(7), 5811–5852. doi: 10.3168/jds.2018-15829.

Ipharraguerre, I. R. and Clark, J. H. 2005. Impacts of the source and amount of crude protein on the intestinal supply of nitrogen fractions and performance of dairy cows. Journal of Dairy Science 88(Suppl. 1), E22–E37. doi: 10.3168/jds.S0022-0302(05)73134-9.

Jouany, J. P. 1996. Effect of rumen protozoa on nitrogen utilization by ruminants. Journal of Nutrition 126(4), 1335S–1346S. doi: 10.1093/jn/126.suppl_4.1335S.

Kamoun, M., Ammar, H., Théwis, A., Beckers, Y., France, J. and López, S. 2014. Comparison of three ^{15}N methods to correct for microbial contamination when assessing in situ protein degradability of fresh forages. Journal of Animal Science 92(11), 5053–5062. doi: 10.2527/jas.2014-7691.

Kand, D., Castro-Montoya, J., Selje-Assmann, N. and Dickhoefer, U. 2021. The effects of rumen nitrogen balance on intake, nutrient digestibility, chewing activity, and milk yield and composition in dairy cows vary with dietary protein sources. Journal of Dairy Science 104(4), 4236–4250. doi: 10.3168/jds.2020-19129.

Kand, D. and Dickhoefer, U. 2021. The effects of rumen nitrogen balance on nutrient intake, nitrogen partitioning, and microbial protein synthesis in lactating dairy cows offered different dietary protein sources. Journal of Dairy Science 104(4), 4223–4235. doi: 10.3168/JDS.2019-18005.

Krawielitzki, K., Schmidt, T., Voigt, J., Kowalczyk, J. and Gabel, M. 2006. Dynamics of microbial contamination of protein during ruminal in situ incubation of feedstuffs. Journal of Animal and Feed Sciences 15(2), 313–328. doi: 10.22358/jafs/66903/2006.

Krizsan, S. J., Ahvenjärvi, S., Volden, H. and Broderick, G. A. 2010. Estimation of rumen outflow in dairy cows fed grass silage-based diets by use of reticular sampling as an alternative to sampling from the omasal canal. Journal of Dairy Science 93(3), 1138–1147. doi: 10.3168/jds.2009-2661.

Krizsan, S. J., Rinne, M., Nyholm, L. and Huhtanen, P. 2015. New recommendations for the ruminal in situ determination of indigestible neutral detergent fibre. Animal Feed Science and Technology 205, 31–41. doi: 10.1016/j.anifeedsci.2015.04.008.

Lapierre, H., Binggeli, S., Sok, M., Pellerin, D. and Ouellet, D. R. 2019. Estimation of correction factors to determine the true amino acid concentration of protein after a 24-hour hydrolysis. Journal of Dairy Science 102(2), 1205–1212. doi: 10.3168/jds.2018-15392.

Lapierre, H., Blouin, J. P., Bernier, J. F., Reynolds, C. K., Dubreuil, P. and Lobley, G. E. 2002. Effect of supply of metabolizable protein on whole body and splanchnic leucine metabolism in lactating dairy cows. Journal of Dairy Science 85(10), 2631–2641. doi: 10.3168/jds.S0022-0302(02)74348-8.

Lapierre, H., Girard, C. L., Matte, J. J. and Lobley, G. E. 2005. Effects of stage of lactation on protein metabolism in dairy cows. Journal of Animal and Feed Sciences 14(1), 53–62. doi: 10.22358/jafs/66967/2005.

Lapierre, H., Larsen, M., Sauvant, D., Van Amburgh, M. E. and Van Duinkerken, G. V. 2018. Review: Converting nutritional knowledge into feeding practices: A case study comparing different protein feeding systems for dairy cows. Animal 12(s2), s457–s466. doi: 10.1017/S1751731118001763.

Lapierre, H., Lobley, G. E. and Ouellet, D. R. 2021. Histidine optimal supply in dairy cows through determination of a threshold efficiency. Journal of Dairy Science 104(2), 1759–1776. doi: 10.3168/jds.2020-19205.

Lapierre, H., Martineau, R., Hanigan, M. D., Van Lingen, H. J., Kebreab, E., Spek, J. W. and Ouellet, D. R. 2020. Review: Impact of protein and energy supply on the fate of amino acids from absorption to milk protein in dairy cows. Animal 14(Suppl. 1), s87–s102. doi: 10.1017/S1751731119003173.

Larsen, M., Galindo, C., Ouellet, D. R., Maxin, G., Kristensen, N. B. and Lapierre, H. 2015. Abomasal amino acid infusion in postpartum dairy cows: Effect on whole-body, splanchnic, and mammary amino acid metabolism. Journal of Dairy Science 98(11), 7944–7961. doi: 10.3168/jds.2015-9439.

Larsen, M., Madsen, T. G., Weisbjerg, M. R., Hvelplund, T. and Madsen, J. 2001. Small intestinal digestibility of microbial and endogenous amino acids in dairy cows. Journal of Animal Physiology and Animal Nutrition 85(1–2), 9–21. doi: 10.1046/j.1439-0396.2001.00297.x.

Liebe, D. M., Firkins, J. L., Tran, H., Kononoff, P. J. and White, R. R. 2018. Technical note: Methodological and feed factors affecting measurement of protein A, B, and C fractions, degradation rate, and intestinal digestibility of rumen-undegraded protein. Journal of Dairy Science 101(9), 8046–8053. doi: 10.3168/jds.2018-14837.

Lopreiato, V., Mezzetti, M., Cattaneo, L., Ferronato, G., Minuti, A. and Trevisi, E. 2020. Role of nutraceuticals during the transition period of dairy cows: A review. Journal of Animal Science and Biotechnology 11(1), 96. doi: 10.1186/s40104-020-00501-x.

Marini, J. C., Fox, D. G. and Murphy, M. R. 2008. Nitrogen transactions along the gastrointestinal tract of cattle: A meta-analytical approach. Journal of Animal Science 86(3), 660–679. doi: 10.2527/jas.2007-0039.

Mariz, L. D. S., Amaral, P. M., Valadares Filho, S. C., Santos, S. A., Detmann, E., Marcondes, M. I., Pereira, J. M. V., Silva Júnior, J. M., Prados, L. F. and Faciola, A. P. 2018a. Dietary protein reduction on microbial protein, amino acid digestibility, and body retention in beef cattle: 2. Amino acid intestinal absorption and their efficiency for whole-body deposition. Journal of Animal Science 96(2), 670–683. doi: 10.1093/jas/sky018.

Mariz, L. D. S., Amaral, P. M., Valadares Filho, S. C., Santos, S. A., Marcondes, M. I., Prados, L. F., Carneiro Pacheco, M. V., Zanetti, D., de Castro Menezes, G. C. and Faciola, A. P. 2018b. Dietary protein reduction on microbial protein, amino acids digestibility, and body retention in beef cattle. I. Digestibility sites and ruminal synthesis estimated by purine bases and ^{15}N as markers. Journal of Animal Science 96(6), 2453–2467. doi: 10.1093/jas/sky134.

McFadden, J. W., Girard, C. L., Tao, S., Zhou, Z., Bernard, J. K., Duplessis, M. and White, H. M. 2020. Symposium review: One-carbon metabolism and methyl donor nutrition in the dairy cow. Journal of Dairy Science 103(6), 5668–5683. doi: 10.3168/jds.2019-17319.

Muck, R. E., Nadeau, E. M. G., McAllister, T. A., Contreras-Govea, F. E., Santos, M. C. and Kung, L. 2018. Silage review: Recent advances and future uses of silage additives. Journal of Dairy Science 101(5), 3980–4000. doi: 10.3168/jds.2017-13839.

National Academies of Science and Medicine. 2016. Nutrient Requirements of Beef Cattle, Eight (Rev. edn.). Washington, DC: National Academies Press. doi: 10.17226/19014.

National Academies of Sciences and Medicine. 2021. Nutrient Requirements of Dairy Cattle: Eighth (Rev. edn.). Washington, DC: The National Academies Press. doi: 10.17226/25806.

NRC. 1985. Ruminant Nitrogen Usage. Washington, DC: National Academy of Sciences.

NRC. 2001. Nutrient Requirements of Dairy Cattle (7th rev. edn.). Washington, DC: National Academy of Sciences.

Offner, A. and Sauvant, D. 2004. Prediction of in vivo starch digestion in cattle from in situ data. Animal Feed Science and Technology 111(1-4), 41-56. doi: 10.1016/S0377-8401(03)00216-5.

Olmos Colmenero, J. J. and Broderick, G. A. 2006. Effect of dietary crude protein concentration on milk production and nitrogen utilization in lactating dairy cows. Journal of Dairy Science 89(5), 1704-1712. doi: 10.3168/jds.S0022-0302(06)72237-8.

Onodera, R. 1993. Methionine and lysine metabolism in the rumen and possible effects of their metabolites on the nutrition and physiology of ruminants. Amino Acids 5(2), 217-232. doi: 10.1007/BF00805984.

Ouellet, D. R., Berthiaume, R., Holtrop, G., Lobley, G. E., Martineau, R. and Lapierre, H. 2010. Effect of method of conservation of timothy on endogenous nitrogen flows in lactating dairy cows. Journal of Dairy Science 93(9), 4252-4261. doi: 10.3168/jds.2010-3085.

Ouellet, D. R., Demers, M., Zuur, G., Lobley, G. E., Seoane, J. R., Nolan, J. V. and Lapierre, H. 2002. Effect of dietary fiber on endogenous nitrogen flows in lactating dairy cows. Journal of Dairy Science 85(11), 3013-3025. doi: 10.3168/jds.S0022-0302(02)74387-7.

Owens, F. N. and Hanson, C. F. 1992. External and internal markers for appraising site and extent of digestion in ruminants. Journal of Dairy Science 75(9), 2605-2617. doi: 10.3168/jds.S0022-0302(92)78023-0.

Paz, H. A., Klopfenstein, T. J., Hostetler, D., Fernando, S. C., Castillo-Lopez, E. and Kononoff, P. J. 2014. Ruminal degradation and intestinal digestibility of protein and amino acids in high-protein feedstuffs commonly used in dairy diets. Journal of Dairy Science 97(10), 6485-6498. doi: 10.3168/jds.2014-8108.

Raggio, G., Lemosquet, S., Lobley, G. E., Rulquin, H. and Lapierre, H. 2006a. Effect of casein and propionate supply on mammary protein metabolism in lactating dairy cows. Journal of Dairy Science 89(11), 4340-4351. doi: 10.3168/jds.S0022-0302(06)72481-X.

Raggio, G., Lobley, G. E., Lemosquet, S., Rulquin, H. and Lapierre, H. 2006b. Effect of casein and propionate supply on whole body protein metabolism in lactating dairy cows. Canadian Journal of Animal Science 86(1), 81-89. doi: 10.4141/a05-055.

Reveneau, C., Karnati, S. K. R., Oelker, E. R. and Firkins, J. L. 2012. Interaction of unsaturated fat or coconut oil with monensin in lactating dairy cows fed twelve times daily. I. Protozoal abundance, nutrient digestibility, and microbial protein flow to the omasum. Journal of Dairy Science 95(4), 2046-2060. doi: 10.3168/jds.2011-4887.

Reynal, S. M., Ipharraguerre, I. R., Liñeiro, M., Brito, A. F., Broderick, G. A. and Clark, J. H. 2007. Omasal flow of soluble proteins, peptides, and free amino acids in dairy cows fed diets supplemented with proteins of varying ruminal degradabilities. Journal of Dairy Science 90(4), 1887-1903. doi: 10.3168/jds.2006-158.

Reynolds, C. K. and Kristensen, N. B. 2008. Nitrogen recycling through the gut and the nitrogen economy of ruminants: An asynchronous symbiosis. Journal of Animal Science 86(14), E293–E305. doi: 10.2527/jas.2007-0475.

Roman-Garcia, Y., White, R. R. and Firkins, J. L. 2016. Meta-analysis of postruminal microbial nitrogen flows in dairy cattle. I. Derivation of equations. Journal of Dairy Science 99(10), 7918–7931. doi: 10.3168/jds.2015-10661.

Sandek, A., Krawielitzki, K., Kowalczyk, J., Kreienbring, F., Schoenhusen, U., Gabel, M., Zebrowska, T., Hagemeister, H. and Voigt, J. 2001. Studies on N-metabolism in different gastrointestinal sections of sheep using the digesta exchange technique. 2. Passage of endogenous nitrogen. Journal of Animal and Feed Sciences 10(4), 605–618. doi: 10.22358/jafs/68012/2001.

Santos, F. A. P., Santos, J. E. P., Theurer, C. B. and Huber, J. T. 1998. Effects of rumen-undegradable protein on dairy cow performance: A 12-year literature review. Journal of Dairy Science 81(12), 3182–3213. doi: 10.3168/jds.S0022-0302(98)75884-9.

Sauvant, D., Cantalapiedra-Hijar, G., Delaby, L., Daniel, J. B., Faverdin, P. and Nozière, P. 2015. Actualisation des besoins protéiques des ruminants et détermination des réponses des femelles laitières aux apports de protéines digestibles dans l'intestin. INRA Productions Animales 28(5), 347–368. doi: 10.20870/productions-animales.2015.28.5.3038.

Sauvant, D. and Nozière, P. 2016. Quantification of the main digestive processes in ruminants: The equations involved in the renewed energy and protein feed evaluation systems. Animal 10(5), 755–770. doi: 10.1017/S1751731115002670.

Schieber, A., Brückner, H. and Ling, J. R. 1999. Gc–ms analysis of diaminopimelic acid stereoisomers and amino acid enantiomers in rumen bacteria. Biomedical Chromatography 13(1), 46–50. doi: 10.1002/(SICI)1099-0801(199902)13:1<46::AID-BMC808>3.0.CO;2-4.

Schwab, C. G. and Broderick, G. A. 2017. A 100-year review: Protein and amino acid nutrition in dairy cows. Journal of Dairy Science 100(12), 10094–10112. doi: 10.3168/jds.2017-13320.

Seo, S., Lanzas, C., Tedeschi, L. O., Pell, A. N. and Fox, D. G. 2009. Development of a mechanistic model to represent the dynamics of particle flow out of the rumen and to predict rate of passage of forage particles in dairy cattle. Journal of Dairy Science 92(8), 3981–4000. doi: 10.3168/jds.2006-799.

Seo, S., Tedeschi, L. O., Lanzas, C., Schwab, C. G. and Fox, D. G. 2006a. Development and evaluation of empirical equations to predict feed passage rate in cattle. Animal Feed Science and Technology 128(1–2), 67–83. doi: 10.1016/j.anifeedsci.2005.09.014.

Seo, S., Tedeschi, L. O., Schwab, C. G., Garthwaite, B. D. and Fox, D. G. 2006b. Evaluation of the passage rate equations in the 2001 dairy NRC model. Journal of Dairy Science 89(6), 2327–2342. doi: 10.3168/jds.S0022-0302(06)72304-9.

Sok, M., Ouellet, D. R., Firkins, J. L., Pellerin, D. and Lapierre, H. 2017. Amino acid composition of rumen bacteria and protozoa in cattle. Journal of Dairy Science 100(7), 5241–5249. doi: 10.3168/jds.2016-12447.

Spanghero, M. and Kowalski, Z. M. 2021. Updating analysis of nitrogen balance experiments in dairy cows. Journal of Dairy Science 104(7), 7725–7737. doi: 10.3168/jds.2020-19656.

St-Pierre, N. R. 2001. Invited review: Integrating quantitative findings from multiple studies using mixed model methodology. Journal of Dairy Science 84(4), 741–755. doi: 10.3168/jds.S0022-0302(01)74530-4.

Stern, M. D., Bach, A. and Calsamiglia, S. 1997. Alternative techniques for measuring nutrient digestion in ruminants. Journal of Animal Science 75(8), 2256–2276. doi: 10.2527/1997.7582256x.

Stern, M. D., Varga, G. A., Clark, J. H., Firkins, J. L., Huber, J. T. and Palmquist, D. L. 1994. Evaluation of chemical and physical properties of feeds that affect protein metabolism in the rumen. Journal of Dairy Science 77(9), 2762–2786. doi: 10.3168/jds.S0022-0302(94)77219-2.

Swanson, E. W. 1977. Factors for computing requirements of protein for maintenance of cattle. Journal of Dairy Science 60(10), 1583–1593. doi: 10.3168/jds.S0022-0302(77)84074-5.

Tamminga, S., Van Straalen, W. M., Subnel, A. P. J., Meijer, R. G. M., Steg, A., Wever, C. J. G. and Blok, M. C. 1994. The Dutch protein evaluation system: The DVE/OEB-system. Livestock Production Science 40(2), 139–155. doi: 10.1016/0301-6226(94)90043-4.

Tedeschi, L. O. 2019. Relationships of retained energy and retained protein that influence the determination of cattle requirements of energy and protein using the California Net Energy System. Translational Animal Science 3(3), 1029–1039. doi: 10.1093/tas/txy120.

Thivierge, M. C., Petitclerc, D., Bernier, J. F., Couture, Y. and Lapierre, H. D. 2002. Variations in mammary protein metabolism during the natural filling of the udder with milk over a 12-h period between two milkings: Leucine kinetics. Journal of Dairy Science 85(11), 2974–2985. doi: 10.3168/jds.S0022-0302(02)74383-X.

Tran, H., Schlageter-Tello, A., Caprez, A., Miller, P. S., Hall, M. B., Weiss, W. P. and Kononoff, P. J. 2020. Development of feed composition tables using a statistical screening procedure. Journal of Dairy Science 103(4), 3786–3803. doi: 10.3168/jds.2019-16702.

Van Amburgh, M. E. 2018. Modeling and integrating metabolizable energy and protein supply and requirements in dry and lactating dairy cattle to optimize nitrogen utilization. In: Canada A. N. C. O. (Ed.), ANAC. Edmonton, AB, pp. 90–104.

Van Amburgh, M. E., Collao-Saenz, E. A., Higgs, R. J., Ross, D. A., Recktenwald, E. B., Raffrenato, E., Chase, L. E., Overton, T. R., Mills, J. K. and Foskolos, A. 2015. The Cornell Net carbohydrate and protein system: Updates to the model and evaluation of version 6.5. Journal of Dairy Science 98(9), 6361–6380. doi: 10.3168/jds.2015-9378.

Van Duinkerken, G., Blok, M. C., Bannink, A., Cone, J. W., Dijkstra, J., Van Vuuren, A. M. and Tamminga, S. 2011. Update of the Dutch protein evaluation system for ruminants: The DVE/OEB2010 system. Journal of Agricultural Science 149(3), 351–367. doi: 10.1017/S0021859610000912.

Walker, N. D., Newbold, C. J. and Wallace, R. J. 2005. Nitrogen metabolism in the rumen. In: Pfeffer, E. and Hristov, A. (Eds.), Nitrogen and Phosphorus Nutrition of Cattle. Cambridge, MA: CABI Publishing, pp. 71–115.

Wallace, R. J. 1983. Digestion of rumen bacteria in vitro. British Journal of Nutrition 49(1), 101–108. doi: 10.1079/BJN19830015.

Wallace, R. J. 1996. Ruminal microbial metabolism of peptides and amino acids. Journal of Nutrition 126(4), 1326S–1334S. doi: 10.1093/jn/126.suppl_4.1326S.

White, R. R., Roman-Garcia, Y., Firkins, J. L., Kononoff, P. J., VandeHaar, M. J., Tran, H., McGill, T., Garnett, R. and Hanigan, M. D. 2017. Evaluation of the 2001 dairy NRC and derivation of new prediction equations. 2. Rumen degradable and undegradable protein. Journal of Dairy Science 100(5), 3611–3627. doi: 10.3168/jds.2015-10801.

Whitehouse, N., Schwab, C., Luchini, D. and Sloan, B. 2010a. A critique of doseresponse plots that relate changes in content and yield of milk protein to predicted concentrations of lysine in metabolizable protein by the NRC (2001), CPM-Dairy (v.3.0.10) and AMTS.Cattle (v.2.1.1) models. Journal of Dairy Science 93: 447. doi: doi.

Whitehouse, N., Schwab, C., Luchini, D. and Sloan, B. 2010b. A critique of doseresponse plots that relate changes in content and yield of milk protein to predicted concentrations of methionine in metabolizable protein by the NRC (2001), CPM-Dairy (v.3.0.10) and AMTS.Cattle (v.2.1.1) models. Journal of Dairy Science 93: 447.

Woods, V. B., Moloney, A. P. and O'Mara, F. P. 2003. The nutritive value of concentrate feedstuffs for ruminant animals. Part II: In situ. Animal Feed Science and Technology 110(1-4), 131-143. doi: 10.1016/S0377-8401(03)00222-0.

Wu, Y. and Wu, G. 2017. Nutritionally nonessential amino acids: A misnomer in nutritional sciences. Advances in Nutrition 8(1), 137-139. doi: 10.3945/an.116.012971.

Chapter 3

Advances in understanding lipid requirements and utilization in dairy cattle

James K. Drackley, University of Illinois at Urbana-Champaign, USA

1 Introduction

Although lipids represent a smaller part of the diet than carbohydrates and protein, lipid metabolism is central to the production of milk and milk components. In many high-producing dairy herds, cows on average consume more than 1 kg of dietary lipid daily and secrete more than 1.5 kg of milk fat daily. The current US record holding cow, Selz-Pralle Aftershock 3918, produced 1406 kg of milk fat in 1 year, equating to 3.85 kg/day. The mammary gland uses the

http://dx.doi.org/10.19103/AS.2022.0117.03

carbohydrate fermentation products acetate and butyrate to synthesize about half of the fat secreted in milk, with the remainder coming from preformed fatty acids consumed in the diet. Lipid nutrition is complicated by the extensive body fat mobilization that occurs in early lactation, with subsequent restoration of body fat later in lactation. During early lactation, milk fat becomes enriched with the long-chain fatty acids mobilized from adipose tissue stores (Palmquist et al., 1993).

No requirement for dietary lipids has been established, except for the so-called essential fatty acids (linoleic and linolenic) that cannot be synthesized by mammals (NASEM, 2021). However, use of supplemental fats and oils in dairy cow diets has developed to the point where it now is a standard practice. Fats are supplemented to increase ration energy density, which ideally will lead to increased energy intake if dry matter intake (DMI) is not decreased extensively. In turn, increased energy intake should improve energy balance and benefit body condition, milk production, reproduction, and health. Specialized lipid sources rich in the essential fatty acids or unique isomers have been developed to aid reproductive success and health.

Dietary lipids have been studied for more than 100 years. Wood (1894) reported that feeding oils from cottonseed, palm, and coconut increased the milk fat percentage. Modern interest in the use of supplemental lipids grew out of research that emerged after the Second World War and began to flourish around the time of publication of a seminal review by Palmquist and Jenkins (1980). Since then, much has been learned about digestive effects, metabolism, and productive benefits of supplemental fats (Palmquist and Jenkins, 2017). Nevertheless, much confusion remains in the dairy industry, starting with the types and forms of lipids and their differential utilization. The purpose of this chapter is to provide an overview of the processes of fat digestion and use in dairy cows and to summarize the effects of dietary lipids on intake, production, milk fatty acid composition, metabolism, reproduction, and health of dairy cows.

2 Lipids in dietary ingredients

The term 'lipid' refers to a broad class of substances that are insoluble in water or other aqueous solvents but are soluble in organic solvents such as ether, chloroform, hexane, acetone, and certain alcohols. Historically, this property has defined the lipid fraction in feeds because it was determined as the residue after ether extraction (i.e. 'ether extract'). Various extraction procedures exist, which yield different amounts of lipid from the same feed. Thus, the term 'crude fat' is a better descriptive factor than ether extract. However, this definition is vague and imprecise, as it includes compounds as diverse as carotenoids, steroids, and bile salts. The definition from Christie (2019) is more accurate for consideration of lipids in dairy cattle nutrition: 'Lipids are fatty acids and their derivatives, and substances related biosynthetically or functionally to these compounds'. Most normal forages, grains, and by-products contain some amount of lipid. The discussion in this chapter will focus on those lipids that are found as part of feedstuffs or as commercial fat supplements and which provide a source of fatty acids as energy-yielding nutrients to cows. The major lipid constituents important in dairy cow nutrition are free fatty acids, triglycerides, glycolipids, and phospholipids. Within each of these classes, there is considerable variability due to chemical makeup that results in different physical characteristics of the fat. These are discussed in more detail below.

2.1 Free fatty acids

Free or nonesterified fatty acids are carboxylic acids not attached to a glycerol molecule. Few fatty acids are found in free form within feeds. Fatty acids as part of complex lipids in feeds consist of a hydrocarbon chain ranging in length from 14 to 18 carbons. Marine oils such as fish oil also contain longer-chain fatty acids with 20–24 carbon atoms in their chains. Saturated fatty acids have all the carbon atoms linked with single bonds, while unsaturated fatty acids have one or more double bonds. The first double bond is found linking the ninth and tenth carbons, while additional double bonds are located farther down the chain away from the acid head group. The most abundant fatty acid in forages is linolenic acid, which contains 18 carbon atoms with three double bonds in the chain. Linolenic acid is abbreviated, therefore, as C18:3, indicating the number of carbons and the number of double bonds. In cereals and oilseeds, the most abundant fatty acid is linoleic acid, or C18:2. Both of these fatty acids are called 'polyunsaturated' fatty acids, or PUFAs, because they contain two or more double bonds. In animal fats, oleic acid or C18:1 is the most common fatty acid. Oleic acid is a 'monounsaturated' fatty acid. In cows' milk fat, palmitic acid (C16:0) is the most abundant fatty acid. This saturated fatty acid is also the most abundant fatty acid in palm oil, from which it gets its name. In ruminant milk and body fat, stearic acid (C18:0) is also abundant because it results from bacterial modification of dietary unsaturated fatty acids in the rumen.

Unsaturated fatty acids are denoted by one of two systems. The International Union of Pure and Applied Chemistry (IUPAC) system counts carbons from the acid (or carboxyl) end of the chain. The IUPAC system denotes the site of a double bond as Δ (delta) with a numeral. For example, linoleic acid is C18:2 $\Delta_{9,12}$, indicating that the two double bonds are on the ninth and twelfth carbons from the carboxyl end. An alternate system denotes the site of the first double bond starting from the methyl end of the chain as ω (omega) or n-. In this case, linoleic acid is C18:2 ω-6 or n-6. Mammalian enzymes are not capable of inserting a double bond closer than 9 carbons from the methyl end of the fatty acid; thus, n-6 and n-3 fatty acids are termed essential fatty acids because they are needed by the body but must be obtained from the diet.

An additional complication in the structure of unsaturated fatty acids is that the double bonds are found in one of two different three-dimensional orientations that result in markedly different physical characteristics. Double bonds can be in either '*cis*' or '*trans*' configurations. The presence of *cis*-configurations makes fatty acids more fluid (a lower melting point) than the presence of *trans*-configured bonds. All unsaturated fatty acids in naturally occurring plant lipids have only *cis*-double bonds. *Trans*-double bonds result from microbial modification of *cis*-double bonds or from chemical hydrogenation. As hydrogens are added to unsaturated double bonds, they are converted to single bonds, resulting in the conversion of unsaturated to saturated fatty acids. The first step in these hydrogenation processes, whether microbial or chemical, is conversion from *cis*- to *trans*-orientation.

Multiple double bonds in PUFA are located three carbons apart, resulting in an intervening methylene group between the two carbons that have double bonds. This is called a 'non-conjugated' arrangement. During hydrogenation, the *trans*-bond that is formed occurs between the preceding carbon atoms, resulting in a 'conjugated' arrangement. For example, in the ruminal biohydrogenation of linoleic acid (non-conjugated C18:2 with double bonds at the ninth and twelfth carbons), the first intermediate is conjugated linoleic acid (CLA), with a *cis*-double bond at carbon 9 and a *trans*-double bond at carbon 11.

In oils, increasing free fatty acid content is a sign of rancidity. In dairy cow nutrition, nearly all long-chain fatty acids found as part of forage-derived glycolipids or the triglycerides in cereals or oilseeds ultimately are converted to free fatty acids before they can be absorbed. In addition, fats supplemented to the diet as triglycerides, calcium soaps or salts of fatty acids, and free saturated fatty acids all will be absorbed as free fatty acids. Like triglycerides, the degree of saturation affects the physical characteristics. Saturated fatty acids are solids at environmental and body temperature, while unsaturated fatty acids are fluid. Nonruminant animals cannot digest saturated free fatty acids very well, while ruminants like dairy cows have evolved elaborate mechanisms to be able to digest saturated fatty acids with high efficiency. This is essential because saturated free fatty acids constitute the main type of lipid reaching the intestine for absorption from 'normal' or 'natural' ruminant diets.

2.2 Triglycerides

The basic unit of triglycerides (triacylglycerols) is a molecule of glycerol (a three-carbon alcohol) combined with three fatty acid molecules by ester linkages. The term 'fat' generally means a triglyceride that is solid or semi-solid at typical environmental temperatures, whereas 'oil' is a triglyceride that is liquid at working temperatures. Whether something is a fat or oil largely depends on the fatty acids that are attached to the glycerol backbone. For example, beef fat trimmed from a cut of meat is a semi-solid at room temperature, but corn oil is liquid. Both are triglycerides, but the beef fat has a higher melting point because of its greater content of saturated fatty acids, and corn oil is liquid because it is very high in unsaturated fatty acids. Chemically, the introduction of double bonds causes a 'kink' in the otherwise straight-chain structure of the fatty acid molecule, and this 'kinking' means that the fatty acids pack less tightly with one another, making the oil liquid at colder temperatures. The appropriate mix of saturated and unsaturated fatty acids gives the triglyceride the correct physical properties for a particular organism. For example, at body temperature, the triglycerides of beef muscle are mostly fluid; seed oils must remain fluid at extremely low environmental temperatures and so contain mostly unsaturated fatty acids.

The purpose of triglycerides in plants or animals is to store energy. As components of dairy cow rations, triglycerides are mostly broken down in the rumen to release the fatty acids and the glycerol, unless the triglyceride is extremely saturated (e.g. a highly hydrogenated tallow, grease, or vegetable oil). In this case, microbial lipases are unable to gain access to the ester bonds, and the saturated triglyceride will pass to the abomasum and small intestine.

2.3 Glycolipids

The lipids of forage stems and leaves are primarily glycolipids, which are similar to triglycerides except they have two or more sugars linked to one position of the glycerol backbone instead of the third fatty acid. The most common are galactolipids, which have galactose (a component sugar of milk lactose) linked to the glycerol. The two fatty acids that make up the glycolipids are generally unsaturated, especially linolenic acid. Glycolipids are structural components of plant tissues. In most forages, whether fresh, dry, or ensiled, glycolipids are almost completely broken down in the rumen of the cow.

2.4 Phospholipids

Phospholipids make up the cell membranes of plants and animals. They consist of a glycerol backbone with two fatty acids attached, with the third position of glycerol attached to a phosphate group that links an organic base such as choline, ethanolamine, serine, or inositol to the molecule. In the rumen, bacteria largely remove the base group and fatty acids from the phospholipids in dietary ingredients. However, the protozoa and bacteria in the rumen also make their own phospholipids for their cell membranes. Usually, therefore, the amount of phospholipid leaving the rumen is more than the amount consumed by the cow in the diet.

3 Lipid supplements

Fat supplements are used to increase the energy density of the diet without resorting to amounts of cereal grains that would cause ruminal acidosis, which can lead to loss of absorptive function of the rumen epithelium, liver abscesses, hoof problems, and poor efficiency of milk production. Dietary lipid supplements are often derived from waste streams of various lipid-processing industries. As such, lipids play a role in enhancing both the biological sustainability of milk production and the economic sustainability of production of fats and oils.

In thinking about fat supplementation to the diet, the provision of fat in basal ingredients and by-products must be accounted for first. Forages typically contain about 2–3% fatty acids, with maize silage being higher than alfalfa or grasses because of the higher oil content of the maize grain. Oilseeds (e.g. cottonseed and soybeans) often are added into rations for purposes other than fat supplementation (e.g. digestible fiber or protein), with the fat providing additional benefit. Of growing importance is that nutritionists know and account for the oil content of various by-product feeds, such as distillers' grains, bakery waste, and fish meal.

A variety of fat supplements are available today for dairy nutritionists to use. These include various commodity fats such as tallow, choice white grease, yellow grease, lard, and animal-vegetable blends. Ruminants might be called 'the original recyclers' because of their ability to convert non-human-edible ingredients into high-quality food for humans. This extends to the commodity fats, although amounts that can be fed are limited without decreasing DMI, interfering with rumen function, or causing milk fat depression (Grummer, 1996).

The rendering industry provides large quantities of lipids back to the livestock industries (Table 1). Fats and oils are classified by their titer, which is defined as the temperature at which the free fatty acids from the fat become solidified. Under accepted US trading laws, inedible fats with a titer of 40°C or above are classified as tallows, and those with titer less than 40°C are classified as greases (National Renderers Association, Inc., 2008). The titer and its converse, the melting point, reflect the degree of unsaturation of the fatty acids attached to glycerol. In general, tallows are derived from ruminants, while greases are derived from swine or poultry. Other specifications, including the maximum amount of free fatty acids, are used to set the grades (Table 1). Increased free fatty acids indicate some hydrolysis during manufacture or storage, which can increase the chances of rancidity and decrease digestibility in nonruminants.

Table 1 American Fats and Oils Association specifications for tallows and greases

	Specifications	
Grades	Titer, minimum °C	Free fatty acids, % maximum
Edible tallow	41.0	0.75
Lard (edible)	38.0	0.50
Top white tallow	41.0	2
All beef packer tallow	42.0	2
Extra fancy tallow	41.0	3
Fancy tallow	40.5	4
Bleachable fancy tallow	40.5	4
Prime tallow	40.5	6
Special tallow	40.0	10
No. 2 tallow	40.0	35
Choice white grease	36.0	4
Yellow grease	Negotiated	Negotiated

Source: Modified from National Renderers Association, Inc., 2008.

The emulsification system of non-ruminants does not deal efficiently with free fatty acids.

Another source of fats is the commercial 'bypass' fat supplements that are designed in some way to be 'inert' in the rumen. The concept of an 'inert fat' is preferable and should be interpreted as a fat supplement that has less effect on rumen fermentation than native fats. The terms 'escape' or 'bypass' fat are less desirable because many fats are altered somewhat in the rumen, and all fats must 'bypass' the rumen for their absorption. Another major advantage of commercial fats is that they are 'dry' fats that can be handled more easily without heated storage tanks and the inconvenience of liquid fats.

There are three main strategies for producing inert fats. One is to increase the saturation of the fatty acids in the product so that the melting point is above the range of environmental temperatures and body temperature. This decreases solubility and the potential interaction of fats with rumen microbes. The second is to complex the fatty acids with calcium to form calcium salts or soaps. Because a free carboxyl group is required for biohydrogenation and interaction with microbes, tying up the fatty acids as soaps prevents this interaction to some degree. Finally, fats may be 'encapsulated' in various ways to physically prevent interaction with the ruminal microbes. These can be non-nutritive compounds (e.g. sodium alginate) or formaldehyde treatment of proteins to prevent rumen access but allow intestinal access. Of these strategies, only the first two are in common use currently. Formaldehyde-treated products are not approved for use in many countries.

Increasing saturation is an effective way to produce a highly digestible free fatty acid product but is not desirable for intact triglycerides. Highly saturated triglycerides such as hydrogenated tallow or hydrogenated soybean oil are very poorly digested because not only microbial cells are denied access but also pancreatic lipase of the cow is unable to split the fatty acids from the triglyceride molecule (Elliott et al., 1999). A moderately saturated mixture of free fatty acids is well digested because of the unique digestive system of the ruminant, as discussed in a later section.

Formation of calcium soaps of fatty acids is another common method of producing dry fats. Such products are easily handled and are more inert in the rumen than native fats because of their decreased solubility and complexed carboxyl group. The unsaturated fatty acids in calcium soap products are only partially protected from rumen biohydrogenation, and this protection is less complete with PUFA than with the oleic acid found as a major component of palm oil fatty acid distillate, a by-product of the palm-processing industry (Tomkins and Drackley, 2010). The fatty acids in calcium soaps of palm fatty acids are highly digestible. The soaps are dissociated in the acid conditions of the abomasum so that the free fatty acid is available for absorption in the small intestine.

4 Fat digestion and absorption in ruminants

4.1 Processes in the rumen

Major differences in processes of fat digestion and absorption exist between ruminant and nonruminant animals, as a result of the profound impact of the rumen on dietary lipids (Jenkins, 1993). As described above, dairy cows consume a diet that predominantly contains PUFA as part of plant triglycerides and glycolipids. Bacteria in the rumen split off the fatty acids (and sugars) from the glycerol backbone. The glycerol and the sugars released from glycolipids are fermented by the rumen microbes to volatile fatty acids.

The breakdown of dietary lipids by rumen bacteria and protozoa generally occurs quite rapidly as the lipids are exposed during rumination and bacterial digestion of feed particles. In addition, the process generally is essentially complete so that few monoglycerides or diglycerides pass to the lower digestive tract. The major exception to this would be when a highly saturated (or hydrogenated) triglyceride is fed. Because of the very high melting point of such fats, and their resulting low solubility, bacterial enzymes cannot gain access to the bonds linking fatty acids to glycerol, and so these pass to the lower digestive tract (Elliott et al., 1999). Unfortunately, the same limitations of solubility and melting point result in poor access of the animal's digestive enzymes in the small intestine and very poor digestibility in that site as well (Elliott et al., 1999).

The fatty acids released in the rumen are not absorbed from the rumen but will pass to the abomasum and then the small intestine, which is the primary site for absorption of fatty acids (Fig. 1). However, the profile of fatty acids that reach the intestine will be very different from what the animal has consumed. This is because of the extensive biohydrogenation that occurs in the rumen as a result of bacterial activity.

Unsaturated fatty acids are toxic to many of the species of rumen bacteria, particularly those that are involved in fiber digestion (Jenkins, 1993). Also, because of the anaerobic environment of the rumen, there is an excess of hydrogen that the microbial population must continually dispose. This is the primary reason why methane is produced within the rumen, as it is an important 'sink' for disposal of hydrogen to keep rumen fermentation running efficiently. Thus, hydrogen for biohydrogenation is readily available, although it is a minor sink for hydrogen disposal relative to methane formation. Consequently, as unsaturated fatty acids are released from the glycerol backbone, they are quickly hydrogenated to saturated fatty acids. In cows fed most typical diets, 60-90% of the unsaturated fatty acids will be biohydrogenated to produce saturated fatty acids that flow to the small intestine (Doreau and Ferlay, 1991).

From the standpoint of rumen carbohydrate fermentation, biohydrogenation is a favorable process because it reduces the potential negative effects of unsaturated fatty acids on the ruminal microbiota (Jenkins, 1993). The resulting negative effect on microbial fiber digestion is the primary reason why large

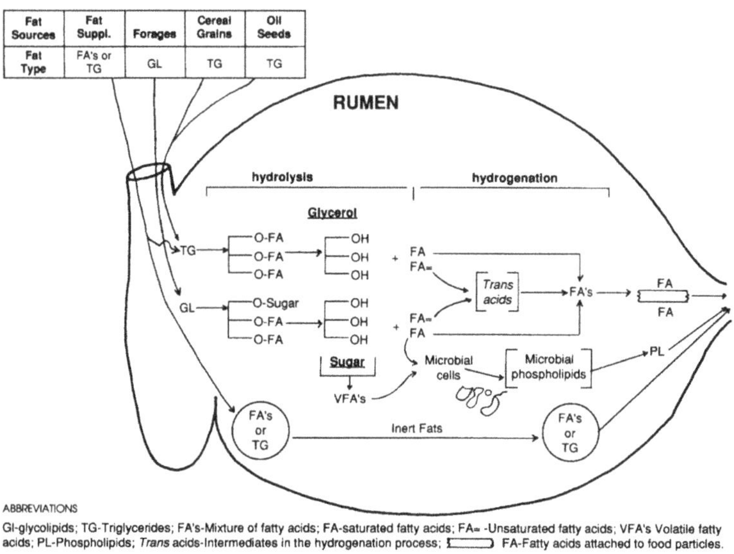

Figure 1 Schematic of lipid metabolism in the rumen. Source: From Davis (1990), used with permission.

amounts of free vegetable oils are not fed to dairy cows and led to the development of the rumen-inert fats (Palmquist and Jenkins, 2017). However, a meta-analysis determined that addition of 3% vegetable oil decreased the total tract neutral detergent fiber (NDF) digestibility by only 1.3%; an equivalent amount of medium-chain fats (containing large amounts of C12:0 and C14:0) decreased NDF digestibility by 8% (Weld and Armentano, 2017). While these saturated fatty acids do not require biohydrogenation, they are more toxic to fibrolytic microbes than the unsaturated long-chain fatty acids. In the same analysis, saturated fats and calcium salts of long-chain fatty acids increased the NDF digestibility by 1.3% and 3.2%, respectively (Weld and Armentano, 2017). An explanation for this effect has not been provided.

During the biohydrogenation process, intermediate compounds with *trans*-double bonds are produced (Fig. 1). Some of these *trans*-intermediates escape from the rumen and are incorporated into body fat and milk fat of ruminants. This accounts for the relatively high content of *trans*-fatty acids in ruminant products. The first step of linoleic acid biohydrogenation produces *cis*-9, *trans*-11 C18:2 or CLA (or rumenic acid), which has received much attention from the medical community because of its potent anti-cancer effects and other health benefits (Ferlay et al., 2017). The next step produces *trans*-11 C18:1 (vaccenic acid), which is the most abundant *trans*-isomer in ruminant milk and meat. Vaccenic acid also is of interest to human nutrition because it can be converted to CLA by the enzyme stearoyl-CoA desaturase, in the mammary gland of cattle but also in human tissues after consumption of milk or meat.

As mentioned above, breakdown of the dietary lipids to free fatty acids occurs rapidly and is a prerequisite for subsequent biohydrogenation. Large amounts of unsaturated oils can 'overwhelm' the biohydrogenation process and result in undesirable effects on the rumen microbial population. Under low rumen pH conditions that may result from excessive grain or insufficient effective fiber in the diet, a different set of *trans*-intermediates may be produced (Fig. 2). Some of these alternate products, particularly *trans*-10, *cis*-12 C18:2, have very powerful inhibitory effects on milk fat synthesis, which can cause milk fat depression (Baumgard et al., 2000). While *trans*-10 C18:1 is produced in larger quantities under the same conditions, it is not as inhibitory to milk fat synthesis. However, it provides a useful analytical marker for the metabolite(s) that depresses milk fat synthesis (Lock et al., 2007).

The rate of duodenal fatty acid flow is consistently less than fatty acid intake, indicating that fatty acids disappear on a net basis across the rumen (Doreau and Ferlay, 1991). This occurs despite the simultaneous synthesis of fatty acids by the rumen microbiota, particularly odd-chain and branched-chain fatty acids used by microbial cells for membrane formation. Disappearance may be due to the catabolism of fatty acids by microbes, although microbial

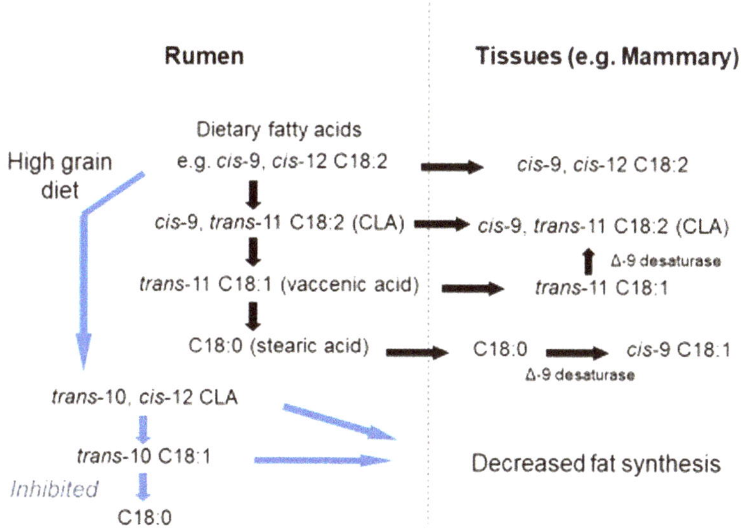

Figure 2 Pathways of normal biohydrogenation of linoleic acid in the rumen (left), with the alternative pathway induced by high-grain or low-fiber diets on the far left. The right-hand portion shows the metabolism of the intermediates in tissues such as the mammary gland. CLA, conjugated linoleic acid.

oxidation of long-chain fatty acids is limited given the highly reduced, anaerobic environment of the rumen. Fatty acids of 14 carbons or less have been demonstrated to disappear both *in vitro* (Wu and Palmquist, 1991) and *in vivo* (Wu et al., 1991). The nature of this disappearance remains poorly defined. Because ruminal appearance of phospholipids occurs as a result of microbial synthesis, total lipid balance across the rumen usually is close to zero.

4.2 Processes in the small intestine

The lipids that leave the rumen are predominantly free fatty acids (80–90%) and phospholipids (10–20%) found as part of microbial cell membranes (Fig. 3). The fatty acids making up the free fatty acid portion will be predominantly saturated (80–90%), consisting of about two-thirds to three-fourths stearic acid and about one-quarter to one-third palmitic acid. In addition, C18:1 is present, consisting primarily of *cis*-9-18:1 but also *trans*-10, *trans*-11, and other *trans*-isomers of C18:1 produced by microbial biohydrogenation. In the rumen, most of the free fatty acids are actually found as potassium, sodium, or calcium salts of fatty acids because of the near-neutral pH in the rumen contents (6.0–6.8). After passing through the acid conditions (pH ~2.0) of the abomasum, however, the fatty acid salts are dissociated and the protonated free fatty acids will be found adsorbed (or 'stuck') to the surface of small feed particles that pass as part

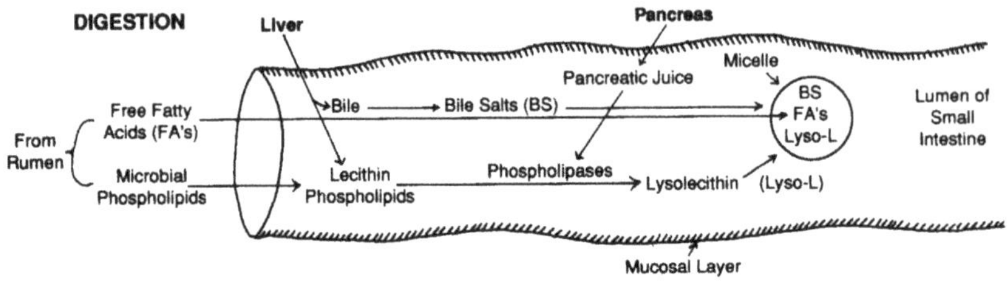

Figure 3 Digestion of lipids in the small intestine of ruminants. Source: From Davis (1990), used with permission.

of the digestive contents (Bauchart, 1993). Nonruminants would have a very difficult time trying to absorb such a profile of high-melting point, insoluble fatty acids, but ruminants have developed processes that result in saturated fatty acids being absorbed nearly as well as unsaturated fatty acids, and with much greater efficiency than in nonruminants.

The key to absorption of fatty acids in ruminants is formation in the intestine of complexes called micelles, which are bilayer disks consisting of bile salts (secreted in bile from the liver into the intestine, by way of the gall bladder) and phospholipids, with the insoluble polar lipids in the interior (Fig. 3). Micelles are needed to move the free fatty acids across the unstirred water layer on the surface of the intestinal cells, where the fatty acids can be absorbed into the intestinal epithelial cells. In nonruminants (and also in pre-ruminant calves), monoglycerides that result from pancreatic lipase-mediated digestion of triglycerides in the small intestine are needed for micelle formation. Bile salts and monoglycerides are amphipathic molecules, meaning they have portions of their molecular structure that can interact with aqueous systems (like the fluid in the intestinal lumen) as well as portions that can interact with lipids, so they form an 'interface' between fat and water. In the absence of monoglycerides, nonruminants are not able to absorb many free fatty acids.

In ruminants, however, a compound called lysolecithin takes the place of monoglycerides. Lysolecithin is produced from lecithin (phosphatidylcholine, the main phospholipid in rumen microbial cells, pancreatic juice, and bile) by action of the enzyme phospholipase-A2 that is secreted from the pancreas of the cow into the upper small intestine. Phospholipase-A2 converts lecithin into lysolecithin, which is an extremely efficient emulsifier particularly for saturated free fatty acids. In fact, a series of studies in the late 1960s demonstrated that lysolecithin was the most potent natural emulsifier for stearic acid and was more potent than any man-made emulsifier available at that time (Freeman, 1969). Other natural emulsifiers such as oleic acid (normally not present in large amounts in the small intestine of ruminants) were effective but not to the degree as was lysolecithin. Consequently, Mother Nature has ensured that dairy cows, like other ruminants, can efficiently absorb the mostly saturated free fatty acids presented to the intestine every day.

The saturated fatty acids produced from biohydrogenation of unsaturated fatty acids in feedstuffs are absorbed efficiently in part because they are finely distributed on the surface of feed particles. Saturated fatty acids, particularly stearic acid (C18:0), in fat supplements are agglomerated and crystallized (Shepardson et al., 2020) and have lower digestibilities in dairy cows. This may be due to a limitation in the biological emulsifier system.

4.3 Intestinal processing and delivery of dietary fat

After the absorption of fatty acids into intestinal cells, the free fatty acids are reconverted to triglycerides by combining with α-glycerol-phosphate produced from glycolytic metabolism of blood glucose (Fig. 4). The triglycerides are packaged into triglyceride-rich lipoprotein particles [chylomicrons and very low-density lipoproteins (VLDL)] in combination with cholesterol, phospholipids, and specific proteins. These proteins (called apoproteins) serve to direct the trafficking and use of the lipoprotein triglycerides. Because chylomicrons are too large to pass directly into the venous bloodstream draining the intestinal cells, they are secreted into the lymph, which is delivered back into the venous bloodstream near the heart. Some VLDL may be absorbed into the portal vein (Bauchart, 1993). After the blood is oxygenated through the lungs, the lipoprotein particles are delivered to various organs of the body such as the mammary gland, muscle, and heart that can use the triglycerides.

Triglycerides in chylomicrons or VLDL are broken down to free fatty acids by the enzyme lipoprotein lipase that is found in the capillaries of these tissues.

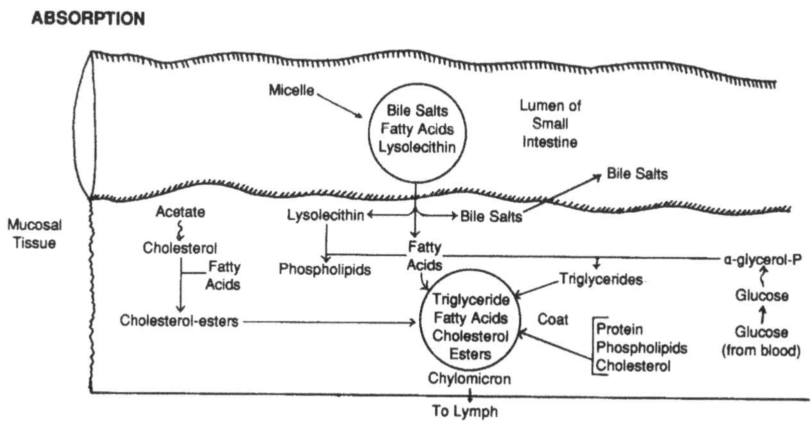

Figure 4 Lipid metabolism in the small intestinal tissue after absorption. Source: From Davis (1990), used with permission.

The free fatty acids then enter the cells where they can be formed back into triglycerides (such as milk fat) or burned to release energy that can fuel cell functions (such as contraction in skeletal or heart muscle). It should be noted from the scheme for lymphatic absorption described here that most dietary fats do not reach the liver directly, in contrast to other nutrients like amino acids or propionate that are absorbed into the portal circulation. The liver does not metabolize those VLDL that may reach the liver from portal absorption because it lacks lipoprotein lipase. Consequently, dietary fats do not contribute appreciably to fat accumulation in the liver (fatty liver) that often is observed around calving. Fatty liver results from the accumulation of triglycerides driven by extensive mobilization of nonesterified fatty acids (NEFAs) from the adipose tissue during negative energy balance and the conversion of NEFA back to triglycerides in the liver (Fig. 5).

Lipoprotein metabolism in ruminants differs from that in nonruminants, as discussed by Bauchart (1993). The remaining remnant particle after lipoprotein lipase action, sometimes called an intermediate-density lipoprotein, is enriched with cholesterol, which is delivered to tissues by binding to low-density

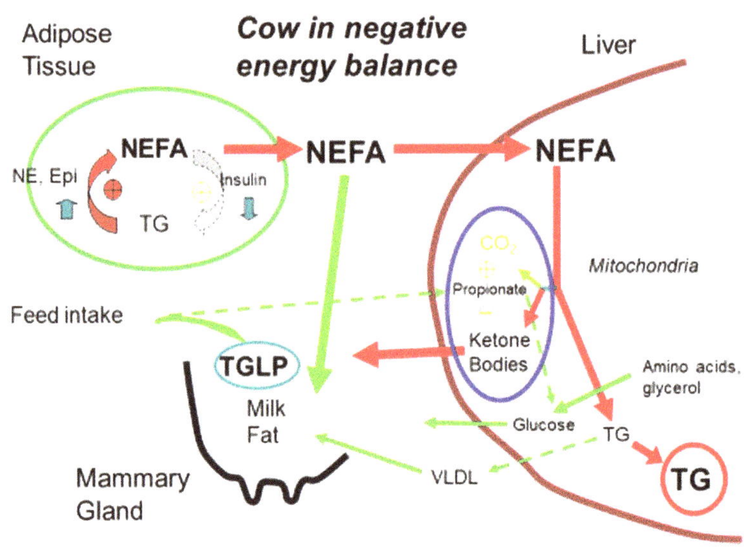

Figure 5 Schematic of lipid metabolism during negative energy balance in early lactation. Adipose tissue mobilization of nonesterified fatty acids (NEFAs) from triglycerides (TGs) in response to norepinephrine (NE) and epinephrine (Epi) leads to increased NEFA concentration in the blood and increased uptake of NEFAs by liver and mammary tissues. Limited feed intake means lower availability of triglyceride-rich lipoproteins (TGLPs) from intestinal absorption and lower supply of propionate to make glucose in the liver. Greater NEFA uptake by the liver leads to the production of ketone bodies, limited secretion of VLDL, and accumulation of TG, leading to fatty liver. Source: Modified from Drackley (1999).

lipoprotein (LDL) receptors (Fig. 6). The LDL particles represent only about 10% of bovine lipoproteins. The high-density lipoproteins (HDL) are the predominant form of lipoproteins in dairy cows. They are secreted as nascent, phospholipid-enriched particles by the small intestine and liver. Cholesterol is picked up from peripheral tissues by HDL and is esterified primarily to linoleic acid by the enzyme lecithin-cholesterol acyltransferase to form cholesterol esters, which accumulate in the core of the HDL particle. Eventually, these enlarged HDLs are cleared by receptors in peripheral tissues or liver, where the phospholipids are recycled. Cholesterol may be used to form new lipoprotein particles or, in the liver, a portion is converted to bile salts.

The triglycerides circulating in intestinally derived chylomicrons and VLDL are rich in saturated fatty acids, approximately 23% C16:0 and 48% C18:0 (Drackley et al., 2007). This is much more highly saturated than triglycerides in adipose tissue, liver, or milk (Douglas et al., 2007). Much of the C18:0 is desaturated to C18:1 by the enzyme stearoyl-CoA desaturase abundant in many tissues, including adipose tissue, liver, and mammary gland.

In ruminant animals, oxidative or fuel use of long-chain fatty acids is more limited than in most nonruminants (Hocquette and Buachart, 1999). This may be due, at least in part, to the abundance of acetic acid from rumen fermentation. Acetate (the salt form of acetic acid present at normal body pH) is the most abundant oxidative fuel in ruminants such as dairy cows. Consequently,

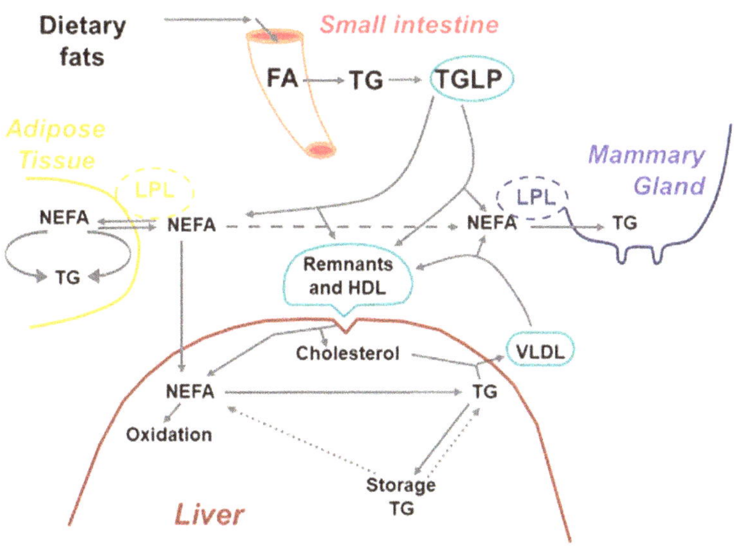

Figure 6 Schematic of lipoprotein metabolism in ruminants. FA, fatty acid; HDL, high-density lipoprotein; LPL, lipoprotein lipase; NEFA, nonesterified fatty acids; TG, triglyceride; TGLP, triglyceride-rich lipoproteins [chylomicrons and very low-density lipoproteins (VLDL)]. Source: Adapted from Drackley (1999).

the main use of fatty acids from dietary fats and oils will be for triglyceride synthesis. For cows in negative energy balance during early lactation, which are mobilizing body fat, the only substantial use of dietary fatty acids will be for milk fat synthesis (Fig. 5). For cows in positive energy balance that are also depositing body fat, some dietary fat can be stored in adipose tissues as well as used for milk fat synthesis (Fig. 7).

4.4 Polyunsaturated fatty acid escape and use

The small amounts of PUFAs that escape through the rumen without being hydrogenated are very important for proper structure of cellular membranes in the animal. The PUFA linoleic acid and linolenic acid cannot be made in the cow's body and so must be absorbed from the intestine. Ruminal escape of linoleic and linolenic acids was reported to average 20% and 8%, respectively, with a range of 5–30% and 0–15% (Doreau and Ferlay, 1991). Estimates for linolenic acid escape meet the proposed tissue requirements. A dietary intake of 1% of DMI as linoleic acid, with a 20% escape rate, would provide 2 g/kg DMI

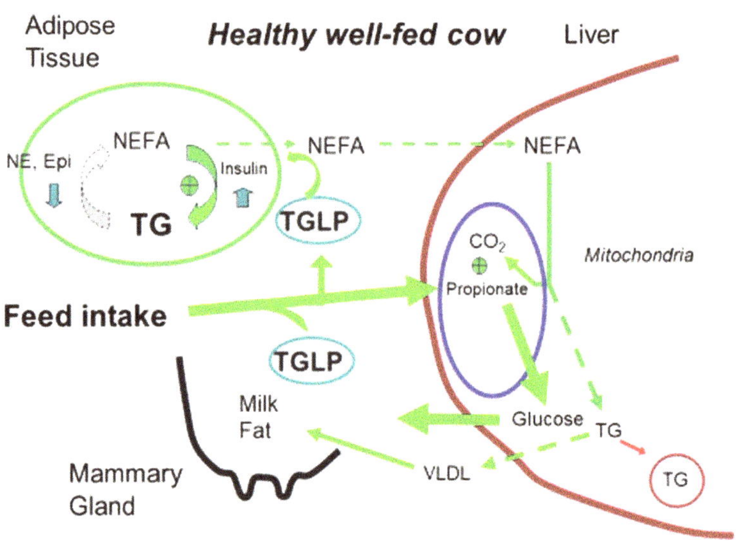

Figure 7 Lipid metabolism in healthy well-fed cows in positive energy balance. Insulin suppresses the mobilization of nonesterified fatty acids (NEFAs) from adipose tissue triglycerides (TGs) in response to norepinephrine (NE) and epinephrine (Epi), and so NEFA concentration in the blood is low. There is limited uptake of NEFAs by the liver and mammary tissues. Adequate feed intake means abundant mammary tissue uptake of TG-rich lipoproteins (TGLP) from intestinal absorption. The abundant supply of propionate is used to make glucose in the liver to drive milk production. The liver has limited secretion of very low-density lipoproteins (VLDL) and little accumulation of TGs. Source: Modified from Drackley (1999).

of linoleic acid to the intestine, easily meeting the estimated tissue requirement of 1 g/kg intake set for growing and reproductive swine (National Research Council [NRC], 2012). From another perspective, a cow producing 1.8 kg of milk fat daily would secrete approximately 45 g of linoleic acid in that fat. If the cow consumes 26 kg of DM, the estimated ruminal escape of linoleic acid would be approximately 52 g/day. That dairy cows meet their tissue requirements for essential fatty acids despite extensive ruminal biohydrogenation is substantiated by the fact that no clinical signs of deficiency are noted (Palmquist, 2009).

Ruminants have evolved ways to conserve the supply of essential fatty acids. Intestinal cells primarily esterify absorbed PUFAs to phospholipids rather than triglycerides. In this way, PUFAs are protected from being burned for energy and instead are incorporated into cell membrane phospholipids. The PUFAs can be released from the membrane lipids and converted into important signaling molecules, such as prostaglandins (PGs) and leukotrienes. The unique structure and metabolism of PUFAs result in their incorporation into cell membranes of tissues all over the body, including immune cells and the reproductive tract. Proposed benefits of partially rumen-protected PUFA supplements on reproduction and health likely arise in this way (Moallem, 2018). The challenge is to minimize their biohydrogenation in the rumen so that an appreciable quantity of additional PUFAs reaches the intestine for absorption.

5 Digestibility of dietary lipids

Determination of lipid digestibility in dairy cows poses several methodological challenges. Early studies reported decreased fiber digestion and increased fat digestibility with fat supplementation. However, much of the fat in feces is present as insoluble calcium salts or soaps of fatty acids, which are not measured by classical ether extraction procedures unless pretreated with acid. Studies in which this step was not carried out will erroneously show high lipid digestibility. In addition, non-fatty acid endogenous lipids will be a lesser amount of fecal lipid as dietary lipid increases, thereby leading to increasing lipid digestibility at high lipid intakes. These limitations are overcome by measuring fatty acid digestibility, which is most meaningful from a nutritional standpoint. Careful technique must be utilized to ensure quantitative recovery, particularly of saturated free fatty acid products (Palmquist and Jenkins, 2003).

Digestibility of fatty acids in basal ingredients was set to 73% by NASEM (2021). Calculated fatty acid digestibilities assigned to various fat supplements were: oilseeds 73%, oils 70%, calcium salts of palm fatty acids 76%, saturated fatty acid-enriched nonesterified fatty acids 69%, and 85%-palmitic acid 73% (NASEM, 2021). Intestinal digestibility of all dietary fatty acids and C18 biohydrogenation products does not differ greatly. However, at high duodenal flows of C18:0, its digestibility declines quadratically (Glasser et al., 2008b) or linearly (Boerman et al., 2015). Western et al. (2020) observed that supplementing dairy cows with a mixed C16:0 plus C18:0 supplement (33.1% C16:0 and 53.3% C18:0) decreased fatty acid digestibility by 9.1% compared with a non-supplemented control. The physiochemical structure of saturated fatty acid prills may play a role in the decreased digestibility (Shepardson et al., 2020).

Low digestibility of supplemented C18:0 is likely attributable to its low solubility in intestinal micelles. Adding emulsifiers to aqueous solutions of bile salts greatly increased the solubility of C18:0 (Freeman, 1969). Prom and Lock (2021) observed that increasing *cis*-9 C18:1 from 10% to 30% in a fatty acid supplement predominantly composed of C18:0 increased fatty acid digestibility by 11.5%. de Souza et al. (2020) abomasally infused a nonionic surfactant (30 g/day of polysorbate-C18:1) and reported that 18-carbon and total fatty acid digestibility increased by 11.3% and 10.2%, respectively. Prom et al. (2022) discovered that the fatty acid attached to the polysorbate influenced the effects on C18 digestibility. Prom et al. (2021) abomasally infused increasing amounts of oleic acid (20, 40, or 60 g/day) and found that C18 digestibility increased linearly. These amounts of oleic acid would be in the range of amounts delivered to the intestine by calcium salts of palm fatty acids (Klusmeyer and Clark, 1991), perhaps accounting for the high digestibility of calcium salts.

6 Mammary metabolism of lipids

Milk fat is a highly complex material, containing over 400 different fatty acids (Jensen, 2002), although just 14 are usually present in amounts more than 1% of total. The fatty acids C4:0, C6:0, C8:0, C10:0. C12:0, and C14:0 are synthesized *de novo* within the mammary gland from acetate, with 3-hydroxybutyrate (BHBA) able to serve as a primer molecule. The fatty acids C16:0 and C16:1 are both synthesized in the gland and taken up from blood in approximately equal portions. All of the C18 and longer fatty acids are taken up preformed from blood, as a result of lipid mobilization from adipose tissue in times of negative energy balance or from absorbed dietary fatty acids. A substantial amount of C18:0 is desaturated to *cis*-9 C18:1 by the enzyme stearoyl-CoA desaturase.

Palmquist (2006) thoroughly reviewed the biochemistry of milk lipid synthesis. Fatty acid synthesis proceeds in repeated cycles of elongation using malonyl-CoA derived from acetate until reaching C16:0. Synthesis of nicotinamide adenine dinucleotide phosphate (NADPH) needed for elongation is derived from glucose metabolism in the pentose phosphate pathway and acetate metabolism through isocitrate dehydrogenase. No elongation of C16:0 to C18:0 occurs in the mammary gland. Milk from dairy cows and other ruminants is unique in having a large proportion of C4:0 through C12:0

fatty acids, which are derived by thioesterase activity terminating the chain elongation process before reaching C16:0. Long-chain fatty acids are taken up from circulating triglycerides by the action of lipoprotein lipase or from plasma NEFA when concentrations are elevated. Synthesis of C18:1 (oleic acid) occurs by desaturation of C18:0 by stearoyl-CoA desaturase.

Interpretation of milk fatty acid composition is very different when considering the molar composition versus the percentages by weight. For example, butyrate (C4:0) makes up only about 4% by weight of total milk fatty acids but is present at about one-fourth or more of fatty acids on a molar basis. The six fatty acids shorter than C16 make up about a quarter of the milk fatty acids by weight but a greater molar proportion of the total. A milk fat triglyceride with C4:0, C16:0, and C18:1 fatty acids contains equimolar amounts of the three fatty acids but if hydrolyzed yields 14% C4:0, 41% C16:0, and 45% C18:1 by weight (NASEM, 2021).

Milk fat triglycerides are not assembled with a random distribution but have positional specificity. Enzymes for triglyceride synthesis are associated with the smooth endoplasmic reticulum and require glycerol-3-phosphate as precursor. The stereospecific number (*sn*)-3 position is preferentially bound to a short-chain fatty acid, most often C4:0. Palmitic acid occurs in almost equal proportions at the *sn*-1 and *sn*-2 positions. Fatty acids C4:0 and C6:0 do not appear at the *sn*-2 position. Oleic acid is esterified preferentially to the *sn*-1 and *sn*-3 positions, thereby competing with the short-chain fatty acids for the *sn*-3 position. Milk fat triglycerides are characterized by carbon number (the total C in the acyl chains), which range from 26 to 54. About 36% of the milk triglycerides contain either C4:0 or C6:0 and two long-chain fatty acids (Jensen, 2002). Palmitic acid plays a predominant role in determining the physical properties of milk fat in dairy products (Ortiz Gonzalez et al., 2007, 2022).

7 Effects of lipids on feed intake

For supplemental lipids to increase energy intake, DMI must not be decreased by the lipid. Effects of fats on DMI were measured in many studies, and recent meta-analyses have summarized these data (Rabiee et al., 2012; Hu et al., 2017; Weld and Armentano, 2017; dos Santos Neto et al., 2021a,b). Table 2 shows the results of these meta-analyses. In general, most fats caused some decrease of DMI, with the exception of the prilled fats consisting of mostly saturated free fatty acids. Much of this effect of the non-saturated fatty acid supplements seems to be due to greater unsaturation of the dietary lipid source, which decreases meal size (Harvatine and Allen, 2006).

A series of studies by the author's research group demonstrated that abomasal infusion of unsaturated fatty acids potently decreased DMI and milk production (Drackley et al., 1992; Christensen et al., 1994; LaCount et al., 1994;

Table 2 Results from meta-analyses of the effects of added fats to diets of lactating dairy cows[a]

			Difference from control diet						
Study	Treatment	*n*	Added fat, % DM	DMI, kg/day	Milk, kg/day	Fat, %	Protein, %	ECM, kg/day	ECM/DMI
Rabiee et al. (2012)	Tallow	15	3.48	-1.06	0.61	-0.03	-0.09	ND	ND
	Ca salts palm	28	2.30	-0.64	1.55	0.10	-0.05	ND	ND
	Oilseeds	29	2.16	-0.56	0.84	-0.10	-0.10	ND	ND
	Prilled fat	4	2.92	-0.09	1.00	0.10	-0.02	ND	ND
	Other Ca salts	10	2.32	-2.11	0.92	-0.47	-0.18	ND	ND
Hu et al. (2017)	Mixed fatty acids	21	≤3.5%	0.17	1.80	0.10	-0.02	ND	ND
	Single fatty acids	9	≤3.5%	-0.28	1.21	0.10	-0.02	ND	ND
dos Santos Neto et al. (2021a)	Ca salts palm	60	2.20	-0.56	1.53	0.03	-0.05	1.12	0.08
dos Santos Neto et al. (2021b)	Mix	28	2.26	0.0	1.20	0.0	-0.05	0.80	0.02
	Palmitic acid	23	1.81	0.1	1.60	0.18	0.02	2.00	0.07

[a]Values in table are differences from the control diet, i.e. if the value is negative the change was a decrease relative to control.
DMI, dry matter intake; ECM, energy-corrected milk; ND, not determined.

Bremmer et al., 1998; Drackley et al., 2007; Litherland et al., 2022). Unsaturated triglycerides decreased DMI to a lesser extent than the same fatty acids in free form (Litherland et al., 2005, 2022), presumably because the triglycerides needed to be hydrolyzed to have the effect of stimulating endocrine cells. The gut hormone glucagon-like peptide 1 was associated with the decreases in intake, but cholecystokinin was not (Benson and Reynolds, 2001; Litherland et al., 2005, 2022). This hormonal response also was observed in cows fed dietary fat supplements of increasing unsaturation (Relling and Reynolds, 2007).

8 Effects of lipid supplements on milk production and composition

An abundant literature supports the supplementation of fats to dairy cow diets. Several meta-analyses have been conducted, and the results of four of these that looked at effects of fats on production are summarized in Table 2 (Rabiee et al., 2012; Hu et al., 2017; dos Santos Neto et al., 2021a,b). In general, supplemental lipids increase milk yield, with the greatest increases occurring with rumen-inert fats and the least with unprotected fats such as tallow and oilseeds, although results for the latter are confounded with changes in protein supply. Milk fat percentage was increased by fat supplementation, except for the unprotected fats, while milk protein percentage often was slightly decreased. Efficiency of milk production, measured as energy-corrected milk/DMI, was improved.

Supplemental fats change the composition of milk fat, which has been reviewed by several authors (Grummer, 1991; Palmquist et al., 1993; Loften and Cornelius, 2004; Glasser et al., 2008a; Shingfield et al., 2013; Loften et al., 2014). In general, milk fat becomes enriched with the long-chain fatty acids supplied by the supplemental lipid, whereas the contents of short- and medium-chain fatty acids in milk fat are decreased. These changes contribute to the improved efficiency of milk production, as the use of preformed rather than *de novo* synthesized fatty acids saves glucose and NADPH normally used in *de novo* synthesis as dietary fatty acids are transferred to milk fat without the energetic cost of synthesis. A major consideration is maintenance of milk fat fluidity at body temperature, with the abundant C16:0 and C18:0 pushing toward a higher melting point that must be offset by incorporation of oleic acid or the short-chain fatty acids.

9 Use of dietary lipids to modify milk fat for human health

Researchers have sought to modify the fatty acid composition of milk and meat lipids with the intent of improving the health of humans consuming the products (Shingfield et al., 2013; Toral et al., 2018). Milk and dairy products as part of a healthy diet are associated with several important health measures (Thorning et al., 2016). Nevertheless, recommendations to decrease consumption of saturated fatty acids persist (U.S. Department of Agriculture and U.S. Department of Health and Human Services, 2020). In particular, the fatty acids C12:0, C14:0, and C16:0 are hypercholesterolemic (Hillbrick and Augustin, 2002). These three fatty acids are abundant in milk fat. Moving toward a more nutritionally desirable milk fat necessitates a decrease in the medium-chain saturated fatty acids and an increase in oleic acid (Jenkins and McGuire, 2006). As discussed previously, an increased supply of intestinally available C18:1 decreases those fatty acids made *de novo* in the mammary gland and increases C18:1 in milk fat. Enrichment of milk fat with monounsaturated or PUFAs at the expense of saturated fatty acids decreased total and LDL cholesterol in the blood of humans that consumed the dairy products in four studies (Livingstone et al., 2012).

Another way the nutritional value of milk fat could be enhanced would be increasing the content of the n-3 fatty acids in milk fat. Enrichment of milk fat with n-3 fatty acids could supply about 30% of the recommended human consumption of 1–1.5 g/day of n-3 fatty acids (Moallem, 2018). However, the greater unsaturation leads to alterations of milk-processing qualities and a propensity for oxidation and off-flavors (Palmquist and Jenkins, 2017).

Various fatty acids in milk fat have beneficial effects on human health even in small doses. One of the best studied is CLA ($C18:2_{cis\text{-}9,\ trans\text{-}11}$), now named rumenic acid. This CLA isomer exerts powerful anti-cancer, anti-atherosclerosis, and anti-obesity effects (den Hartigh, 2019). Feeding small amounts of vegetable oils or oilseeds generally increases the amount of CLA in milk fat (Plata-Pérez et al., 2022). Other lipids in milk fat, such as vaccenic acid and sphingomyelin, also have beneficial effects in humans (Cichosz et al., 2020). Butyric acid (C4), an abundant fatty acid in milk fat, has beneficial effects on intestinal health (Cichosz et al., 2020).

10 Effects of lipids on reproduction

Energy balance has profound effects on fertility in dairy cows. Given this, substantial research has been conducted on the use of supplemental fats to improve energy balance and reproduction. This topic has been summarized by Palmquist and Jenkins (2017). Supplemental fats may improve reproduction by several methods: increasing energy intake and improving energy balance; stimulating cholesterol synthesis, needed for synthesis of reproductive steroid hormones; increasing follicular size; and altering the synthesis and secretion of PGs, thereby prolonging the life of the corpus luteum (Grummer and Carroll, 1991; Thatcher et al., 2006). While nearly any fat source could cause the first three effects, the effect on PGs would require specific PUFA (Moallem, 2018).

Modifying intakes of n-6 and n-3 fatty acids regulates the balance of two- and three-series PGs. High concentrations of $PGF_{2\alpha}$ cause regression of the corpus luteum, resulting in the loss of pregnancy before the conceptus is implanted in the uterus. Increasing the availability of linoleic acid or linolenic acid decreases the synthesis of arachidonic acid (C20:4), which is the precursor for $PGF_{2\alpha}$. In contrast, linolenic acid increases the synthesis of eicosapentaenoic acid, which is the precursor for the three-series PGs that counter the effects of $PGF_{2\alpha}$. This biology led to the development of a phased reproductive program in which a calcium salt of linoleic acid was fed from 28 days before to 28 days after parturition, followed by the supplementation of high n-3 source such as fish oil (Thatcher et al., 2006). In sum, the studies conducted have shown that feeding fats enriched with specific essential fatty acids can positively benefit the health and reproduction of dairy cows (Moallem, 2018). These programs are difficult to manage on farms, however, as the timing of the two fats may not correspond to grouping strategies often used on large farms.

Rodney et al. (2015) conducted a meta-analysis of studies that determined the effects of fat supplementation on calving to pregnancy interval and the proportion pregnant to service. Only 17 studies with 26 treatment comparisons were available for inclusion. The analysis showed that fat supplementation resulted in a 27% increase in pregnancy to service and decreased the calving to pregnancy interval.

11 Role of dietary lipids in the regulation of metabolism

Lipids not only serve roles in energy storage and transport as well as cell structure but also play important roles as regulators of metabolism. Regulatory molecules include the well-known PGs, leukotrienes, and steroid hormones. Fatty acids themselves can activate peroxisome proliferator-activated receptors (PPARs) and other nuclear receptors, with downstream roles such as increased fatty acid oxidation or increased lipogenesis depending on which PPAR is activated. Other lipid mediators are the topic of current research, including acyl-carnitines, diacylglycerols, and sphingolipids (McFadden, 2020).

Sphingolipids are components of cellular membranes but also act in the regulation of cell growth, differentiation, and apoptosis (Hannun and Obeid, 2008). One type, the ceramides, has created recent interest because of its role as a biomarker of metabolic diseases such as type 2 diabetes mellitus, nonalcoholic fatty liver disease, and cardiovascular disease in rodent models and humans (Borodzicz et al., 2015). Ceramides are components of all sphingolipids and are synthesized *de novo* from palmitoyl-CoA and serine, followed by the addition of a second long-chain or very long-chain fatty acid. Ceramides interfere with insulin signaling, leading to the development of insulin resistance.

In dairy cows, ceramide increases in the plasma during the transition to lactation, regardless of the body condition score (Rico et al., 2015, 2017). Plasma ceramides were greater in cows with greater body condition score, which mobilized more fatty acids from adipose tissue and were positively correlated with hepatic lipid content (Rico et al., 2017). These data support a possible role of ceramides in development of the insulin resistance of early lactation (Rico et al., 2015, 2017).

In mid-lactation dairy cows, feeding a diet enriched with 4% supplemental palmitic acid increased plasma ceramides and milk fat yields without affecting DMI. The palmitic acid supplement also decreased the disappearance of plasma NEFA following an intravenous glucose challenge, suggesting the development of adipose tissue insulin resistance (Rico et al., 2016). Palmitic acid also prevented the gradual decrease of plasma ceramides in cows as lactation progressed. Changes in circulating ceramides over the lactation cycle suggest that ceramides are inversely related to insulin concentrations.

Although this area of investigation is still in early stages, it appears that ceramides may inhibit adipose and muscle insulin sensitivity, which in early lactation could enhance lipolysis in adipose tissue and lead to uptake of NEFA by liver. In later lactation, palmitic acid feeding could be used to help partition glucose and fatty acids to milk production and decrease the rate of body fat accrual (McFadden and Rico, 2019).

12 Important considerations for selection of commercial inert fats

Fat supplements are added in an attempt to increase energy intake because they are more energy dense than carbohydrate-providing feeds such as grains and forages. The energetic advantages of replacing 0.5 kg of a carbohydrate source (e.g. corn) with 0.5 kg of fat equate to a potential gain of 2-3 kg of milk if DMI is not affected. Rumen-inert dry fats are used to avoid interference with rumen function that in turn could decrease nutrients supplied by the rest of the ration ingredients or that could result in low milk fat tests. Most of the commercial fats that are still on the market have acceptable handling characteristics. To actually increase usable energy by the cow, however, requires that the fats are digestible and do not decrease DMI more than the increased energy density can compensate for. Consequently, the most important aspects of evaluating the commercial inert fats are their digestibility and their effects on feed intake.

The low digestibility of hydrogenated triglycerides has resulted in a fall in their use. The two most popular types of inert fats used today in the USA are prilled saturated free fatty acids and calcium soaps of palm fatty acids, both of which are highly digestible. Data from a collection of research studies indicate only a small difference in digestibility between the two types of fats, with a small advantage for the calcium soaps. The high digestibility of the saturated free fatty acid product is attributable not only to the unique digestive processes in ruminants as discussed already but also to the fact that the particles (prills) are small and so have extensive surface area for the digestive secretions to suspend and emulsify the fatty acids. Larger particles of similar composition are less digestible (Elliott et al., 1994).

Of equal or greater importance in evaluating different fats is their effect on DMI. The primary reason for feeding supplemental fats is to increase digestible energy intake; if DMI is decreased by the supplemental fat, digestible energy intake may not increase or may even decrease. The body of scientific evidence clearly shows that the calcium soaps consistently have a negative impact on DMI. This was stated by the NRC (2001) in its 2001 version of *Nutrient Requirements of Dairy Cattle* as well as by more recent meta-analyses (Weld and Armentano, 2017; dos Santos Neto et al., 2021a). The NRC concluded that calcium soaps decreased DMI by 2.5% for each percentage unit added to the diet, while hydrogenated fatty acids or triglycerides did not affect DMI. The relative importance of digestibility and effects on DMI can be shown by the following calculations.

Saturated free fatty acid products are 80–98% fatty acids and calcium soaps are no more than 84% fatty acids. Let us assume that both products contain 82% fatty acids and that saturated free fatty acids are 69% digestible versus 76% digestible for the calcium soaps at production intakes, as specified by NASEM (2021). Calculated net energy of lactation (NEL) values, then, are 4.36 and 4.80 Mcal/kg fat source, respectively, showing the digestibility advantage for the calcium soaps. However, for cows consuming 25 kg of DM daily when 0.25 kg of fat (1% of DM) is added from either source, the NRC predicted that cows fed calcium soaps would decrease DMI by 2.5% or 0.625 kg/day. In this case, total NE_L intake falls by more than 1 Mcal, enough for 1.45 kg/day of milk. In contrast, at equal intakes, the slightly lower digestibility of the saturated fatty acids would mean only 0.6 kg less milk for the saturated fatty acids at an intake of 0.25 kg/day. Clearly, modest effects of fats on DMI may have more important effects on energy available for milk production than modest differences in fat digestibility.

13 Practical recommendations for supplementing fats

General philosophies about the use of supplemental fats have gone through several swings of the pendulum over the last half century. The initial description of the negative effects of free unsaturated oils on rumen fermentation led to active interest in the use of calcium soaps of fatty acids first and then other inert fat products, as they were developed. Subsequent demonstration of relative 'inertness' of tallow, at least in diets high in alfalfa, coupled with the higher cost of commercial fats led to commercial fats falling somewhat out of favor during the early to late 1990s. However, interest in commercial fats has been rekindled recently due to the increased problems with commodity fat use in diets based more heavily on maize silage (Ruppert et al., 2003) and containing large amounts of by-products such as bakery waste and high-oil distillers' grains that are high in oil, particularly in very high-producing herds.

Recommendations for maximum amounts of fat have not changed much over the years. A practical rule of thumb, originally proposed by Maynard (1937), is that the total amount of fat fed (from basal ingredients and supplemental fat sources) should equal the amount of fat secreted in milk. Suboptimal amounts of fat decrease the milk production and decrease the energetic efficiency of milk production. Kronfeld (1976) estimated that fat should constitute 16% of the total metabolizable energy (ME) in the diet of dairy cows for maximum metabolic efficiency of milk synthesis. If fat is fed at amounts similar to that secreted in milk fat, the energetic contribution of fat increases from 8% of ME at milk production of 10 kg/day to more than 18% of ME at 50 kg/day of 3.5% fat-corrected milk production (Palmquist and Jenkins, 2017). The content of C16:0 in milk fat is higher than normal when dietary fat intake is suboptimal (Banks et al., 1976).

Total dietary fat should not exceed 7% of the DM (NASEM, 2021) and usually will be adequate at lesser amounts. In general, 2–3% added fat should be the most needed under typical conditions. The first 1.5–2% (0.35–0.4 kg) can be from good-quality tallow (highly saturated, low free fatty acid content), if the diet is based primarily on alfalfa or grass hay or silage (Ruppert et al., 2003). Amounts in addition to that should be from a commercial inert fat product. When diets are based on forages that are composed of more than 2/3 (DM basis) maize silage, or when TMR particle size is borderline, no more than 0.25 kg/day of tallow or grease should be supplemented. When rations also contain cottonseed or soybeans and high-oil distillers' grains, then the amount of tallow that will be tolerated without some decrease in feed intake or milk fat content will be even less (Ruppert et al., 2003).

Highly rumen-available sources of unsaturated fatty acids can decrease the milk fat concentration, depending on the dietary effective fiber content. Jenkins and Harvatine (2014) attempted to provide guidance by proposing an equation to calculate the amount of unsaturated fat that can be added safely to diets of dairy cows. The equation is:

Fat to include in diet (%) = (4 × %NDF)/sum of % unsaturated fatty acids in the supplement

As always, the cows should be the final authority. If milk fat content decreases or is low, the amount of vegetable oil should be decreased by, for example, decreasing the amount of distillers' grains or decreasing the amount of cottonseed or whole soybeans. Commodity fats should be removed or replaced incrementally by commercial inert fats until milk fat concentration returns to the desired target. Although commercial fat prices may be high, the economics of the gain in two- to three-tenths of a percentage unit in milk fat concentration plus a milk production increase generally will be favorable for the addition of a good commercial fat product.

14 Future trends in research

Dairy cows have the ability to utilize a wide variety of fatty acids obtained from bioprocessing waste streams. Research to improve utilization of these low-quality streams should be beneficial in helping to create sustainability both in the processing industries and in dairy production.

Research should continue to seek ways to improve the digestibility of C18:0 because this fatty acid is widely available for use in fat supplements. Possible areas of emphasis include providing exogenous emulsifiers or learning how to stimulate the secretion of the endogenous emulsifiers such as lysolecithin.

The area of lipid nutrition is hampered by not being quantitative. Research must seek to define 'requirements' or adequate intakes of individual fatty acids, analogous to amino acids. In particular, given that C16:0 is both synthesized *de novo* and taken up from blood, it seems likely that there is an optimal amount of C16:0 to supplement to meet demands for milk fat synthesis with optimal efficiency. Likewise, adequate amounts of 18-carbon fatty acids need to be supplemented to replenish adipose tissue fatty acids, which are rich in *cis*-9 18:1, as well as meeting lactational demands. The need for quantitative recommendations also extends to the use of specific fatty acid supplements to alter the composition of milk fat for human use and health.

Research should continue to explore the profiles of fatty acids that negatively impact feed intake by dairy cows and to understand the mechanism involved in the suppression of DMI. This is critical to meet the goal of improving energy intake and balance.

Finally, the roles of lipids in modulating or regulating reproduction and metabolism will be an area of continued research. Roles of acyl-carnitines, ceramides and other sphingolipids, and diglycerides will continue to be defined.

15 Conclusion

Digestion, absorption, and metabolism of dietary lipids in dairy cows differ from the corresponding processes in non-ruminants. Increased understanding of these areas should improve the sustainability of milk production. Amounts of the nutritionally important C16:0, C18:0, and C18:1 *cis*-9 needed for optimal production efficiency remain to be defined. Researchers must continue to work toward establishing actual requirements or adequate intakes of these fatty acids needed for efficient milk production by dairy cows. Lipid nutrition is an important area of the sustainability of milk production.

16 Where to look for further information

- Linn, J. (2021). Fats in Dairy Cattle Nutrition. A Guide to Feeding Fats and Fatty Acids in Dairy Cow Rations. Chemistry, Sources, Digestion, Absorption. Eden Prairie, MN: Milk Specialties Global Animal Nutrition. https://www.milkspecialties.com/animal-nutrition/animal-nutrition-research/.
- Searchable Proceedings of Animal Conferences (S-PAC). https://spac.adsa.org/.
- Palmquist, D. L. and Jenkins, T. C. (1980). Fat in lactation rations: Review, J. Dairy Sci. 63(1), 1–14.
- Palmquist, D. L. and Jenkins, T. C. (2017). 'A 100-year review: Fat feeding of dairy cows', J. Dairy Sci. 100(12), 10061–10077.

17 References

Banks, W., Clapperton, J. L. and Ferrie, M. E. (1976). Effect of feeding fat to dairy cows receiving a fat-deficient basal diet. II. Fatty acid composition of the milk fat, *J. Dairy Res.* 43(2), 219–227.

Bauchart, D. (1993). Lipid absorption and transport in ruminants, *J. Dairy Sci.* 76(12), 3864–3881.

Baumgard, L. H., Corl, B. A., Dwyer, D. A., Saebø, A. and Bauman, D. E. (2000). Identification of the conjugated linoleic acid isomer that inhibits milk fat synthesis, *Am. J. Physiol. Regul. Integr. Comp. Physiol.* 278(1), R179–R184.

Benson, J. A. and Reynolds, C. K. (2001). Effects of abomasal infusion of long-chain fatty acids on splanchnic metabolism of pancreatic and gut hormones in lactating dairy cows, *J. Dairy Sci.* 84(6), 1488–1500.

Boerman, J. P., Firkins, J. L., St-Pierre, N. R. and Lock, A. L. (2015). Intestinal digestibility of long-chain fatty acids in lactating dairy cows: A meta-analysis and meta-regression, *J. Dairy Sci.* 98(12), 8889–8903.

Borodzicz, S., Czarzasta, K., Kuch, M. and Cudnoch-Jedrzejewska, A. (2015). Sphingolipids in cardiovascular diseases and metabolic disorders, *Lipids Health Dis.* 14, 55.

Bremmer, D. R., Ruppert, L. D., Clark, J. H. and Drackley, J. K. (1998). Effects of chain length and unsaturation of fatty acid mixtures infused into the abomasum of lactating dairy cows, *J. Dairy Sci.* 81(1), 176–188.

Christensen, R. A., Drackley, J. K., LaCount, D. W. and Clark, J. H. (1994). Infusion of four long-chain fatty acid mixtures into the abomasum of lactating dairy cows, *J. Dairy Sci.* 77(4), 1052–1069.

Christie, W. W. (2019). The lipid web. Available at: https://www.lipidhome.co.uk/index.html.

Cichosz, G., Czeczot, H. and Bielecka, M. (2020). The anticarcinogenic potential of milk fat, *Ann. Agric. Environ. Med.* 27(4), 512–518.

Davis, C. L. (1990). *Fats in Animal Nutrition*. Dundee, IL: Milk Specialties Co.

de Souza, J., Westerrn, M. and Lock, A. L. (2020). Abomasal infusion of an exogenous emulsifier improves fatty acid digestibility and milk fat yield of lactating dairy cows, *J. Dairy Sci.* 103(7), 6167–6177.

den Hartigh, L. J. (2019). Conjugated linoleic acid effects on cancer, obesity, and atherosclerosis: A review of pre-clinical and human trials with current perspectives, *Nutrients* 11(2), 370.

Doreau, M. and Ferlay, A. (1991). Digestion and utilization of fatty acids by ruminants, *Anim. Feed Sci. Technol.* 45, 379–396.

dos Santos Neto, J. M., de Souza, J. and Lock, A. L. (2021a). Effects of calcium salts of palm fatty acids on nutrient digestibility and production responses of lactating dairy cows: A meta-analysis and meta-regression, *J. Dairy Sci.* 104(9), 9752–9768.

dos Santos Neto, J. M., de Souza, J. and Lock, A. L. (2021b). Nutrient digetibility and production responses of lactating dairy cows when saturated free fatty acid supplements are included in diets: A meta-analysis, *J. Dairy Sci.* 104(12), 12628–12646.

Douglas, G. N., Rehage, J., Beaulieu, A. D., Bahaa, A. O. and Drackley, J. K. (2007). Prepartum nutrition alters fatty acid composition in plasma, adipose tissue, and liver lipids of periparturient dairy cows, *J. Dairy Sci.* 90(6), 2941–2959.

Drackley, J. K. (1999). ADSA Foundation Scholar Award. Biology of dairy cows during the transition period: The final frontier? *J. Dairy Sci.* 82(11), 2259–2273.

Drackley, J. K., Klusmeyer, T. H., Trusk, A. M. and Clark, J. H. (1992). Infusion of long-chain fatty acids varying in saturation and chain length into the abomasum of lactating dairy cows, *J. Dairy Sci.* 75(6), 1517–1526.

Drackley, J. K., Overton, T. R., Ortiz Gonzalez, G., Beaulieu, A. D., Barbano, D. M., Lynch, J. M. and Perkins, E. G. (2007). Responses to increasing amounts of high-oleic sunflower fatty acids infused into the abomasum of lactating dairy cows, *J. Dairy Sci.* 90(11), 5165–5175.

Elliott, J. P., Drackley, J. K., Beaulieu, A. D., Aldrich, C. G. and Merchen, N. R. (1999). Effects of saturation and esterification of fat sources on site and extent of digestion in steers: Digestion of fatty acids, triglycerides, and energy, *J. Anim. Sci.* 77(7), 1919–1929.

Elliott, J. P., Overton, T. R. and Drackley, J. K. (1994). Digestibility and effects of three forms of mostly saturated fatty acids, *J. Dairy Sci.* 77(3), 789–798.

Ferlay, A., Bernard, L., Meynadier, A. and Malpuech-Brugère, C. (2017). Production of trans and conjugated fatty acids in dairy ruminants and their putative effects on human health: A review, *Biochimie* 141, 107–120.

Freeman, C. P. (1969). Properties of fatty acids in dispersions of emulsified lipid and bile salt and the significance of these properties in fat absorption in the pig and the sheep, *Br. J. Nutr.* 23(2), 249–263.

Glasser, F., Ferlay, A., Doreau, M., Schmidely, P., Sauvant, D. and Chilliard, Y. (2008a). Long-chain fatty acid metabolism in dairy cows: A meta-analysis of milk fatty acid yield in relation to duodenal flows and de novo synthesis, *J. Dairy Sci.* 91(7), 2771–2785.

Glasser, F., Schmidely, P., Sauvant, D. and Doreau, M. (2008b). Digestion of fatty acids in ruminants: A meta-analysis of flows and variation factors: 2. C18 fatty acids, *Animal* 2(5), 691–704.

Grummer, R. R. (1991). Effect of feed on the composition of milk fat, *J. Dairy Sci.* 74(9), 3244–3257.

Grummer, R. R. (1996). Chapter 8: Utilizing rendered products: Ruminants. In: *The Original Recyclers*. Alexandria, VA: The Animal Protein Producers Industry, The Fats & Proteins Research Foundation, and The National Renderers Association.

Grummer, R. R. and Carroll, D. J. (1991). Effects of dietary fat on metabolic disorders and reproductive performance of dairy cattle, *J. Anim. Sci.* 69(9), 3838-3852.

Hannun, Y. A. and Obeid, L. M. (2008). Principles of bioactive lipid signaling: Lessons from sphingolipids, *Nat. Rev. Mol. Cell Biol.* 9(2), 139-150.

Harvatine, K. J. and Allen, M. S. (2006). Effects of fatty acid supplements on feed intake, and feeding and chewing behavior of lactating dairy cows, *J. Dairy Sci.* 89(3), 1104-1112.

Hillbrick, G. and Augustin, M. A. (2002). Milkfat characteristics and functionality: Opportunities for improvement, *Austr. J. Dairy Tech.* 57, 45-51.

Hocquette, J. F. and Buachart, D. (1999). Intestinal absorption, blood transport and hepatic and muscle metabolism of fatty acids in preruminant and ruminant animals, *Reprod. Nutr. Dev.* 39(1), 27-48.

Hu, W., Boerman, J. P. and Aldrich, J. M. (2017). Production responses of Holstein dairy cows when fed supplemental fat containing saturated free fatty acids: A meta-analysis, *Asian-Australas. J. Anim. Sci.* 30(8), 1105-1116.

Jenkins, T. C. (1993). Lipid metabolism in the rumen, *J. Dairy Sci.* 76(12), 3851-3863.

Jenkins, T. C. and Harvatine, K. J. (2014). Lipids feeding and milk fat depression, *Vet. Clin. North Am. Food Anim. Pract.* 30(3), 623-642.

Jenkins, T. C. and McGuire, M. A. (2006). Major advances in nutrition: Impact on milk composition, *J. Dairy Sci.* 89(4), 1302-1310.

Jensen, R. G. (2002). The composition of bovine milk lipids: January 1995 to December 2000, *J. Dairy Sci.* 85(2), 295-350.

Klusmeyer, T. H. and Clark, J. H. (1991). Effects of dietary fat and protein on fatty acid flow to the duodenum and in milk produced by dairy cows, *J. Dairy Sci.* 74(9), 3055-3067.

Kronfeld, D. S. (1976). Bovine ketosis in high yielding dairy cows. The potential importance of the proportions of glucogenic, lipogenic and aminogenic nutrients in regard to the health and productivity of dairy cows, *Fortschr. Tierphysiol. Tierernahr.* 7(7), 3-26.

LaCount, D. W., Drackley, J. K., Laesch, S. O. and Clark, J. H. (1994). Secretion of oleic acid in milk fat in response to abomasal infusions of canola or high-oleic sunflower fatty acids, *J. Dairy Sci.* 77(5), 1372-1385.

Litherland, N. B., Beaulieu, A. D., Reynolds, C. K. and Drackley, J. K. (2022). Effects of esterification, saturation and amount of fatty acids infused into the rumen or abomasum in lactating dairy cows, *J. Anim. Physiol. Anim. Nutr. (Berl).*

Litherland, N. B., Thire, S., Beaulieu, A. D., Reynolds, C. K., Benson, J. A. and Drackley, J. K. (2005). Dry matter intake is decreased more by abomasal infusion of unsaturated free fatty acids than by unsaturated triglycerides, *J. Dairy Sci.* 88(2), 632-643.

Livingstone, K. M., Lovegrove, J. A. and Givens, D. I. (2012). The impact of substituting SFA in dairy products with MUFA or PUFA on CVD risk: Evidence from human intervention studies, *Nutr. Res. Rev.* 25(2), 193-206.

Lock, A. L., Tyburczy, C., Dwyer, D. A., Harvatine, K. J., Destaillats, F., Mouloungui, Z., Candy, L. and Bauman, D. E. (2007). Trans-10 octadecenoic acid does not reduce milk fat synthesis in dairy cows, *J. Nutr.* 137(1), 71-76.

Loften, J. R. and Cornelius, S. G. (2004). Responses of supplementary dry, rumen-inert fat sources in lactating dairy cow diets, *Appl. Anim. Sci.* 20, 461-469.

Loften, J. R., Linn, J. G., Drackley, J. K., Jenkins, T. C., Soderholm, C. G. and Kertz, A. F. (2014). Invited review: Palmitic and stearic acid metabolism in lactating dairy cows, *J. Dairy Sci.* 97(8), 4661-4674.

Maynard, L. A. (1937). *Animal Nutrition* (1st edn.). New York, NY: McGraw-Hill.

McFadden, J. W. (2020). Review: Lipid biology in the periparturient dairy cow: Contemporary perspectives, *Animal* 14(S1), s165-s175.

McFadden, J. W. and Rico, J. E. (2019). Invited review: Sphingolipid biology in the dairy cow: The emerging role of ceramide, *J. Dairy Sci.* 102(9), 7619-7639.

Moallem, U. (2018). Invited review: Roles of dietary n-3 fatty acids in performance, milk fat composition, and reproductive and immune systems in dairy cattle, *J. Dairy Sci.* 101(10), 8641-8661.

National Academies of Sciences, Engineering, and Medicine (NASEM). (2021). *Nutrient Requirements of Dairy Cattle* (8th rev. edn.). Washington, DC: The National Academies Press.

National Renderers Association, Inc. (2008). *Pocket Information Manuel: A Buyer's Guide to Rendered Products*. Alexandria, VA: National Renderers Association.

National Research Council. (2001). *Nutrient Requirements of Dairy Cattle* (7th rev. edn.). Washington DC: National Academy Press.

National Research Council. (2012). Nutrient Requirements of Swine (11th rev. edn.). Washington, DC: National Academy Press.

Ortiz Gonzalez, G., Jimenez Flores, R., Bremmer, D. R., Clark, J. H., DePeters, E. J., Schmidt, S. J. and Drackley, J. K. (2007). Functional properties of butter oil made from bovine milk with experimentally altered fat composition, *J. Dairy Sci.* 90(11), 5018-5031.

Ortiz Gonzalez, G., Jimenez Flores, R., Bremmer, D. R., Clark, J. H., DePeters, E. J., Schmidt, S. J. and Drackley, J. K. (2022). Functional properties of cream from dairy cows with experimentally altered milk fat composition, *J. Dairy Sci.* 105(5), 3861-3870.

Palmquist, D. L. (2006). Chapter 2: Milk fat: Origin of fatty acids and influence of nutritional factors thereon. In: Fox, P. F. and McSweeney, P. L. H. (Eds.), *Advanced Dairy Chemistry:* Lipids (Vol. 2, 3rd edn.). New York: Springer.

Palmquist, D. L. (2009). Omega-3 fatty acids in metabolism, health, and nutrition and for modified animal product foods, *Prof. Anim. Sci.* 25(3), 207-249.

Palmquist, D. L., Beaulieu, A. D. and Barbano, D. M. (1993). 'ADSA Foundation Symposium: Milk fat synthesis and modification. Feed and animal factors influencing milk fat composition', *J. Dairy Sci.* 76(6), 1753-1771.

Palmquist, D. L. and Jenkins, T. C. (1980). Fat in lactation rations: Review, *J. Dairy Sci.* 63(1), 1-14.

Palmquist, D. L. and Jenkins, T. C. (2003). Challenges with fats and fatty acid methods, *J. Anim. Sci.* 81(12), 3250-3254.

Palmquist, D. L. and Jenkins, T. C. (2017). A 100-year review: Fat feeding of dairy cows, *J. Dairy Sci.* 100(12), 10061-10077.

Plata-Pérez, G., Angeles-Hernandez, J. C., Morales-Almaráz, E., Del Razo-Rodríguez, O. E., López-González, F., Peláez-Acero, A., Campos-Montiel, R. G., Vargas-Bello-Pérez, E. and Vieyra-Alberto, R. (2022). Oilseed supplementation improves milk composition and fatty acid profile of cow milk: A meta-analysis and meta-regression, *Animals (Basel)* 12(13), 1642.

Prom, C. M., dos Santos Neto, J. M. and Lock, A. L. (2022). Abomasal infusion of different exogenous emulsifiers alters fatty acid digestibility and milk fat yield of lactating dairy cows, *J. Dairy Sci.* 105(4), 3102-3112.

Prom, C. M., dos Santos Neto, J. M., Newbold, J. R. and Lock, A. L. (2021). Abomasal infusion of oleic acid increases fatty acid digestibility and plasma insulin of lactating dairy cows, *J. Dairy Sci.* 104(12), 12616–12627.

Prom, C. M. and Lock, A. L. (2021). Replacing stearic acid with oleic acid in supplemental fat blends improves fatty acid digestibility of lactating dairy cows, *J. Dairy Sci.* 104(9), 9956–9966.

Rabiee, A. R., Breinhild, K., Scott, W., Golder, H. M., Block, E. and Lean, I. J. (2012). Effect of fat additions to diets of dairy cattle on milk production and components: A meta-analysis and metaregression, *J. Dairy Sci.* 95(6), 3225–3247.

Relling, A. E. and Reynolds, C. K. (2007). Feeding rumen-inert fats differing in their degree of saturation decreases intake and increases plasma concentrations of gut peptides in lactating dairy cows, *J. Dairy Sci.* 90(3), 1506–1515.

Rico, J. E., Bandaru, V. V., Dorskind, J. M., Haughey, N. J. and McFadden, J. W. (2015). Plasma ceramides are elevated in overweight Holstein dairy cows experiencing greater lipolysis and insulin resistance during the transition from late pregnancy to early lactation, *J. Dairy Sci.* 98(11), 7757–7770.

Rico, J. E., Mathews, A. T., Lovett, J., Haughey, N. J. and McFadden, J. W. (2016). Palmitic acid feeding increases ceramide supply in association with increased milk yield, circulating nonesterified fatty acids, and adipose tissue responsiveness to a glucose challenge, *J. Dairy Sci.* 99(11), 8817–8830.

Rico, J. E., Saed Samii, S., Mathews, A. T., Lovett, J., Haughey, N. J. and McFadden, J. W. (2017). Temporal changes in sphingolipids and systemic insulin sensitivity during the transition from gestation to lactation, *PLoS ONE* 12(5), e0176787.

Rodney, R. M., Celi, P., Scott, W., Breinhild, K. and Lean, I. J. (2015). Effects of dietary fat on fertility of dairy cattle: A meta-analysis and meta-regression, *J. Dairy Sci.* 98(8), 5601–5620.

Ruppert, L. D., Drackley, J. K., Bremmer, D. R. and Clark, J. H. (2003). Effects of tallow in diets based on corn silage or alfalfa silage on digestion and nutrient use by lactating dairy cows, *J. Dairy Sci.* 86(2), 593–609.

Shepardson, R. P., Bazilevskaya, E. A. and Harvatine, K. J. (2020). Physical characterization of fatty acid supplements with varying enrichments of palmitic and stearic acid by differential scanning calorimetry, *J. Dairy Sci.* 103(10), 8967–8975.

Shingfield, K. J., Bonnet, M. and Scollan, N. D. (2013). Recent developments in altering the fatty acid composition of ruminant-derived foods, *Animal* 7 (Suppl 1), 132–162.

Thatcher, W. W., Bilby, T. R., Bartolome, J. A., Silvestre, F., Staples, C. R. and Santos, J. E. P. (2006). Strategies for improving fertility in the modern dairy cow, *Theriogenology* 65(1), 30–44.

Thorning, T. K., Raben, A., Tholstrup, T., Soedamah-Muthu, S. S., Givens, I. and Astrup, A. (2016). Milk and dairy products: Good or bad for human health. An assessment of the totality of scientific evidence, *Food Nutr. Res.* 60, 32527.

Tomkins, T. and Drackley, J. K. (2010). Applications of palm oil in animal nutrition, *J. Oil Palm Res.* 22, 835–845.

Toral, P. G., Monahan, F. J., Hervás, G., Frutos, P. and Moloney, A. P. (2018). Review: Modulating ruminal lipid metabolism to improve the fatty acid composition of meat and milk. Challenges and opportunities, *Animal* 12(s2), s272–s281.

U.S. Department of Agriculture and U.S. Department of Health and Human Services. (2020). *Dietary Guidelines for Americans, 2020-2025* (9th edn.). Available at: DietaryGuidelines.gov.

Weld, K. A. and Armentano, L. E. (2017). The effects of adding fat to diets of lactating cows on total-tract neutral detergent fiber digestibility: A meta-analysis, *J. Dairy Sci.* 100(3), 1766–1779.
Western, M. M., deSouza, J. and Lock, A. L. (2020). Effects of commercially available palmitic and stearic acid supplements on nutrient digestibility and production responses of lactating dairy cows, *J. Dairy Sci.* 103(6), 5131–5142.
Wood, A. H. (1894). Effect of food on milk. Feeding with fats. *Univ New Hampshire Agric. Expt. Sta. Bull*, 20 March. Durham: University of New Hampshire.
Wu, Z., Ohajuruka, O. A. and Palmquist, D. L. (1991). Ruminal synthesis, biohydrogenation, and digestibility of fatty acids by dairy cows, *J. Dairy Sci.* 74(9), 3025-3034.
Wu, Z. and Palmquist, D. L. (1991). Synthesis and biohydrogenation of fatty acids by ruminal microorganisms in vitro, *J. Dairy Sci.* 74(9), 3035–3046.

Part 2

The use of dietary supplements to optimise dairy cattle nutrition: plant extracts

Chapter 4

The use of plant extracts as dietary supplements in dairy cow nutrition: plant essential oils

Sergio Calsamiglia and María Rodríguez-Prado, Animal Nutrition and Welfare Service, Universitat Autònoma de Barcelona, Spain; Gonzalo Fernández-Turren, Instituto de Producción Animal, Facultad de Veterinaria, Universidad de la República, Uruguay; and Lorena Castillejos, Animal Nutrition and Welfare Service, Universitat Autònoma de Barcelona, Spain

1 Introduction

Plants produce an extensive variety of small volatile organic compounds derived from their secondary metabolism that are obtained by steam or hydro distillation (Gershenzon and Croteau, 1991a). These molecules play a relevant role in the protection of plants against environmental or biological threats or act as chemical messengers between the plant and its ecosystem (Balandrin and Klocke, 1985). The word 'essential' derives from 'essence' that means smell or taste and provides plants and spices with their characteristic odor and flavor. Essential oils extracted from plants are composed of a mix of different molecules where up to four of them represent the largest proportion and determine their main activity (Benchaar et al., 2008a). Essential oils can be classified based on their chemical structure in two groups: terpenoids and

http://dx.doi.org/10.19103/AS.2022.0117.04

phenylpropanoids. Terpenoids are the more numerous and diversified group and are synthesized through the mevalonate and deoxyxylulose pathways (Fig. 1). Terpenoids derive from an isoprene unit of five carbons (Gershenzon and Croteau, 1991b), and the most relevant present in essential oils are monoterpenoid (up to 90%) and sesquiterpenoid (Figs. 1 and 2; Gershenzon and Croteau, 1991b). Monoterpenoids are further classified as phenolic (i.e. thymol, carvacrol, carvone, terpinene, limonene) and noncyclic (i.e. linalool, geraniol, citral). Gingerol and piperine are examples of sesquiterpenoids. Capsicum oleoresin is a plant extract that contains lipophilic substrates, pigments, essential oils and other compounds, and it is often classified as an essential oil. Capsaicin is the main active compound of capsicum oleoresin, and it is a volatile molecule obtained by steam distillation with a sesquiterprenoid-like structure often classified as an alkaloid, and it is a product of the secondary metabolism synthesized from vanillylamide and fatty acids. Phenylpropanoids are less common and are synthesized through the shikimate metabolic pathway (Figs. 1 and 2). Phenylpropanoids are composed of a six-carbon aromatic ring with a three-carbon side chain. Cinnamaldehyde, eugenol and anethol are examples of phenylpropanoids (Fig. 2). A different group of essential oils is the organosulfur-derivate compounds found in garlic and wasabi oils and have a completely different structure that derives from the metabolism of cysteine.

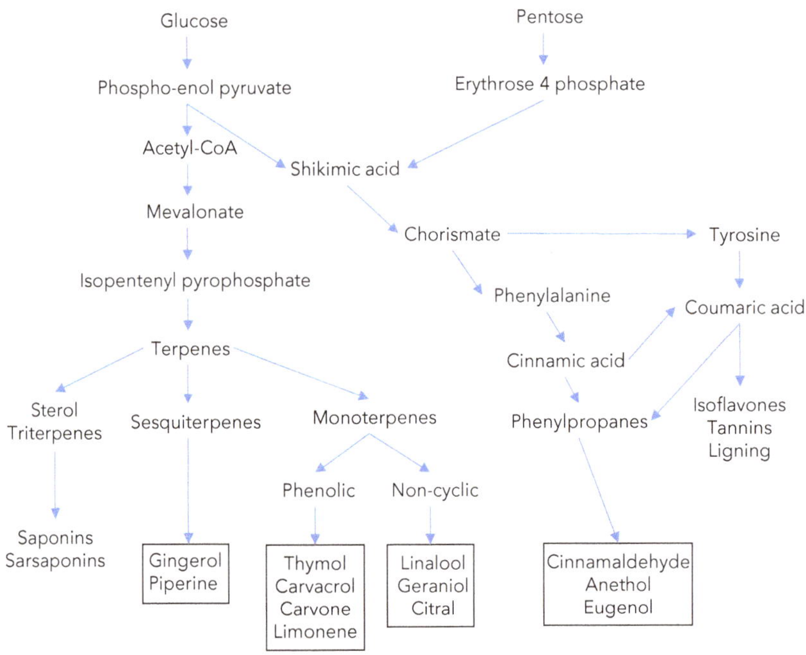

Figure 1 Metabolic pathways of the biosynthesis of main plant extract active compounds.

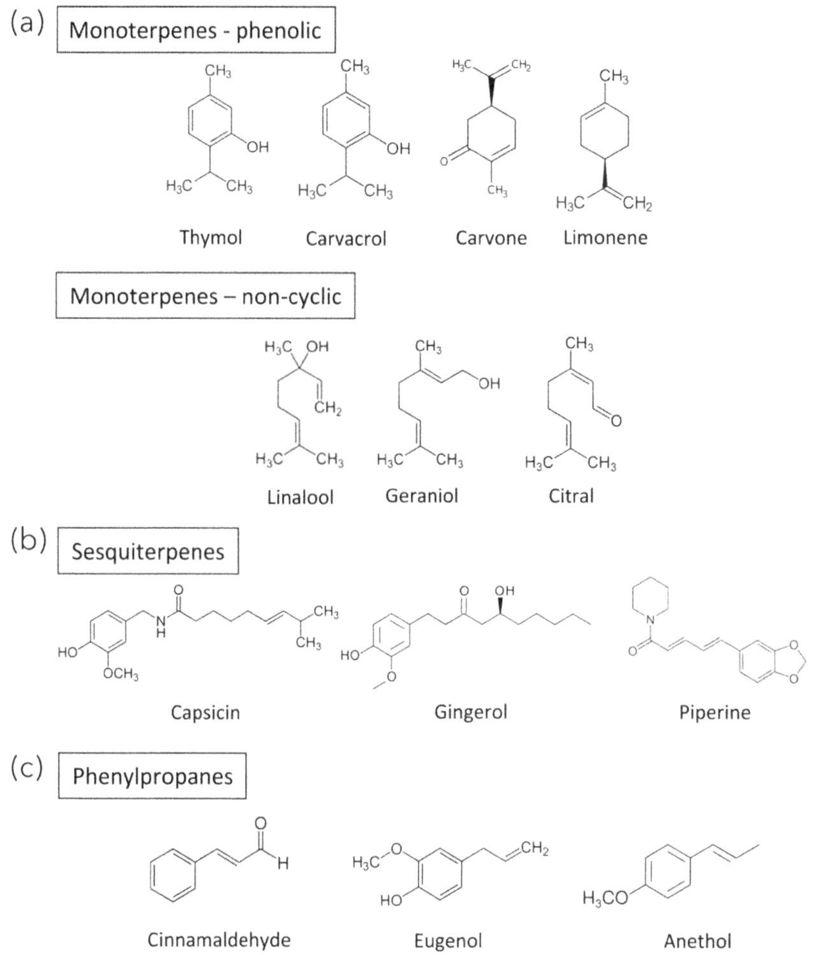

Figure 2 Main monoterpenes (a), sesquiterpenes (b) and phenylpropanes (c) compounds of essential oils.

Essential oils have been used for centuries in traditional medicine for their effects on cardiovascular diseases, inflammatory processes, immunity and as antioxidants (Davidson and Naidu, 2000; Reddy et al., 2003; Trouillas et al., 2003; Giannenas et al., 2020). However, the antimicrobial and antiseptic activity is their best-known use, although it was not until the first part of the twentieth century that this activity was demonstrated scientifically (Hoffmann and Evans, 1911).

The interest in the use of essential oils in animal nutrition did not develop until the ban on the use of antibiotics growth promoters in animal feeds was announced in the European Union (Directive 1831/2003/CEE). In the last 20 years, there has been intense research interest in developing these products as natural alternatives to antibiotic growth promoters, initially by modulating the rumen microbiome (Calsamiglia et al., 2007) but also for their potential postruminal effects (Oh et al., 2017b). In the present chapter, the impact of essential oils on ruminal fermentation, postruminal effects and dairy cattle performance will be reviewed.

2 Ruminal effects of essential oils

2.1 Mechanisms of action of essential oils

There are two main types of mechanism of action to explain the antimicrobial activity of essential oils: (1) the interaction with the microbial cell membrane; and (2) the interaction with proteins or other molecules inside the bacterial cell (Griffin et al., 1999; Davidson and Naidu, 2000; Dorman and Deans, 2000). Terpenoids and phenylpropanoids are lipophilic molecules that accumulate in the lipidic membrane of bacteria resulting in changes in its structure and function (Griffin et al., 1999; Ultee et al., 1999). The result is an imbalance in the flow of ions across the membrane that bacteria will counterbalance dedicating energy resources to maintain its homeostasis (Griffin et al., 1999; Ultee et al., 1999). In these conditions, bacteria will slow their growth and eventually reduce the size of the population. Ultee et al. (2002) proposed an alternative mechanism by which some essential oils can affect bacterial cell membrane permeability. In this case, the hydroxyl group of phenolic compounds (i.e. thymol, carvacrol, carvone) function as transmembrane ion carriers similar to ionophore antibiotics. The mechanisms of action related to the interaction with the cell membrane are generally more effective against gram-positive bacteria, where these hydrophobic molecules can interact directly with the cell membrane (Chao et al., 2000; Cimanga et al., 2002). In contrast, gram-negative bacteria have a hydrophilic external cell wall that makes it difficult for lipophilic substances to reach the cell membrane. In fact, this cell wall barrier is responsible for the selective effect of monensin and other ionophores, because their large molecular weight (670 g/mol) does not allow them to cross the cell wall of gram-negative bacteria, which makes monensin selective, affecting mostly gram-positive bacteria. In contrast, phenolic compounds such as thymol, carvacrol and carvone are lipophilic but of smaller molecular weight (around 150 g/mol), and can cross the external cell wall through small pores and interact with the bacterial cell membrane (Griffin et al., 1999; Dorman and Deans, 2000). The small molecular weight of some essential oil compounds makes them active against both gram-positive and negative bacteria. Therefore, these molecules are strong antimicrobials active against a wide range of bacteria but, at the same time, are less specific, making the modulation of the rumen bacterial fermentation profile more difficult.

The other type of mechanism of action is related to the capacity of some essential oils to interact and modify the activity of different bacterial proteins or enzymes (Juven et al., 1994; Gustafson and Bowen, 1997). Essential oils may interact with these molecules through hydrogen bridges, ionic interactions or carbonyl groups, among others (Prescott et al., 2004). For example, cinnamaldehyde does not interact with the membrane and its antimicrobial activity is related to the reactivity of its carbonyl group against plasmatic proteins (Helander et al., 1998). The mechanism of action of garlic on the rumen microbial population was described by Busquet et al. (2005a,b). Methanogenic archea have unique membrane lipids that contain glycerol linked to long-chain isoprenoid alcohols that provide stability to the cell membrane (De Rosa et al., 1986). The synthesis of the isoprenoid units in archea is catalyzed by a hydroxymethylglutaryl coenzyme A (HMG-CoA) reductase. Garlic oil and its active compounds inhibit this HMG-CoA reductase (Gebhardt and Beck, 1996). As a result, the membrane of archea becomes unstable and cells die. This hypothesis was confirmed by Miller and Wolin (2001) when demonstrated that lovastatin and mevastatin, specific HMG-CoA reductase inhibitors, drastically reduce the rumen archea.

The mechanism of action of essential oils on N metabolism has been explained through two different pathways. Busquet et al. (2005a) reported that cinnamaldehyde and garlic oil reduced ammonia-N concentration and that change was paralleled with a reduction in *Prevotella* spp., a common rumen bacterium with proteolytic and deamination activity (Ferme et al., 2004). However, the reduction in ammonia-N concentration can also be achieved by controlling a small but very active deaminating microbial population known as hyper-ammonia-producing bacteria (i.e. *Peptostreptococcus* and *Clostridium* spp.; Wallace et al., 2002). These bacteria use the carbon fraction of amino acids as the main energy source, releasing ammonia-N into the media. McIntosh et al. (2003) demonstrated that the reduction in ammonia-N concentration after the supplementation of a blend of essential oils in a batch fermentation study was due to the reduction in hyper-ammonia-producing bacteria counts.

2.2 Targets to modulate ruminal fermentation to improve efficiency

The rumen ecosystem is a complex symbiosis between the ruminant and its microbiome. The ruminant provides feed and environmental conditions such as anaerobiosis, pH, temperature and redox to allow microbial growth. In return, the microbial population provides the ability to degrade fiber and supplies microbial protein to the cow. The rumen microbial ecosystem is characterized by its functional diversity and redundancy (Weimer, 2015), making it very robust and resilient to changes in media conditions. Therefore, while changes in the microbial population structure are possible, the resulting fermentation profile only fluctuates within a relatively narrow range. Changes in the fermentation profile are restricted by the anaerobic nature of the system and the required biochemical balance. Therefore, the first step in the selection of useful essential oils to modify rumen microbial fermentation is to define the target and the expected magnitude of the changes to be achieved in the fermentation profile.

Carbohydrates, namely fiber and starch, are the main source of energy for rumen microbes and the cow. These carbohydrates are digested in the rumen into glucose and glucose into pyruvate. This process generates energy for microbial growth but also metabolic hydrogens in the form of reduced nicotinamide adenine dinucleotide ($NADH_2$). These $NADH_2$ need to be recycled back into NAD by transferring the H to other acceptors. In the rumen, the main H acceptors are methane, propionate and, to a lesser extent, butyrate (Fig. 3). The balance of H is essential to maintain the metabolic pathway active, but also the retention of H in volatile fatty acids (VFA) is essential to maximize the energy supply to the cow. Methane plays an essential role in maintaining this balance because it is able to accept four H per unit of carbon. Unfortunately, methane is a gas lost through eructation and represents a 3–13% loss of energy for the cow (Van Nevel and Demeyer, 1988). Reducing methane and retaining the energy requires that the H are effectively transferred to propionate or, to a lesser extent, butyrate. Compared with pyruvate, energy retained in propionate (109%) is higher than that retained in butyrate (78%) or acetate (63%; Richardson et al., 1976). Other alternative H sinks are either unsaturated fatty acids, which are not quantitatively relevant (Nagaraja et al., 1997), or the production of H_2 gas. For example, the inhibition of methane production by

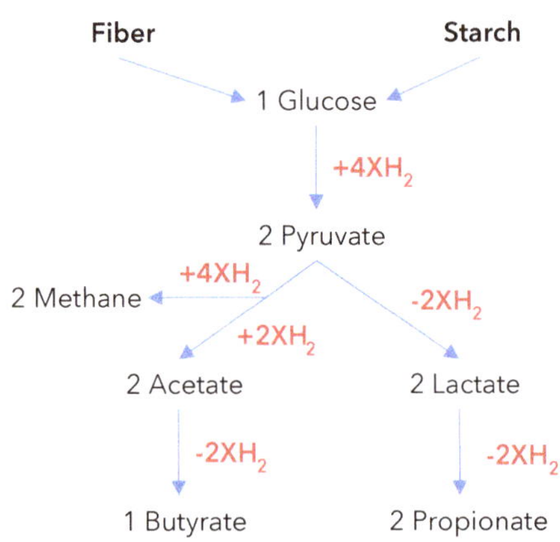

Figure 3 Rumen microbial fermentation pathways and metabolic hydrogens balance.

anthraquinone occurred together with an increase in the concentration of hydrogen (Garcia-Lopez et al., 1996; Kung et al., 2003). While the process regenerates $NADH_2$ into NAD, the energy is lost as H_2 gas through eructation. Therefore, the increase in propionate production should be the main target in the modulation of rumen fermentation not only because it retains the largest amount of energy but also because it is a precursor for neoglucogenesis, which may contribute to improve performance and health in high-producing dairy cows in early lactation. Alternatively, although acetate and butyrate retain less energy and would be less desirable from the energy point of view, they play a key role as precursors for milk fat synthesis and their production is also relevant in dairy cattle. Therefore, the initial target for dairy cattle would be to enhance the production of propionate while maintaining a reasonable production of acetate and butyrate.

On the other hand, dietary protein is degraded to peptides, amino acids and, finally, deaminated into ammonia-N. Ammonia-N plays an important role in supplying N for microbial protein synthesis. However, in normal feeding conditions, proteolysis and deamination proceed faster that the capacity of the rumen microbial population to capture it. This excess ammonia-N is absorbed, transformed into urea in the liver and excreted through the urine, which implies a loss of N for the cow and the release of N into the environment (Tamminga, 1996; Weimer, 1998; Bach et al., 2005). The decrease in deamination decreases losses of ammonia-N and increases free AA that will be used more efficiently by most rumen bacteria (Russell et al., 1992). Therefore, the target of essential oils should also be the reduction of proteolysis and deamination. In most cases, improving the efficiency of energy and protein retention is directly associated with a reduction in the emission of pollutants. In fact, antibiotic growth promoters like ionophores have been very successful in reducing these energy and protein losses in the rumen and the emission of pollutants to the environment through the modulation of fermentation toward more propionate and less ammonia-N production (Van Nevel and Demeyer, 1988; Duffield et al., 2008).

2.3 Effects of essential oils on rumen fermentation in vitro

There has been extensive research developed in the last 20 years on the effects of essential oils on rumen microbial fermentation. Most of the research has been conducted *in vitro* using short-term batch and long-term continuous culture fermentation studies and has been reviewed elsewhere (Calsamiglia et al., 2007; Benchaar et al., 2008a; Hart et al., 2008; Cobellis et al., 2016). Interpreting and summarizing the results of such studies is difficult because some studies report the effects of pure active compounds while others use essential oils that are mixes of different active compounds with variable concentrations that are often not reported. Furthermore, the potential impact of synergies and antagonisms, as well as the influence of the type of diet and fermentation conditions, makes the world of essential oils very complex. Hart et al. (2008) summarized the effect of 84 different *in vitro* tests with essential oils and reported that total VFA was reduced in 22% of the tests, propionate concentration only increased in 5% of the tests and ammonia-N concentration decreased in 32% of the tests. This reflects the difficulty in identifying essential oils and doses that meet the criteria to become useful for modulating ruminal microbial fermentation profile. However, there are some common outcomes that are worth summarizing:

2.3.1 Promising products

There is a general consensus that the most promising products to modulate rumen fermentation are garlic oil and three phenylpropanoids (cinnamaldehyde, present in cinnamon oil; eugenol, present in clove bud oil; and anethol, present in anise oil) (Calsamiglia et al., 2007). This observation agrees with the review by Cobellis et al. (2016) who, in addition to the previous essential oils, also included some phenolic terpenes as carvacrol (present in oregano oil), thymol (present in thyme oil), carvone (present in spearmint and dill oils) and gingerol (present in ginger). Garlic oil, cinnamaldehyde and eugenol at optimal doses frequently reduce the proportion of acetate and increase the proportions of propionate and butyrate without reducing total VFA production (Busquet et al., 2005a,b,c, 2006; Foskolos et al., 2015). These changes in the fermentation profile are similar to those observed with the use of antimethanogenic molecules (Hino and Russell, 1985; Martin and Macy, 1985). Busquet et al. (2005b) demonstrated that garlic reduced the methane (μmol):VFA (μmol) ratio and provided evidence for the antimethanogenic effect through the inhibition of archea. The fermentation profile of cinnamaldehyde and eugenol also suggest an antimethanogenic effect, but the exact mechanism of action has not been elucidated. Phenolic compounds (i.e. thymol, carvacrol and carvone) are strong antimicrobial with a wide spectrum activity against gram-positive and negative bacteria. Previous studies have shown that in many cases, even at moderate doses, the change in the fermentation profile occurs at doses higher than those that reduce total VFA production, which is counterproductive for dairy cattle performance, being too strong and nonspecific against microorganisms to have selective effects on rumen fermentation (Cardozo et al., 2005; Busquet et al., 2005c, 2006; Zhou et al., 2020). In contrast, anethol had no effect on VFA profile (Cardozo et al., 2004; Benchaar et al., 2007; Chaves et al., 2008).

The other target to modify rumen microbial fermentation is the reduction of proteolysis and deamination. Eugenol and phenolic compounds at low doses reduce ammonia-N concentration (Cardozo et al., 2005; Busquet et al.,

2005a,c, 2006; Castillejos et al., 2006). In contrast, the effects of garlic oil, cinnamaldehyde and anethol have been small and inconsistent (Cardozo et al., 2004; Busquet et al., 2005a,b,c). As discussed earlier, this effect is mediated through the modulation of bacterial populations involved in proteolysis and deamination (Wallace et al., 2002; Ferme et al., 2004). Because data indicate that the active compounds that modulate energy and protein metabolism are different, combining essential oils or their active compounds in a way that affects the production of VFA and ammonia-N in the desired direction would allow the development of optimal products.

Many other essential oils or their active compounds have also been tested for their impact on rumen fermentation, but their effects were small and inconsistent. For example, the supplementation of citral or essential oils containing citral resulted in either small or negative effects (Hristov et al., 2008; Nanon et al., 2015; Joch et al., 2016; Temmar et al., 2021). Similar lack or negative effects have been reported for limonene (Castillejos et al., 2006; Joch et al., 2016; Temmar et al., 2021), terpinene (Busquet et al., 2006; Hristov et al., 2008; Temmar et al., 2021), linalool (Joch et al., 2016; Temmar et al., 2021), geraniol oil (Joch et al., 2017; Temmar et al., 2021), capsaicin (Cardozo et al., 2004, 2005; Busquet et al., 2005c), and black pepper, turmeric and ginger oils (Busquet et al., 2006; Nanon et al., 2015; Temmar et al., 2021).

2.3.2 Combination of essential oils

Because the effects and mechanisms of action of essential oils are very diverse, Calsamiglia et al. (2007) suggested that a careful combination may result in synergistic effects that would enhance ruminal fermentation. In fact, these synergies may already occur among different active compounds in essential oils. For example, the antimicrobial activity of garlic and cinnamon oil has been reported to be different than the use of their active compounds individually (Reuter et al., 1996; O'Gara et al., 2000; Ross et al., 2001; Busquet et al., 2005b), which suggests that minor active compounds of the essential oils interact with their main active component modulating its activity. Bassolé and Juliani (2012) reported a synergy between linalool and eugenol. In contrast, Galucci et al. (2009) reported that the combination of thymol and carvacrol resulted in antagonistic effects. However, all these studies were conducted in the field of cosmetics and food processing, and the evidence for these interactions on rumen microbial fermentation is very limited. Most commercial products in the market (i.e. CRINA-Ruminants®, DSM, Switzerland; AGOLIN®, Agolin Sa, Switzerland; XTRACT®, Pancosma, Switzerland) are combinations of two or more different essential oils or active compounds, but their potential synergies or antagonisms have not been reported. Fandiño et al. (2020) conducted an *in vitro* study to test for these interactions among essential oils in a beef cattle

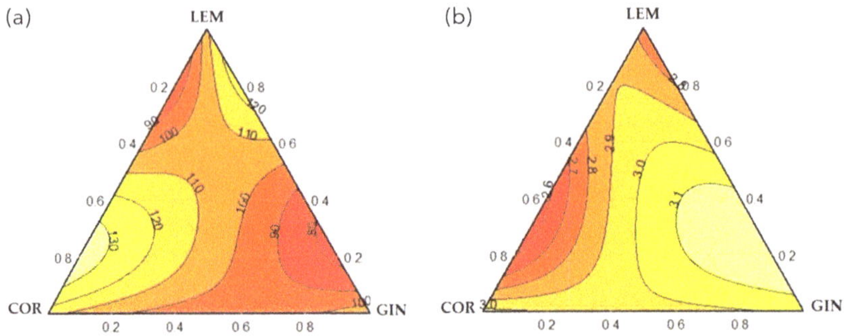

Figure 4 Contour plots of the effects of mixtures of lemongrass (LEM), coriander (COR) and ginger (GIN) on total VFA (mM) (a) and the acetate-to-propionate ratio (b). The vertices represent individual essential oils and the lines follow the mixtures that have the same response (Temmar et al., 2021).

dietary environment and reported that thyme and oregano oils incubated alone reduced acetate, increased propionate and reduced the acetate-to-propionate ratio. However, when they were incubated alone or together mixed with clove bud (containing eugenol as the main active ingredient) resulted in an antagonistic effect and no changes in the rumen fermentation profile were observed. In the same study, when oregano, thyme or clove bud oils were incubated alone, the fermentation profile changed in the desired direction. However, when mixed with tea tree oil, there were no changes in the rumen fermentation profile, confirming the negative interaction. Temmar et al. (2021) used a different approach. They selected three essential oils (coriander, lemongrass and ginger oils) and used a simplex centroid design to determine potential interactions among them. Authors reported that a combination of 60–80% coriander oil with the remaining being lemongrass oil optimized total VFA concentration and the acetate-to-propionate ratio (Fig. 4). However, as the amount of ginger oil increased in the mix, total VFA decreased and the acetate-to-propionate ratio increased, suggesting that ginger oil had an antagonistic effect on the lemongrass and coriander oils mix. Therefore, it is always necessary to evaluate the potential interactions when different essential oils or active compounds are mixed.

2.3.3 Optimal doses

Determining the doses at which essential oils have positive effects on fermentation is not easy. As indicated previously, when the oils are used, there are three elements that complicate the task: (a) essential oils are a mix of active compounds and it is always difficult to attribute the effects to a specific one; (b) the actual concentration of each active compound in the essential oil is often not reported, and there is considerable variation within oils. For example, oregano is rich in thymol and carvacrol, but their concentration ranged from 0.44% to 31.8% and 0.43% to 79.6%, respectively (Sivropoulou et al., 1996). This variability may be partially responsible for the inconsistent responses reported in literature; and (c) there are potential interactions among active compounds within an essential oil that may increase or decrease the effect of the most relevant one. Therefore, it is necessary to identify the range of doses of the pure active compounds that are effective in modifying rumen microbial fermentation first and to evaluate synergies and antagonisms thereafter. Previous studies indicate that changes considered positive for the optimization of rumen fermentation *in vitro* occurred at doses in the range of 50–500 mg/L (Cardozo et al., 2005; Busquet et al., 2005c; Castillejos et al., 2005; Temmar et al., 2021). It is relevant to indicate that these suggested concentrations relate to the active component and should not be confused with the dose of the corresponding essential oil.

2.3.4 The magnitude of changes

Another important point is to determine the magnitude of the expected changes. To try to provide some objective measure of this potential, eight *in vitro* studies from our laboratory where different essential oils or their active compounds were tested in a dairy cattle environment were selected. From these studies, we only considered treatments and doses where differences were significant and in the desired direction without decreasing total VFA and used these values to obtain an average. Therefore, results represent the maximum expected change. From this analysis, we observed that the expected positive response was a decrease in the acetate proportion from 63.3 mol/100 mol to 60.5 mol/100 mol (−4.4%), an increase in the propionate proportion from 20.1 mol/100 mol to 25.0 mol/100 mol (+24%) and a decrease in the ammonia-N concentration from 15.9 mg N/dl to 11.5 mg N/dl (−28%). The overall balance reflects that the decrease in acetate corresponded closely to the increase in propionate, as expected. If we use the energy retained in each of these products (Richardson et al., 1976) and assume an intake of 22 kg of DM with a 50% rumen fermentability, one can conclude that diets supplemented with essential oils would have a maximum of 2.5% increase in energy supply. If we transfer this increase in energy supply into energy for milk yield in a cow producing 35 kg, it will imply an increase of approximately 0.8 kg/day. This is an amount close to some reports where the supplementation of essential oils increased milk production (1.12 kg/day; Belanche et al., 2020), although no effects were observed in many other studies.

2.3.5 Variability

Finally, there is considerable variation among studies and repeatability is difficult. This variable response can be attributed to different experimental conditions, variability in the concentration of active compounds in essential oils, the dose used, the media pH, the rumen fluid donor and diet composition, among others (Calsamiglia et al., 2007; Klevenhusen et al., 2012).

2.4 Effects of essential oils on rumen fermentation in vivo

In contrast to *in vitro*, *in vivo* studies are limited and the interpretation is more difficult because other factors add variability. That includes the level of production and intake, the flow and absorption of endproducts of fermentation and rumen sampling time. At higher intakes, rumen retention time is reduced and the effect of essential oils on rumen bacteria may be reduced. Ammonia-N and VFA concentrations are used as makers of changes in rumen fermentation. However, because they are absorbed through the rumen wall, the concentration does not reflect actual production, which may hide potential changes. Finally, the concentration of total and individual VFA and the different protein fractions change with time after feeding either increasing or decreasing. Therefore, sampling time plays a relevant role. *In vitro*, samples are taken either at the end of the incubation period in batch fermentation studies or at specific times after feeding in continuous culture systems (Cardozo et al., 2004; Busquet et al., 2005a,b). In contrast, *in vivo* it is common to collect ruminal samples along the day and composite them. In this case, differences that may be significant a few hours after feeding may be diluted with the flow of digesta along the day, making the difference often difficult to detect.

In spite of the difficulties, data from 17 papers published in peer-reviewed journals that compared 29 essential oils versus a control on rumen microbial fermentation profile *in vivo* were collected. Doses tested ranged from 0.04 g/cow/day to 10 g/cow/day, with an average of 1.39 g/cow/day. Cows in the study had an average milk yield of 35.2 kg/day and a DMI of 23.1 kg/day. There were no differences in any of the items reported (Table 1).

Table 1 Effects of supplementing essential oils on rumen microbial fermentation profile *in vivo* (summary of 29 comparisons *in vivo*)

Item	Average	Difference	SE	*P*-value
pH	6.16	+0.03	0.07	0.68
Total VFA, mM	126.1	+0.09	4.70	0.19
Acetate, mol/100 mol	65.7	−0.08	0.84	0.25
Propionate, mol/100 mol	23.7	+0.30	0.75	0.28
Butyrate, mol/100 mol	12.9	−0.03	0.67	0.36
Acetate:propionate ratio	2.84	−0.04	0.13	0.15
NH_3, mg N/ml	14.7	−0.44	2.34	0.74

A meta-analysis of *in vivo* studies also reported no effect of essential oils on rumen fermentation (Khiaosa-Ard and Zebeli, 2013). There are several reasons that may justify this lack of effect. The dose *in vivo* is consistently lower than the dose used *in vitro*. For example, taking the average daily dose of 1.39 g/cow/day and assuming that the daily flow through the rumen is around 200 L/day, the actual concentration in the rumen fluid would be 7 mg/L, when *in vitro* data indicate that the effective dose is in the range of 50–500 mg/L (Calsamiglia et al., 2007). Calsamiglia et al. (2007) argue that *in vitro* models use diluted rumen fluid, and the effective concentration of the essential oils needs to be much higher because the probability of interacting with the active site (microbial cell membrane) is much lower. This is not a problem specific to essential oils. Similar differences in doses between *in vitro* and *in vitro* occur with monensin, where the effective dose for dairy cattle (185–660 mg/day) would correspond to 0.9–3.3 mg/L in an estimated flow of 200 L/day of fluid through the rumen, a concentration much lower than the one reported to have an effect *in vitro* (12 mg/L, Busquet et al., 2005a,b). Another possible reason to justify the lack of effect may be related to the samples used for the measurements. The *in vitro* batch fermentation measures the accumulation of the effects after 24 h, while *in vivo*, it is common to use spot samples at a specific time after feeding. In fact, research conducted in a continuous culture system showed that the changes in the fermentation profiles were significant when samples were collected 3-6 h after feeding, but these differences disappeared when samples were collected from the 24 h overflow. Finally, there may also be a reason related to the absorption of nutrients. In *in vitro* systems, there is no absorption, and the concentration of each VFA or ammonia-N reflects its production. In contrast, because absorption occurs only in the undissociated form of the acid and the pK of individual VFA is different, the concentration of VFA in the rumen does not necessarily reflect the actual production. These effects may be quantitatively minor, but because the expected changes are also minor, they may be sufficient to reflect the lack of effect on the rumen fermentation profile. Therefore, the lack of changes in the concentration of VFA and/or ammonia-N *in vivo* does not necessarily imply that the fermentation was not affected at some point in the daily feeding cycle.

3 Postruminal effects of essential oils

Essential oils have also been successfully used in monogastric animals resulting in changes in digestion, health and performance (Trouillas et al., 2003; Reyes-Camacho et al., 2020; Montoya et al., 2021). Many *in vivo* studies in ruminants have reported changes in performance at doses that did not affect ruminal fermentation, suggesting that the changes could also be attributed to a postruminal effect due to their partial ruminal escape (Oh et al., 2015). These postruminal effects can be classified into three areas: impact on the intestinal ecosystem, effects on immunity and hormonal regulation (Oh et al., 2017b; Zeng et al., 2015).

3.1 Intestinal microbiome and gut morphology

There has been considerable research on the relevance of the equilibrium of the intestinal microbial ecosystem in monogastric. This equilibrium is established between the microbial population, the digestion and absorption processes, the morphology of the intestinal wall and the local immunity. In the intestinal ecosystem, there is competition for specific niches between the local microbiome and pathogens, and the maintenance of adequate equilibria is essential for intestinal health (Oviedo-Rondón et al., 2006; Tiihonen et al., 2010; Reyes-Camacho et al., 2020; Montoya et al., 2021). The lactobacilli to coliforms ratio has been proposed as an indicator of a healthy microbial population, which reduces the negative impact of the microbiome on intestinal epithelial cells (Mourão et al., 2006). For example, essential oils reduced the counts of *E. coli* and increased the counts of lactobacilli in weaned pigs (Huang et al., 2010; Li et al., 2012). Montoya et al. (2021) also reported that a blend of trans-cinnamaldehyde, eugenol, carvacrol, thymol and diallylsidulfutre supplemented to piglets challenged with enterotoxigenic *E. coli* F4 improved the microbiota balance, increasing the fecal lactobacilli to coliforms ratio. Zou et al. (2016) reported better villi morphology increased expression of genes related to epithelia integrity and a reduced inflammatory response in pigs supplemented with oregano oil, although it was not clear to what extent the effect was mediated through changes in the microbiome or a direct effect of the essential oil on mucosa cells. However, the selection of the type of essential oils and the dose is important, and negative side effects are also possible. Muhl and Liebert (2007) reported that some essential oils reduced the bacterial counts of some beneficial bacteria at similar or even lower doses required to reduce the counts of pathogenic bacteria. Similar results were also reported by Horošová et al. (2006), where oregano oil reduced the counts of lactobacilli in fecal samples of chickens. Unfortunately, there are no reports on such effects in ruminants. In fact, the intestinal ecosystem in ruminants is more robust because of the buffering effect of the rumen microbial ecosystem, and external factors tend to have a smaller influence on intestinal health. However, some intestinal dysbiosis may occur in cases of ruminal acidosis, and different stressors like heat or metabolic stress may affect intestinal mucosa permeability, what has been named 'leaky gut'. There is growing evidence of the implication of lipopolysaccharides (LPS) produced by bacteria in the small intestine as a key element of these problems (Horst et al., 2019). The control of the microbiome and the resulting inflammatory reaction may be relevant in the development of the 'leaky gut'. Only recently, the microbiome of the lower gut in ruminants has been investigated (Mao et al., 2015), but the effect of essential oils on this microbial ecosystem has not been explored. However, there is a challenging opportunity to identify which factors associated with the microbiome may affect the gastrointestinal function, how to modulate them and their implications for cow's health and performance.

3.2 Impact of essential oils on immunity

Animals are exposed to risks and threats that are controlled by the immune system. The protection measures range from the physical barriers of mucosa to a complex immune reaction. The digestive tract is an important first barrier that contains an extensive array of immune structures prepared to prevent any external threat. The first step in the immune function is the identification of the causing agent. When the causing agent is identified, an intracellular signaling cascade results in the activation of transcription factors (i.e. the AP-1 protein, the NF-κB factor). These factors regulate the expression of genes responsible for the inflammatory response, releasing pro-inflammatory cytokines such as interleukins 1β and 6 (IL-1β; IL-6) and the tumor necrosis factor-α (TNF-α). These cytokines participate in the process of inflammation and the recruitment of white blood cells to neutralize and destroy the causing agent. During the process, macrophages release nitric oxide that further exacerbates the inflammatory process (MacMicking et al., 1997), and there is extensive production of reactive oxygen species (ROS) that can damage proteins, lipids and DNA (McCall and Frei, 1999). The body has developed an endogenous antioxidant system (i.e. glutathione peroxidase, superoxide dismutase or alfa-lipoic acid) that stops the chain reaction inhibiting lipid peroxidation and formations of other free radicals. However, excessive inflammatory cytokine production can lead to tissue damage, which may become harmful to the host. Therefore, the inflammatory reaction needs to be carefully controlled by a precise equilibrium between pro and anti-inflammatory regulatory molecules. Some inflammatory cytokines, such as IL-1β, IL-6 and TNF-α, and oxidative stress indicators can be used as biomarkers of inflammation.

There is evidence that some essential oils can interfere with different pathways involved in the inflammation cascade. An in-depth review of the immune response in ruminants has been recently published by Oh et al. (2017b). Most of the preliminary research was conducted *in vitro* or using animal models. For example, cinnamaldehyde, eugenol, capsaicin and allicin have been shown to activate some specific transient receptor potential channels that result in the regulation of pro- or anti-inflammatory cytokines and oxidative stress (Takano et al., 2007; Holzer, 2011; Nevius et al., 2012; Zeng et al., 2015). Several essential oil mixes based on cinnamaldehyde, thymol, oregano oil and/or capsicum oleoresin increased lymphocyte proliferation and phagocytosis rate and the production of immunoglobulins (Li et al., 2012; Liu et al., 2013). The immune stimulation resulting from the supplementation with essential oils improved the response of chickens to a vaccination challenge (Lee et al., 2011). While immune stimulation may prepare the animal to confront pathogenic challenges, in most production conditions oxidative stress and chronic inflammation become a problem. The inflammation requires a considerable amount of energy that is diverted away from production objectives. In this context, an anti-inflammatory modulation may be beneficial. Some essential oils can have this anti-inflammatory response (Lang et al., 2004). In fact, the same essential oils may respond differently depending on the inflammatory status of the animal. For example, Liu et al. (2012) reported that anethol, capsicum oleoresin, carvacrol, eugenol, garlic, cinnamaldehyde and turmeric had a pro-inflammatory effect when used in nonactivated porcine macrophages *in vitro*. However, when macrophages were challenged with LPS, the effect was anti-inflammatory. The anti-inflammatory activity is mediated through the blocking of the NF-κB pathway that results in a regulation of several genes involved in the inflammatory responses, resulting in the reduction of the concentration of pro-inflammatory cytokines and acute phase immune molecules (i.e. intraepithelial lymphocytes, IL-1β and TNF-α) (Xie and Nathan, 1994). Nofrarías et al. (2006) and Manzanilla et al. (2006, 2009) reported similar effects with a mixture of carvacrol, capsicum oleoresin and cinnamaldehyde in weaned piglets. Xun et al. (2015) reported that the supplementation with curcumin decreased several inflammation markers (i.e. intraepithelial lymphocytes, IL-1β and TNF-α) and improved the intestinal mucosal barrier integrity in piglets challenged with *E. coli*. In fact, curcumin has been reported to block the synthesis of several proteins involved in the inflammation process (Jobin et al., 1999). Similar anti-inflammatory response was observed in chickens challenged with necrotic enteritis after the supplementation of a mix of capsicum oleoresin and turmeric, resulting in an improved body weight gain (Lee et al., 2013). Overall, the anti-inflammatory effects reduce the energy cost of the inflammatory reaction, allowing to divert energy and other nutrients to production targets, which explains the improved performance. For example, the supplementation of a blend of eucalyptol, p-cymene, linalool, anethole and thymol to hyperprolific sows during gestation and lactation increased the number of piglets born alive and improved the colostrum and milk composition, provided bacteriostatic effects in milk and modified the antioxidant status of sows and their offspring as well as the suckling piglets' gut health gene expression (Reyes-Camacho et al., 2020). These results confirmed the prenatal and postnatal maternal transfer of several volatile compounds of these essential oils that reached the placental fluid and milk.

Another relevant side-effect of the inflammatory process is the excessive production of ROS. If the production of ROS surpasses the capacity of the endogenous antioxidant system to control them, an oxidative stress develops. Some essential oils may play a relevant role in these processes (Frankič et al., 2010; Amorati et al., 2013). For example, thymol and carvacrol contain OH groups that can supply hydrogens to free radicals and delay the formation of peroxides (Farag et al., 1989; Djeridane et al., 2006). However, the free radical scavenging potential of phenolic compounds at the doses fed does not fully explain their antioxidant activity observed *in vivo*. Masella et al. (2005) reported that some phenolic compounds stimulate the production of superoxide dismutase, catalase and glutathione peroxidase. Other essential oils that have shown antioxidant capacity include garlic, curcumin and capsaicin (Manjunatha and Srinivasan, 2006; Mirunalini et al., 2010; Nencini et al., 2010). The capacity of essential oils or their active components to enhance the antioxidant capacity may be relevant in conditions of metabolic stress commonly found in critical periods of the production cycle of cattle.

Unfortunately, the evidence of the effects of essential oils on the immunity of cattle is more limited. The first evidence in dairy cattle was reported by Oh et al. (2013), where the abomasal infusion of garlic oil, curcuma and capsicum oleoresin improved some indicators of adaptive immunity without affecting the intestinal microbial populations in dairy cows. When capsicum oleoresin was supplemented in the diet in a rumen-protected form, there was an increase in neutrophils and eosinophils counts, a decrease in lymphocytes counts and a reduction in the neutrophil to lymphocyte ratio, changes consistent with improvements in the immunity status (Oh et al., 2015). This pro-inflammatory effect of capsicum oleoresin may be beneficial in dairy cows around calving, where immunosuppression is common and there is a high risk of infections. However, when cows were challenged with LPS, capsicum oleoresin resulted in an anti-inflammatory and antioxidant response reflected by a reduction in the concentration of haptoglobin, cortisol and thiobarbituric acid reactive substances (TBARS; Oh, et al., 2017a). A mix of cinnamaldehyde, eugenol, curcumin and carnosic acid reduced somatic cell counts in milk (Hashemzadeh-Cigari et al., 2014), and a mix of rosemary, cinnamon bark, turmeric and clove bud decreased blood TBARS concentration postpartum (Hashemzadeh-Cigari et al., 2015). Overall, reducing inflammation will not only prevent the development of a pathological condition but will also redirect nutrient resources to production purposes.

3.3 Hormonal and metabolic regulation

Plant extracts, in general, and essential oils, in particular, have been used for more than 6000 years in traditional medicine (Giannenas et al., 2020). In addition to their well-known antimicrobial and anti-inflammatory activity, the effects of essential oils on systemic metabolism have also been studied. One such effect is the glucose-insulin regulation (van de Wall et al., 2005; Qin et al., 2010). Furness et al. (2013) indicated that the intestinal mucosa is a sensory organ containing receptors that can be activated by some specific molecules and result in a neuroendocrine response. Some essential oils can interact with these receptors and activate transient receptor potential channels that result in a secondary intracellular response. Then, the intestinal secretory cells release signals that modulate systemic metabolism. For example, some of these receptors release GLP-1 from the intestinal secretory cells that act on the pancreatic β-cells and release insulin (Moran et al., 2014). A different pathway is the release of neuroendocrine substances by enterocytes that activate receptors in the neuronal terminals in the lamina propria of the intestinal wall. These pathways can be used to modulate insulin secretion and/or sensitivity (Holzer, 2011). Most of the preliminary research has been conducted *in vitro* or in animal models. Cinnamon oil and cinnamaldehyde reduced the development of insulin resistance resulting in a hypoglycemic effect (Sheng et al., 2008; Anan et al., 2010; Qin et al., 2010). Garlic and turmeric oils are also hypoglycemic through increasing insulin release and/or modifying insulin sensitivity (Nishiyama et al., 2005; Thomson et al., 2007; Xie et al., 2011). In contrast, capsaicin reduced plasma insulin concentration during a glucose tolerance test without affecting plasma glucose concentrations (van de Wall et al., 2005, 2006). While most peripheral tissues are insulin dependent, the mammary gland utilizes glucose through the insulin-independent glucose transporter 1 (Zhao and Keating, 2007). Therefore, when an insulin resistance condition occurs, the availability of glucose to the mammary gland increases, which enhances milk production. Alternatively, a reduction in insulin secretion will also increase the availability of glucose for the mammary gland. Therefore, milk production may be regulated by insulin resistance or a decrease in insulin secretion. In fact, bovine somatotropin increases milk production by causing insulin resistance that will result in the redirection of plasma glucose to the mammary gland for lactose synthesis (Bauman and Currie, 1980). Similar effects of some essential oils on glucose-insulin metabolism have been reported in cattle. Hashemzadeh-Cigari et al. (2015) reported that feeding a mix of rosemary, cinnamon bark, turmeric and clove bud oils to dairy cows increased insulin sensitivity. Increasing insulin sensitivity will direct glucose toward peripheral tissues and away from the mammary gland, which would not be desirable for dairy cows. In contrast, an encapsulated mix of cinnamaldehyde and diallyl disulfide, an active component of garlic oil, increased insulin plasma concentration without affecting glucose levels, suggesting that insulin sensitivity was decreased (Serbester et al., 2012). Oh et al. (2017a) reported that feeding a rumen-protected capsicum oleoresin decreased serum insulin concentration without affecting glucose levels during a glucose tolerance test, suggesting also an insulin resistance type of effect. It is interesting to observe that while the average effect of rumen modifiers like monensin, essential oils or yeast increase milk production by 2-4% attributed to changes in rumen fermentation (Duffield et al., 2008; Robinson and Erasmus, 2009; Belanche et al., 2020), recent work with rumen-protected capsicum oleoresin resulted in larger increases in milk production. In three trials conducted on dairy cows, supplementation with a rumen-protected capsicum oleoresin resulted in an average increase in milk yield of 6.6% (Oh et al., 2015; Stelwagen et al., 2016; Wall and Bravo, 2016). Because the impact of capsicum oleoresin on rumen microbial fermentation in dairy cattle is negligible (Calsamiglia et al., 2007), these effects can only be attributed to a postruminal effect where glucose is redirected to the mammary gland under the effect of a reduction in insulin concentration. It should be noted that up to 33% of capsaicin may escape rumen microbial degradation (Oh et al., 2016), which may also explain some improvements in performance when fed unprotected. While further research is necessary to elucidate this mechanism of action, these results suggest that there is potential to develop new applications for essential oils to improve production through the modulation of systemic metabolism.

4 Impact of essential oils on dairy cattle performance

The effect of feeding essential oils on dairy cow performance is highly variable. At least part of this variation can be attributed to the essential oil used, the dose, the type of diet, the level of production and the physiological conditions of cows. For example, the supplementation with cinnamaldehyde (Benchaar et al., 2008b; Benchaar, 2016) or a mix of cinnamaldehyde and eugenol (Tekippe et al., 2013) had no effect on milk yield and composition. Feeding oregano or thyme oil and their active compounds also failed to result in improvements in performance (Benchaar, 2019, 2020). In contrast, Wall et al. (2014) reported that a mix of cinnamaldehyde and eugenol increased milk yield in primiparous cows but not in multiparous. More recently, a series of studies supplementing rumen-protected capsicum oleoresin reported consistent increases in milk yield. Stelwagen et al. (2016) reported a 6.6% increase in milk yield in a pasture production system. Other studies also reported that supplementation of rumen-protected capsicum oleoresin improved milk production by 9.1% (Wall and Bravo, 2016), 6.2% (Oh et al., 2015) and 4.4% (Oh et al., 2017a).

Evaluating the impact of supplementing essential oils on dairy cattle performance is complex because there are few studies, they use different active compounds and/or essential oils and experimental conditions are different. Sixteen peer-review papers on experiments designed to evaluate cow performance were selected, and data were subject to a meta-analysis. Small Latin Square designs with a small number of cows and abstracts were excluded. The studies provided 28 different treatment comparisons.

Table 2 Summary of the effects of supplementing essential oils on dairy cattle performance

Item	Average	Difference	SE	*P*-value
DMI, kg/d	22.0	+0.01	0.81	0.84
Milk, kg/d	31.1	+1.3	1.41	0.10
FE, kg milk/kg DMI	1.57	+0.00	0.04	0.29
Fat, %	3.63	−0.07	0.067	0.09
Protein, %	3.04	+0.06	0.040	0.01

On average, milk production tended to increase an average of 1.3 kg/day, with no change in dry matter intake. Milk fat content tended to decrease (−0.07), and milk protein content increased (+0.06) (Table 2). These results are close to the improved performance reported by Belanche et al. (2020), who observed that the supplementation of a specific blend of essential oils containing coriander seed oil, eugenol, geranyl acetate and geraniol for more than 28 days increased milk yield (+3.6%, equivalent to 1.12 kg/d), and fat and protein corrected milk (+4.1%). In contrast, Khiaosa-Ard and Zebeli (2013) conducted a meta-analysis to evaluate the effect of feeding essential oils on dairy cattle performance and reported no effects on milk yield but tended to increase milk protein content.

5 Current problems in the use of essential oils in dairy cattle feeding

In the last 20 years, there has been a significant advancement in the understanding and uses of essential oils in ruminants. While most research has been conducted *in vitro*, there are still numerous questions and opportunities in their application *in vivo* that need to be addressed. Some of these limitations include:

1. Identification of the optimal dose *in vivo*. There is strong evidence that the doses used *in vitro* are much larger than those that have effects *in vivo* and may be related to the effective ratio between the concentration of the active component and the concentration of bacteria in the media (Calsamiglia et al., 2007). The calculation of such dose is also complicated by the incomplete reporting of the actual concentration of the active compounds in essential oils. Therefore, it is recommended to use pure active compounds or to report the actual content of the active component in the essential oils used. Finally, there seems also to be inconsistencies between the doses used in research reports compared with those recommended for commercial purposes. For example, the amount of cinnamaldehyde, eugenol, garlic or capsicum oleoresin used in research

reports has been between 100 mg/cow/day and 2000 mg/cow/day (Benchaar et al., 2008b; Benchaar, 2020; Oh et al., 2013, 2015, 2017a; Wall et al., 2014; Foskolos et al., 2020), while the amount recommended in commercial products are consistently ten times lower. The solution to this problem requires the development of dose-response studies *in vivo*.

2 Potential adaptation of the rumen microbial population to the effects of essential oils has been discussed for a long time (Molero et al., 2004; Cardozo et al., 2004; Castillejos et al., 2007). The use of short-term *in vitro* batch fermentation studies may exclude essential oils that may need a period of adaptation. However, essential oils selected based on this short-term effect may not work after a period of adaptation (Cardozo et al., 2004). Therefore, *in vivo* studies should be conducted with enough adaptation time to study the persistency of the effects. Belanche et al. (2020) recently conducted a meta-analysis to evaluate the impact of supplementing a blend of essential oils on performance and reported that significant increases in milk yield occurred only after 3 weeks of adaptation.

3 There is a challenging opportunity to explore the potential postruminal effects of essential oils on dairy cattle performance and health. Preliminary data obtained with the use of capsicum oleoresin is promising (Oh et al., 2017b). However, if a postruminal effect is sought, the degradation of the active compounds in the rumen and, if appropriate, the development of rumen-protected products, is necessary. There is very limited research in this area. Some products like thymol and carvacrol appear to be relatively resistant to ruminal degradation (>50%); others are intermediate like cineol, menthol and capsicum oleoresin (30–40%), while others like limonene, linalool, pinene and geraniol are mostly degraded (>90%) in the rumen, and only small amounts would reach the postruminal tract (Franz et al., 2010; Oh et al., 2016).

6 Case study

A major problem in the development of an essential oil-based additive is that, in spite of the diversity of products available, only a few and at specific doses are able to provide positive and consistent results (Hart et al., 2008). The first step in the process is a screening test where different essential oils and active compounds are evaluated at different doses using simple *in vitro* batch fermentation studies (Busquet et al., 2006; Castillejos et al., 2007). These studies are relatively inexpensive and fast, and provide a first view of the potential candidates. These preliminary studies pointed out that garlic, cinnamaldehyde, eugenol and anethol were potential candidates to modulate ruminal microbial fermentation. However, the *in vitro* batch fermentation method has several shortcomings that need to be evaluated. We reported that some products may require adaptation before an effect is observed or may lose their effectiveness after the microbial population adapts to the product (Cardozo et al., 2004). For this reason, we followed up the *in vitro* batch fermentation studies with long-term dual flow continuous culture fermentation studies (Busquet et al., 2005a,b; Castillejos et al., 2006; Foskolos et al., 2015). These studies are longer and more expensive but allow us to study the potential impact of adaptation and the study of mechanisms of action by following the dynamics of fermentation over time (Busquet et al., 2005a; Castillejos et al., 2006). Although we had several potential candidates, the most consistent results were obtained with garlic oil and we pursued the study of this oil. Once the product was selected, we aimed to identify the main mechanisms of action on protein and energy metabolism. The study of the dynamics of large peptides, small peptides, free amino acids and ammonia-N concentrations in the ruminal fluid allowed us to suggest that the main action of garlic oil on protein metabolism was associated with a reduction of deamination (Busquet et al., 2005a). Furthermore, the analysis of the changes in the microbial population allowed us to identify species of the genus Prevotella as responsible for such reduction (*P. bryantii* and *P. ruminicola*; Ferme et al., 2004). These bacteria are common in the rumen, have a deaminating activity and are responsible for the reduction of the ammonia-N concentration. On the energy side, the increase in the proportion of propionate and butyrate and the decrease in acetate suggested that the mechanism of action was the inhibition of methanogenesis (Busquet et al., 2005a,b). This hypothesis was confirmed with the reduction of total methane and the ratio of methane to total VFA concentrations after supplementing garlic oil (Busquet et al., 2006). We further proposed that the mechanism of garlic oil was mediated through the reduction of methanogenic archea population. The stability of the cell membrane of archea depends on an isoprenoid molecule synthesized through a metabolic pathway where a hydroxymethylglutaryl coenzyme A (HMG-CoA) reductase is essential. Miller and Wolin (2001) confirmed this hypothesis by reporting a reduction in methane and archea population in a rumen fermentation study using lovastatin and mevastatin, two specific HMG-CoA reductase inhibitors. Finally, we were interested in identifying the main active component in garlic oil. For that purpose, we selected four main compounds of garlic oil (i.e. diallyl sulfide, diallyl disulfide, allicin and allyl mercaptan) and tested their effects using an *in vitro* batch fermentation model followed by a long-term dual flow continuous culture system. Results showed that diallyl disulfide and allyl mercaptan, but not allicin or diallyl sulfide, had effects similar to or stronger than garlic oil (Busquet et al., 2005b). Then, garlic active compounds were mixed with other essential oils (cinnamaldehyde and eugenol) to produce a patented additive (Bravo and Calsamiglia, 2010). While the impact of this essential oil mix requires *in vivo* performance confirmation, this is an example of the process necessary to follow for the development of an effective essential oil-based additive for ruminants.

7 Conclusion and future trends

In vitro research has shown that the use of moderate doses of some essential oils or their active compounds can modify the rumen microbial fermentation profile in a direction supportive of more efficient energy and protein utilization. The reduction observed in the acetate-to-propionate ratio and methane, and the increase in butyrate concentration suggest that their main mechanism of action is the inhibition of methanogenesis. The reduction in protein degradation and deamination is mediated through the reduction of a specific population of rumen microorganisms. The combination of different essential oils needs to be evaluated for potential synergies or antagonisms that may increase the effects or allow the use of lower doses. Among essential oils, garlic, cinnamaldehyde, eugenol and anethol appear to be the best candidates. Other essential oils may be relevant if synergies are found. The evidence for postruminal benefits in ruminants is currently limited to the use of capsicum oleoresin, where improved immunity and glucose utilization for milk production has been demonstrated. However, in this area there is a challenging opportunity to explore the potential benefits of using other essential oils or active compounds on the lower gut microbiome, immunity and/or regulation of metabolism.

8 Where to look for further information

8.1 Further reading

- A good introduction to the use of essential oils in animal nutrition is presented in the book 'Aromatic Plants and Herbs in Animal Nutrition and Health' by Florou-Paneri et al. (2019).
- Some review papers on the effects of essential oils on rumen microbial fermentation are Calsamiglia et al. (2007); Benchaar et al. (2008a); Cobellis et al. (2016).
- A good introduction to the potential use of essential oils for postruminal effects in ruminants is that of Oh et al. (2017b).

8.2 Key journals/conferences

- Journal of Dairy Science publishes high-quality research papers in the area of dairy cattle nutrition, including numerous studies on essential oils.
- Animal Feed Science and Technology publishes high-quality research papers associated with feeding and feed technology, including numerous studies on essential oils.

8.3 Major international projects

- RUMEN UP: https://cordis.europa.eu/project/id/QLK5-CT-2001-00992 was an EU-funded project that explored the effect of a large number of plant extracts on rumen microbial fermentation.
- REPLACE: https://cordis.europa.eu/project/id/506487 was an EU-funded project designed to evaluate plant extracts as candidates to substitute antibiotic growth promoters in animal feeds.

9 References

Amorati, R., Foti, M. C. and Valgimigli, L. (2013). Antioxidant activity of essential oils. *Journal of Agricultural and Food Chemistry* 61(46), 10835–10847. https://doi.org/10.1021/jf403496k.

Anan, F., Masaki, T., Kikuchi, H., Iwao, T., Shimomura, T., Umeno, Y., Eshima, N., Saikawa, T. and Yoshimatsu, H. (2010). Association between plasma high-sensitivity C-reactive protein and insulin resistance and white matter lesions in Japanese type 2 diabetic patients. *Diabetes Research and Clinical Practice* 87(2), 233–239. https://doi.org/10.1016/j.diabres.2009.10.017.

Bach, A., Calsamiglia, S. and Stern, M. D. (2005). Nitrogen metabolism in the rumen. *Journal of Dairy Science* 88 (Suppl. 1), E9–21. https://doi.org/10.3168/jds.S0022-0302(05)73133-7.

Balandrin, M. F. and Klocke, J. A. (1985). Natural plant chemicals: sources of industrial and medicinal materials. *Science* 228(4704), 1154–1160.

Bassolé, I. H. N. and Juliani, H. R. (2012). Essential oils in combination and their antimicrobial properties. *Molecules* 17(4), 3989–4006. https://doi.org/10.3390/molecules17043989.

Bauman, D. E. and Currie, W. B. (1980). Partitioning of nutrients during pregnancy and lactation: a review of mechanisms involving homeostasis and homeorhesis. *Journal of Dairy Science* 63(9), 1514–1529. https://doi.org/10.3168/jds.S0022-0302(80)83111-0.

Belanche, A., Newbold, C. J., Morgavi, D. P., Bach, A., Zweifel, B. and Yáñez-Ruiz, D. R. (2020). A meta-analysis describing the effects of the essential oils blend agolin ruminant on performance, rumen fermentation and methane emissions in dairy cows. *Animals: An Open Access Journal from MDPI* 10(4). https://doi.org/10.3390/ani10040620.

Benchaar, C., Petit, H. V., Berthiaume, R., Ouellet, D. R., Chiquette, J. and Chouinardt, P. Y. (2007). Effects of essential oils on digestion, ruminai fermentation, rumen microbial populations, milk production and milk composition in dairy cows fed alfalfa silage or corn silage. *Journal of Dairy Science* 90(2), 886–897.

Benchaar, C., Calsamiglia, S., Chaves, A. V., Fraser, G. R., Colombatto, D., McAllister, T. A. and Beauchemin, K. A. (2008a). A review of plant-derived essential oils in ruminant nutrition and production. *Animal Feed Science and Technology* 145(1–4), 209–228. https://doi.org/10.1016/j.anifeedsci.2007.04.014.

Benchaar, C., McAllister, T. A. and Chouinard, P. Y. (2008b). Digestion, ruminal fermentation, ciliate protozoal populations and milk production from dairy cows fed cinnamaldehyde, quebracho condensed tannin, or Yucca schidigera saponin extracts. *Journal of Dairy Science* 91(12), 4765–4777. https://doi.org/10.3168/jds.2008-1338.

Benchaar, C. (2016). Diet supplementation with cinnamon oil, cinnamaldehyde, or monensin does not reduce enteric methane production of dairy cows. *Animal* 10(3), 418–425. https://doi.org/10.1017/S175173111500230X.

Benchaar, C. (2019). Feeding oregano oil and its main component carvacrol does not affect ruminal fermentation, nutrient utilization, methane emissions, milk production, or milk fatty acid composition of dairy cows. *Journal of Dairy Science* 103(2), 1516–1527. https://doi.org/10.3168/jds.2019-17230.

Benchaar, C. (2020). Diet supplementation with thyme oil and its main component thymol failed to favorably alter rumen fermentation, improve nutrient utilization, or enhance milk production in dairy cows. *Journal of Dairy Science* 104(1), 324–336. https://doi.org/10.3168/jds.2020-18401.

Bravo, D. and Calsamiglia, S. (2010). *Food Additive Containing Eugenol, Cinnamaldehyde and an Alliaceous Extract*. US patent number: 20120171323.

Busquet, M., Calsamiglia, S., Ferret, A., Cardozo, P. W. and Kamel, C. (2005a). Effects of cinnamaldehyde and garlic oil on rumen microbial fermentation in a dual flow continuous culture. *Journal of Dairy Science* 88(7), 2508–2516. https://doi.org/10.3168/jds.S0022-0302(05)72928-3.

Busquet, M., Calsamiglia, S., Ferret, A., Carro, M. D. and Kamel, C. (2005b). Effect of garlic oil and four of its compounds on rumen microbial fermentation. *Journal of Dairy Science* 88(12), 4393–4404. https://doi.org/10.3168/jds.S0022-0302(05)73126-X.

Busquet, M., Calsamiglia, S., Ferret, A. and Kamel, C. (2005c). Screening for effects of plant extracts and active compounds of plants on dairy cattle rumen microbial fermentation in a continuous culture system. *Animal Feed Science and Technology 123–124 PA*, 597–613. https://doi.org/10.1016/j.anifeedsci.2005.03.008.

Busquet, M., Calsamiglia, S., Ferret, A. and Kamel, C. (2006). Plant extracts affect in vitro rumen microbial fermentation. *Journal of Dairy Science* 89(2), 761–771. https://doi.org/10.3168/jds.S0022-0302(06)72137-3.

Calsamiglia, S., Busquet, M., Cardozo, P. W., Castillejos, L. and Ferret, A. (2007). Invited review: essential oils as modifiers of rumen microbial fermentation. *Journal of Dairy Science* 90(6), 2580–2595. https://doi.org/10.3168/jds.2006-644.

Cardozo, P. W., Calsamiglia, S., Ferret, A. and Kamel, C. (2004). Effects of natural plant extracts on ruminal protein degradation and fermentation profiles in continuous culture. *Journal of Animal Science* 82(11), 3230–3236. https://doi.org/10.2527/2004.82113230x.

Cardozo, P. W., Calsamiglia, S., Ferret, A. and Kamel, C. (2005). Screening for the effects of natural plant extracts at different pH on in vitro rumen microbial fermentation of a high-concentrate diet for beef cattle. *Journal of Animal Science* 83(11), 2572–2579. https://doi.org/10.2527/2005.83112572x.

Castillejos, L., Calsamiglia, S., Ferret, A. and Losa, R. (2005). Effect a specific blend of essential oil compounds and the type of diet on rumen microbial fermentation and nutrient flow from a continuous culture system. *Animal Feed Science and Technology* 119(1–2), 29–41. https://doi.org/10.1016/j.anifeedsci.2004.12.008.

Castillejos, L., Calsamiglia, S. and Ferret, A. (2006). Effect of essential oil active compounds on rumen microbial fermentation and nutrient flow in in vitro systems. *Journal of Dairy Science* 89(7), 2649–2658. https://doi.org/10.3168/jds.S0022-0302(06)72341-4.

Castillejos, L., Calsamiglia, S., Ferret, A. and Losa, R. (2007). Effect of dose and adaptation time of a specific blend of essential oil compounds on rumen fermentation. *Animal Feed Science and Technology* 132(3–4), 186–201. https://doi.org/10.1016/j.anifeedsci.2006.03.023.

Chao, S. C., Young, D. G. and Oberg, C. J. (2000). Screening for inhibitory activity of essential oils on selected bacteria, fungi and viruses. *Journal of Essential Oil Research* 12(5), 639–649. https://doi.org/10.1080/10412905.2000.9712177.

Chaves, A. V., He, M. L., Yang, W. Z., Hristov, A. N., McAllister, T. A. and Benchaar, C. (2008). Effects of essential oils on proteolytic, deaminative and methanogenic activities of mixed ruminal bacteria. *Canadian Journal of Animal Science* 88(1), 117–122. https://doi.org/10.4141/CJAS07061.

Cimanga, K., Apers, S., De Bruyne, T., Van Miert, S., Hermans, N., Totté, J., Pieters, L., Vlietinck, A. J., Kambu, K. and Tona, L. (2002). Chemical composition and antifungal activity of essential oils of some aromatic medicinal plants growing in the Democratic Republic of Congo. *Journal of Essential Oil Research* 14(5), 382–387. https://doi.org/10.1080/10412905.2002.9699894.

Cobellis, G., Trabalza-Marinucci, M. and Yu, Z. (2016). Critical evaluation of essential oils as rumen modifiers in ruminant nutrition: a review. *Science of the Total Environment* 545–546, 556–568. https://doi.org/10.1016/j.scitotenv.2015.12.103.

Davidson, P. M. and Naidu, A. S. (2000). Phyto-phenols. Pages 265–293. In: *Natural Food Antimicrobial Systems*. A. S. Naidu (Eds). CRC Press, Boca Raton, FL

De Rosa, M., Gambacorta, A. and Gliozzi, A. (1986). Structure, biosynthesis and physicochemical properties of archaebacterial lipids. *Microbiological Reviews* 50(1), 70–80. https://doi.org/10.1128/mr.50.1.70-80.1986.

Djeridane, A., Yousfi, M., Nadjemi, B., Boutassouna, D., Stocker, P. and Vidal, N. (2006). Antioxidant activity of some Algerian medicinal plants extracts containing phenolic compounds. *Food Chemistry* 97(4), 654–660. https://doi.org/10.1016/j.foodchem.2005.04.028.

Dorman, H. J. D. and Deans, S. G. (2000). Antimicrobial agents from plants: antibacterial activity of plant volatile oils. *Journal of Applied Microbiology* 88(2), 308–316. https://doi.org/10.1046/j.1365-2672.2000.00969.x.

Duffield, T. F., Rabiee, A. R. and Lean, I. J. (2008). A meta-analysis of the impact of monensin in lactating dairy cattle. Part 2. Production effects. *Journal of Dairy Science* 91(4), 1347–1360. https://doi.org/10.3168/jds.2007-0608.

Fandiño, I., Fernandez-Turren, G., Ferret, A., Moya, D., Castillejos, L. and Calsamiglia, S. (2020). Exploring additive, synergistic or antagonistic effects of natural plant extracts on in vitro beef feedlot-type rumen microbial fermentation conditions. *Animals: An Open Access Journal from MDPI* 10(1). https://doi.org/10.3390/ani10010173.

Farag, R. S., Daw, Z. Y., Hewedi, F. M. and El-Baroty, G. S. A. (1989). Antimicrobial activity of some Egyptian spice essential oils. *Journal of Food Protection* 52(9), 665–667. https://doi.org/10.4315/0362-028X-52.9.665.

Ferme, D., Banjac, M., Calsamiglia, S., Busquet, M., Kamel, C. and Avguštin, G. (2004). The effects of plant extracts on microbial community structure in a rumen-simulating continuous-culture system as revealed by molecular profiling. *Folia Microbiologica* 49(2), 151–155. https://doi.org/10.1007/BF02931391.

Florou-Paneri, P., Christaki, E. and Giannenas, I. (2019). *Aromatic Plants and Herbs in Animal Nutrition and Health*. Academic Press.

Foskolos, A., Siurana, A., Rodriguez-Prado, M., Ferret, A., Bravo, D. and Calsamiglia, S. (2015). The effects of a garlic oil chemical compound, propyl-propane thiosulfonate, on ruminal fermentation and fatty acid outflow in a dual-flow continuous culture system. *Journal of Dairy Science* 98(8), 5482–5491. https://doi.org/10.3168/jds.2014-8674.

Foskolos, A., Ferret, A., Siurana, A., Castillejos, L. and Calsamiglia, S. (2020). Effects of capsicum oleoresin and propyl-propane thiosulfonate on rumen fermentation, digestion, and milk production and composition in dairy cows. *Animals: An Open Access Journal from MDPI* 10(5), 859. https://doi.org/10.3390/ani10050859.

Frankič, T., Levart, A. and Salobir, J. (2010). The effect of vitamin E and plant extract mixture composed of carvacrol, cinnamaldehyde and capsaicin on oxidative stress induced by high PUFA load in young pigs. *Animal* 4(4), 572–578. https://doi.org/10.1017/S1751731109991339.

Franz, C., Baser, K. and Windisch, W. (2010). Essential oils and aromatic plants in animal feeding–A European perspective. A review. *Flavour and Fragrance Journal* 25(5), 327–340. https://doi.org/10.1002/ffj.1967.

Furness, J. B., Rivera, L. R., Cho, H. J., Bravo, D. M. and Callaghan, B. (2013). The gut as a sensory organ. *Nature Reviews. Gastroenterology and Hepatology* 10(12), 729–740. https://doi.org/10.1038/nrgastro.2013.180.

Galucci, M. N., Oliva, M., Casero, C., Dambolena, J., Luna, A., Zygadlo, J. and Demo, M. (2009). Antimicrobial combined action of terpenes against the food-borne microorganisms *Escherichia coli*, *Staphylococcus aureus* and *Bacillus cereus*. *Flavour and Fragrance Journal* 24(6), 348–354. https://doi.org/10.1002/ffj.1948.

Garcia-Lopez, P. M., Kung, L. and Odom, J. M. (1996). In vitro inhibition of microbial methane production by 9,10-anthraquinone. *Journal of Animal Science* 74(9), 2276–2284. https://doi.org/10.2527/1996.7492276x.

Gebhardt, R. and Beck, H. (1996). Differential inhibitory effects of garlic-derived organosulfur compounds on cholesterol biosynthesis in primary rat hepatocyte cultures. *Lipids* 31(12), 1269–1276. https://doi.org/10.1007/BF02587912.

Gershenzon, J. and Croteau, R. (1991a). Regulation of monoterpene biosynthesis in higher plants. Pages 99–159. In: *Biochemistry of the Mevalonic Acid Pathway to Terpenoids* Towers, G. H. N. and Stafford, H. A. (Eds). Plenum Press, New York, NY.

Gershenzon, J. and Croteau, R. (1991b). Terpenoids. Pages 165–219. In: *Herbivores: Their Interactions with Secondary Plant Metabolites* (vol. 1) Rosenthal, G. A. and Berenbaum, M. R. (Eds). Academic Press, San Diego, CA.

Giannenas, I., Sidiropoulou, E., Bonos, E., Christaki, E. and Florou-Paneri, P. (2020). The history of herbs, medicinal and aromatic plants and their extracts: past, current situation and future perspectives. In: *Feed Additives: Aromatic Plants and Herbs in Animal Nutrition and Health*. Elsevier Inc. https://doi.org/10.1016/B978-0-12-814700-9.00001-7

Griffin, S. G., Wyllie, S. G., Markham, J. L. and Leach, D. N. (1999). The role of structure and molecular properties of terpenoids in determining their antimicrobial activity. *Flavour and Fragrance Journal* 14(5), 322–332. https://doi.org/10.1002/(SICI)1099-1026(199909/10)14:5<322::AID-FFJ837>3.0.CO;2-4.

Gustafson, R. H. and Bowen, R. E. (1997). Antibiotic use in animal agriculture. *Journal of Applied Microbiology* 83(5), 531–541. https://doi.org/10.1046/j.1365-2672.1997.00280.x.

Hart, K. J., Yáñez-Ruiz, D. R., Duval, S. M., McEwan, N. R. and Newbold, C. J. (2008). Plant extracts to manipulate rumen fermentation. *Animal Feed Science and Technology* 147(1-3), 8-35. https://doi.org/10.1016/j.anifeedsci.2007.09.007.

Hashemzadeh-Cigari, F., Khorvash, M., Ghorbani, G. R., Kadivar, M., Riasi, A. and Zebeli, Q. (2014). Effects of supplementation with a phytobiotics-rich herbal mixture on performance, udder health and metabolic status of Holstein cows with various levels of milk somatic cell counts. *Journal of Dairy Science* 97(12), 7487-7497. https://doi.org/10.3168/jds.2014-7989.

Hashemzadeh-Cigari, F., Ghorbani, G. R., Khorvash, M., Riasi, A., Taghizadeh, A. and Zebeli, Q. (2015). Supplementation of herbal plants differently modulated metabolic profile, insulin sensitivity and oxidative stress in transition dairy cows fed various extruded oil seeds. *Preventive Veterinary Medicine* 118(1), 45-55. https://doi.org/10.1016/j.prevetmed.2014.10.013.

Helander, I. M., Alakomi, H. L., Latva-Kala, K., Mattila-Sandholm, T., Pol, I., Smid, E. J., Gorris, L. G. M. and Von Wright, A. (1998). Characterization of the action of selected essential oil components on Gram-negative bacteria. *Journal of Agricultural and Food Chemistry* 46(9), 3590-3595. https://doi.org/10.1021/jf980154m.

Hino, T. and Russell, J. B. (1985). Effect of reducing-equivalent disposal and NADH/NAD on deamination of amino acids by intact rumen microorganisms and their cell extracts. *Applied and Environmental Microbiology* 50(6), 1368-1374. https://doi.org/10.1128/aem.50.6.1368-1374.1985.

Hoffmann, C. and Evans, A. C. (1911). The use of spices as preservatives. *Journal of Industrial and Engineering Chemistry* 3(11), 835-838.

Holzer, P. (2011). TRP channels in the digestive system. *Current Pharmaceutical Biotechnology* 12(1), 24-34.

Horošová, K., Bujňáková, D. and Kmeť, V. (2006). Effect of oregano essential oil on chicken Lactobacilli and *E. coli*. *Folia Microbiologica* 51(4), 278-280. https://doi.org/10.1007/BF02931812.

Horst, E. A., Mayorga, E. J., Rodriguez-Jimenez, S., Abeyta, A., Goetz, B. M., Carta, S., Al-Qaisi, M., Kvidera, S. K. and Baumgard, L. H. (2019). Causes and metabolic consequences of leaky gut. *Journal of Chemical Information and Modeling* 53(9), 1689-1699.

Hristov, A. N., Ropp, J. K., Zaman, S. and Melgar, A. (2008). Effects of essential oils on in vitro ruminal fermentation and ammonia release. *Animal Feed Science and Technology* 144(1-2), 55-64. https://doi.org/10.1016/j.anifeedsci.2007.09.034.

Huang, Y., Yoo, J. S., Kim, H. J., Wang, Y., Chen, Y. J., Cho, J. H. and Kim, I. H. (2010). Effects of dietary supplementation of modified zinc oxide on growth performance, nutrient digestibility, blood profiles, fecal microbial shedding and fecal score in weanling pigs. *Animal Science Journal* 86(6), 617-623. https://doi.org/10.1111/asj.12329.

Jobin, C., Bradham, C. A., Russo, M. P., Juma, B., Narula, A. S., Brenner, D. A. and Sartor, R. B. (1999). Curcumin blocks cytokine-mediated NF-kappa B activation and proinflammatory gene expression by inhibiting inhibitory factor I-kappa B kinase activity. *Journal of Immunology* 163(6), 3474-3483. http://www.ncbi.nlm.nih.gov/pubmed/10477620.

Joch, M., Cermak, L., Hakl, J., Hucko, B., Duskova, D. and Marounek, M. (2016). In vitro screening of essential oil active compounds for manipulation of rumen fermentation and methane mitigation. *Asian-Australasian Journal of Animal Sciences* 29(7), 952-959. https://doi.org/10.5713/ajas.15.0474.

Joch, M., Kudrna, V. and Hučko, B. (2017). Effects of geraniol and camphene on in vitro rumen fermentation and methane production. *Scientia Agriculturae Bohemica* 48(2), 63–69. https://doi.org/10.1515/sab-2017-0012.

Juven, B. J., Kanner, J., Schved, F. and Weisslowicz, H. (1994). Factors that interact with the antibacterial action of thyme essential oil and its active constituents. *Journal of Applied Bacteriology* 76(6), 626–631. https://doi.org/10.1111/j.1365-2672.1994.tb01661.x.

Khiaosa-Ard, R. and Zebeli, Q. (2013). Meta-analysis of the effects of essential oils and their bioactive compounds on rumen fermentation characteristics and feed efficiency in ruminants. *Journal of Animal Science* 91(4), 1819–1830. https://doi.org/10.2527/jas.2012-5691.

Klevenhusen, F., Muro-Reyes, A., Khiaosa-ard, R., Metzler-Zebeli, B. U. and Zebeli, Q. (2012). A meta-analysis of effects of chemical composition of incubated diet and bioactive compounds on in vitro ruminal fermentation. *Animal Feed Science and Technology* 176(1–4), 61–69. https://doi.org/10.1016/j.anifeedsci.2012.07.008.

Kung, L., Smith, K. A., Smagala, A. M., Endres, K. M., Bessett, C. A., Ranjit, N. K. and Yaissle, J. (2003). Effects of 9,10 anthraquinone on ruminal fermentation, total-tract digestion and blood metabolite concentrations in sheep. *Journal of Animal Science* 81(1), 323–328. https://doi.org/10.2527/2003.811323x.

Lang, A., Lahav, M., Sakhnini, E., Barshack, I., Fidder, H. H., Avidan, B., Bardan, E., Hershkoviz, R., Bar-Meir, S. and Chowers, Y. (2004). Allicin inhibits spontaneous and TNF-α induced secretion of proinflammatory cytokines and chemokines from intestinal epithelial cells. *Clinical Nutrition* 23(5), 1199–1208. https://doi.org/10.1016/j.clnu.2004.03.011.

Lee, S. H., Lillehoj, H. S., Jang, S. I., Lee, K. W., Bravo, D. and Lillehoj, E. P. (2011). Effects of dietary supplementation with phytonutrients on vaccine-stimulated immunity against infection with Eimeria tenella. *Veterinary Parasitology* 181(2–4), 97–105. https://doi.org/10.1016/j.vetpar.2011.05.003.

Lee, S. H., Lillehoj, H. S., Jang, S. I., Lillehoj, E. P., Min, W. and Bravo, D. M. (2013). Dietary supplementation of young broiler chickens with Capsicum oleoresin and turmeric oleoresins increases resistance to necrotic enteritis. *British Journal of Nutrition* 110(5), 840–847. https://doi.org/10.1017/S0007114512006083.

Li, S. Y., Ru, Y. J., Liu, M., Xu, B., Péron, A. and Shi, X. G. (2012). The effect of essential oils on performance, immunity and gut microbial population in weaner pigs. *Livestock Science* 145(1–3), 119–123. https://doi.org/10.1016/j.livsci.2012.01.005.

Liu, Y., Song, M., Che, T. M., Bravo, D. and Pettigrew, J. E. (2012). Anti-inflammatory effects of several plant extracts on porcine alveolar macrophages in vitro. *Journal of Animal Science* 90(8), 2774–2783. https://doi.org/10.2527/jas.2011-4304.

Liu, Y., Song, M., Che, T. M., Almeida, J. A., Lee, J. J., Bravo, D., Maddox, C. W. and Pettigrew, J. E. (2013). Dietary plant extracts alleviate diarrhea and alter immune responses of weaned pigs experimentally infected with a pathogenic Escherichia coli. *J. Anim. Sci.* 91:5294–5306. doi: 10.2527/jas.2012-6194

MacMicking, J., Xie, Q. W. and Nathan, C. (1997). Nitric oxide and macrophage function. *Annual Review of Immunology* 15(1), 323–350. https://doi.org/10.1146/annurev.immunol.15.1.323.

Manjunatha, H. and Srinivasan, K. (2006). Protective effect of dietary curcumin and capsaicin on induced oxidation of low-density lipoprotein, iron-induced hepatotoxicity and carrageenan-induced inflammation in experimental rats. *FEBS Journal* 273(19), 4528–4537. https://doi.org/10.1111/j.1742-4658.2006.05458.x.

Manzanilla, E. G., Nofrarías, M., Anguita, M., Castillo, M., Perez, J. F., Martín-Orúe, S. M., Kamel, C. and Gasa, J. (2006). Effects of butyrate, avilamycin and a plant extract combination on the intestinal equilibrium of early-weaned pigs. *Journal of Animal Science* 84(10), 2743–2751. https://doi.org/10.2527/jas.2005-509.

Manzanilla, E. G., Pérez, J. F., Martín, M., Blandón, J. C., Baucells, F., Kamel, C. and Gasa, J. (2009). Dietary protein modifies effect of plant extracts in the intestinal ecosystem of the pig at weaning. *Journal of Animal Science* 87(6), 2029–2037. https://doi.org/10.2527/jas.2008-1210.

Mao, S., Zhang, M., Liu, J. and Zhu, W. (2015). Characterizing the bacterial microbiota across the gastrointestinal tracts of dairy cattle: membership and potential function. *Scientific Reports* 5, 16116. https://doi.org/10.1038/srep16116.

Martin, S. A. and Macy, J. M. (1985). Effects of monensin, pyromellitic diimide and 2-bromoethanesulfonic acid on rumen fermentation in vitro. *Journal of Animal Science* 60(2), 544–550. https://doi.org/10.2527/jas1985.602544x.

Masella, R., Di Benedetto, R., Varì, R., Filesi, C. and Giovannini, C. (2005). Novel mechanisms of natural antioxidant compounds in biological systems: involvement of glutathione and glutathione-related enzymes. *Journal of Nutritional Biochemistry* 16(10), 577–586. https://doi.org/10.1016/j.jnutbio.2005.05.013.

McCall, M. R. and Frei, B. (1999). Can antioxidant vitamins materially reduce oxidative damage in humans? *Free Radical Biology and Medicine* 26(7–8), 1034–1053. https://doi.org/10.1016/S0891-5849(98)00302-5.

McIntosh, F. M., Williams, P., Losa, R., Wallace, R. J., Beever, D. A. and Newbold, C. J. (2003). Effects of essential oils on ruminal microorganisms and their protein metabolism. *Applied and Environmental Microbiology* 69(8), 5011–5014. https://doi.org/10.1128/AEM.69.8.5011-5014.2003.

Miller, T. L. and Wolin, M. J. (2001). Inhibition of growth of methane-producing bacteria of the ruminant forestomach by hydroxymethylglutaryl~SCoA reductase inhibitors. *Journal of Dairy Science* 84(6), 1445–1448. https://doi.org/10.3168/jds.S0022-0302(01)70177-4.

Mirunalini, S., Dhamodharan, G. and Karthishwaran, K. (2010). A natural wonder drug helps to prevent cancer: garlic oil. *Notulae Scientia Biologicae* 2(1), 14–19. https://doi.org/10.15835/nsb213593.

Molero, R., Ibars, M., Calsamiglia, S., Ferret, A. and Losa, R. (2004). Effects of a specific blend of essential oil compounds on dry matter and crude protein degradability in heifers fed diets with different forage to concentrate ratios. *Animal Feed Science and Technology* 114(1–4), 91–104. https://doi.org/10.1016/j.anifeedsci.2003.11.011.

Montoya, D., D'Angelo, M., Martín-Orúe, S. M., Rodríguez-Sorrento, A., Saladrigas-García, M., Araujo, C., Chabrillat, T., Kerros, S. and Castillejos, L. (2021). Effectiveness of two plant-based in-feed additives against an Escherichia coli F4 oral challenge in weaned piglets. *Animals: An Open Access Journal from MDPI* 11(7), 2024. https://doi.org/10.3390/ani11072024.

Moran, B. M., Abdel-Wahab, Y. H. A., Flatt, P. R. and McKillop, A. M. (2014). Activation of GPR119 by fatty acid agonists augments insulin release from clonal β-cells and isolated pancreatic islets and improves glucose tolerance in mice. *Biological Chemistry* 395(4), 453–464. https://doi.org/10.1515/hsz-2013-0255.

Mourão, J. L., Pinheiro, V., Alves, A., Guedes, C. M., Pinto, L., Saavedra, M. J., Spring, P. and Kocher, A. (2006). Effect of mannan oligosaccharides on the performance, intestinal morphology and cecal fermentation of fattening rabbits. *Animal Feed Science and Technology* 126(1-2), 107-120. https://doi.org/10.1016/j.anifeedsci.2005.06.009.

Muhl, A. and Liebert, F. (2007). Growth and parameters of microflora in intestinal and faecal samples of piglets due to application of a phytogenic feed additive. *Journal of Animal Physiology and Animal Nutrition* 91(9-10), 411-418. https://doi.org/10.1111/j.1439-0396.2006.00668.x.

Nagaraja, T. G., Newbold, C. J., van Nevel, C. J. and Demeyer, D. I. (1997). Manipulation of rumen fermentation. Pages 524-632. In: *The Rumen Microbial Ecosystem* Hobson, P. N. and Stewart, C. S. (Eds). Blackie Academic and Professional, London.

Nanon, A., Suksombat, W. and Yang, W. Z. (2015). Original Article 168 Nanon A. et al. / Thai J Vet Med. 45(2), 167-180. Available at: http://www.thaiscience.info/journals/Article/TJVM/10961970.pdf.

Nencini, C., Franchi, G. G. and Micheli, L. (2010). Cardiovascular receptor binding affinity of aqueous extracts from Allium species. *International Journal of Food Sciences and Nutrition* 61(4), 433-439. https://doi.org/10.3109/09637481003591608.

Nevius, E., Srivastava, P. K. and Basu, S. (2012). Oral ingestion of capsaicin, the pungent component of chili pepper, enhances a discreet population of macrophages and confers protection from autoimmune diabetes. *Mucosal Immunology* 5(1), 76-86. https://doi.org/10.1038/mi.2011.50.

Nishiyama, T., Mae, T., Kishida, H., Tsukagawa, M., Mimaki, Y., Kuroda, M., Sashida, Y., Takahashi, K., Kawada, T., Nakagawa, K. and Kitahara, M. (2005). Curcuminoids and sesquiterpenoids in turmeric (Curcuma longa L.) Suppress an increase in blood glucose level in type 2 diabetic KK-Aγ mice. *Journal of Agricultural and Food Chemistry* 53(4), 959-963. https://doi.org/10.1021/jf0483873.

Nofrarías, M., Manzanilla, E. G., Pujols, J., Gibert, X., Majó, N., Segalés, J. and Gasa, J. (2006). Effects of spray-dried porcine plasma and plant extracts on intestinal morphology and on leukocyte cell subsets of weaned pigs. *Journal of Animal Science* 84(10), 2735-2742. https://doi.org/10.2527/jas.2005-414.

O'Gara, E. A., Hill, D. J. and Maslin, D. J. (2000). Activities of garlic oil, garlic powder and their diallyl constituents against Helicobacter pylori. *Applied and Environmental Microbiology* 66(5), 2269-2273. https://doi.org/10.1128/AEM.66.5.2269-2273.2000.

Oh, J., Hristov, A. N., Lee, C., Cassidy, T., Heyler, K., Varga, G. A., Pate, J., Walusimbi, S., Brzezicka, E., Toyokawa, K., Werner, J., Donkin, S. S., Elias, R., Dowd, S. and Bravo, D. (2013). Immune and production responses of dairy cows to postruminal supplementation with phytonutrients. *Journal of Dairy Science* 96(12), 7830-7843. https://doi.org/10.3168/jds.2013-7089.

Oh, J., Giallongo, F., Frederick, T., Pate, J., Walusimbi, S., Elias, R. J., Wall, E. H., Bravo, D. and Hristov, A. N. (2015). Effects of dietary Capsicum oleoresin on productivity and immune responses in lactating dairy cows. *Journal of Dairy Science* 98(9), 6327-6339. https://doi.org/10.3168/jds.2014-9294.

Oh, J., Bravo, D. M., Wall, E. H. and Hristov, A. N. (2016). Rumen disappearance of capsaicin and dihydrocapsaicin in lactating dairy cows. *Journal of Dairy Science* 99(E-Suppl. 1), 788.

Oh, J., Harper, M., Giallongo, F., Bravo, D. M., Wall, E. H. and Hristov, A. N. (2017a). Effects of rumen-protected Capsicum oleoresin on productivity and responses to a glucose tolerance test in lactating dairy cows. *Journal of Dairy Science* 100(3), 1888-1901. https://doi.org/10.3168/jds.2016-11665.

Oh, J., Wall, E. H., Bravo, D. M. and Hristov, A. N. (2017b). Host-mediated effects of phytonutrients in ruminants: A review. *Journal of Dairy Science* 100(7), 5974–5983. https://doi.org/10.3168/jds.2016-12341.

Oviedo-Rondón, E. O., Hume, M. E., Hernández, C. and Clemente-Hernández, S. (2006). Intestinal microbial ecology of broilers vaccinated and challenged with mixed Eimeria species and supplemented with essential oil blends. *Poultry Science* 85(5), 854–860. https://doi.org/10.1093/ps/85.5.854.

Prescott, L. M., Harley, J. P.and Klein, D. A. (2004). Control de microorganismos por agentes físicos y químicos. Pages 145–162. In: *Microbiología*. McGraw-Hill-Interamericana de España, Madrid, Spain.

Qin, B., Polansky, M. M., Harry, D. and Anderson, R. A. (2010). Green tea polyphenols improve cardiac muscle mrna and protein levels of signal pathways related to insulin and lipid metabolism and inflammation in insulin-resistant rats. *Molecular Nutrition and Food Research* 54 (Suppl. 1), S14–S23. https://doi.org/10.1002/mnfr.200900306.

Reddy, L., Odhav, B. and Bhoola, K. D. (2003). Natural products for cancer prevention: A global perspective. *Pharmacology and Therapeutics* 99(1), 1–13. https://doi.org/10.1016/S0163-7258(03)00042-1.

Reuter, H. D., Koch, J. P. and Lawson, L. (1996). Therapeutic effects and applications of garlic and its preparations. Pages 135–212. In: *Garlic: the Science and Therapeutic Application of Allium sativum L. and Related Species* Koch, H. P. and Lawson, L. D. (Eds). Williams and Wilkins, Baltimore, MD.

Reyes-Camacho, D., Vinyeta, E., Pérez, J. F., Aumiller, T., Criado, L., Palade, L. M., Taranu, I., Folch, J. M., Calvo, M. A., Van der Klis, J. D. and Solà-Oriol, D. (2020). Phytogenic actives supplemented in hyperprolific sows: effects on maternal transfer of phytogenic compounds, colostrum and milk features, performance and antioxidant status of sows and their offspring, and piglet intestinal gene expression. *Journal of Animal Science* 98(1), 1–03. https://doi.org/10.1093/jas/skz390.

Richardson, L. F., Raun, A. P., Potter, E. L., Cooley, C. O. and Rathmacher, R. P. (1976). Effect of monensin on rumen fermentation in vitro and in vivo. *Journal of Animal Science* 43(3), 657–664.

Robinson, P. H. and Erasmus, L. J. (2009). Effects of analyzable diet components on responses of lactating dairy cows to Saccharomyces cerevisiae based yeast products: A systematic review of the literature. *Animal Feed Science and Technology* 149(3–4), 185–198. https://doi.org/10.1016/j.anifeedsci.2008.10.003.

Ross, Z. M., O'Gara, E. A., Hill, D. J., Sleightholme, H. V. and Maslin, D. J. (2001). Antimicrobial properties of garlic oil against human enteric bacteria: evaluation of methodologies and comparisons with garlic oil sulfides and garlic powder. *Applied and Environmental Microbiology* 67(1), 475–480. https://doi.org/10.1128/AEM.67.1.475-480.2001.

Russell, J. B., O'Connor, J. D., Fox, D. G., Van Soest, P. J. and Sniffen, C. J. (1992). A net carbohydrate and protein system for evaluating cattle diets: I. Ruminal fermentation. *Journal of Animal Science* 70(11), 3551–3561. http://www.ncbi.nlm.nih.gov/pubmed/1459918.

Serbester, U., Çinar, M., Ceyhan, A., Erdem, H., Görgülü, M., Kutlu, H. R., Çelik, L. B., Yücelt, O. and Cardozo, P. W. (2012). Effect of essential oil combination on performance, milk composition, blood parameters and pregnancy rate in early lactating dairy cows during heat exposure. *Journal of Animal and Plant Sciences* 22(3), 556–563.

Sheng, X., Zhang, Y., Gong, Z., Huang, C. and Zang, Y. Q. (2008). Improved insulin resistance and lipid metabolism by cinnamon extract through activation of peroxisome proliferator-activated receptors. *PPAR Research* 2008, 581348. https://doi.org/10.1155/2008/581348.

Sivropoulou, A., Papanikolaou, E., Nikolaou, C., Kokkini, S., Lanaras, T. and Arsenakis, M. (1996). Antimicrobial and cytotoxic activities of Origanum essential oils. *Journal of Agricultural and Food Chemistry* 44(5), 1202–1205. https://doi.org/10.1021/jf950540t.

Stelwagen, K., Wall, E. H. and Bravo, D. M. (2016). 1394 Dose-dependent effects of a sensory additive on the eating behavior of TMR-fed dairy cows. *Journal of Animal Science* 94(suppl_5), 674–675. https://doi.org/10.2527/jam2016-1394.

Takano, F., Yamaguchi, M., Takada, S., Shoda, S., Yahagi, N., Takahashi, T. and Ohta, T. (2007). Capsicum oleoresin ethanol extracts and capsaicin enhance interleukin-2 and interferon-gamma production in cultured murine Peyer's patch cells ex vivo. *Life Sciences* 80(17), 1553–1563. https://doi.org/10.1016/j.lfs.2007.01.031.

Tamminga, S. (1996). A review on environmental impacts of nutritional strategies in ruminants. *Journal of Animal Science* 74(12), 3112–3124. https://doi.org/10.2527/1996.74123112x.

Tekippe, J. A., Tacoma, R., Hristov, A. N., Lee, C., Oh, J., Heyler, K. S., Cassidy, T. W., Varga, G. A. and Bravo, D. (2013). Effect of essential oils on ruminal fermentation and lactation performance of dairy cows. *Journal of Dairy Science* 96(12), 7892–7903. https://doi.org/10.3168/jds.2013-7128.

Temmar, R., Rodríguez-Prado, M., Forgeard, G., Rougier, C. and Calsamiglia, S. (2021). Interactions among natural active ingredients to improve the efficiency of rumen fermentation in vitro. *Animals: An Open Access Journal from MDPI* 11(5), 1-12. https://doi.org/10.3390/ani11051205.

Thomson, M., Al-Amin, Z. M., Al-Qattan, K. K., Shaban, L. H. and Ali, M. (2007). Anti-diabetic and hypolipidaemic properties of garlic (Allium sativum) in streptozotocin-induced diabetic rats. *International Journal of Diabetes and Metabolism* 15(3), 108–115.

Tiihonen, K., Kettunen, H., Bento, M. H. L., Saarinen, M., Lahtinen, S., Ouwehand, A. C., Schulze, H. and Rautonen, N. (2010). The effect of feeding essential oils on broiler performance and gut microbiota. *British Poultry Science* 51(3), 381–392. https://doi.org/10.1080/00071668.2010.496446.

Trouillas, P., Calliste, C. A., Allais, D. P., Simon, A., Marfak, A., Delage, C. and Duroux, J. L. (2003). Antioxidant, anti-inflammatory and antiproliferative properties of sixteen water plant extracts used in the Limousin countryside as herbal teas. *Food Chemistry* 80(3), 399–407. https://doi.org/10.1016/S0308-8146(02)00282-0.

Ultee, A., Bennik, M. H. J. and Moezelaar, R. (2002). The phenolic hydroxyl group of carvacrol is essential for action against the food-borne pathogen Bacillus cereus. *Applied and Environmental Microbiology* 68(4), 1561–1568. https://doi.org/10.1128/AEM.68.4.1561-1568.2002.

Ultee, A., Kets, E. P. W. and Smid, E. J. (1999). Mechanisms of action of carvacrol on the food-borne pathogen Bacillus cereus. *Applied and Environmental Microbiology* 65(10), 4606–4610. https://doi.org/10.1128/AEM.65.10.4606-4610.1999.

van de Wall, E. H. E. M., Gram, D. X., Strubbe, J. H., Scheurink, A. J. W. and Koolhaas, J. M. (2005). Ablation of capsaicin-sensitive afferent nerves affects insulin response during an intravenous glucose tolerance test. *Life Sciences* 77(11), 1283–1292. https://doi.org/10.1016/j.lfs.2005.03.011.

van de Wall, E. H. E. M., Wielinga, P. Y., Strubbe, J. H. and van Dijk, G. (2006). Neonatal capsaicin causes compensatory adjustments to energy homeostasis in rats. *Physiology and Behavior* 89(1), 115-121. https://doi.org/10.1016/j.physbeh.2006.03.018.

Van Nevel, C. J. and Demeyer, D. I. (1988). Manipulation of rumen fermentation. Pages 387-443. In: *The Rumen Microbial Ecosystem* Hobson, P. N. (Ed.). Elsevier Applied Science, New York, NY.

Wall, E. H., Doane, P. H., Donkin, S. S. and Bravo, D. (2014). The effects of supplementation with a blend of cinnamaldehyde and eugenol on feed intake and milk production of dairy cows. *Journal of Dairy Science* 97(9), 5709-5717. https://doi.org/10.3168/jds.2014-7896.

Wall, E. H. and Bravo, D. M. (2016). Supplementation with a blend of capsicum oleoresin and artificial sweetener alters milk yield and nutrient partitioning in lactating dairy cows. *Journal of Animal Science* 94(suppl_5), 754-755. https://doi.org/10.2527/jam2016-1553.

Wallace, R. J., McEwan, N. R., McIntosh, F. M., Teferedegne, B. and Newbold, C. J. (2002). Natural products as manipulators of rumen fermentation. *Asian-Australasian Journal of Animal Sciences* 15(10), 1458-1468. https://doi.org/10.5713/ajas.2002.1458.

Weimer, P. J. (1998). Manipulating ruminal fermentation: A microbial ecological perspective. *Journal of Animal Science* 76(12), 3114-3122. https://doi.org/10.2527/1998.76123114x.

Weimer, P. J. (2015). Redundancy, resilience and host specificity of the ruminal microbiota: implications for engineering improved ruminal fermentations. *Frontiers in Microbiology* 6(Apr), 1-16. https://doi.org/10.3389/fmicb.2015.00296.

Xie, Q. W. and Nathan, C. (1994). The high-output nitric oxide pathway: role and regulation. *Journal of Leukocyte Biology* 56(5), 576-582. https://doi.org/10.1002/jlb.56.5.576.

Xie, X., Li, W., Lan, T., Liu, W., Peng, J., Huang, K., Huang, J., Shen, X., Liu, P. and Huang, H. (2011). Berberine ameliorates hyperglycemia in alloxan-induced diabetic C57BL/6 mice through activation of Akt signaling pathway. *Endocrine Journal* 58(9), 761-768. https://doi.org/10.1507/endocrj.K11E-024.

Xun, W., Shi, L., Zhou, H., Hou, G., Cao, T. and Zhao, C. (2015). Effects of curcumin on growth performance, jejunal mucosal membrane integrity, morphology and immune status in weaned piglets challenged with enterotoxigenic Escherichia coli. *International Immunopharmacology* 27(1), 46-52. https://doi.org/10.1016/j.intimp.2015.04.038.

Zeng, Z., Zhang, S., Wang, H. and Piao, X. (2015). Essential oil and aromatic plants as feed additives in non-ruminant nutrition: a review. *Journal of Animal Science and Biotechnology* 6(1), 7. https://doi.org/10.1186/s40104-015-0004-5.

Zhao, F. Q. and Keating, A. F. (2007). Functional properties and genomics of glucose transporters. *Current Genomics* 8(2), 113-128. https://doi.org/10.2174/138920207780368187.

Zhou, R., Wu, J., Lang, X., Liu, L., Casper, D. P., Wang, C., Zhang, L. and Wei, S. (2020). Effects of oregano essential oil on in vitro ruminal fermentation, methane production and ruminal microbial community. *Journal of Dairy Science* 103(3), 2303-2314. https://doi.org/10.3168/jds.2019-16611.

Zou, Y., Wang, J., Peng, J. and Wei, H. (2016). Oregano essential oil induces SOD1 and GSH expression through Nrf2 activation and alleviates hydrogen peroxide-induced oxidative damage in IPEC-J2 Cells. *Oxidative Medicine and Cellular Longevity* 2016, 5987183. https://doi.org/10.1155/2016/5987183.

Chapter 5

The use of tannins as dietary supplements in dairy cattle nutrition

Michael Kreuzer, ETH Zurich, Switzerland

1 Introduction

Tannins are phenolic compounds with a complex structure (Crozier, 2003; Naumann et al., 2017; Aboagye and Beauchemin, 2019; Mueller-Harvey et al., 2019; Kelln et al., 2021) and form an important group of secondary plant compounds that are prevalent in a number of plants. They are formed during plant metabolism and contribute to the plant's defence system among other functions (Swanson, 2003; Kelln et al., 2021). There are two classes of tannins that are determined by their difference in size and complexity, as well as degradability, known as

- larger-sized condensed tannins (CT); and
- hydrolysable tannins (HT).

Apart from the basic differences in molecular weight, there is also a large structural diversity within CT and HT. Of all secondary plant compounds, tannins have received the most attention in dairy cow nutrition research as they are considered to have potential beneficial effects in the diet at concentrations where possible adverse effects are either absent or less

http://dx.doi.org/10.19103/AS.2022.0117.05

important than the beneficial effects. The research on tannins and tanniferous feeds/supplements has been discussed in a number of review articles (e.g. Frutos et al., 2004; Patra and Saxena, 2011; Jayanegara et al., 2012a; Piluzza et al., 2013; Morales and Ungerfeld, 2015; Naumann et al., 2017; Min and Solaiman, 2018; Aboagye and Beauchemin, 2019; Mueller-Harvey et al., 2019; Nawab et al., 2020; Kelln et al., 2021). It should be noted that many other review articles include the effects of tannins when describing the influence of various diet supplements. Due to the plethora of information and reviews, this chapter focuses on results from *in vivo* studies. *In vitro* work is very important for screening and for researching modes of action, but the proof of principle and determination of the level of effect *in vivo* is the ultimate step prior to implementation in farm practice. As the number of *in vivo* studies carried out exclusively in dairy cows is limited, evidence from growing cattle and small ruminants is also considered. Emphasis is also placed on important, less well-researched aspects of tannin supplementation, such as the influence on emissions from the manure of tannin-supplemented animals or the level of transfer of such polyphenols into milk. As tannins are highly complex and largely differ in structure, their effect is difficult to predict and requires specific analytical and experimental techniques, which are aspects also considered in this chapter.

2 Sources of tannins for supplementation

Feed sources of tannins are often the entire above ground plant material (legumes and herbs) or their leaves (shrubs and trees), but tannins are also prevalent in roots, fruits, flowers, buds, wood and bark (Piluzza et al., 2013; Kelln et al., 2021). In order to identify sources of tannins with beneficial effects as feeds or supplements in ruminants, a clear programme has to be followed, for instance, by using the five-step procedure suggested by Flachowsky and Lebzien (2012). Hence, the key steps in tannin research would involve:

1. characterising the botanical and chemical composition of the sources;
2. analytical description of the tannins;
3. *in vitro* screening;
4. *in vivo* confirmation with recording of palatability and side effects; and
5. long-term studies confirming use and efficacy in farm practice.

Extensive *in vitro* screenings (Step 3), including the search for effective sources of tannins and tanniferous plants as supplements to ruminant diets, have been carried out by a number of researchers (e.g. Bodas et al., 2008; García-González et al., 2008; Kamra et al., 2008; Staerfl et al., 2010; Jayanegara et al., 2011, 2012b; Terranova et al., 2018; Kapp-Bitter et al., 2021b). Sources of tannins may originate from the tropics or temperate climates. In this chapter, the most well-researched sources are presented, and their effects are described.

2.1 Tannin sources from the tropics

Two sources of tannins (especially CT), which are subject to a lot of *in vivo* studies, are extracts of the bark of an acacia species (*Acacia mearnsii*; mimosa; also called 'Black wattle') and of quebracho (*Schinopsis quebracho-colorado*). These extracts are produced in large amounts in South America for the industrial tanning of leather and are used in tannin research worldwide. In addition to these two sources, leaves from tropical shrubs rich in tannins (sometimes up to 40% of dry matter) have also been investigated and used as feeds in ruminant diets in the regions where these shrubs are grown as forage. Among them, the most widespread is *Leucaena leucocephala*, which has exceptional growth performance under good cultivation conditions (Tiemann et al., 2009b). Its high tannin concentration (similar to leaves from the related *Leucaena diversifolia*) does not seem to impair its feeding value as much as that of other tanniferous shrub species, like the leaves of *Calliandra calothyrsus* and *Flemingia macrophylla*. All of these plants are rich in CT and less so in HT. By-products from the food industry in arid and semi-arid countries, such as HT-rich pomegranate by-products, have recently also gained interest. Canul-Solis et al. (2020) and Nawab et al. (2020) specify further tropical feed resources rich in tannins, such as *Acacia nilotica*, *Acacia pennatula* and *Lespedeza cuneata*.

2.2 Tannin sources from temperate climates

Tannin sources originating from plants that only grow in tropical areas have to be imported when used in countries with a temperate climate. In these countries, the use of sources from on-farm cultivation would, therefore, be more sustainable. Tannin-rich (mainly CT) forages that can be grown on arable land under temperate climatic conditions include, in particular, the legumes sainfoin (*Onobrychis viciifolia*; Fig. 1) and birdsfoot trefoil (*Lotus corniculatus*), but big trefoil (*Lotus pedunculatus*) and sulla (*Hedysarum coronarium*) are also of interest. A number of herbs, among them small burnet (*Sanguisorba minor*), rosebay willow (*Epilobium angustifolium*) and wood avens (*Geum urbanum*), could also be future options as dietary tannin sources. Among the woody plants, leaves from hazel (*Corylus avellana*) and green grapevine (*Vitis vinifera*) are promising as dietary supplements, with hazel leaves containing both HT and CT. Other tannin sources from woody plants for ruminant nutrition, though not nutritious feeds, are HT-rich extracts from chestnut wood (*Castanea sativa*) and valonea oak (*Quercus valonea*). Interesting tanniferous pomaces from temperate climate plants that are by-products of the food industry include grape pomace.

Figure 1 Sainfoin (*Onobrychis viciifolia*), one of the most promising tanninferous forage plants growing under temperate climatic conditions. Photograph by Anna Scharenberg.

3 Effects of tannins in dairy cows

3.1 Effects on palatability, digestion and performance

The utility of tannins and tanniferous feeds in dairy cow nutrition depends on the presence or absence of adverse side effects of their dietary inclusion on performance. The key to maintaining high performance is a sufficient nutrient supply, which may be reduced by the low palatability of the diet and reduced digestibility of the nutrients. The majority of reports on the effects of tanniferous feeds point towards a lower or unchanged milk yield (e.g. Frutos et al., 2004).

Due to their astringent taste, tannins may substantially impair feed intake by reducing the palatability of the entire diet or part of the diet, forage or concentrate, that contains the tannins. An example of substantial intake decline was a diet supplemented with *C. calothyrsus* (Hindrichsen et al., 2004). To date, only a few studies have been carried out with dietary tanniferous supplements that have been specifically designed to test palatability, for instance, by offering choice situations (e.g. Terranova et al., 2020). Due to a lack of such studies, a decline in feed intake under both *ad libitum* and restricted feeding conditions is considered to be an indicator of reduced palatability. Attempts to avoid reduced nutrient supply include limiting dietary tannin concentration to below 5% and using palatable tannin sources, in addition to considering the nutritional value of the non-tannin constituents in these sources (Frutos et al., 2004). Adding molasses or other palatable supplements may also be useful. In addition, it is important to note that, when offered the choice, animals select their own diet and thus decide about the quantity of less palatable components, such as tanniferous supplements, they will ingest. For instance, intake may be promoted when animals experience positive metabolic feedback, as is the case when combatting parasites or illness (Villalba et al., 2010). This points to situational responses in the palatability of tanniferous supplements. Accordingly, promising tanniferous supplements have to be tested for palatability under various conditions, including target animal species and production purpose (e.g. milk production by cows), dietary proportion and complementary dietary components.

Concerning digestion, inhibiting the growth of cellulolytic and proteolytic bacteria is known to be mediated via the formation of complexes of tannins and the bacterial cell wall (Bodas et al., 2012), resulting from their complex and large structures complete with side chains that have binding properties. However, the greater influence is probably due to the capacity of tannins to bind nutrients, making them inaccessible to microbes. For this reason, the tannin structure and side-chain composition determine the binding activity of tannins (Frutos et al., 2004; Tiemann et al., 2010a). At ruminal pH, binding by tannins is most pronounced with dietary protein, but nutrients such as fibre might also be bound or otherwise become inaccessible to rumen microbes (Vasta et al., 2019). Accordingly, Orlandi et al. (2015) found a substantially increased duodenal flux of amino acids in the presence of the *A. mearnsii* extract. The conditions in the lower gut, where pH substantially declines in the abomasum, might help cleave these bonds, but pH increases again in the small intestine. The extent to which protein–tannin complexes are actually cleaved in the lower gut determines how much metabolizable protein is generated from the amount of protein escaping ruminal degradation. Otherwise, only a shift of nitrogen (N) excretion from urine to faeces can be achieved, which is a goal in itself. Cortés et al. (2009) developed an *in vitro* model combining ruminal degradation with post-ruminal enzyme treatment and found highly variable post-ruminal soybean protein release levels when incubating the protein with pure CT extracts from tropical shrubs. This model was again applied by Tiemann et al. (2010b) who found effects of cultivation conditions of the shrubs on the level of post-ruminal protein release. Declines in ruminal protein degradability and increases in protein bypassing the ruminal degradation under simulated conditions of the lower gut were also found by Lavrenčič and Levart (2021) in the case of chestnut and quebracho tannins and by Abarghueia et al. (2021) with pomegranate peel, which indicates substantial cleaving activity. However, *in vivo* studies on the post-ruminal digestion of tanniferous diets are widely lacking, and there is no indirect evidence for a better metabolic protein supply of tanniferous diets, because no positive effects on milk N secretion or body N retention with tannin supplements have been reported (e.g. dairy cows: Grosse Brinkhaus et al., 2016; Gerlach et al., 2018; sheep: Hindrichsen et al., 2002, 2004; Carulla et al., 2005; Tiemann et al., 2008b). An exception is Pathak et al.

(2017), where lambs fed tropical tanniferous feeds exhibited a clearly higher body N retention. Lower protein digestibility and higher ruminal escape of protein often compensate each other (Naumann et al., 2017). Both outcomes, more metabolisable protein or lower urine N losses or both, are only helpful with diets containing excessive rumen-degraded protein. In tropical countries where diets are very often limited in rumen-degraded protein, the protein and energy supply is further reduced by tanniferous supplements, even when provided as high-crude protein–high-tannin legume shrub leaves (Tiemann et al., 2008b). The tannins still limit the amount of rumen-degraded protein, and a deficiency in degraded protein is particularly detrimental for fibre-degrading microbes. This hampers the protein supply from microbial matter and energy supply from volatile fatty acids.

The class of tannins gets important when provided at higher doses as, compared with CT, HT or its breakdown products may be absorbed to some extent and could be toxic (Reed, 1995). However, some bacteria are able to degrade HT not only to tannic acid but also to volatile fatty acids in further steps with the help of tannase (Bhat et al., 1998; Vasta et al., 2019); thus, the risk of toxicity could be lower than that assumed by Reed (1995). Different from the general assumption, Bhat et al. (1998) claimed that even CT might be degradable by certain microbes to a limited extent.

Overall, it seems that there are a few tannin sources that are highly palatable and have no or very low adverse effects on performance. Independent of the actual effects of tannins, their lack of (in tannin extracts), or low supply with, energy (in tanniferous plants) has to be balanced by high-energy ingredients when feeding tannins to high-yielding ruminants.

3.2 Use of tannins to mitigate environmentally relevant emissions

The two most important environmental concerns of dairy husbandry are emissions of N (as ammonia, nitrous oxides and nitrate) from manure and methane from the cow and its manure. In experimental research, N and methane emissions from the manure of tannin-fed cows are rarely measured. Indicators of potential N emissions from manure are the excretion of easily volatilised N from urine and, earlier in the digestive system, ruminal ammonia formation as a precursor of urine N. The milk urea content is another useful indicator of the N-emitting potential of manure. The considerations in Section 3.2.1 concerning N emissions mainly rely on these indicator traits, but the available experimental evidence on actual N emissions from manure is also described.

Tannins represent the most well-researched class of secondary plant compounds concerning urine N and methane mitigation. They are among the few dietary supplements that present a win–win situation in terms of mitigating both types of emission simultaneously. An important, but not always reported, measure of the efficiency of tannins in mitigation is the emission intensity per unit of milk (protein). The efficiency may be small when tannins would impair feed intake and digestibility, and more animals are needed to secure food demand, ultimately leading to constant or even higher, rather than lower, system emissions.

3.2.1 Nitrogen emissions

The mode of action of tannins in reducing ammonia formation in the rumen has two components. The most important component is the binding of dietary protein, making it inaccessible to ruminal degradation; the other is the direct adverse effect of tannins against (certain) rumen microbes, as described in Section 3.1.

There is abundant evidence for the efficiency of various tannin sources to reduce the predicted N emission potential of manure, and one well-investigated source is the *A. mearnsii* extract. Carulla et al. (2005) fed 4% of this extract in the diet dry matter to growing sheep, which reduced urine N losses from 60% to 52% of N intake. In the study of Orlandi et al. (2015), feeding 0%, 0.9%, 1.8% and 2.7% *A. mearnsii* extract reduced urine N losses in growing steers from 67% to 46% of N intake. In these two studies, carried out with growing ruminants, the proportion of dietary N lost with urine was high compared with that expected from dairy cows. The urine N loss was reduced from 39% to 26% and 22% of N intake with 1% and 2% *A. mearnsii* tannins in the dairy cow study of Grainger et al. (2009). Gerlach et al. (2018) found no clear reduction of urinary N with 3% *A. mearnsii* extract fed to dairy cows despite trends towards a lower milk urea content. Combining quebracho tannins with chestnut tannins (0.75% each) was more efficient in reducing urine N (from 55 g/day to 43 g/day; an estimated decline from 26% to 20% of N intake) than supplementing 1.5% chestnut tannins alone (down to 51 g/day; 24% of N intake) (Aboagye et al., 2018), indicating the greater efficiency of quebracho tannins compared to chestnut tannins. There are also several *in vivo* studies investigating the effect of temperate climate forage legumes on potential N emissions. In the study by Grosse Brinkhaus et al. (2016), urine N excretion of dairy cows declined with 16% sainfoin in the diet from 21% to 18% of N intake but not with the same amount of birdsfoot trefoil, which may be explained by the 10-fold higher concentration of tannins analysed in the sainfoin compared to the birdsfoot trefoil. In the study by Huyen et al. (2016), a diet containing 30% sainfoin did not affect urine N excretion in dairy cows. In beef heifers fed exclusively sainfoin as a diet, absolute urine N losses decreased, but those relative to N intake were less clearly affected (Chung et al., 2013), illustrating the limited potential of temperate climate forage legumes in mitigating N emissions. A recent *in vitro* screening identified several particularly promising temperate climate shrub and herb species that effectively mitigated ruminal ammonia formation (Terranova et al., 2018), of which four turned out to be highly palatable to dairy cows, namely, the leaves from hazel and green grapevine and the herbs rosebay willow and wood avens (Terranova et al., 2020). Hazel leaves were particularly efficient in reducing the urine N proportion of N intake, decreasing from 30% to 15% in sheep (Wang et al., 2018) and from 50% to 15% in dairy cows (Terranova et al., 2021) fed diets containing up to 50% hazel leaves. Part of the reduction may have resulted from certain concomitant declines in dietary N content. As in the study of Terranova et al. (2021) the milk yield of mid-to-late lactating cows remained unchanged, this mitigating effect also included urine N emission intensity. It should be noted that the maximum levels of hazel leaves tested were in a range not widely applicable in farm practice. Different from hazel leaves, vine leaves unexpectedly did not mitigate urine N proportion of intake in dairy cows (Birkinshaw et al., 2022). A certain but low efficiency in reducing urine N excretion was found for relatively low levels of chestnut extract by Wischer et al. (2014) as well as Aboagye et al. (2018 and 2019) and for valonea oak extract (Wischer et al. 2014). Highly tanniferous leaves of tropical plants also seem to be less efficient than hazel leaves in shifting N excretion from urine to faeces. For instance, *L. diversifolia* increased faecal N in sheep, but a decrease in urine N excretion was only found in one study (Hindrichsen et al., 2004) and not in the other (Hindrichsen et al., 2002). Similarly, *C. calothyrsus* and *F. macrophylla* enhanced faecal N excretion, but the reduction in urine N was not significant though numerically present (Tiemann et al., 2008b).

As outlined above, the proof of principle requires the actual measurement of N emissions from manure, which is not necessarily associated with urine N losses in a linear manner. Tannins may mitigate N emission from manure via two mechanisms. One is reducing the proportion of highly volatile urine N (as outlined above); the second is a direct effect of the tannins remaining undigested and therefore being present in manure. Indeed, in recently it was reported that adding 4% quebracho tannins or chestnut tannins directly to manure reduced nitrous oxide emissions (Min et al., 2022). This indicates that tannins act in a similar manner against microbial protein degradation in manure as they do in the rumen. However, the use of tannins in the diet of animals, other than adding them directly to manure, may also provide extra substrates, especially protein protected from digestion by tannins, which might get available to manure microbes and thus might promote N emissions. The few studies directly investigating N emissions from the manure of tannin-supplemented animals suggest that an effect either is dose dependent or may need time to be exhibited or both. In lambs, Śliwiński et al. (2002) found no effect of 0.1% and 0.2% of chestnut wood extract in the diet on urine N losses and ammonia emission from manure within 2 weeks of storage. In contrast,

Śliwiński et al. (2004) noted that gaseous N losses from the manure of cows fed a diet containing 0.5% of the same extract were reduced by half within 8 weeks of storage even though urine N losses had not been significantly lowered. An amount of 2.5% *A. mearnsii* extract in the diet of beef cattle had no clear effect on ammonia emission from manure in the study of Koenig et al. (2018), but the method used (horizontal flux technique with passive ammonia samplers from a few pens) might have added uncertainty to the measurement. It may also be important that, when inhibiting ammonia formation in manure, undigested tannins are likely to also impair the fertiliser value of manure. This was investigated by Tiemann et al. (2009a) feeding 15% and 30% of high-CT leaves of tropical shrubs (*C. calothyrsus* and *F. macrophylla*) in the diet to sheep. No clear effect on the N fertiliser value was found, but the soil used was probably not first limited by N as stated by the authors. Therefore, further studies are needed to clarify the conditions under which tannins excreted by cows are active in manure and may affect the fertiliser value of manure. It is also unclear if there is a difference in the efficacy between HT (chestnut) and CT. Hao et al. (2011) found no effect of a diet with 2.5% *A. mearnsii* tannins on nitrous oxide emissions during the composting of manure of growing cattle. By contrast, 15% of HT-containing pomegranate peel per kilogram of the diet of sheep reduced nitrous oxide emissions from manure (Yurtseven et al., 2018), which was estimated to be equivalent to about 1.5% HT when considering the values for pomegranate pomace made by Giller et al. (2022). Another approach was chosen by Duval et al. (2016), who measured total on-farm emissions when dairy cows were fed diets containing 0.45% and 1.8% of a 2:1 mix of quebracho and chestnut tannins. They found that emissions of ammonia and nitrous oxide were numerically lower per unit of milk than without tannin supplementation. The authors of one study (Clemensen et al., 2019) went even further and investigated whether grazing of a tannin-containing forage (sainfoin) influences soil N turnover and reported a reduction of soil nitrate concentration, pointing towards a mitigating effect of tannins against N leaching.

The experimental evidence illustrates that the most reliable environmental effect of tannins consists of mitigating urine N emissions and, thus ultimately, N emissions as ammonia, nitrous oxide and nitrate from their excreta during excretion, storage and distribution on the field.

3.2.2 Methane emissions

The mode of action of a tannin-caused mitigation of enteric (and manure-derived) methane has two components. One is the binding activity of the tannins, which also includes a reduction in ruminal fibre degradation (see Section 3.1). This phenomenon will also decrease the metabolic energy supply to the animals and thus only reduce absolute emissions and probably not contribute to the more important mitigation of emission intensity. The second component is a direct action of tannins against ruminal microbes, which is explained by binding to microbes (especially by CT) as well as bacteriostatic effects or direct effects against methanogens (especially by HT) (Bodas et al., 2012; Aboagye and Beauchemin, 2019). Vasta et al. (2019) reported that the expenditure of hydrogen associated with ruminal degradation of HT to volatile fatty acids might provide another way of explaining the effect of HT against methanogenesis, but this offers no explanation for the frequently reported CT effects and seems to be too small in magnitude as a sole explanation.

Tannins appear to have an underlying general effect against methanogenesis. A meta-analysis demonstrated a clear dose-response relationship between tannin concentration and the level of *in vivo* methane mitigation (Jayanegara et al., 2012a). In addition to the general effect, there are clear differences between tannin sources in mitigation efficiency, with the majority of tannin sources remaining limited in their efficacy of methane mitigation. Among the sources currently investigated *in vivo*, the *A. mearnsii* extract is the most well researched and is one of the most efficient. This extract was found to be methane mitigating in sheep, growing cattle and dairy cows (Carulla et al., 2005; Grainger et al., 2009; Staerfl et al., 2012; Denninger et al., 2020). The level of methane mitigation obtained by 3% of the *A. mearnsii* extract ranges from 10% to 30%. Although rarely adversely affecting feed intake, digestion and performance, there is a tendency for lower values with *A. mearnsii*, and therefore, the efficiency to reduce methane emission intensity per kilogram of milk might be slightly lower than that of absolute emission. Less reliable than *A. mearnsii* in terms of methane mitigation in the animal, despite ample *in vitro* evidence, are extracts of chestnut (e.g. Śliwiński et al., 2002; Wischer et al., 2014; Aboagye et al., 2019), quebracho (e.g. Beauchemin et al., 2007; Aboagye et al., 2018) and valonia oak (Wischer et al., 2014). Leaves of the tropical shrubs *C. calothyrsus* and *F. macrophylla*, which are extremely rich in tannins, fed in dietary proportions of 30% reduced absolute methane emission by 20–25%, but not per unit of digested organic matter, an indicator of emission intensity (Tiemann et al., 2008b). This means that the decline in methane simply followed the lower level of available fermentable organic matter. Differing to these shrubs, beef cattle fed *L. leucocephala* emitted less methane and exhibited increased growth under subtropical and tropical conditions when pastures were enriched with this shrub (Harrison et al., 2015). For that study, open path lasers were applied to estimate methane emissions at herd scale in grazing cattle. Controlled *in vivo* studies to confirm the methane-suppressing activity of *L. leucocephala* are still necessary. In addition, other tanniferous parts of tropical plants could be of interest. As an example extract of *Terminalia chebula* seeds was reported to slightly reduce sheep methane emissions in the study by Patra et al. (2011), but *in vitro* results indicated a 90%

reduction of methane formation. Some tanniferous temperate climate forage legumes were shown to have potential in mitigating methane emission. After demonstrating this for big trefoil (Woodward et al., 2001) and sulla (Waghorn et al., 2002) in sheep, New Zealand researchers were probably the first to show a methane-declining effect of a tanniferous forage legume (birdsfoot trefoil) in dairy cows (Woodward et al., 2004). Huyen et al. (2016) described a similar effect of sainfoin in dairy cows. In both cases, methane yield per unit of feed intake was significantly reduced; however, the effects were minimal and Chung et al. (2013) did not find clear sainfoin effects in beef heifers. Small burnet (*Sanguisorba minor*) was found by Stewart et al. (2019) to be more efficient in mitigating methane emission in cows and heifers than sainfoin and birdsfoot trefoil, but digestibility was also impaired. Substantial methane-mitigating activity of other tanniferous herbs such as wood avens and rosebay willow which were effective *in vitro* (Terranova et al., 2020), may be possible in live animals. Effective woody plants from temperate climates include the leaves from hazel and green grapevine which are highly palatable (Terranova et al., 2020). Hazel leaves reduced methane emission in sheep by up to 35%; this high level of mitigation was achieved with a dietary hazel proportion of 50% (Wang et al., 2018). Lower dosages of hazel leaves, which are easier to implement in farm practice, were also effective, and in dairy cattle a close dose-response relationship was recently described (Terranova et al., 2021). Grapevine leaves, provided at 13% of dietary dry matter, reduced methane emission intensity per unit of milk yield by 12% in dairy cows (Birkinshaw et al., 2022). Grape pomace (marc) was found to be efficient in reducing methane emissions by 20% when added to the diet of low-yielding dairy cows at 25% (Moate et al., 2014), a level probably not applicable to diets of high-yielding cows. Caetano et al. (2019) found a 14% reduction in daily methane production with this feed, but emission intensity was probably less clearly mitigated due to the low energy value of the pomace. The use of pomegranate pomace to mitigate methane, as demonstrated *in vitro*, still awaits confirmation *in vivo* (Giller et al., 2022).

Concerning methane emission from manure, the same aspects are valid as for N emissions, namely that undigested tannins may inhibit methanogens whereas any extra undigested fibre might provide additional substrates for manure methanogens. The first was demonstrated by adding 4% quebracho tannins directly to manure, which caused a reduction in methane emissions, whereas 4% chestnut tannins remained ineffective (Min et al., 2022); however, the second component, extra fibre, also appears to be important. Indeed, the results to date would indicate a low (net) efficiency of fed tannins, if any, against methane formation in manure. Accordingly, including 2.5% *A. mearnsii* in the diet of growing cattle did not affect methane emissions during the composting of manure (Hao et al., 2011). Furthermore, adding 15% of pomegranate peel to the diet did not affect the methane emission from sheep manure (Yurtseven et al., 2018). In this context, it is fortunate that most methane is formed in the rumen of the animal. Therefore, the absence or low tannin effects in manure are less detrimental than for N emissions, where tannin effects are restricted to those in manure. Accordingly, in the study of Staerfl et al. (2012), 3% of *A. mearnsii* tannin extract in the diet unexpectedly caused a non-significant increase in slurry-derived methane (+14% and +39% at storage temperatures of 14°C and 27°C, respectively). However, total system methane emissions (enteric and manure derived) were still substantially mitigated by tannin supplementation in that study. Supplementing 1.5% and 2.5% of CT from an *A. mimosa* extract mitigated absolute methane emission in adult cattle, but not methane yield per kilogram feed intake, despite a reduction of archaeal counts, and did not influence methane formation in biogas produced from their manure (Fagundes et al., 2020). The total on-farm methane emissions of dairy cows fed diets with 0.45% and 1.8% of a 2:1 mix of quebracho and chestnut tannins in the study of Duval et al. (2016) also numerically declined per unit of milk, like those of N emissions. It should be noted that the observed general lack of a clear methane-mitigating effect in the manure of tannin-fed cows would be beneficial when manure is used for biogas production.

In conclusion, there appears to be an underlying general tannin effect against methanogenesis, but only a few tannin sources exert substantial mitigating effects *in vivo*. It is, therefore, not possible to predict the efficiency of not well-researched sources simply by analogy.

3.3 Effects of tannins on milk quality

Tannins do not have a substantial effect on the gross composition of milk (e.g. Śliwiński et al., 2004; Woodward et al., 2004; Grainger et al., 2009; Huyen et al., 2016; Gerlach et al., 2018; Terranova et al., 2020, 2021). This would be different if tannins severely affected protein supply and/or the profile of volatile fatty acids. Then, changes in milk protein and fat content could not be excluded, but this would represent an indirect and no direct effect of tannins on rumen microbes.

3.3.1 Effects on ruminal biohydrogenation and milk fatty acid composition

Unlike the gross composition of milk, the fatty acid composition of milk fat can be substantially influenced by tannin supplementation (e.g. Buccioni et al., 2015; Morales and Ungerfeld, 2015; Girard et al., 2016a; Grosse Brinkhaus et al., 2016). Many microbial species are involved in the ruminal biohydrogenation of polyunsaturated fatty acids, such as linoleic acid and α-linolenic acid, but *Butyrivibrio fibrisolvens* is particularly important and can be inhibited by tannins (Min and Solaiman, 2018). Tannins alter fatty acid composition by protecting distinct fatty acids, considered beneficial for human health, from ruminal biohydrogenation, which promotes their transfer to milk (Patra and Saxena, 2011). In addition, even an only partial biohydrogenation of fatty acids in the presence of tannins provides valuable intermediates such as C18:1 *trans*-11 and C18:2 *cis*-9, *trans*-11 (the latter being a conjugated linoleic acid and often generated from C18:1 *trans*-11 in the mammary gland). This effect again seems to depend on tannin source. For instance, Buccioni et al. (2015) found such effects using a quebracho extract but not with a chestnut extract. The *A. mearnsii* extract (Khiaosa-ard et al., 2009) and quebracho tannins (Vasta et al., 2010) were both shown to partially specifically inhibit the last step of ruminal biohydrogenation, whereas tanniferous sainfoin remained ineffective in this respect (Khiaosa-ard et al., 2009). It was, therefore, unexpected that the *A. mearnsii* extract did not change the fatty acid composition of the perirenal adipose tissue of beef cattle (Staerfl et al., 2011), whereas that of the intramuscular fat of lamb meat was observed to be richer in *n*-3 fatty acids when using sainfoin (Girard et al., 2016b). The same was found for sainfoin in dairy cows, where a 17% increase of the proportion of α-linolenic acid was recorded in both milk and cheese lipids when feeding sainfoin (Girard et al., 2016a). With birdsfoot trefoil, the increase in the proportion of α-linolenic acid was only 3%. *In vitro*, α-linolenic acid biohydrogenation was reduced with sainfoin, but it remained unclear if this was an effect of lower nutrient degradation or a tannin effect on ruminal biohydrogenation (Khiaosa-ard et al., 2009). Feeding vine leaves resulted in a trend towards higher proportions of C18:2 *cis*-9, *trans*-11 and α-linolenic acid in the milk fat of dairy cows (Birkinshaw et al., 2022). Overall, the results in terms of tannin effects on the fatty acids are highly variable and predictions are difficult, as many factors influence the outcome, including tannin dosage and source, other dietary ingredients, metabolic function of the body lipids and, possibly, animal species. Further studies are required to quantify the importance of the different factors (Vasta et al., 2019).

Accordingly, research evidence to date points towards a specific property of tannins that affects the fatty acid profile of foods from ruminants via their influence on ruminal biohydrogenation, but the effects are variable in terms of which fatty acids are influenced and they not very pronounced.

3.3.2 Transfer of tannins to milk

It would be of interest if bioactive substances such as tannins could also be transferred to human food, i.e. ruminant milk; however, whether or not this is beneficial needs to be clarified in individual cases. As CT are considered to be widely indigestible, no substantial transfer would be expected, unless there is a not yet quantified partial degradation of CT molecules in the gut with some monomers being absorbed. A higher potential for transfer is given with HT, where the possibility of absorption (probably as breakdown products such as tannic acid) is demonstrated by their potential toxicity (Section 3.1). Studies specifically investigating the transfer of tannins into milk are scarce. Leparmarai et al. (2021) followed the recovery of total phenols from the diet in the milk of cattle and camels grazing tropical wood pastures with accessibility of the animals to various tanniferous shrub and tree leaves; elevated levels of phenols were indeed detected in milk. The transfer was, however, extremely low with around 0.1% of intake in cattle and not related to intake in camels. Camels seem to have developed coping mechanisms against tannins such as specific salivary binding properties. Feeding tanniferous hazel leaves (containing both CT and HT) to sheep resulted in an increased total antioxidant capacity of blood plasma which pointed towards the absorption of bioactive phenols, but unexpectedly total phenol concentrations in plasma were not elevated (Wang et al., 2019). In addition, hazel leaves did not significantly elevate the milk phenol concentration of dairy cows, whereas the tanniferous wood avens did but only by a small magnitude (Birkinshaw et al., 2020). Certain non-tannin phenols, like those from grapeseed extract, may be more efficiently transferred into milk as was shown for the milk of goats and sheep, but even in that case most phenols were excreted via urine (Leparmarai et al., 2019). Overall, limited data suggest that there is no great potential of tannins to beneficially elevate phenol concentrations of milk.

3.4 Anthelmintic properties of tannins

Intestinal parasites like nematodes mainly occur in small ruminants, but there are reports by veterinarians regarding an increasing prevalence in cattle, especially when grazing highly infected pastures. Anthelmintics are chemicals to which resistance is becoming more frequent (Barone et al., 2019) and, therefore, should be replaced by natural effective ingredients. Various studies point towards an anthelmintic activity of tannins and feeds rich in tannins. *In vivo* studies targeting infections with *Haemonchus contortus*, as a major nematode species, included the forage legumes birdsfoot trefoil (Lonngren et al., 2020) and sainfoin (Heckendorn et al., 2006, 2007; Azuhnwi et al., 2013b) as well as *Mimosa caesalpiniifolia* leaves (Brito et al., 2018) and *A. mearnsii* extract (Lima et al., 2019). Sulla is also among the tanniferous plants known for their antiparasitic properties (Ramírez-Restrepo and Barry, 2005). The structure of tannin plays an important role in antiparasitic function (Mueller-Harvey et al., 2019). Apart from varying efficacy among cultivars (Barone et al., 2019) and sometimes a low *in vivo* recovery of *in vitro* results (e.g. Lonngren et al., 2020), in general, the complete extinction of nematode activity is unachievable. However, slowing down the life cycle may still be an important element of an on-farm strategy to combat nematode infections.

4 Factors influencing the effects of tannins in animals

4.1 Variability among plants and cultivars as well as within cultivars

One of the difficulties in recovering the effects of tannins on ruminants reported in previous studies is due to their variability, which concerns both the concentration and composition of the very complex tannin polymers. It is generally accepted that different plant species differ in their tannin properties as they may also vary in the proportions of CT and HT; cultivars may also differ in this respect [shown for birdsfoot trefoil by Barone et al. (2019) and for sainfoin by Azuhnwi et al. (2013a)]. This phenomenon got more even apparent from recent research triggered by the regained interest on temperate climate legumes for their usefulness as forage, with plant breeders searching for the most appropriate cultivars (Mueller-Harvey et al., 2019). Also the difficulty in efforts to reintroduce higher tannin levels in high-performance forage legumes (Kelln et al., 2021) emphasised this phenomenon. Therefore, expectations of reliable effects have concentrated on the same cultivar; however, environmental conditions during cultivation remain a major source of variability (Frutos et al., 2004; Aboagye and Beauchemin, 2019). This was experimentally demonstrated for the leaves of tropical shrubs where the same cultivar was grown in different countries (Hess et al., 2006) and on different soils with or without drought conditions and with or without mineral fertilisation (Tiemann et al., 2009b, 2010a,b). These studies revealed a repeatable pattern showing that tannin concentrations decline with improved plant growing conditions. It cannot be excluded that this is simply the consequence of a dilution of the tannins by non-tannin plant matter promoted by better environmental conditions. In addition, decreased environmental stress may lower the need to maintain the associated defence mechanisms. Even with this knowledge, it is difficult to predict tannin concentration and composition under defined cultivation conditions. Such differences in growing conditions may also affect the adoption of tanniferous plants by farmers. For instance, in Colombia, *C. calothyrsus* is considered to be low-quality forage, whereas it is called 'dairy shrub' in parts of Kenya, with this indigenous knowledge being confirmed by corresponding differences in quality found when obtaining the same cultivar of *C. calothyrsus* from the two countries (Hess et al., 2006). Another form of variability should also not be underestimated. Minimal concentrations seem to be needed to obtain noticeable bioactivity, as concluded from the meta-analysis by Jayanegara et al. (2012a) where extreme variability in the antimethanogenic effects of tannins was found across *in vivo* studies at tannin concentrations below 2%, whereas results from studies with higher dosages were much nearer to the line of regression. For cost reasons, recommendations for the supply of commercial tannin supplements are often below this threshold, explaining the variable and inconsistent effects.

4.2 Importance of analytical and experimental techniques

The variability of tannin supplement effects is also caused by analytical limitations. Naumann et al. (2017) consider tannin diversity and analytical deficits as the main cause for minimal progress in knowledge regarding tannin-animal interactions. In particular, limitations concern the determination of structural properties such as molecular weight and subunit structure, which could be crucial for a tannin effect to be exhibited (Aboagye and Beauchemin, 2019; Mueller-Harvey et al., 2019), but even tannin concentrations have substantial analytical uncertainties. The standard method is outlined in the laboratory manual of Makkar (2003a), according to which the analysis of total tannins is performed by the Folin-Ciocalteu method and CT are analysed using the butanol-hydrochloric acid-iron method, sometimes complemented by distinguishing between free and bound tannins/CT. The HT are estimated from the difference between total tannins and CT, where the commonly different tannin standards used for total tannins and CT add to the already high uncertainty. The tannin standards used in these analyses may not even exist in the tannin supplement analysed. The limitations of this simple analysis are outlined by Ramsay et al. (2015). Few researchers create their own standards from the test plant species or cultivar via time-consuming extraction (Piluzza et al. 2013), as realised by Azuhnwi et al. (2013a). This would substantially improve the accuracy of quantitative values. Although considerable progress in analytical methods has been made (Ramsay et al., 2015; Mueller-Harvey et al., 2019), methods for tannin structural analysis are rarely applied [such as differentiation for delphinidin, prodelphinidin, cyanidin, procyanidin and pelargonidin (Tiemann et al., 2010a; Azuhnwi et al., 2013a)], and difficulty remains in predicting which structure will promote the desired effects of tannins in the animal or which have adverse effects. The bioactivity of tannins may, however, be assessed by other methods both in the animal and *in vitro*. The most widespread approach is to add polyethylene glycol (PEG) to diets (inert to the animal) containing tannin supplements, because it has been known since the 1970s that this compound widely inactivates tannins (Makkar 2003b). Tiemann et al. (2008a) reviewed issues associated with the PEG approach including that the amount of PEG needed for neutralisation depends on tannin type (plant species) and should probably exceed a ratio of 1:1 of PEG:tannins. In this context, it remains unclear if the efficacy of PEG to inactivate tannins is different for purified tannins and tannin extracts compared with tannins still embedded in the plant matrix where they may have already formed bonds with protein. As tannins replace other nutrients in plant matter and their synthesis and deposition in the plant might be associated with changes in nutritional quality of the non-tannin matter, it remains unclear whether effects are (exclusively) from tannins or not. Tiemann et al. (2008c), therefore, developed a method to purify the fibre of tanniferous leaves of tropical shrubs and was able to separate fibre and tannin effects with this approach. Another laboratory-based indicator of tannin bioactivity (especially CT) is astringency, a measure of the protein-binding capacity of tannins (Tiemann et al., 2010a).

4.3 Verification of tannin effects on farms

The results to date suggest that the implementation of supplements of certain tannin sources on dairy farms is timely. The question remains, how can the effects of tannins be confirmed and controlled? A Swiss group of researchers identified and tested a set of variables to be quantified or sampled for analysis on farm and attempted to confirm their applicability with diets supplemented either with a chestnut tannin extract (Kapp-Bitter et al., 2020) or with sainfoin (Kapp-Bitter et al., 2021a). The focus was on N emission-related variables. Here, a particularly easily accessible variable is the routinely assessed milk urea content. Another approach is based on the analysis of faeces and urine spot samples using indicators for total faeces and urine volumes and N concentrations. Changes in milk protein content in cows fed tannin supplements would indicate variations in the supply of metabolic protein, resulting from concomitant changes in fermentable organic matter. Confirmation of the applicability of these indicators is still awaited as the variables investigated did not clearly respond to the two sources of tannins tested (Kapp-Bitter et al., 2020, 2021a); conversely, a decline in palatability was clearly obvious on farm (Kapp-Bitter et al., 2020). Milk is a type of matrix where samples are easily obtained on farm, allowing tannin effects on fat composition to be determined via gas chromatography.

5 Onset of tannin effects and long-term efficiency

An important aspect concerning the efficiency of tannin supplementation is the duration of the delay until effects begin and, more importantly, whether or not rumen microbes adapt with time to the tannins with time thus counteracting their beneficial effects. Indeed, although tannins may inhibit the growth of many microbes, Bhat et al. (1998) outlined that there are also species (especially from the Streptococcus group; Patra and Saxena, 2011) that are resistant to tannins or have developed such resistance when exposed to tannins, mainly by degrading tannins. Smith et al. (2005) described several other resistance-enhancing mechanisms which bacteria might develop including the secretion of exo-polysaccharides forming a protective layer around the cell wall, or modification of the cell wall, or modes of dissociation of tannin–substrate complexes. If these resistant microbes represent key niches in the rumen and develop their ability under constant exposure to tannins, the tannin effects might be short-lived; hence, there are considerable concerns regarding the transiency of tannin effects. Studies are rare for both questions, i.e. the time for onset and long-term efficiency. It was recently shown that the effect of tannins seems to occur very rapidly, with only 20 minutes needed for significant methane mitigation to be exhibited when feeding 3% of an *A. mearnsii* tannin extract to dairy cows (Denninger et al., 2020). Birkinshaw et al. (2020; testing various shrubs and herbs) and Denninger et al. (2020; testing *A. mearnsii* extract) investigated how quickly tannin causes changes in the fatty acid composition of milk. They found modifications of the fatty acid profile after 3 days, but this included some fatty acids probably not affected in the longer term, and it seems that the key fatty acids of ruminal biohydrogenation were less or differently affected than would be expected with ongoing tannin supplementation. Concerning long-term efficiency, a breakthrough was the demonstration by Staerfl et al. (2012) of the sustained efficiency of the *A. mearnsii* extract at 3% in the diet of fattening bulls across more than 6 months (Fig. 2), a phenomenon that awaits confirmation for most other tannin sources. Wischer et al. (2014) also investigated the long-term

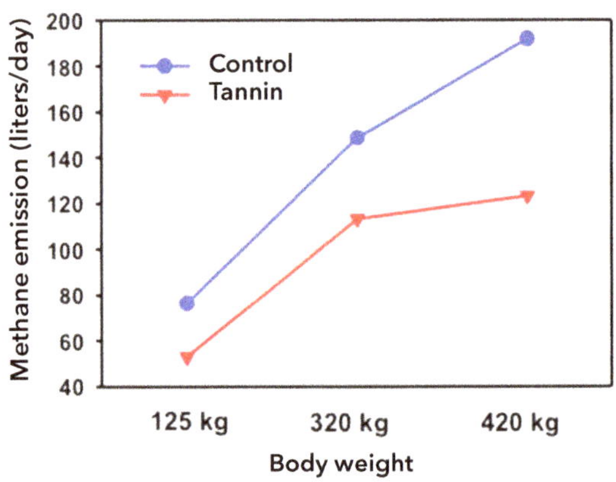

Figure 2 Long-term effect of *Acacia mearnsii* tannins against methane emission found in growing Limousin crossbred bulls. Adapted from Staerfl et al. (2012).

effects of tannins (chestnut and valonea extracts) in dairy cows. The shift from faecal to urinary N was still present 6 months after start of feeding at a level similar to that measured first approximately after 1 month of feeding. They also followed methane emission, but the tannin sources used had no effect on methane. The trend for lower total on-farm methane emissions in dairy cows fed a mix of quebracho and chestnut tannin extracts found by Duval et al. (2016) after 1.5 months of feeding was also recovered after 3 months of supplementation. Therefore, a limited number of studies suggest that, despite mechanisms of adaptation by ruminal microbes, tannin effects seem to be quite long-lasting in ruminants.

6 Strategies to maximise beneficial tannin effects

6.1 Optimal tannin dosage

The optimal dosage is one where the desired effects are achieved and adverse effects on performance and factors determining performance such as palatability, digestion and metabolic side effects are absent or clearly outweighed by the beneficial effects. However, this dosage is very difficult to define and often cannot be generalised as it not only depends on the basic effects of the respective tannins but also on the matrix they are embedded in and on other feed ingredients counteracting or promoting the tannin effects. Furthermore, the optimal dosage also depends on costs and cost-effectiveness, which is increased when, for instance, compensatory payments can be obtained balancing extra costs for supplements or performance declines or both. If tannin sources are homegrown, the farmer can determine the optimal and most cost-effective dosage. The thresholds of 2% tannins of dietary dry matter as minimal dosages for clear effects and 5% specified for the maximal dosage where no adverse effects are expected can probably not be generalised and should be determined separately for each tannin source.

6.2 Optimal form of tannin supplementation

Another decision to be made for farm practice is the form in which tannins should be supplemented. Basically, supplementation could consist of pure or highly purified compounds, extracts where tannins are enriched or entire tanniferous plants (or the respective tannin-rich plant parts). Pure compounds would offer the best repeatable effects, but they are too expensive to be applied on farm and costs are prohibitive even for experimental *in vivo* research. However, they are very helpful in determining the basic efficiency and mechanistic background of the effects in the rumen. They may even point towards the most effective monomers within the complex and large tannin molecules, the abundance of which could be a criterion for selecting efficient tannin sources. Such *in vitro* research has, for instance, been carried out with various pure compounds from a chemical supplier including tannic acid (Sinz et al., 2018). With highly purified tannic acid (purity: 95%) and gallic acid (purity: 99%), Aboagye et al. (2019) carried out a beef cattle study. An approach where pure compounds are used also offers the opportunity to quantify interactions among tannins or of tannins with other bioactive compounds in a way that remains unbiased by other factors. An important restriction of this approach is that the effects of pure compounds may differ from those of extracts or whole plants, where a number of bioactive substances in various concentrations may be present and the efficiency of the tannins will also depend on how easily they are liberated from the plant matrix. The latter, along with the often large natural variation in tannin concentration and composition described in Section 4.1, make extracts possibly the best alternative as the technical effort is limited, compared with the synthesis of pure compounds, and the variation is considerably reduced compared with unprocessed plants. In research, extracts also allow quantifying interactions among secondary plant compounds. For instance, Sinz et al. (2019a) determined the interactions between extracts from *A. mearnsii* and from grapeseed in growing lambs. Nevertheless, when grown on their own land, farmers would prefer to use unprocessed tanniferous plants.

6.3 Combination with other tanniferous or non-tannin supplements

Research is widely in its infancy concerning the effects of combinations of bioactive compounds in dairy cattle nutrition. These effects could be additive or even associative (i.e. non-linear) but also compensatory. As there may be more interactions with the composition of the basal diet, such research is very complicated. Examples of *in vivo* research for interactions, in particular, include the studies by Canadian researchers where interactions of HT and CT were closely followed, one with extracts rich in either HT or CT (Aboagye et al., 2018) and the other with pure compounds (Aboagye et al., 2019). Williams et al. (2020) reported that tannins and fats act additively against methane in dairy cows, pointing towards different modes of action against methanogens. Instead, if the mechanisms for an effect are quite similar, combining extracts could be of lower efficiency as, for instance, shown by Sinz et al. (2019b), for some extracts concerning *in vitro* methane and ammonia formation. Other research has concentrated on alleviating the potentially adverse effects of the typically low energy content of tanniferous plants, which hampers introduction in farm practice even when no direct adverse effects of tannins would occur. Jayanegara et al. (2013) detected

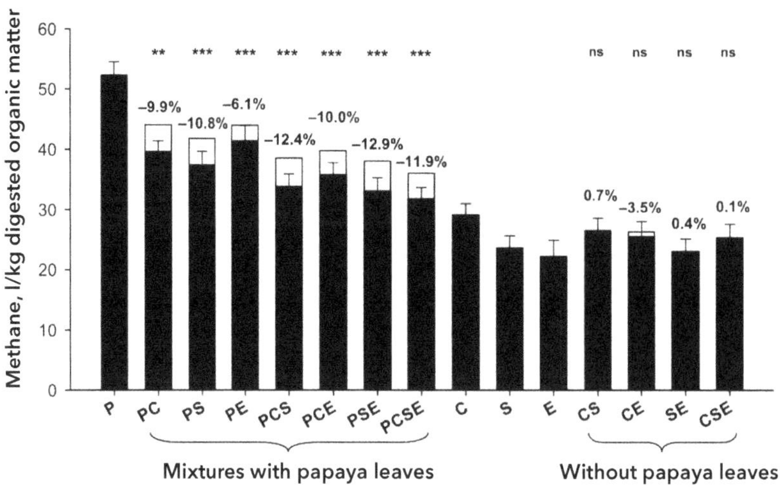

Figure 3 Positive associative effects of the combination of tanniferous tropical forage leaves (C, *Clidemia hirta*; S, *Swietenia mahagoni*; E, *Eugenia aquea*) and leaves from a high-quality forage (P, *Carica papaya*). The open parts of the columns illustrate the gap between methane emissions expected from the plants incubated alone (upper end) and that measured in the combinations (lower end of the open part). ***, significant at $P < 0.001$; **, significant at $P < 0.01$; n.s., not significant. Adapted from Jayanegara et al. (2013).

positive associative effects when combining various tanniferous forages with high-quality papaya leaves (Fig. 3). However, no corresponding positive associative effects were found when combining tanniferous hazel leaves with lucerne in dairy cow diets, where effects were linear or even of a negative associative nature (Terranova et al., 2021). All this points to considerable demand for further research.

7 Conclusion and future trends

This chapter demonstrates that feeding tannins to dairy cows can be strategically used to fulfil goals related to environmental sustainability and quality of the food produced. Several tannin sources have been intensively investigated for their effects, with respect to a number of aspects. Figure 4 illustrates the beneficial effects an optimal tannin supplementation to dairy cows should have and which supplements might be useful to reach the corresponding goals. The research evidence generated to date clearly shows that there is a delicate balance between avoiding adverse effects on performance and being able to profit from the beneficial effects. In addition, the presence and extent of beneficial effects is highly variable; therefore, considerations for the best trade-off are necessary. In this context, Grainger and Beauchemin (2011) asked the

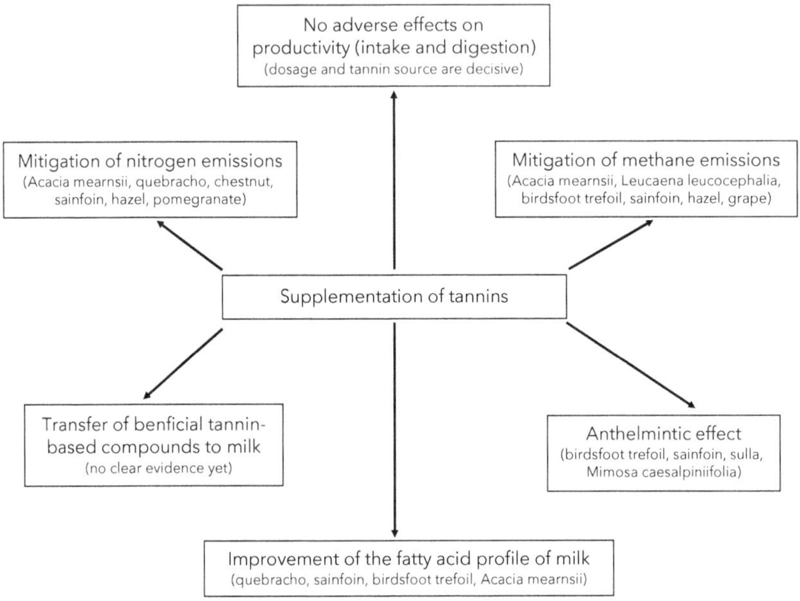

Figure 4 Anticipated effects of the optimal tannin supplementation strategy in dairy cows and tannin sources the use of which at least occasionally exhibits such effects *in vivo*.

question whether enteric methane emissions from ruminants can be mitigated at all without lowering their productivity and concluded that it is possible when certain boundary conditions are considered. An important issue is also the costs associated with feeding tannins. The economic result is influenced by the costs of purchasing the sources and by the indirect costs resulting from growing plants with lower yield and lower energetic value. To compensate for this, these plants either have to be combined with costly higher quality feeds or a lower income from a lower performance of the animals has to be accepted, or both. Growing the tannin sources on farm would be preferable, but only when the effect is as high as that of currently imported sources. Future political and food industrial developments such as compensation payments, fines and high-price labelling of foods produced with environmental and product quality benefits will also be decisive for the question of whether or not tannin supplementation becomes affordable and attractive to the individual farmer. Further research is needed in parallel to the promotion of on-farm implementation of tannin supplements. Large gaps in knowledge include proper tannin analytics and the relation of their structural properties to the recovery of desired tannin effects. This could lead to indicators that allow the quick determination (best *in vitro*) of the suitability of batches of tannin sources before their use on farm.

8 Where to look for further information

In 2006, Mueller-Harvey described the chemistry of the tannins and their role in animal nutrition in more detail than it was possible in this chapter (see: Unravelling the conundrum of tannins in animal nutrition and health, *J. Sci. Food Agric.* 86 (10), 2010–37). A recent article [Arndt et al. (2022). Full adoption of the most effective strategies to mitigate methane emissions by ruminants can help meet the 1.5°C target by 2030 but not 2050, *PNAS* 119, e2111294119] compares the efficiency of dietary measures, including the use of tannins, against methane emissions and their potential quantitative role in assisting to be able to meet the current climate change mitigation targets.

9 References

Abarghueia, M. J., Rouzbehan, Y., Salem, A. Z. M. and Zamiric, M. J. (2021). Effects of pomegranate peel extract on ruminal and post-ruminal *in vitro* degradation of rumen inoculum of the dairy cow, *Anim. Biotechnol.* 32(3): 366–74.

Aboagye, I. A. and Beauchemin, K. A. (2019). Potential of molecular weight and structure of tannins to reduce methane emissions from ruminants: A review, *Animals (Basel)* 9(11): 856.

Aboagye, I. A., Oba, M., Castillo, A. R., Koenig, K. M., Iwaasa, A. D. and Beauchemin, K. A. (2018). Effects of hydrolyzable tannin with or without condensed tannin on methane emissions, nitrogen use, and performance of beef cattle fed a high-forage diet, *J. Anim. Sci.* 96(12): 5276–86.

Aboagye, I. A., Oba, M., Koenig, K. M., Zhao, G. Y. and Beauchemin, K. A. (2019). Use of gallic acid and hydrolyzable tannins to reduce methane emission and nitrogen excretion in beef cattle fed a diet containing alfalfa silage, *J. Anim. Sci.* 97(5): 2230–44.

Azuhnwi, B. N., Boller, B., Dohme-Meier, F., Hess, H. D., Kreuzer, M., Stringano, E. and Mueller-Harvey, I. (2013a). Exploring variation in proanthocyanidin composition and content of sainfoin (*Onobrychis viciifolia*), *J. Sci. Food Agric.* 93(9): 2102–9.

Azuhnwi, B. N., Hertzberg, H., Arrigo, Y., Gutzwiller, A., Hess, H. D., Mueller-Harvey, I., Torgerson, P. R., Kreuzer, M. and Dohme-Meier, F. (2013b). Investigation of sainfoin (*Onobrychis viciifolia*) cultivar differences on nitrogen balance and fecal egg count in artificially infected lambs, *J. Anim. Sci.* 91(5): 2343–54.

Barone, C. D., Zajac, A. M., Ferguson, S. M., Brown, R. N., Reed, J. D., Krueger, C. G. and Petersson, K. H. (2019). *In vitro* screening of 51 birdsfoot trefoil (*Lotus corniculatus* L.; Fabaceae) strains for anti-parasitic effects against *Haemonchus contortus*, *Parasitology* 146(6): 828–36.

Beauchemin, K. A., McGinn, S. M., Martinez, T. F. and McAllister, T. A. (2007). Use of condensed tannin extract from quebracho trees to reduce methane emissions from cattle, *J. Anim. Sci.* 85(8): 1990–6.

Bhat, T. K., Singh, B. and Sharma, O. P. (1998). Microbial degradation of tannins – A current perspective, *Biodegradation* 9(5): 343–57.

Birkinshaw, A., Kreuzer, M., Sutter, M., Reidy, B. and Terranova, M. (2022). Effects of early herbage cutting and vine leaves on methane emission, urine N losses and the milk fatty acid profile of dairy cows, *J. Dairy Sci.* 105(9): 7416–31.

Birkinshaw, A., Schwarm, A., Marquardt, S., Kreuzer, M. and Terranova, M. (2020). Rapid responses in bovine milk fatty acid composition and phenol content to various tanniferous forages, *J. Anim. Feed Sci.* 29(4): 297–305.

Bodas, R., López, S., Fernández, M., García-González, R., Rodríguez, A. B., Wallace, R. J. and González, J. S. (2008). *In vitro* screening of the potential of numerous plant species as antimethanogenic feed additives for ruminants, *Anim. Feed Sci. Technol.* 145(1-4): 245–58.

Bodas, R., Prieto, N., García-González, R., Andrés, S., Giráldez, F. J. and López, S. (2012). Manipulation of rumen fermentation and methane production with plant secondary metabolites, *Anim. Feed Sci. Technol.* 176(1-4): 78–93.

Brito, D. R. B., Costa-Júnior, L. M., Garcia, J. L., Torres-Acosta, J. F. J., Louvandini, H., Cutrim-Júnior, J. A. A., Araújo, J. F. M. and Soares, E. D. S. (2018). Supplementation with dry *Mimosa caesalpiniifolia* leaves can reduce the *Haemonchus contortus* worm burden of goats, *Vet. Parasitol.* 252: 47–51.

Buccioni, A., Pauselli, M., Viti, C., Minieri, S., Pallara, G., Roscini, V., Rapaccini, S., Marinucci, M. T., Lupi, P., Conte, G. and Mele, M. (2015). Milk fatty acid composition, rumen microbial population, and animal performances in response to diets rich in linoleic acid supplemented with chestnut or quebracho tannins in dairy ewes, *J. Dairy Sci.* 98(2): 1145–56.

Caetano, M., Wilkes, M. J., Pitchford, W. S., Lee, S. J. and Hynd, P. I. (2019). Effect of ensiled crimped grape marc on energy intake, performance and gas emissions of beef cattle, *Anim. Feed Sci. Technol.* 247: 166–72.

Canul-Solis, J., Campos-Navarrete, M., Piñeiro-Vázquez, A., Casanova-Lugo, F., Barros-Rodríguez, M., Chay-Canul, A., Cárdenas-Medina, J. and Castillo-Sánchez, L. (2020). Mitigation of rumen methane emissions with foliage and pods of tropical trees, *Animals (Basel)* 10(5): 843.

Carulla, J. E., Kreuzer, M., Machmüller, A. and Hess, H. D. (2005). Supplementation of *Acacia mearnsii* tannins decreases methanogenensis and urinary nitrogen in forage-fed sheep, *Aust. J. Agric. Res.* 56(9): 961–70.

Chung, Y. H., Mc Geough, E. J., Acharya, S., McAllister, T. A., McGinn, S. M., Harstad, O. M. and Beauchemin, K. A. (2013). Enteric methane emission, diet digestibility, and nitrogen excretion from beef heifers fed sainfoin or alfalfa, *J. Anim. Sci.* 91(10): 4861–74.

Clemensen, A. K., Villalba, J. J., Rottinghaus, G. E., Lee, S. T., Provenza, F. D. and Reeve, J. R. (2019). Do plant secondary metabolite-containing forages influence soil processes in pasture systems?, *Agron. J.* 112(5): 3744–57.

Cortés, J. E., Moreno, B., Pabón, M. L., Avila, P., Kreuzer, M., Hess, H. D. and Carulla, J. E. (2009). Effects of purified condensed tannins extracted from Calliandra, Flemingia and Leucaena on ruminal and postruminal degradation of soybean meal as estimated *in vitro*, *Anim. Feed Sci. Technol.* 151(3-4): 194–204.

Crozier, A. (2003). Classification and biosynthesis of secondary plant products: An overview. In: Goldberg, G. (Ed.), *Diet and Health: Plants*. Oxford, UK: Blackwell Publishing Science, pp. 27–48, doi: 10.1002/9780470774465.ch2.

Denninger, T. M., Schwarm, A., Birkinshaw, A., Terranova, M., Dohme-Meier, F., Münger, A., Eggerschwiler, L., Babst, B., Wegmann, S., Clauss, M. and Kreuzer, M. (2020). Immediate effect of *Acacia mearnsii* tannins on methane emissions and milk fatty acid profiles of dairy cows, *Anim. Feed Sci. Technol.* 261: 114388.

Duval, B. D., Aguerre, M., Wattiaux, M., Vadas, P. A. and Powell, J. M. (2016). Potential for reducing on-farm greenhouse gas and ammonia emissions from dairy cows with prolonged dietary tannin additions, *Water Air Soil Pollut.* 227(9): 329.

Fagundes, G. M., Benetel, G., Welter, K. C., Melo, F. A., Muir, J. P., Carriero, M. M., Souza, R. L. M., Meo-Filho, P., Frighetto, R. T. S., Berndt, A. and Bueno, I. C. S. (2020). Tannin as a natural rumen modifier to control methanogenesis in beef cattle in tropical systems: Friend or foe to biogas energy production?, *Res. Vet. Sci.* 132: 88-96.

Flachowsky, G. and Lebzien, P. (2012). Effects of phytogenic substances on rumen fermentation and methane emissions: A proposal for a research process, *Anim. Feed Sci. Technol.* 176(1-4): 70-7.

Frutos, P., Hervás, G., Giráldez, F. J. and Mantecón, A. R. (2004). Review: Tannins and ruminant nutrition, *Span. J. Agric. Res.* 2(2): 191-202.

García-González, R., López, S., Fernández, M., Bodas, R. and González, J. S. (2008). Screening the activity of plants and spices for decreasing ruminal methane production *in vitro*, *Anim. Feed Sci. Technol.* 147(1-3): 36-52.

Gerlach, K., Pries, M., Tholen, E., Schmithausen, A. J., Büscher, W. and Südekum, K. H. (2018). Effect of condensed tannins in rations of lactating dairy cows on production variables and nitrogen use efficiency, *Animal* 12(9): 1847-55.

Giller, K., Bossut, L., Eggerschwiler, L. and Terranova, M. (2022). *In vitro* ruminal fermentation, methane production and nutrient degradability as affected by fruit and vegetable pomaces in differing concentrations, *J. Anim. Physiol. Anim. Nutr. (Berl)* 106(5): 957-67.

Girard, M., Dohme-Meier, F., Wechsler, D., Goy, D., Kreuzer, M. and Bee, G. (2016a). Ability of 3 tanniferous forage legumes to modify quality of milk and Gruyère-type cheese, *J. Dairy Sci.* 99(1): 205-20.

Girard, M., Dohme-Meier, F., Silacci, P., Ampuero Kragten, S., Kreuzer, M. and Bee, G. (2016b). Forage legumes rich in condensed tannins may increase n-3 fatty acid levels and sensory quality of lamb meat, *J. Sci. Food Agric.* 96(6): 1923-33.

Grainger, C. and Beauchemin, K. A. (2011). Can enteric methane emissions from ruminants be lowered without lowering their production?, *Anim. Feed Sci. Technol.* 166-167: 308-20.

Grainger, C., Clarke, T., Auldist, M. J., Beauchemin, K. A., McGinn, S. M., Waghorn, G. C. and Eckard, R. J. (2009). Potential use of *Acacia mearnsii* condensed tannins to reduce methane emissions and nitrogen excretion from grazing dairy cows, *Can. J. Anim. Sci.* 89(2): 241-51.

Grosse Brinkhaus, A., Bee, G., Silacci, P., Kreuzer, M. and Dohme-Meier, F. (2016). Effect of exchanging *Onobrychis viciifolia* and *Lotus corniculatus* for *Medicago sativa* on ruminal fermentation and nitrogen turnover in dairy cows, *J. Dairy Sci.* 99(6): 4384-97.

Hao, X., Benke, M. B., Li, C., Larney, F. J., Beauchemin, K. A. and McAllister, T. A. (2011). Nitrogen transformations and greenhouse gas emissions during composting of manure from cattle fed diets containing corn dried distillers grains with solubles and condensed tannins, *Anim. Feed Sci. Technol.* 166-167: 539-49.

Harrison, M. T., McSweeney, C., Tomkins, N. W. and Eckard, R. J. (2015). Improving greenhouse gas emissions intensities of subtropical and tropical beef farming systems using *Leucaena leucocephala*, *Agric. Syst.* 136: 138-46.

Heckendorn, F., Häring, D. A., Maurer, V., Senn, M. and Hertzberg, H. (2007). Individual administration of three tanniferous forage plants to lambs artificially infected with *Haemonchus contortus* and *Cooperia curticei*, *Vet. Parasitol.* 146(1-2): 123-34.

Heckendorn, F., Häring, D. A., Maurer, V., Zinsstag, J., Langhans, W. and Hertzberg, H. (2006). Effect of sainfoin (*Onobrychis viciifolia*) silage and hay on established populations of *Haemonchus contortus* in lambs, *Vet. Parasitol.* 142(3-4): 293-300.

Hess, H. D., Tiemann, T. T., Noto, F., Franzel, S., Lascano, C. E. and Kreuzer, M. (2006). The effects of cultivation site on forage quality of *Calliandra calothyrsus* var. Patulul, *Agroforest. Syst.* 68(3): 209-20.

Hindrichsen, I. K., Osuji, P. O., Odenyo, A. A., Madsen, J. and Hvelplund, T. (2002). The effect of supplementation of a basal diet of maize stover fed to sheep with different amounts of *Leucaena diversifolia* on intake, digestibility, nitrogen metabolism and rumen parameters in sheep, *Anim. Feed Sci. Technol.* 98(3-4): 131-42.

Hindrichsen, I. K., Osuji, P. O., Odenyo, A. A., Madsen, J. and Hvelplund, T. (2004). Effect of supplementation of maize stover with foliage of various tropical multipurpose trees and *Lablab purpureus* on intake, rumen fermentation, digesta kinetics and microbial protein supply of sheep, *Anim. Feed Sci. Technol.* 113(1-4): 83-96.

Huyen, N. T., Desrues, O., Alferink, S. J. J., Zandstra, T., Verstegen, M. W. A., Hendriks, W. H. and Pellikaan, W. F. (2016). Inclusion of sainfoin (*Onobrychis viciifolia*) silage in dairy cow rations affects nutrient digestibility, nitrogen utilization, energy balance, and methane emissions, *J. Dairy Sci.* 99(5): 3566-77.

Jayanegara, A., Kreuzer, M., Wina, E. and Leiber, F. (2011). Significance of phenolic compounds in tropical forages for the ruminal bypass of polyunsaturated fatty acids and the appearance of biohydrogenation intermediates as examined *in vitro*, *Anim. Prod. Sci.* 51(12): 1127-36.

Jayanegara, A., Leiber, F. and Kreuzer, M. (2012a). Meta-analysis of the relationship between dietary tannin level and methane formation in ruminants from *in vivo* and *in vitro* experiments, *J. Anim. Physiol. Anim. Nutr. (Berl)* 96(3): 365-75.

Jayanegara, A., Kreuzer, M. and Leiber, F. (2012b). Ruminal disappearance of polyunsaturated fatty acids and appearance of biohydrogenation products when incubating linseed oil with alpine forage plant species *in vitro*, *Livest. Sci.* 147(1-3): 104-12.

Jayanegara, A., Marquardt, S., Wina, E., Kreuzer, M. and Leiber, F. (2013). *In vitro* indications for favourable non-additive effects on ruminal methane mitigation between high-phenolic and high-quality forages, *Br. J. Nutr.* 109(4): 615-22.

Kamra, D. N., Patra, A. K., Chatterjee, P. N., Kumar, R., Agarwal, N. and Chaudhary, L. C. (2008). Effect of plant extracts on methanogenesis and microbial profile of the rumen of buffalo: A brief overview, *Aust. J. Exp. Agric.* 48(2): 175-8.

Kapp-Bitter, A. N., Dickhoefer, U., Suglo, E., Baumgartner, L., Kreuzer, M. and Leiber, F. (2020). Graded supplementation of chestnut tannins to dairy cows fed protein-rich spring pasture: Effects on indicators of protein utilization, *J. Anim. Feed Sci.* 29(2): 97-104.

Kapp-Bitter, A. N., Dickhoefer, U., Kaptijn, G., Pedan, V., Erler, E., Kreuzer, M. and Leiber, F. (2021a). On-farm examination of sainfoin supplementation effects in dairy cows in a roughage-based feeding system: Indicators of protein utilization, *Livest. Sci.* 248: 104509.

Kapp-Bitter, A. N., Dickhoefer, U., Kreuzer, M. and Leiber, F. (2021b). Mature herbs as supplements to ruminant diets: Effects on *in vitro* ruminal fermentation and ammonia production, *Anim. Prod. Sci.* 61(5): 470-9.

Kelln, B. M., Penner, G. B., Acharya, S. N., McAllister, T. A. and Lardner, H. A. (2021). Impact of condensed tannin-containing legumes on ruminal fermentation, nutrition, and performance in ruminants: A review, *Can. J. Anim. Sci.* 101(2): 210-23.

Khiaosa-ard, R., Bryner, S. F., Scheeder, M. R. L., Wettstein, H. R., Leiber, F., Kreuzer, M. and Soliva, C. R. (2009). Evidence for inhibition of the terminal step of ruminal α-linolenic acid biohydrogenation by condensed tannins, *J. Dairy Sci.* 92(1): 177-88.

Koenig, K. M., Beauchemin, K. A. and McGinn, S. M. (2018). Feeding condensed tannins to mitigate ammonia emissions from beef feedlot cattle fed high-protein finishing diets containing distillers grains, *J. Anim. Sci.* 96(10): 4414-30.

Lavrenčič, A. and Levart, A. (2021). *In vitro* dry matter and crude protein rumen degradation and abomasal digestibility of soybean meal treated with chestnut and quebracho wood extracts, *Food Sci. Nutr.* 9(2): 1034-9.

Leparmarai, P. T., Kunz, C., Mwangi, D. M., Gluecks, I., Kreuzer, M. and Marquardt, S. (2021). Camels and cattle respond differently in milk phenol excretion and milk fatty acid profile to free ranging conditions in East-African rangelands, *Sci. Afr.* 13: e00896.

Leparmarai, P. T., Sinz, S., Kunz, C., Liesegang, A., Ortmann, S., Kreuzer, M. and Marquardt, S. (2019). Transfer of total phenols from a grapeseed-supplemented diet to dairy sheep and goat milk, and effects on performance and milk quality, *J. Anim. Sci.* 97: 1840-51.

Lima, P. M. T., Crouzoulon, P., Sanches, T. P., Zabre, G., Kabore, A., Niderkorn, V., Hoste, H., Amarante, A. F. T. D., Costa-Junior, L. M., Abdalla, A. L. and Louvandin, H. (2019). Effects of *Acacia mearnsii* supplementation on nutrition, parasitological, blood parameters and methane emissions in Santa Ines sheep infected with *Trichostrongylus colubriformis* and *Haemonchus contortus*, *Exp. Parasitol.* 207: 107777.

Lonngren, K. J., Barone, C. D., Zajac, A. M., Brown, R. N., Reed, J. D., Krueger, C. G. and Petersson, K. H. (2020). Effect of birdsfoot trefoil cultivars on exsheathment of *Haemonchus contortus* in fistulated sheep, *Vet. Parasitol.* 287: 109271.

Makkar, H. P. S. (2003a). *Quantification of Tannins in Tree and Shrub Foliage: A Laboratory Manual*. Dordrecht, The Netherlands: Kluwer Academic Publishers.

Makkar, H. P. S. (2003b). Effects and fate of tannins in ruminant animals, adaptation to tannins, and strategies to overcome detrimental effects of feeding tannin-rich feeds, *Small Rumin. Res.* 49(3): 241-56.

Min, B. R., Willis, W., Parker, D., Casey, K., Castleberry, L., Waldrip, H. and Parker, D. (2022). Condensed and hydrolyzable tannins for reducing methane and nitrous oxide emissions in dairy manure - A laboratory incubation study, *Animals*, 12: 2876.

Min, B. R. and Solaiman, S. (2018). Comparative aspects of plant tannins on digestive physiology, nutrition and microbial community changes in sheep and goats: A review, *J. Anim. Physiol. Anim. Nutr. (Berl)* 102(5): 1181-93.

Moate, P. J., Williams, S. R. O., Torok, V. A., Hannah, M. C., Ribaux, B. E., Tavendale, M. H., Eckard, R. J., Jacobs, J. L., Auldist, M. J. and Wales, W. J. (2014). Grape marc reduces methane emissions when fed to dairy cows, *J. Dairy Sci.* 97(8): 5073-87.

Morales, R. and Ungerfeld, E. M. (2015). Use of tannins to improve fatty acids profile of meat and milk quality in ruminants: A review, *Chil. J. Agric. Res.* 75(2): 239-48.

Mueller-Harvey, I., Bee, G., Dohme-Meier, F., Hoste, H., Karonen, M., Kölliker, R., Lüscher, A., Niderkorn, V., Pellikaan, W. F., Salminen, J.-P., Skøt, L., Smith, L. M. J., Thamsborg, S. M., Totterdell, P., Wilkinson, I., Williams, A. R., Azuhnwi, B. N., Baert, N., Grosse

Brinkhaus, A. G., Copani, G., Desrues, O., Drake, C., Engström, M., Fryganas, C., Girard, M., Huyen, N. T., Kempf, K., Malisch, C., Mora-Ortiz, M., Quijada, J., Ramsay, A., Ropiak, H. M. and Waghorn, G. C. (2019). Benefits of condensed tannins in forage legumes fed to ruminants: Importance of structure, concentration, and diet composition, *Crop Sci.* 59(3): 861-85.

Naumann, H. D., Tedeschi, L. O., Zeller, W. E. and Huntley, N. F. (2017). The role of condensed tannins in ruminant animal production: Advances, limitations and future directions, *Rev. Bras. Zootec.* 46(12): 929-49.

Nawab, A., Li, G., An, L., Nawab, Y., Zhao, Y., Xiao, M., Tang, S. and Sun, C. (2020). The potential effect of dietary tannins on enteric methane emission and ruminant production, as an alternative to antibiotic feed additives - a review, *Ann. Anim. Sci.* 20(2): 355-88.

Orlandi, T., Kozloski, G. V., Alves, T. P., Mesquita, F. R. and Ávila, S. C. (2015). Digestibility, ruminal fermentation and duodenal flux of amino acids in steers fed grass forage plus concentrate containing increasing levels of *Acacia mearnsii* tannin extract, *Anim. Feed Sci. Technol.* 210: 37-45.

Pathak, A. K., Dutta, N., Pattanaik, A. K., Chaturvedi, V. B. and Sharma, K. (2017). Effect of condensed tannins from *Ficus infectoria* and *Psidium guajava* leaf meal mixture on nutrient metabolism, methane emission and performance of lambs, Asian-Australas. *J. Anim. Sci.* 30(12): 1702-10.

Patra, A. K., Kamra, D. N., Bhar, R., Kumar, R. and Agarwal, N. (2011). Effect of *Terminalia chebula* and *Allium sativum* on *in vivo* methane emission by sheep, *J. Anim. Physiol. Anim. Nutr. (Berl)* 95(2): 187-91.

Patra, A. K. and Saxena, J. (2011). Exploitation of dietary tannins to improve rumen metabolism and ruminant nutrition, *J. Sci. Food Agric.* 91(1): 24-37.

Piluzza, G., Sulas, L. and Bullitta, S. (2013). Tannins in forage plants and their role in animal husbandry and environmental sustainability: A review, *Grass Forage Sci.* 69(1): 32-48.

Ramírez-Restrepo, C. A. and Barry, T. N. (2005). Alternative temperate forages containing secondary compounds for improving sustainable productivity in grazing ruminants, *Anim. Feed Sci. Technol.* 120(3-4): 179-201.

Ramsay, A., Drake, C., Grosse Brinkhaus, A., Girard, M., Copani, G., Dohme-Meier, F., Bee, G., Niderkorn, V. and Mueller-Harvey, I. (2015). Sodium hydroxide enhances extractability and analysis of proanthocyanidins in ensiled sainfoin (*Onobrychis viciifolia*), *J. Agric. Food Chem.* 63(43): 9471-79.

Reed, J. D. (1995). Nutritional toxicology of tannins and related polyphenols in forage legumes, *J. Anim. Sci.* 73(5): 1516-28.

Sinz, S., Kunz, C., Liesegang, A., Braun, U., Marquardt, S., Soliva, C. R. and Kreuzer, M. (2018). *In vitro* bioactivity of various pure flavonoids in ruminal fermentation, with special reference to methane formation, *Czech J. Anim. Sci.* 63(8), 293-304.

Sinz, S., Liesegang, A., Kreuzer, M. and Marquardt, S. (2019a). Do supplements of *Acacia mearnsii* and grapeseed extracts alone and in combination alleviate metabolic nitrogen load and manure nitrogen emissions of lambs fed a high crude protein diet?, *Arch. Anim. Nutr.* 73(4): 306-23.

Sinz, S., Marquardt, S., Soliva, C. R., Braun, U., Liesegang, A. and Kreuzer, M. (2019b). Phenolic plant extracts are additive in their effects against *in vitro* ruminal methane and ammonia formation, Asian-Australas. *J. Anim. Sci.* 32(7): 966-76.

Śliwiński, B. J., Kreuzer, M., Sutter, F., Machmüller, A. and Wettstein, H. R. (2004). Performance, body nitrogen conversion and nitrogen emission from manure of dairy cows fed diets supplemented with different plant extracts, *J. Anim. Feed Sci.* 13(1): 73-91.

Śliwiński, B. J., Kreuzer, M., Wettstein, H. R. and Machmüller, A. (2002). Rumen fermentation and nitrogen balance of lambs fed diets containing plant extracts rich in tannins and saponins, and associated emissions of nitrogen and methane, *Arch. Anim. Nutr.* 56(6): 379-92.

Smith, A. H., Zoetendal, E. and Mackie, R. I. (2005). Bacterial mechanisms to overcome inhibitory effects of dietary tannins, *Microb. Ecol.* 50(2): 197-205.

Staerfl, S. M., Kreuzer, M. and Soliva, C. R. (2010). *In vitro* screening of unconventional feeds and various natural supplements for their ruminal methane mitigation potential when included in a maize-silage based diet, *J. Anim. Feed Sci.* 19(4): 651-64.

Staerfl, S. M., Soliva, C. R., Leiber, F. and Kreuzer, M. (2011). Fatty acid profile and oxidative stability of the perirenal fat of bulls fattened on grass silage and maize silage supplemented with tannins, garlic, maca and lupines, *Meat Sci.* 89(1): 98-104.

Staerfl, S. M., Zeitz, J. O., Kreuzer, M. and Soliva, C. R. (2012). Methane conversion rate of bulls fattened on grass or maize silage as compared with the IPCC default values, and the long-term methane mitigation efficiency of adding acacia tannin, garlic, maca and lupine, *Agric. Ecosyst. Environ.* 148: 111-20.

Stewart, E. K., Beauchemin, K. A., Dai, X., MacAdam, J. W., Christensen, R. G. and Villalba, J. J. (2019). Effect of tannin-containing hays on enteric methane emissions and nitrogen partitioning in beef cattle, *J. Anim. Sci.* 97(8): 3286-99.

Swanson, B. G. (2003). Tannins and polyphenols. In: Caballero, B., Trugo, L. C. and Finglas, P. M. (Eds.), *Encyclopaedia of Food Sciences and Nutrition*. London, UK: Academic Press, pp. 5729-33.

Terranova, M., Eggerschwiler, L., Ortmann, S., Clauss, M., Kreuzer, M. and Schwarm, A. (2021). Increasing the proportion of hazel leaves in the diet of dairy cows reduced methane yield and excretion of nitrogen in volatile form, but not milk yield, *Anim. Feed Sci. Technol.* 276: 114790.

Terranova, M., Kreuzer, M., Braun, U. and Schwarm, A. (2018). *In vitro* screening of temperate climate forages from a variety of woody plants for their potential to mitigate ruminal methane and ammonia formation, *J. Agric. Sci.* 156(7): 929-41.

Terranova, M., Wang, S., Eggerschwiler, L., Braun, U., Kreuzer, M. and Schwarm, A. (2020). Dose-response effects of woody and herbaceous forage plants to *in vitro* ruminal methane and ammonia formation, and their short-term palatability in lactating cows, *Animal* 14(3): 538-48.

Tiemann, T. T., Avila, P., Ramízrez, G., Lascano, C. E., Kreuzer, M. and Hess, H. D. (2008a). *In vitro* ruminal fermentation of tanniniferous tropical plants: Plant-specific tannin effects and counteracting efficiency of PEG, *Anim. Feed Sci. Technol.* 146(3-4): 222-41.

Tiemann, T. T., Lascano, C. E., Wettstein, H. R., Mayer, A. C., Kreuzer, M. and Hess, H. D. (2008b). Effect of the tropical tannin-rich shrub legumes *Calliandra calothyrsus* and *Flemingia macrophylla* on methane emission, nitrogen and energy balance in growing lambs, *Animal* 2(5): 790-9.

Tiemann, T. T., Lascano, C. E., Kreuzer, M. and Hess, H. D. (2008c). The ruminal degradability of fibre explains part of the low nutritional value and reduced methanogenesis in highly tanniniferous tropical legumes, *J. Sci. Food Agric.* 88(10): 1794-803.

Tiemann, T. T., Hincapie, B., Frossard, E., Kreuzer, M. and Hess, H. D. (2009a). Effect of supplementing tropical tannin-free and tanniniferous legumes to grass-fed sheep on the utility of their manure as nitrogen fertilizer, *Livest. Res. Rural Dev.* 21(3). Available at: www.lrrd.org/lrrd21/3/tiem21041.htm.

Tiemann, T. T., Franco, L. H., Peters, M., Frossard, E., Kreuzer, M., Lascano, C. E. and Hess, H. D. (2009b). Effect of season, soil type and fertilizer on the biomass production and chemical composition of five tropical shrub legumes with forage potential, *Grass Forage Sci.* 64(3): 255-65.

Tiemann, T. T., Franco, L. H., Ramirez, G., Kreuzer, M., Lascano, C. E. and Hess, H. D. (2010a). Influence of cultivation site and fertilisation on the properties of condensed tannins and *in vitro* ruminal nutrient degradation of *Calliandra calothyrsus*, *Flemingia macrophylla* and *Leucaena leucocephala*, *Anim. Feed Sci. Technol.* 157(1-2): 30-40.

Tiemann, T. T., Cortés, J. E., Pabón, M. L., Hess, H. D., Kreuzer, M. and Carulla, J. E. (2010b). *In vitro* evidence for the importance of cultivation conditions on the effects of Calliandra tannins on ruminal escape of soybean protein and its postruminal degradability, *J. Anim. Physiol. Anim. Nutr. (Berl)* 94(6): e225-30.

Vasta, V., Daghio, M., Cappucci, A., Buccioni, A., Serra, A., Viti, C. and Mele, M. (2019). Invited review: Plant polyphenols and rumen microbiota responsible for fatty acid biohydrogenation, fiber digestion, and methane emission: Experimental evidence and methodological approaches, *J. Dairy Sci.* 102(5): 3781-804.

Vasta, V., Yáñez-Ruiz, D. R., Mele, M., Serra, A., Luciano, G., Lanza, M., Biondi, L. and Priolo, A. (2010). Bacterial and protozoal communities and fatty acid profile in the rumen of sheep fed a diet containing added tannins, *Appl. Environ. Microbiol.* 76(8): 2549-55.

Villalba, J. J., Provenza, F. D. and Manteca, X. (2010). Links between ruminants' food preference and their welfare, *Animal* 4(7): 1240-7.

Waghorn, G. C., Tavendale, M. H. and Woodfield, D. R. (2002). Methanogenesis from forages fed to sheep, *Proc. N. Z. Soc. Anim. Prod.* 62: 159-65.

Wang, S., Giller, K., Hillmann, E., Marquardt, S. and Schwarm, A. (2019). Effect of supplementation of pelleted hazel (*Corylus avellana*) leaves on blood antioxidant activity, cellular immune response and heart beat parameters in sheep, *J. Anim. Sci.* 97(11): 4496-502.

Wang, S., Terranova, M., Kreuzer, M., Marquardt, S., Eggerschwiler, L. and Schwarm, A. (2018). Supplementation of pelleted hazel (*Corylus avellana*) leaves decreases methane and urinary nitrogen emissions by sheep at unchanged forage intake, *Sci. Rep.* 8(1): 5427.

Williams, S. R. O., Hannah, M. C., Eckard, R. J., Wales, W. J. and Moate, P. J. (2020). Supplementing the diet of dairy cows with fat or tannin reduces methane yield, and additively when fed in combination, *Animal* 14(S3): S464-72.

Wischer, G., Greiling, A. M., Boguhn, J., Steingass, H., Schollenberger, M., Hartung, K. and Rodehutscord, M. (2014). Effects of long-term supplementation of chestnut and valonea extracts on methane release, digestibility and nitrogen excretion in sheep, *Animal* 8(6): 938-48.

Woodward, S. L., Waghorn, G. C. and Laboyrie, P. G. (2004). Condensed tannins in birdsfoot trefoil (*Lotus corniculatus*) reduce methane emissions from dairy cows, *Proc. N. Z. Soc. Anim. Prod.* 64: 160-4.

Woodward, S. L., Waghorn, G. C., Ulyatt, M. J. and Lassey, K. R. (2001). Early indications that feeding Lotus will reduce methane emissions from ruminants, *Proc. N. Z. Soc. Anim. Prod.* 61: 23-6.

Yurtseven, S., Avci, M., Çetin, M., Öztürk, I. and Boğa, M. (2018). Emissions of some greenhouse gases from the manure of ewes fed on pomegranate peel, yucca extract, and thyme oil, *Appl. Ecol. Environm. Res.* 16(4): 4217-28.

Chapter 6

The use of plant extracts as dietary supplements in dairy cow nutrition: saponins

Chaouki Benchaar, Sherbrooke Research and Development Centre, Agriculture and Agri-Food Canada, Canada

1 Introduction

In livestock production, feed additives are routinely used to increase animal productivity by improving efficiency of feed conversion into animal products (e.g. milk, meat), animal performance, health and preventing metabolic disorders (e.g. acidosis). Most nutrients supplied to the ruminant are produced from microbial metabolism of carbohydrates (i.e. structural and non-structural), protein and non-protein nitrogen. Therefore, the main objectives of nutritionists are to optimize specific rumen functions for the benefit of the animal. Feed additives are primarily used to improve nutrient utilization and efficiency of rumen fermentation and enhance animal productivity. For example, feed additives can be used to: (i) increase propionate production in the rumen to improve the energy balance of the animal; (ii) reduce dietary protein degradation in the rumen to increase availability of amino acids (AA) in the small intestine; (iii) reduce degradation rate of rapidly fermentable carbohydrates (e.g. starch,

http://dx.doi.org/10.19103/AS.2022.0117.06
Published by Burleigh Dodds Science Publishing Limited, 2023.

sucrose) and control lactic acid concentration; and (iv) enhance fibre digestion in the rumen (Jouany and Morgavi, 2007). Based on these considerations, several feed additives have been developed and used to achieve one or more of these four targets. Recent advances in understanding rumen function and its microbiome and the increasing concern over the contribution from cattle in greenhouse gas (GHG) emissions has resulted in the development of feed additives to mitigate enteric methane (CH_4) production.

Among the feed additives used in ruminant nutrition, the ionophore monensin have been extensively used as a rumen modifier. Monensin is known for reducing protein degradation in the rumen (Tedeschi et al., 2003), thereby increasing the flow of AA to the small intestine (Yang and Russell, 1993); increasing propionate production (H_2-consuming pathway) at the expense of that of acetate (H_2-producing pathway) thereby reducing H_2 availability for CH_4 synthesis by methanogens (Callaway et al., 1997); and slowing the rate of acid production and helping to prevent lactic acidosis (Osborne et al., 2004). Monensin has also been shown to be helpful in reducing the incidence of bloat in cattle grazing legume pastures (Lowe et al., 1991).

Despite benefiting animal production and productivity, the use feed additives (i.e. chemical and antibiotics) has become increasingly controversial due to the potential appearance of chemical residues in food and potential emergence of multi-drug-resistant bacteria that pose a risk to human health, if the antibiotics are also used in human health (Benchaar et al., 2009; Benchaar and Greathead, 2011). Feed antibiotics (e.g. monesin) are used at subtherapeutic levels and are not currently used in human medicine and thus, may not contribute to antibiotic resistance. Despite this, they have been banned in Europe from use in livestock production (European Union, 2003). Accordingly, the livestock feed additive industry and scientists have been actively seeking for alternatives to feed antibiotics. Plants and their secondary metabolites offer a tremendous opportunity for development in this regard (Wallace 2004; Benchaar et al., 2007; Tedeschi et al., 2021). Many plants produce secondary metabolites that have antimicrobial properties (Conner 1993; Greathead, 2003) and many of these secondary metabolites have been assessed for their potential to favourably alter rumen microbial fermentation (Benchaar et al., 2007; Tedeschi et al., 2021). As a consequence of this intense research and development work, several feed additives based on plants, plant extracts or synthetic secondary metabolites are commercially available for use in livestock nutrition and production. When feed additives are of plant origin (i.e. phytogenic), they are collectively referred as phytochemicals (Tedeschi et al., 2021).

The most known plant-derived additives are saponins, tannins and essential oils. The interest in the use of tannins and saponins in ruminant nutrition dates back to several decades. However, interest for essential oils has grown rapidly and significantly over the last decade (Benchaar et and Greathead, 2011; Tedeschi et al., 2021). In the last two decades, advancements research in phytochemicals has been stimulated by the increasing pressure from governments and the consumers on the livestock industry to mitigate greenhouse gas (i.e. CH_4, N_2O) and other (e.g. NH_3) emissions and reduce nutrient (e.g. nitrogen) excretion in the environment.

This chapter aims to provide a concise information on the potential use of saponins as a feed additive in dairy cow nutrition. It summarizes available literature information on sources, chemistry and antimicrobial properties (i.e. mode of action) of saponins, their effects on rumen microbial fermentation, their potential to mitigate enteric methane (CH_4) emissions and how their use may affect animal performance. Although the emphasis of this chapter is particularly on saponins for dairy cows, data from *in vitro* studies and those generated from experiments involving other ruminant species (e.g. sheep) were used for discussion.

2 Sources, chemistry, biosynthesis pathway and antimicrobial properties

Saponins are secondary metabolites naturally occurring in plants, few marine animals (e.g. starfish and sea cucumber) and some bacteria (Sparg et al., 2004; Güçlü-Üstündag and Mazza, 2007). Saponins form a stable foam in an aqueous solution hence the origin of their name 'saponins', which is derived from the Latin word 'sapo', meaning soap. This surfactant activity has been the source of humankind interest since ancient times. Saponins are present in wild and cultivated crops (Francis et al., 2002). They have been reported present in more than 100 plant families (Sparg et al., 2004; Güçlü-Üstündag and Mazza, 2007) and may occur in different parts of the plant such as roots, tubes, barks, leaves, seeds and fruits. The most widely used commercial sources of saponins in ruminant nutrition are *Yucca schidigera* and *Quillaja saponaria*. *Yucca schidigera* is a desert plant native to the southwestern United States and Mexico whereas *Quillaja saponaria* (soapbark tree) is a tree native to Chile (Cheeke, 1999). *Yucca schidigera* is rich in steroidal saponins whereas *Quillaja saponaria* contains triterpene saponins. According to Wina et al. (2005a), yucca extract contains 4.4% of steroid saponins, which consist of 28 different structures of spirostanol and furastonal glycosides. *Quillaja saponaria* extract contains 10% total saponins, and more than 20 different structures of triterpenoid saponin have been identified in the extract. The U.S. Food and Drug Administration has classified saponins of *Y. schidigera* and *Q. saponaria* under GRAS (generally recognised as safe) for human consumption (Fenwick et al., 1991).

Many legume plants contain saponins. With the exception of *Trigonella foenum-graecum* which contains saponins of steroidal type, most legume plants contain triterpenoid saponins. Wina et al. (2005a) suggested that despite only one main saponin or sapogenin structure is reported in legume plants, it actually occurs not as a single compound, but as various compounds with different sugar moieties. For instance, in different parts of alfalfa plant (leaves, seeds), 29 structures of saponins have been determined with medicagenic acid being the main sapogenin (Wina et al., 2005a). Like other plant secondary metabolites (e.g. essential oils), it appears that saponins play an important role by providing the plant with protection against microbial (e.g. bacterial and fungal) and insect attacks (Wallace, 2004; Benchaar et al., 2009). In several plant species, saponin production is often induced by abiotic stressors (e.g. humidity, temperature, light intensity, etc.), which influence the amount and the composition of saponin produced (Szakiel et al., 2011; De Costa et al., 2013). Thus, saponins may be considered a part of the defence system of the plants (Francis et al., 2002) and they have been included in a wide group of protective molecules found in plants named 'phytoanticipins' or 'phytoprotectants' (Morrissey and Osburn, 1999). Other factors such as the plant maturity, the environment and agronomic conditions may also affect saponin level in the plant (Yoshiki et al., 1998; Francis et al., 2002).

Saponins consist of a sugar moiety (i.e. oligosaccharide chain) with glycosidic links to a hydrophobic aglycone (sapogenin). Depending on the type of the aglycone, plants mainly synthesize two categories of saponins: steroidal or triterpenoid saponins (Fig. 1; Bruneton, 1999. Steroidal saponins comprise a steroidal aglycone and are almost exclusively found in monocotyledonous angiosperms. Triterpenoid saponins possess triterpene as the aglycone core and most commonly occur in dicotyledonous angiosperms. Both forms of saponins possess biological activities that are contingent upon the types, development stages and tissues of the plants. Podolak et al. (2010) suggested that the presence of different components (e.g. hydroxyl, hydroxymethyl, carboxyl and acyl groups) in the sapogenin, the difference in the composition, the linkage and the number of sugar chains may account for significant variation in their structure and thus, their bioactivity.

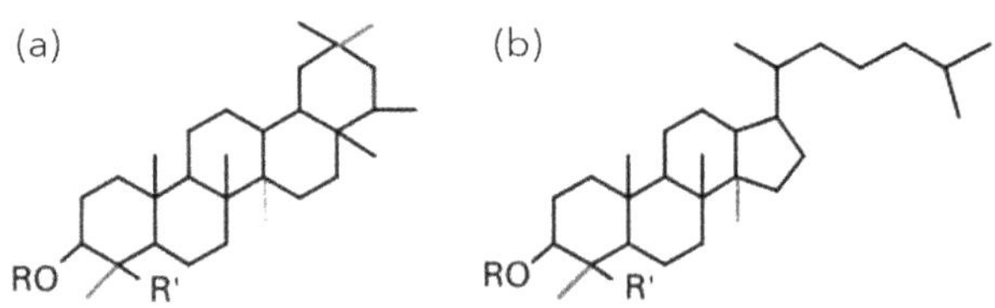

Figure 1 Chemical structure of sapogenins: (a) steroidal and (b) triterpenoid saponins. Source: Bruneton (1999).

Saponins are glycosides of triterpenes and steroids. Steroidal glycoalkaloids are sometimes also referred to as saponins. The triterpene and steroid backbones are both derived from the mevalonic acid (MVA) pathway, the common precursor being 2,3-oxidosqualene (Moses et al., 2014; Mugford and Osbourn, 2021). Steroidal saponins are synthesized via two pathways: the mevalonate (MVA) pathway and methylerythritol 4-phosphate (MEP) pathway (Fig. 2). Acetyl coA and glyceraldehyde 3-phosphate act as precursors for MVA and MEP pathways, respectively. In plants, the cyclization of 2,3-oxidosqualene is mediated by cycloartenol synthase and oxidosqualene cyclase (OSC) to synthesize steroidal and triterpenoid saponins (Abe, 2007; Naoumkina et al., 2010). The isoprenoid is the main pathway of biosynthesis of triterpenoid saponins and 2,3-oxidosqalene is the precursor of these saponins (Haralampidis et al., 2002; Kalinowska et al., 2005).

There is a general consensus that the biological effect of saponins is on the cell membrane. Saponins have the ability to form complexes with the lipid membrane of bacteria, thereby increasing their permeability, which generates an imbalance, and consequently, the lysis and the death of the microorganism (Makkar et al., 1995; Francis et al., 2002). The haemolytic effect of saponins appears to be related to the affinity of the aglycone moiety for membrane sterols, particularly cholesterol (Glauert et al., 1962), with which they form insoluble complexes (Bangham and Horne, 1962). Wallace et al. (2002) suggested that the sensitivity of protozoa to saponins is likely due to the interaction with the sterol moiety present the membrane of protozoa (Williams and Coleman, 1992). The sterol-binding property of saponins (Hostettmann and Marston, 1995) most likely causes the destruction of protozoal cell membranes.

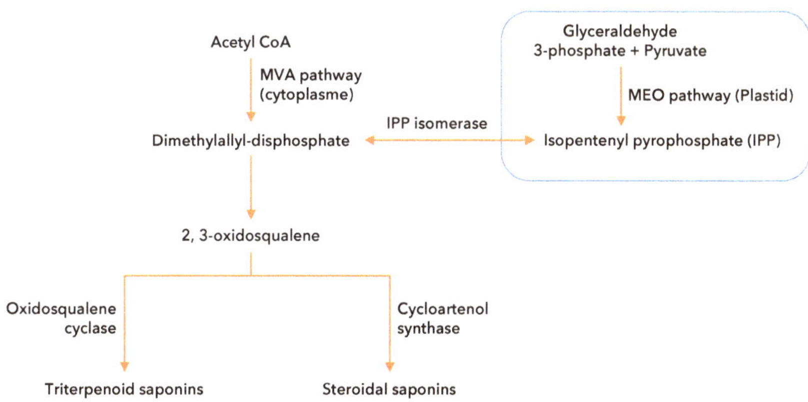

Figure 2 Simplified schematic representation of biosynthesis pathway of saponins. Source: Adapted from: Kavya et al. (2021).

3 Effects on rumen microbial populations

3.1 Protozoa

In the rumen, protozoa are able to engulf and digest bacteria thereby reducing the efficiency of nitrogen (N) use. Protozoa also have proteolytic and deamination activities (Williams and Coleman, 1992). Therefore, reducing (i.e. partial defaunation) or eliminating (i.e. total defaunation) of protozoa allows the reduction of N recycling between protozoa and bacteria and increases microbial protein flow from the rumen to the intestines (Teferedegne, 2000). A meta-analysis by Newbold et al. (2015) revealed that the suppression of protozoa from the rumen could increase microbial protein supply to the host by up to 30% and reduce CH_4 production by up to 11%. In addition, improving efficient utilization of N in the rumen may decrease N excretion into the environment, which may reduce potential nitrous oxide (N_2O), a potent greenhouse gas and ammonia (NH_3) emissions. The symbiotic (exo- and endo-) relationship between protozoa and methanogens in the rumen is well established (Newbold et al., 2015). This symbiosis occurs because of the ability of protozoa to produce H_2 in their hydrogenesomes (Fenchel and Finlay, 2006; Morgavi et al., 2012). Substantial methanogenic activity in the rumen has been linked to protozoa-related methanogens (Newbold et al., 1995; Guyader et al., 2014). It has been reported that up to 25% of rumen methanogens are associated with protozoa (Newbold et al., 1995). Thus, removal of ciliate protozoa from the rumen can help to mitigate enteric CH_4 production in the rumen and decrease energy losses associated with ruminal methanogenesis. However, elimination or reduction of ruminal ciliate protozoa should be performed with caution because some protozoa species are cellulolytic and their suppression would have adverse effects on fibre breakdown in the rumen (Newbold et al., 2015). In addition, protozoa, particularly, the entodiniomorphs protozoa have the capacity to rapidly engulf starch and avoiding its fermentation by ruminal bacteria to acids, which contributes to stabilizing ruminal pH (Williams and Coleman, 1992). It appears therefore, that a partial defaunation rather than a total elimination of protozoa is preferable so as not to impact the positive role of protozoa in ruminal metabolism. However, partial defaunation has not been always associated with positive impact on rumen metabolism and animal productivity (Wallace et al., 2002; Patra et al., 2006).

Several studies have assessed the effects of different sources of saponins on ruminal ciliate protozoa. However, very few studies have determined such effects in dairy cows (Table 1). From a compilation of 46 reports of *in vivo* and *in vitro* data on the effects of saponins or plants containing saponins on ruminal protozoa (Wina et al., 2005a), 28 reports revealed a decrease in protozoa

Table 1 Effects of feeding saponin extracts to dairy cows on rumen protozoa numbers, fermentation and methane production (Examples)

Extract (saponin concentration)	Dosage (g/kg DMI)[b]	Diet (F:C)	Effect[a] Protozoa	NH_3	TVFA	C_2/C_3	CH_4 (g/d)	CH_4 (g/kg DMI)	Reference
Y. schidigera (sarsaponins)	0.077	40:60	ND	=	=	=	ND	ND	Valdez et al. (1986)
Sarsaponin	0.044	67:33	ND	=	ND	ND	ND	ND	Goestch and Owens (1985)
Y. schidigera	0.03–0.12	38:62	ND	=	=	=	ND	ND	Wu et al. (1994)
Y. schidigera	0.38	51:49	ND	=	ND	ND	ND	ND	Wilson et al. 1998)
Y. schidigera (10%)	2.8	60:40	=	=	=	=	ND	ND	Benchaar et al. (2008)
Sarsaponin (9.5%)	0.20, 0.41, 0.62	36:64	=	-	=	=	ND	ND	Singer et al. (2008)
Y. schidigera (6%)	10	51:49	=	-	=	=	=	=	Holtshausen et al. (2009)
Q. saponaria (3%)	10	51:50	=	=	=	=	=	=	Holtshausen et al. (2009)
Tea (6.89%)	0.52%	50:50	=	=	+	=	=	=	Guyader et al. (2017)
Tea	0.03%	60:40	ND	ND	=	=	=	=	Kolling et al. (2018)

[a] =, no effect; −, significant decrease; +, significant increase; ND, not determined; DMI, dry matter intake; F:C, forage:concentrate ratio of the diet (on dry matter basis); TVFA, total volatile fatty acid concentration.
[b] Dose: as reported in the reference cited or calculated based on the reported dry matter intake and inclusion level of the extract.

numbers, 8 reports showed an inhibition of protozoal activity, 7 reports showed no effects and 3 showed an increase. The anti-protozoal effect of saponins or saponin-containing plants was mostly demonstrated *in vitro* and these *in vitro* observations were not always confirmed *in vivo* (Wina et al., 2005a), suggesting that caution must be taken when extrapolating *in vitro* findings to *in vivo* conditions. Literature data reveal inconsistent and variable responses of ruminal protozoa to saponins. Presumably, these variable results are due to several factors including the experimental approach (*in vitro* vs. *in vivo*), the dose of saponins, the source of saponin-containing plant/extract, the type of saponins (e.g. steroidal vs. terpinoidal) and composition of the diet fed to the animal.

Among the different sources of saponins assessed for their antiprotozal effect, steroidal saponins from *Y. schidigera* are probably the most investigated ones. *In vitro* supplementation with *Y. schidigera* saponins decreased protozoal count (Makkar et al., 1998; Pen et al., 2006) and *in vivo* (Valdez et al., 1986; Hristov et al., 1999). In contrast, other *in vitro* studies (Hristov et al., 2003; Śliwiński et al., 2002b) and *in vivo* (Pen et al., 2007; Baah et al., 2007) reported no effects of saponin-containing plants on ruminal protozoal populations. Likewise, Benchaar et al. (2008) reported that feeding 60 g/d of *Y. schidigera* extract (10% saponins) to dairy cows had no effect on total number and distribution by genera of ruminal ciliate protozoa.

It seems that the inhibitory effect of saponins is more pronounced when animals are fed high-concentrate versus high-forages diets. Lu and Jorgensen (1987) observed a larger decrease in protozoal number when pure alfalfa saponins were infused in the rumen of sheep fed concentrate-based diet than in the rumen of sheep fed forage-based diet (76% vs. 47% concentrate feeds). The anti-protozoal activity of alfalfa saponins (i.e. triterpenoid saponins) has also been demonstrated in other studies. For example, Kozłowska et al. (2020) observed *in vitro* a linear reduction in protozoa count as saponin level extracted from two cultivars (fresh or ensiled alfalfa) increased. Likewise, Klita et al. (1996) reported a decrease in protozoa number in sheep intra-ruminally infused with saponins extracted from alfalfa. Other saponin-containing plants such as seeds of fenugreek (*Trigonella foenum-graecum* L.) or leaves from Sesbania (*Sesbania sesban*) also possess anti-protozoal properties (Goel and Makkar, 2012).

Tea saponins have been assessed for their potential to inhibit ruminal protozoa. Guyader et al. (2017) reported *in vitro* a linear decrease (up to 50%) in protozoa counts with increasing addition of tea saponins (up to 0.50 mg/ml), but no change was observed when dairy cows were fed diets supplemented with 0.52% of tea saponins. This discrepancy between *in vitro* and *in vivo* approach was also observed by the same authors in a previous study (Guyader et al., 2015) when similar doses of tea saponins were added in non-lactating dairy cow diets. Popova et al. (2019) observed no change in the community structure of protozoal populations in the rumen of dairy cows fed diets supplemented with 0.5% of tea saponin.

The lack of effects of saponins on protozoa very often reported *in vivo* can be attributed to several factors that include among others, the capacity of rumen bacteria to hydrolyse saponins into sapogenins (Flåøyen et al., 2001; Flåøyen et al., 2002), compounds that cannot affect protozoa (Newbold et al., 1997; Ramos-Morales et al., 2017); salivary inactivation of saponins (Teferedegne, 2000); and the ability of protozoa to protect their membrane from its degradation by saponins by producing extracellular polysaccharides around the membrane (Wina et al., 2005a). In addition, adaptation of protozoa to ruminal exposure to saponins has also been reported (Newbold et al., 1997; Teferedegne, 2000; Ramos-Morales et al., 2017). Changing the chemical structure of saponins has been suggested as an option to prevent saponin degradation by rumen microbes and strengthen their anti-protozoal effects (Ramos-Morales et al., 2017).

3.2 Bacteria

Compared with protozoa, few studies have investigated the effects of saponins on ruminal bacteria. It has been suggested that gram-positive bacteria are more sensitive than gram-negative bacteria to saponins (Wallace et al., 1994; Wang et al., 2000b). Wallace et al. (1994) assessed the influence of *Y. schidigera* extract on growth of pure cultures of ruminal bacteria. The addition of the yucca extract to the growth medium at the rate of 1% (vol/vol) stimulated the growth of *Prevotella (Bacteroides) ruminicola*, had no effect on that of *Selenomonas ruminantium* whereas it inhibited the growth of *Butyrivibrio fibrisolvens* and *Streptococcus bovis*. In another study, saponins extracted from *Y. schidigera* inhibited the growth of *Streptococcus bovis*, *Prevotella bryantii* and *Ruminobacter amylophilus*, but stimulated that of *Selenomonas ruminantium* (Wang et al., 2000b). In the same study, the cellulolytic activity of *Ruminococcus albus* and *Ruminococcus flavefaciens* was markedly inhibited by saponins, whereas that of *Fibrobacter succinogenes* was less impacted (Wang et al., 2000b). This limited effect on *Fibrobacter succinogenes* can be attributed to their resistance to saponins as reported by Wang et al. (2000b) and Vinogradov et al. (2001). Wang et al. (2000b) observed that saponins extracted from *Y. schidigera* substantially and consistently inhibited the growth of cellulolytic bacteria but their effects on amylolytic bacteria were variable and species dependant. However, Wang et al. (1998) reported no change in celluloytic bacteria count upon the addition of *Y. schidigera extract* (0.5 mg/ml) in the RUSITEC fermenter. Using another source of saponins, Muetzel et al. (2003) observed *in vitro* that the addition of *Sesbania pachycarpa* leaves had minor effect on the growth of *Fibrobacter* spp. but enhanced that of *Ruminococcus* spp. (i.e. *albus* and *flavefaciens*). In contrast, *in vivo* studies (Diaz et al., 1993; Thalib et al., 1996) reported increases in cellulolytic and total bacteria in the rumen of sheep fed saponins from S. *saponaria* fruit or a methanol extract of *S. rarak*. Collectively, these results suggest that the effect of saponins on rumen bacteria varies with the source of saponins and is species dependant.

3.3 Fungi

Despite they represent only a small proportion of the rumen microbes, fungi fill an important niche in the ruminal microbial ecosystem as they contribute to the breakdown of fibre (William and Wither, 1991; Williams and Coleman, 1992). Information on the effects of saponins on ruminal fungi is limited but it is generally accepted that saponins are toxic to fungi (Wang et al., 2000a). It is thought that the anti-fungal activity of saponins is via their interaction with membrane sterols (Francis et al., 2002). Wang et al. (2000a) observed that the ruminal anaerobic fungi *Neocallimastix frontalis* and *Piromyces rhizinflata* are highly sensitive to *Y. schidigera* saponins being completely inhibited by concentrations as low as 2.25 μg/ml in the medium. Goel et al. (2008b) assessed *in vitro* the effects of saponins extracted from *Carduus pycnocephalus*, *Sesbania sesban* and *Knautia arvensis* leaves and *Trigonella foenum-graecum* L (fenugreek) seeds on fungi. The saponin-rich fractions decreased the total anaerobic fungal population with the fenugreek being the most inhibitory (65%) followed by *Sesbania* (38%). Likewise, Muetzel et al. (2003) observed that saponin-containing *Sesbania pachycarpa* leaves inhibited the growth of ruminal fungi (*Chytridiomycetes*). An adaptation of fungi may occur during long-term exposure, and it seems that fungi produce certain carbohydrates (Mountfort, 1989), which may degrade saponins, but more research is warranted to confirm such effect (Wina et al., 2005a).

Collectively, these results show the sensitivity of ruminal fungi to saponins. Fungi are metabolically important in rumen function as they contribute to the disruption of plant cell walls both physically and enzymatically (Orpin and Joblin, 1997). As a consequence, possible implications on fibre breakdown may occur if fungi are severely inhibited upon feeding saponins-containing plants/extracts to ruminant. Such adverse effects should be taken into account particularly in animals fed poor quality forages or crop by-products.

3.4 Methanogens

Studies on the effects of saponins on pure cultures of methanogens are scarce. Guo et al. (2008) observed that tea saponins (400 mg/l medium) did not influence the growth and the expression of the methyl coenzyme-M reductase subunit A (*mcrA*) of pure cultures of *Methanobrevibacter ruminantium*. Using mixed cultures of rumen bacteria in the same study, methanogen numbers were not affected, but the activity of the *mcrA* gene was strongly inhibited. Liu et al. (2019) observed no change in methanogen density in ruminal fluid of sheep fed diets supplemented with 2.0 g/d of tea saponins. Of the saponin sources evaluated, Goel et al. (2008b) found that *Sesbania* saponins were the most effective against methanogens (78% decrease) followed by fenugreek and *Knautia saponins* (78%, 22% and 21% decrease, respectively). Patra and Yu (2013) assessed *in vitro* the effect of quillaja saponins on methanogens at two dose levels and observed that abundance (*rrs* gene copies/ml of culture samples) of methanogens was not affected by low saponin dose (0.6 mg/l) but decreased at higher dose (1.2 mg/l). This finding indicates a dose-effect response of methanogens to quillaja saponins. Likewise, Kozłowska et al. (2020) observed *in vitro* that total methanogens, methanobacteriales and methanomicrobiales counts linearly decreased with increasing levels of saponins extracted from two alfalfa cultivars (fresh or ensiled). Narvaez et al. (2013) observed *in vitro* a decrease in methanogens (16S rRNA gene copies) when 650 mg/l of *Y. schidigera* extract (i.e. 100 mg/l of steroidal saponins) when added in the incubation medium. Wina et al. (2005b) observed that saponins of *Sapindus rarak* fruits decreased methanogen RNA concentration at the highest saponins concentration (4 mg/ml) whereas no change was observed at lower doses (0.25 mg/l, 0.5 mg/l, 1.0 mg/l and 2.0 mg/l). In an *in vivo* study, Popova et al. (2019) observed no effect of feeding tea saponins (0.5%) to non-lactating dairy cows on concentration and taxonomic composition of the archaeal community. Taken together, literature findings, mostly from *in vitro* studies clearly demonstrate the inhibitory effect of saponins on rumen methanogens, although the effect is dose dependent and varies with the sources of saponins used. Also, more *in vivo* data are needed to clearly determine the potential of saponins to inhibit the activity and/or the growth or rumen archaea when these microbes are exposed to saponins longer than under *in vitro* conditions.

4 Enteric methane production

The livestock sector contributes to greenhouse gas (GHG) emissions, mainly through the production of methane (CH_4) from enteric fermentation. Methane is a potent GHG and its emission into the atmosphere is directly related to ruminant production systems, mainly dairy and beef cattle. The global warming of CH_4 is at least 28 times that of carbon dioxide (Forster et al., 2007). In addition, enteric CH_4 is a loss of productive energy (4% to 7% of gross energy intake) for lactating dairy cows (Kebreab et al., 2008). Thus, mitigation of enteric CH_4 is beneficial from both nutritional and environmental standpoints. In this regard, the use of plant-derived secondary metabolites such as saponins have been suggested as a dietary strategy to mitigate enteric CH_4 emissions (Beauchemin et al., 2022).

Because of the demonstrated, at least *in vitro*, anti-protozoal and/or anti-methanogenic activities of saponins, and given the symbiotic association between protozoa and methanogens, a number of studies, most of them *in vitro*, have assessed the potential of saponins to inhibit ruminal methanogenesis. Literature data show inconsistent effects of saponins, saponin-containing plants or extracts containing saponins on CH_4 production. The efficacy of saponins in supressing CH_4 production have been found to vary with the source of saponins, their chemical structure, the dose applied and the diet fed to the animal.

Using the RUSITEC system with ruminal content obtained from a defaunated and faunated dairy cow, Hess et al. (2003a) assessed the effects on rumen fermentation of saponins from three saponins-rich tropical fruits (*Sapindus saponaria*, *Enterolobium cyclocarpum*, *Pithecellobium saman*) supplemented to forage-based diets. Of the three sources evaluated, only *S. saponaria* decreased (by 20%) CH_4 production and this decrease was associated with reduction (54%) in protozoa count but without affecting that of methanogens. The extent of CH_4 reduction was more pronounced in defaunated ruminal fluid than in faunated ruminal content (29% vs. 14% decrease, respectively). Since CH_4 inhibition occurred in both defaunated and faunated ruminal fluid, it was concluded that this inhibitory effect was not solely due to the observed depression in protozoa count (Hess et al., 2003a).

Saponins effect on CH_4 production seems to be dose- and saponins source-dependant. Pen et al. (2006a) observed *in vitro* that adding liquid extract of *Y. schidigera* at 2-6 ml/l, reduced in a dose-dependent manner the rate (up to 42% decrease) and the extent (up to 32% reduction) of CH_4 production, whereas *Q. saponaria* had no effect on CH_4 production. In two separate studies, Wang et al. (1998) and Lila et al. (2003) evaluated *in vitro* supply of *Y. schidigera* using mixed-forage diets. At the dose of 0.022 g/l of saponins, no effect on CH_4 production was observed by Wang et al. (1998) whereas Lila et al. (2003) reported a decrease in CH_4 production at much higher doses (1.2-3.2 g of saponins/l). The inhibitory effect of some saponins on methanogenesis has been shown to vary with composition of the diet. For instance, based on *in vitro* findings, Goel et al. (2008a) concluded that saponins from *Sesbania sesban* and fenugreek are expected to be more effective if animals are fed concentrate-based than forage-based diets. In contrast, *Carduus pycnocephalus* saponins could be effective in animals fed both forage- and concentrate-based diets.

Although several *in vitro* studies have reported an inhibition of saponins, CH_4 production, on the other hand, few *in vivo* studies, more particularly in dairy cows, have confirmed such inhibitory effect (Table 1). A good example illustrating this discrepancy between *in vitro* and *in vivo* findings is the study of Holtshausen et al. (2009) who assessed both *in vitro* and *in vivo* the potential of saponin-containing *Y. schidigera* or *Q. saponaria* on CH_4 production. Both saponin sources linearly reduced CH_4 concentration (mg/kg DM) when added *in vitro* at 15 mg/kg, 30 mg/kg and 45 mg/kg of substrate DM (i.e. 0.9 mg, 1.8 mg and 2.7 mg of saponins/kg DM). In that study, CH_4 reduction was related to an inhibition of feed digestion as reflected by decreases in VFA concentration and NDF digestibility. Such adverse effects of saponins on DM and NDF digestion also occurred in other studies where decreases in CH_4 production were observed upon the addition of saponin-containing extracts (Hess et al., 2003a; Lila et al., 2003). The effect of saponin extracts on CH_4 production was sometimes, but not always, explained by a decrease in fibre digestibility. Among studies that recorded a decrease in CH_4 production, Hess et al. (2004) reported a decrease in apparent total-tract digestibility of NDF and ADF with supplementation of *S. saponaria*. Santoso et al. (2004) reported no effect on NDF and ADF digestibility with *Y. schidigera*, and Pen et al. (2007) reported an increase in NDF digestibility with *Q. saponaria*. Despite the observed decreases in CH_4 observed *in vitro* by Holtshausen et al. (2009), no change in enteric CH_4 was observed *in vivo* upon supplementation of dairy cow diets with 10 g/kg dietary DM (i.e. 6% saponins, wt/wt) of *Y. schidigera* or *Q. saponaria* (3% saponins, wt/wt) extracts. The unaffected CH_4 production was in agreement with the lack of effects on VFA (total concentration and pattern), protozoa and fibre digestibility in that study. The contrasting *in vitro* versus *in vivo* effects of saponins on CH_4 production also occurred in other studies. For example, Guyader et al. (2017) observed that increasing tea saponins doses (0 mg/ml, 0.04 mg/ml, 0.08 mg/ml, 0.12 mg/ml, 0.15 mg/ml, 0.30 mg/ml and 0.50 mg/ml) decreased (up to 29%) CH_4 production *in vitro*, but no effects were observed on CH_4 emissions (g/d) from lactating dairy cows when tea saponins were included in the diet at the rate of 0.52% of DM. In that study, the lack of effects was consistent with the absence of changes in VFA pattern (i.e. acetate:propionate ration), protozoa density and total-tract nutrient digestibility. Although daily CH_4 emissions (g/d) was not affected, CH_4 yield (g/kg DMI) increased upon feeding tea saponins mainly because of the depressed of DMI and the unchanged CH_4 emissions (g/d).

Such discrepancies between *in vivo* and *in vitro* findings in CH_4 production response to saponin supplementation was not observed in other studies. For instance, CH_4 suppression observed *in vitro* by Hess et al. (2003a, b) was confirmed by Hess et al. (2004) in a subsequent study in lambs-fed diets supplemented with saponin-containing fruits of *Sapindus saponaria*. In that study, the inhibition of methanogenesis was mainly the result of the anti-protozoal activity of *S. saponaria* saponins. This anti-protozoal effect contrasts with earlier *in vitro* findings by the same authors. Hess et al. (2003a) demonstrated that the inhibitory effect of *S. saponaria* was not related to the presence of protozoa in ruminal fluid and concluded that saponins or other constituents of *S. saponaria* may have directly influenced the activity of methanogens. Nevertheless, in both studies (Hess et al., 2003a; Hess et al., 2004), suppression of methanogenesis by *S. saponaria* was not associated with reduced methanogen counts. Hess et al. (2004) speculated that the inhibition of methanogenesis was rather related to a suppressed activity per methanogen cell. Śliwiński et al. (2002b) reported no change in CH_4 production with the addition of increasing concentrations of *Y. Schidigera* saponins (1, 20 and 100 mg sarsaponin/kg DM) in the RUSITEC system. This result was consistent with the lack of effects on protozoa number in that study. Likewise, similar findings were reported by the same authors (Śliwiński et al., 2002a) in lambs-fed diets supplemented with *Y. Schidigera* saponins (2 mg/kg and 30 mg/kg DM).

A reduction in protozoa and/or methanogens due to feeding saponins does not always account for the observed reduction in CH_4 production. Despite a decrease in protozoa numbers, Klita et al. (1996) observed no change in enteric CH_4 production when saponin extract (27.8% saponins) from alfalfa root was intra-ruminally administered at the rates of 0%, 1%, 2% and 4% of DMI. Pen et al. (2006) observed that the addition (up to 6 ml/l) of *Quillaja saponaria in vitro* had no effect on the rate and extent of CH_4 production although protozoa number declined by up to 41%. Goel et al. (2008a) reported no effect of saponins extracted from *S. sesban* leaves, fenugreek seeds and *Knautia* leaves on CH_4 production *in vitro* despite substantial decreases in total archaeal and protozoal populations. Such findings may be explained by potential changes in the rate of methanogenesis as a result of changing fermentation pattern and microbial diversity (Patra and Saxena, 2010).

Taken together, literature data reveal that the reduction in enteric CH_4 production has not always been associated with an inhibition of the growth and/or activity of protozoa. Also, a clear discrepancy exists between *in vivo* and *in vitro* findings, suggesting that extrapolation of results from *in vitro* to *in vivo* experiments has its limitations.

5 Nitrogen metabolism

Ruminants have a relatively low efficiency of dietary N utilization (Castillo et al., 2000). In a meta-analysis, Hristov et al. (2005) found that the efficiency of transfer of feed N into milk protein averaged 24.7 ± 0.14% (minimum =13.7, maximum = and 39.8%). Such poor utilization of N results in increased production costs and represents an energy loss to the animal, which excretes excess N in the urine, where it becomes an environmental burden. Therefore, improving N utilization has a positive impact both on the efficiency of production of the animal and on the environment. On the other hand, ruminants have the unique advantage of being capable of incorporating sources of non-protein nitrogen into bacterial protein within the rumen. The microbial proteins flowing from the rumen to the intestines represent an excellent source of AA for the host animal. However, within the rumen, protozoa have a negative impact on the supply of microbial N to the animal because they engulf and digest rumen bacteria thereby contributing to overall N inefficiency. Thus, removal of ciliate protozoa from the rumen (i.e. defaunation) prevents the recycling of N between bacteria and protozoa, which increases the efficiency of N metabolism in the rumen. The net result is increased flow of microbial protein from the rumen (Wallace et al., 2002; Patra and Saxena, 2010). Furthermore, urinary N excretion into the environment is decreased because of a more efficient capture of ruminal nitrogen.

5.1 Ammonia

Ammonia (NH_3) is an end-product of protein degradation in the rumen. A decrease in NH_3 may result from an increase of its uptake by rumen microbes for protein synthesis and/or a reduction of feed or microbial protein breakdown. Due to the well-known effect of saponins on the elimination of protozoa and the negative role of protozoa on the efficiency of N use in the rumen, saponins are expected to reduce NH_3 production in the rumen. Wina et al. (2005a) summarized 51 reports on the effects of saponins on ruminal NH_3 concentration. Of the 51 reports, 14 revealed no effect, whereas 17 showed a negative effect of saponins on rumen NH_3 concentration. Wallace et al. (1994) observed a small but a significant reduction (6%) in NH_3 concentration when *Y. schidigera* extract was added (1%, vol/vol) to strained rumen fluid *in vitro*. The authors attributed this decrease to the suppression of ciliate protozoa and reduced proteolysis rather than to ammonia-binding properties as suggested by Headon et al. (1991). Wina et al. (2005a) and Wallace et al. (2002) suggested that the effect of saponins on rumen metabolism is modulated via the reduction in protozoa population rather than an inhibition of protein degradation.

Adding 0.5 mg/ml of *Yucca schidigera* extract (4.4% saponins, w/w) in the RUSITEC, Wang et al. (1998) observed a reduction in NH_3 concentration, but this effect was transient and was explained by a possible adaptation of the mixed rumen microbes to saponins exposure, and also a degradation of saponins by rumen microbes. Contrarily to Wang et al. (1998) findings, Śliwiński et al. (2002b) observed a sustained reduction (up to 21%) in NH_3 concentration only when a high dose of *Y. schidigera* extract (100 mg sarsaponin/kg feed DM) was supplied in the RUSITEC. However, at lower doses (1 mg or 20 mg sarsaponin/kg DM), no changes were observed in NH_3 concentration. In the study of Śliwiński et al. (2002b), the reduced NH_3 concentration was achieved without a decrease in protozoa. The authors suggested that the decrease in NH_3 concentration may be due to the inhibitory effect of saponins on microbial urease as reported in other studies (Makkar, 1993; Hussain and Cheeke, 1995). There are indications that saponins (steroidal saponins) of *Y. schidigera* extract are more effective in reducing ruminal NH_3 concentration than *Q. saponaria* saponins (triterpenoidal saponins). Pen et al. (2006) observed *in vitro* that *Y. schidigera* extract markedly decreased (up to 48%) NH_3 concentration in a dose-dependent manner, whereas *Q. saponaria* extract did not significantly affect NH_3 concentration.

In several *in vivo* studies, feeding *Y. schidigera* and/or *Q. saponaria* extracts to dairy cows failed to alter NH_3 concentration in ruminal fluid (Table 1). Holtshausen et al. (2009) reported that including (10 g/kg DM) *Y. schidigera* (6% saponins, wt/wt) decreased ruminal NH_3 concentration whereas feeding the same amount of *Q. Saponaria* extracts (3% saponins, wt/wt) did not affect ruminal NH_3 concentration. These different responses in NH_3 concentration between the two sources of saponins may be related to the lower saponins concentration in *Q. Saponaria* versus *Y. schidigera* (3% vs. 6% saponins, respectively) and may also indicate that steroidal saponins (i.e. *Y. schidigera)* are more effective than triterpenoid saponins (i.e. *Q. Saponaria).* In contrast, Benchaar et al. (2008) observed no change in ruminal NH_3 concentration in dairy cows fed *Y. schidigera* extract (60 g/d; 10% saponins). In both of these studies, the lack of change in NH_3 concentration was consistent with the unaffected protozoa counts (total and by genera). Other studies with dairy cows also reported no effect of *Y. schidigera* and *Q. Saponaria* extracts on ruminal NH_3 concentrations (Goetsch and Owens, 1985; Valdez et al., 1986; Wu et al., 1994; Wilson et al., 1998). Guyader et al. (2017) observed that including tea saponins in dairy cow diets did not affect ruminal NH_3 concentration consistently with no change in protozoa counts. In one study (Singer et al., 2008), ruminal NH_3 concentration tended to decrease linearly with increasing inclusion of *Y. schidigera* extract (0 g, 5 g, 10 g or 15 g of sarsaponin/d) in dairy cow diets. The lack of influence of saponins on ruminal NH_3 concentration was also observed with other sources of saponins as in the study of Guyader et al. (2017) upon including tea saponins extract at the rate of 0.76% of dietary DM (0.52% of pure saponins).

Based on the results of the literature, it is concluded that the effects of saponins on NH_3 concentration greatly vary with the source of saponins (i.e. chemical structure), the dose and the experimental model used (*in vitro* vs. *in vivo*). In general, *in vitro* experiments show that saponins are effective in reducing NH_3 concentration, but *in vivo* studies failed to confirm such effects. Reasons for such discrepancy between the two methodologies are divers and include the relatively higher doses of saponins applied *in vitro* versus *in vivo*; and the absence of saponins effects on rumen protozoa *in vivo*, and the possible adaptation of rumen microbes to longer exposure *in vivo* compared to shorter exposure *in vitro*.

5.2 Microbial protein synthesis

As the main effect of saponins in the rumen is via their inhibition of rumen protozoa, efficiency of microbial protein synthesis and protein flow to the duodenum are expected to increase with saponin use. *In vitro* addition (0.4–1.2 mg/l) of quillaja saponins to a hay substrate linearly increased efficiency of microbial synthesis (Makkar and Becker, 1996). Likewise, Makkar et al. (1998) also observed that supplementing hay (i.e. substrate) with yucca, quillaja and acacia saponins (0.6 mg/ml and 1.2 mg/ml) increased efficiency of microbial protein synthesis *in vitro* and this improvement was associated with marked decreases in NH_3 concentration and protozoa counts. However, the use of high doses of saponins can negatively affect efficiency of microbial protein synthesis as showed by Wang et al. (2000b). These authors reported that microbial protein synthesis increased *in vitro* with the addition of *Y. schidigera* saponins at the rate of 15 mg/l, whereas it decreased at higher doses (i.e. 75 mg/l or 225 mg/l).

Few studies have assessed the impact of feeding saponinins to ruminants, particularly dairy cows on ruminal microbial synthesis and duodenal protein flow and the effects have been variable. For example, Wu et al. (1994) observed no change in synthesis or duodenal flow of microbial protein when *Y. schidigera* extract was administered into the rumen of dairy cows at rate of 1.2 g/d. Hristov et al. (1999) also observed no effect on microbial protein synthesis when heifers were intra-ruminally dosed with 20 g/d or 60 g/d of *Y. schidigera* (4.4% saponins, wt/wt) despite a decrease in NH_3 concentration (only at 2 h after dosing) and protozoa count. This lack of effect on microbial protein synthesis was in line with the unchanged duodenal flow of microbial protein estimated using urinary excretion of purine derivatives (Hristov et al., 1999). Goetsch and Owens (1985) reported a 36% decrease in the efficiency of microbial protein synthesis in cattle fed *Y. schidigiera* extract (44 ppm). This decrease was the result of unchanged duodenal flow of microbial N and an increased degradation of organic matter in the rumen. In that study, the authors reported that the effects of saponins may vary with the diet. In fact, Lu and Jorgensen (1987) reported decreased bacterial protein synthesis and duodenal bacterial protein when alfalfa saponins (2–4% of dietary DM) were administered into the rumen of sheep fed a roughage-based diet, whereas no effects were observed in sheep receiving a concentrate-based diet. Likewise, efficiency of microbial of protein synthesis was not affected by Y. *schidigera* in sheep were fed timothy silage-based diets while it increased in sheep receiving timothy hay-based diets Santoso et al. (2004).

6 Performance

Productivity of ruminants (e.g. dairy cows) is expected to increase upon feeding saponins mainly because of the anti-protozoal (i.e. defaunation) activity of saponins and its potential to reduce predation of bacteria by protozoa, thereby improving protein metabolism in the rumen and increasing the microbial protein supply to the animal. However, such positive effects can be counterbalanced by a reduction in ruminal fibre degradation given the positive role of protozoa on fibre digestion in the rumen. In addition, the expected inhibition of enteric CH_4 by saponins can positively impact animal productivity if the gain in energy achieved is used for production purpose (e.g. milk). The possible adaptation of rumen microbes to saponins exposure and the potential degradation of these secondary metabolites by the rumen microbes may influence animal productivity response.

Information on the effects of feeding saponins on milk production, milk composition and feed efficiency of dairy cows are scarce. Benchaar et al. (2008) observed no change in milk performance (i.e. production, composition, efficiency) of Holstein cows fed a diet supplemented with 60 g/d of *Y. schidigera* extract (10% saponins, DM basis). This lack of effect was consistent with the absence of effects on DMI and nutrient digestibility. In another study with lactating dairy cows, Holtshausen et al. (2009) fed *Y. schidigera* extract at the rate of 10 g/kg DM (6% saponins, wt/wt) and reported no effects on milk yield, milk composition and nutrient digestibility. However, in that study, feed efficiency (kg milk/kg DMI) was depressed because of the increased DMI when the diet was supplemented with *Y. schidigera* extract. Contrary to findings of Holtshausen et al. (2009), Lovett et al. (2006) observed that supplementing dairy cow diets with *Y. schidigera* extract at the rates of 1.46 g/kg and 3.10 g/kg DMI decreased linearly DMI without affecting milk yield, leading to improved feed efficiency. Other studies reported no change in milk production and milk composition of dairy cows fed diets supplemented with *Y. schidigera* extract (Valdez et al., 1986; Wu et al., 1994; Wilson et al., 1998; Singer et al., 2008; Śliwiński et al., 2004). In these studies, the lack of effect was consistent with the lack of changes in DMI and nutrient digestibility upon diet supplementation with *Y. schidigera* extract.

Contrasting results were reported on the effects of feeding tea saponins on dairy cows' performance. Guyader et al. (2017) observed that feeding diet supplemented with tea saponins (0.52%, DM basis) decreased milk yield, DMI and feed efficiency. In contrast, Kolling et al. (2018) reported neither milk yield nor DMI or nutrient digestibility was affected by including green tea extract (0.028%) in dairy cow diets. Wang et al. (2017) fed 20 g/d, 30 g/d and 40 g/d of tea saponins (56% purity of triterpenoid saponins) to dairy cows and observed quadratic responses of DMI and milk yield with cows receiving 40 g/d consuming less and producing the lowest amount of milk. The same quadratic response was observed for yields of fat and protein. However, feed efficiency (kg milk/kg DMI) was unaffected by dietary addition of tea saponins. Francis et al. (2002) speculated that the adverse effect of saponins on feed intake may be related to poor palatability (i.e. astringent and irritating taste of saponins), reduction in intestinal motility, reduction in protein digestibility, damage to the intestinal membrane and inhibition of nutrient transport. Further research is warranted to elicit reasons of feed intake depression upon feeding tea saponins to dairy cows. Nevertheless, it appears that feeding high amounts of tea saponins may impair animal productivity and there is a need to find optimal tea saponins doses to avoid such negative effect.

Literature findings suggest that milk performance is not affected upon feeding saponin-containing plants/extracts to dairy cows, which is in line with the absence of changes in feed intake, ruminal fermentation characteristics and nutrient digestibility. Such inefficacy calls into question the usefulness of feeding saponin-containing extracts to improve dairy cow productivity.

7 Toxicity

Plant extracts are generally perceived safe because they are produced naturally. However, plant secondary metabolites (e.g. saponins) play an important role in the protection of the plant against herbivores, pathogens, insects and hostile environmental factors. Therefore, the positive perception with regard to plant secondary metabolites must be taken with caution, because toxicity may occur if animals consume specific plants and/or high amounts of plant secondary metabolites.

One of the main known toxic effects of saponins in animals is photosensitization, although liver and kidney degeneration and gut complications such as gastroenteritis and diarrhea have also been reported (Kumar et al., 1992; Cheek al., 1995; Flåøyen and Frøslie, 1997). According to Wina et al. (2005), photosensitization, reflected by skin lesions, is caused by the reaction of an exogenous compound circulating in the blood with ultraviolet light, producing free radicals that react with proteins of tissues. This is classified as a primary photosensitization. The affected animals develop photophobia and severe dermatitis. Secondary photosensitization is a consequence of damages of the liver the bile ducts, thus impeding the excretion of bile, causing chlorophyll metabolites such as phylloerythrin to circulate in the blood. It seems that saponins toxicity towards the animal occurred mainly in rangelands in the United States, where most of the toxic plants containing saponins are mainly found. For instance, plants such as corn cockle, soapwort, cow cockle and broomweed cause serious toxicity problems for grazing livestock (Wina et al., 2005a). Apparently, seeds are the primary cause of photosensitization although, other parts of plants such as leaves and fruits may also cause photosensitization (Miles et al., 1992; Knight and Walter, 2003). Most saponins that have been identified in the toxic plants are steroidal saponins (Wina et al., 2005a). To the best of our knowledge, no toxicity has been reported in animals fed *Y. schidigera* extract, most likely because of the low saponins concentration (3–10%; Table 1) and the low extract feeding rates (0.03–10 g/kg DM; Table 1).

8 Conclusion

Most of the literature available to date on the potential of saponins or saponin-containing plants/extracts to manipulate rumen fermentation have been obtained *in vitro* (batch cultures or continuous cultures). With the *in vitro* data currently available, for most of the plants and extracts, the effects elicited depend on the source of the plant/extract and hence, the chemical structure of saponins (e.g. steroidal versus triterpenoid), the concentration of saponins in the plant/extract, the amount of the plant/extract (i.e. amount of saponins) used, and the composition of the substrate used. Bases on these considerations, it is not then surprising that the *in vitro* effects of saponins on rumen microbial fermentation vary among studies.

Perhaps, one of the main demonstrated effects of saponins is their inhibition of ruminal protozoa. Such effect has been consistently reported in several *in vitro* studies using saponins-containing extracts such as extracts of *Yucca schidigera*, tea, *Sapindus rarak* and *Sesbania sesban*. However, this anti-protozoal activity of saponins did not always result in the inhibition of methanogenesis *in vitro* more likely because other factors such as the composition of the incubated substrate, the source and the dose of saponins may have influenced the rumen fermentation process.

In vivo, most of the available data are from experiments using small ruminants, mainly sheep and to a lesser extent beef cattle. Studies involving dairy cattle are very scarce. Nevertheless, the review of the literature reveals that feeding saponins to ruminants failed to modulate rumen microbial fermentation, nutrient utilization and consequently, animal performance. Several factors have been suggested to explain the discrepancy between *in vitro* and *in vivo* studies. Many of the saponins sources evaluated seem to have a transient effect, mainly on ruminal protozoa. Rumen microbes have a remarkable capacity to adapt to exposure to several types of secondary metabolites and this seems also applies to saponins. Degradation and inactivation of saponins by rumen microbes also contributes to the lack of effects *in vivo*. Although saponins may be considered as safe when fed to animals, toxicity may occur especially when animals are raised on rangelands or grazing pastures containing plants high in saponins.

In conclusion, based on *in vitro* data, saponins or saponins-containing plants/extracts exhibit antimicrobial activities mainly towards ruminal protozoa, making them attractive feed additives in ruminant nutrition to improve efficiency of N and energy utilization, and enhance animal productivity. However, literature data reveal that such positive effects were rarely confirmed in animals fed saponins. Animal productivity, whether milk production by dairy cows or daily gain in beef or sheep were in most cases not improved, which questions the use of saponins and saponin-containing plants/extract as feed additives in ruminant nutrition.

9 References

Abe, I. 2007. Enzymatic synthesis of cyclic triterpenes. *Nat. Prod. Rep.* 24(6):1311-1331.

Baah, J., Ivan, M., Hristov, A. N., Koenig, K. M., Rode, L. M. and McAllister, T. A. 2007. Effects of potential dietary antiprotozoal supplements on rumen fermentation and digestibility in heifers. *Anim. Feed Sci. Technol.* 137(1-2):126-137.

Bangham, A. D., Horne, R. W., Glauert, A. M., Dingle, J. T. and Lucy, J. A. 1962. Action of saponins on biological cell membranes. *Nature* 196:952-955.

Beauchemin, K. A., Ungerfeld, E. M., Abdalla, A. L., Alvarez, C., Arndt, C., Becquet, P., Benchaar, C., Berndt, A., Mauricio, R. M., McAllister, T. A., Oyhantcabal, W., Salami, S. A., Shalloo, L., Sun, Y., Tricarico, J., Uwizeye, A., De Camillis, C., Bernoux, M., Robinson, T. and Kebreab, E. 2022. Invited review: Current enteric methane mitigation options. *J. Dairy Sci.* 105(12):9297-9326.

Benchaar, C. and Greathead, H. 2011. Essential oils and opportunities to mitigate enteric methane emissions from ruminants. *Anim. Feed Sci. Technol.* 166-167:338-355.

Benchaar, C., Hristov, A. N. and Greathead, H. 2009. Essential oils as feed additives in animal nutrition. In T. Steiner (Ed.), *Phytogenics in Animal Nutrition*. Nottingham, UK: Nottingham University Press, pp. 111-146.

Benchaar, C., McAllister, T. A. and Chouinard, P. Y. 2008. Digestion, ruminal fermentation, ciliate protozoal populations, and milk production from dairy cows fed cinnamaldehyde, quebracho condensed tannin, or *Yucca schidigera saponin* extracts. *J. Dairy Sci.* 91(12):4765-4777.

Benchaar, C., Wang, Y., Chaves, A. V., McAllister, T. A. and Beauchemin, K. A. 2007. Use of plant extracts in ruminant nutrition. In S. N. Acharya and J. E. Thomas (Eds.), *Advances in Medicinal Plant Research*. Kerala, India: Research Signpost, pp. 465-489.

Bruneton, J. 1999. *Pharmacognosy, Phytochemistry, Medicinal Plants* (2nd edn.). London, UK: Intercept Ltd.

Callaway, T. R., Carneiro De Melo, A. M. and Russell, J. B. 1997. The effect of nisin and monensin on ruminal fermentations in vitro. *Curr. Microbiol.* 35(2):90-96.

Castillo, A., Kebreab, E., Beever, D. and France, J. 2000. A review of efficiency of nitrogen utilisation in lactating dairy cows and its relationship with environmental pollution. *J. Anim. Feed Sci.* 9(1):1-32.

Cheeke, P. R. 1999. Actual and potential applications of *Yucca schidigera* and *Quillaja saponaria* saponins in human and animal nutrition. *J. Anim. Sci.* 77:1-10.

Conner, D. E. 1993. Naturally occurring compounds. In P. M. Davidson and A. L. Branen (Eds.), *Antimicrobials in Foods*. New York: Dekker, pp. 441-468.

De Costa, F., Yendo, A. C., Fleck, J. D., Gosmann, G. and Fett-Neto, A. G. 2013. Accumulation of a bioactive triterpene saponin fraction of *Quillaja brasiliensis* leaves is associated with abiotic and biotic stresses. *Plant Physiol. Biochem.* 66:56-62.

Diaz, A., Avendano, M. and Escobar, A. 1993. Evaluation of *Sapindus saponaria* as a defaunating agent and its effects on different rumen digestion parameters. *Livest. Res. Rural Dev.* 5:1-6.

European Union. 2003. Regulation (EC) no 1831/2003 of the European Parliament and the Council of 22 September 2003 on additives for use in animal nutrition. *Off. J. L 268/36*.

Fenchel, T. and Finlay, B. J. 2006. The diversity of microbes: Resurgence of the phenotype. *Philos. Trans. R. Soc. Lond. B Biol. Sci.* 361(1475):1965-1973.

Fenwick, S. B., Price, K. R., Tsukamoto, C. and Okubo, K. 1991. Saponins. In J. P. F. D'Mello, C. M. Duffus and J. H. Duffus (Eds.), *Toxic Substances in Crop Plants*. London, UK: The Royal Society of Chemistry, pp. 285-327.

Flåøyen, A. and Frøslie, A. 1997. Photosensitization disorders. D'Mello, J. P. F. (Ed.), *Handbook of Plant and Fungal Toxicants* (Vol. 1997). Boca Raton: CRC Press, pp. 191-204.

Flåøyen, A., Wilkins, A. L., Deng, D. and Brekke, T. 2001. Ovine metabolism of saponins: Evaluation of a method for estimating the ovine uptake of steroidal saponins from *Narthecium ossifragum*. *Vet. Res. Commun.* 25(3):225-238.

Flåøyen, A., Wilkins, A. L. and Sandvik, M. 2002. Ruminal metabolism in sheep of saponins from *Yucca schidigera*. *Vet. Res. Commun.* 26(2):159-169.

Forster, P., Ramaswamy, V., Artaxo, P., Berntsen, T., Betts, R., Fahey, D. W., Haywood, J., Lean, J., Lowe, D. C., Myhre, G., Nganga, J., Prinn, R., Raga, G., Schulz, M., and Van Dorland, R. 2007. The physical science basis: Contribution of Working Group I to the Fourth Assessment Report of the Intergovernmental Panel on Climate Change [Solomon, S., Qin, D., Manning, M., Chen, Z., Marquis, M., Averyt, K. B, Tignor, M., and Miller H. L. (eds.)]. Changes in atmospheric constituents and in radiative forcing. In *Climate Change*. Cambridge, United Kingdom and New York, NY: Cambridge University Press.

Francis, G., Keem, Z., Makkar, H. P. S. and Becker, K. 2002. The biological action of saponins in animal systems: A review. *Br. J. Nutr.* 88(6):587-605.

Glauert, A. M., Dingle, J. T. and Lucy, J. A. 1962. Action of saponin on biological membranes. *Nature* 196(4858):953-955.

Goel, G. and Makkar, H. P. S. 2012. Methane mitigation from ruminants using tannins and saponins. *Trop. Anim. Health Prod.* 44(4):729-739.

Goel, G., Makkar, H. P. S. and Becker, K. 2008a. Effect of *Sesbania sesban* and Carduus pycnocephalus leaves and fenugreek (*Trigonella foenum-Graecum* L.) seeds and their extracts on partitioning of nutrient from roughage and concentrate based feeds to methane. *Anim. Feed Sci. Technol.* 147(1-3):72-89.

Goel, G., Makkar, H. P. S. and Becker, K. 2008b. Changes in microbial community structure, methanogenesis and rumen fermentation in response to saponin-rich fractions from different plant materials. *J. Appl. Microbiol.* 105(3):770-777.

Goetsch, A. L. and Owens, F. N. 1985. Effects of sarsaponin on digestion and passage rates in cattle fed medium to low concentration. *J. Dairy Sci.* 68(9):2377-2384.

Published by Burleigh Dodds Science Publishing Limited, 2023.

Greathead, H. 2003. Plant and plant extract for improving animal productivity. *Proc. Nutr. Soc.* 62(2):279–290.

Güçlü-Üstündag, O. and Mazza, G. 2007. Saponins: Properties, applications and processing. *Crit. Rev. Food Sci. Nutr.* 47(3):231–258.

Guo, Y. Q., Liu, J. X., Lu, Y., Zhu, W. Y., Denman, S. E. and McSweeney, C. S. 2008. Effect of tea saponin on methanogenesis, microbial community structure and expression of *mcrA* gene, in cultures of rumen micro-organisms. *Lett. Appl. Microbiol.* 47(5):421–426.

Guyader, J., Eugene, M., Doreau, M., Morgavi, D. P., Gérard, C., Loncke, C. and Martin, C. 2015. Nitrate but not tea saponin feed additives decreased enteric methane emissions in nonlactating cows. *J. Anim. Sci.* 93(11):5367–5377.

Guyader, J., Eugène, M., Doreau, M., Morgavi, D. P. , Gérard, C. and Martin, C. 2017. Tea saponin reduced methanogenesis in vitro but increased methane yield in lactating dairy cows. *J. Dairy Sci.* 100(3):1845–1855.

Guyader, J., Eugène, M., Nozière, P., Morgavi, D. P., Doreau, M. and Martin, C. 2014. Influence of rumen protozoa on methane emissions in ruminants: A meta-analysis approach. *Animal* 8(11):1816–1825.

Haralampidis, K., Trojanowska, M. and Osbourn, A. E. 2002. Biosynthesis of triterpenoid saponins in plants. *Adv. Biochem. Eng. Biotechnol.* 75:31–49.

Hess, H. D., Beuret, R. A., M. Lötscher, Hindrichsen, I. K., Machmüller, A., Carulla, J. E., Lascano, C. E. and Kreuzer, M. 2004. Ruminal fermentation, methanogenesis and nitrogen utilization of sheep receiving tropical grass hay-concentrate diets offered with *Sapindus saponaria* fruits and *Cratylia argentea* foliage. *Anim. Sci.* 79(1):177–189.

Hess, H. D., Kreuzer, M., Diaz, T. E., Lascano, C. E., Carulla, J. E., Soliva, C. R. and Machmüller, A. 2003a. Saponin rich tropical fruits affect fermentation and methanogenesis in faunated and defaunated rumen fluid. *Anim. Feed Sci. Technol.* 109(1-4):79–94.

Hess, H. D., Monsalve, L. M., Lascano, C. E., Carulla, J. E., Diaz, T. E. and Kreuzer, M. 2003b. Supplementation of a tropical grass diet with forage legumes and *Sapindus saponaria* fruits: Effects on in vitro ruminal nitrogen turnover and methanogenesis. *Aust. J. Agric. Res.* 54(7):703–713.

Holtshausen, L., Chaves, A. V., Beauchemin, K. A., McGinn, S. M., McAllister, T. A., Odongo, N. E., Cheeke, P. R. and Benchaar, C. 2009. Feeding saponin-containing *Yucca schidigera* and *Quillaja saponaria* to decrease enteric methane production in dairy cows. *J. Dairy Sci.* 92(6):2809–2821.

Hostettmann, K. and Marston, A. 1995. *Saponins*. Cambridge, UK: Cambridge University Press.

Hristov, A. N., Ivan, M., Neill, L. and McAllister, T. A. 2003. Evaluation of several potential bioactive agents for reducing protozoal activity in vitro. *Anim. Feed Sci. Technol.* 105(1-4):163–184.

Hristov, A. N., McAllister, T. A., Van Herk, F. H., Cheng, K. J., Newbold, C. J. and Cheeke, P. R. 1999. Effect of *Yucca schidigera* on ruminal fermentation and nutrient digestion in heifers. *J. Anim. Sci.* 77(9):2554–2563.

Hristov, A. N., Price, W. J. and Shafii, B. 2005. A meta-analysis on the relationship between intake of nutrients and body weight with milk volume and milk protein yield in dairy cows. *J. Dairy Sci.* 88(8):2860–2869.

Hussain, I. and Cheeke, P. R. 1995. Effect of dietary *Yucca shidigera* extract on rumen and blood profiles of steers fed concentrate- or roughage-based diets. *Anim. Feed Sci. Technol.* 51(3-4):231-242.

Jouany, J. P. and Morgavi, D. P. 2007. Use of 'natural' products as alternatives to antibiotic feed additives in ruminant production. *Animal* 1(10):1443-1466.

Kalinowska, M., Zimowski, J., Pączkowski, C. and Wojciechowski, Z. A. 2005. The formation of sugar chains in triterpenoid saponins and glycoalkaloids. *Phytochem. Rev.* 4(2-3):237-257.

Kavya, N. M., Adil, L. and Senthilkumar, P. A. 2021. Review on Saponin Biosynthesis and its Transcriptomic Resources in Medical Plants. *Plant Mol.Biol. Rep.* 39:833-840.

Kebreab, E., Johnson, K. A., Archibeque, S. L., Pape, D. and Wirth, T. 2008. Model for estimating enteric methane emissions from United States dairy and feedlot cattle. *J. Anim. Sci.* 86(10):2738-2748.

Klita, P. T., Mathison, G. W., Fenton, T. W. and Hardin, R. T. 1996. Effect of alfalfa root saponins on digestive function in sheep. *J. Anim. Sci.* 74(5):1144-1156.

Knight, A. P. and Walter, R. G. 2003. Plants affecting the skin and liver (part II). In A. P. Knight and R. G. Walter (Eds.), *A Guide to Plant Poisoning of Animals in North America*. Jackson, WY: Teton New Media.

Kolling, G. J., Stivanin, S. C. B., Gabbi, A. M., Machado, F. S., Ferreira, A. L., Campos, M. M., Tomich, T. R., Cunha, C. S., Dill, S. W., Pereira, L. G. R. and Fischer., V. 2018. Performance and methane emissions in dairy cows fed oregano and green tea extracts as feed additives. *J. Dairy Sci.* 101(5):4221-4234.

Kozłowska, M., Cieślak, A., Jóźwik, A., El-Sherbiny, M., Stochmal, A., Oleszek, W., Kowalczyk, M., Filipiak, W. and Szumacher-Strabel, M. 2020. The effect of total and individual lucerne saponins on rumen methane production. *J. Sci. Food Agric.* 100(5):1922-1930.

Kumar, R. 1992. Anti-nutritional factors, the potential risks of toxicity and methods to alleviate them. In A. Speedy and P. L. Pugliese (Eds.), *Legume Trees and Other Fodder Trees as Protein Sources for Livestock; FAO Animal Production and Health Review 102 (Section 9)*. Rome: Food and Agriculture Organization.

Lila, Z. A., Mohammed, N., Kanda, S., Kamada, T. and Itabashi, H. 2003. Effect of sarsaponin on ruminal fermentation with particular reference to methane production in vitro. *J. Dairy Sci.* 86(10):3330-3336.

Liu, Y., Ma, T., Chen, D., Zhang, N., Si, B., Deng, K., Tu, Y. and Diao, Q. 2019. Effects of tea saponin supplementation on nutrient digestibility, methanogenesis, and ruminal microbial flora in Dorper crossbred ewe. *Animals (Basel)* 9(1):29.

Lovett, D. K., Stack, L., Lovell, S., Callan, J., Flynn, B., Hawkins, M. and O'Mara, F. P. 2006. Effect of feeding *Yucca schidigera* extract on performance of lactating dairy cows and ruminal fermentation parameters in steers. *Livest. Sci.* 102(1-2):23-32.

Lowe, L. B., Ball, G. J., Carruthers, V. R., Dobos, R. C., Lynch, G. A., Moate, P. J., Poole, P. R. and Valentine, S. C. 1991. Monensin controlled-release intraruminal capsule for control of bloat in pasture dairy cows. *Aust. Vet. J.* 68(1):17-20.

Lu, C. D. and Jorgensen, N. A. 1987. Alfalfa saponins affect site and extent of nutrient digestion in ruminants. *J. Nutr.* 117(5):919-927.

Makkar, H. P. S. 1993. Antinutritional factors in foods for livestock. *BSAP Occas. Publ.* 16:69-85.

Makkar, H. P. S. and Becker, K. 1996. Effect of quillaja saponins on in vitro rumen fermentation. In G. R. Waller and Y. Yamasaki (Eds.), *Saponins Used in Food and Agriculture*. New York: Plenum Press, pp. 387-394.

Makkar, H. P. S., Blümmel, M. and Becker, K. 1995. In vitro effects of and interactions between tannins and saponins and fate of tannins in the rumen. *J. Sci. Food Agri.* 69(4):481-493.

Makkar, H. P. S., Sen, S., Blümmel, M. and Becker, K. 1998. Effects of fractions containing saponins from *Yucca schidigera*, *Quillaja saponaria*, and *Acacia auriculoformis* on rumen fermentation. *J. Agric. Food Chem.* 46(10):4324-4328.

Miles, C. O., Wlikins, A. L., Munday, S. C., Holland, P. T., Smith, B. L., Lancaster, M. J. and Embling, P. P. 1992. Identification of the calcium salt of epismilagenin â-D-glucuronide in the bile crystals of sheep affected by *Panicum dichotomiflorum* and *Panicum schinzii* toxicoses. *J. Agric. Food Chem.* 40(9):1606-1609.

Morgavi, D. P., Martin, C., Jouany, J. P. and Ranilla, M. J. 2012. Rumen protozoa and methanogenesis: Not a simple cause effect relationship. *Br. J. Nutr.* 107(3):388-397.

Morrissey, J. P. and Osbourn, A. E. 1999. Fungal resistance to plant antibiotics as a mechanism of pathogenesis. *Microbiol. Mol. Biol. Rev.* 63(3):708-724.

Mountfort, D. O. and Asher, R. A. 1989. Production of xylanase from anaerobic fungus, neocallimastix frontalis. *Appl. Environ. Microbiol.* 55(4):1016-1022.

Muetzel, S., Hoffmann, E. M. and Becker, K. 2003. Supplementation of barley straw with *Sesbania pachycarpa* leaves in vitro, effects on fermentation variables and rumen microbial concentration structure quantified by ribosomal RNA-targeted probes. *Br. J. Nutr.* 89(4):445-453.

Mugford, S. T. and Osbourn, A. 2012. Saponin synthesis and function. In T. Bach and M. Rohmer (Eds.), *Isoprenoid Synthesis in Plants and Microorganisms*. New York, NY: Springer, pp. 405-424.

Naoumkina, M. A., Modolo, L. V., Huhman, D. V., Urbanczyk-Wochniak, E., Tang, Y., Sumner, L. W. and Dixon, R. A. 2010. Genomic and coexpression analyses predict multiple genes involved in triterpene saponin biosynthesis in *Medicago truncatula*. *Plant Cell* 22(3):850-866.

Narvaez, N., Wang, Y. and McAllister, T. A. 2013. Effects of extracts of *Humulus lupulus* (hops) and *Yucca schidigera* applied alone or in combination with monensin on rumen fermentation and microbial populations in vitro. *J. Sci. Food Agric.* 93(10):2517-2522.

Newbold, C. J., de la Fuente, G., Belanche, A., Ramos-Morales, E. and McEwan, N. R. 2015. The Role of ciliate protozoa in the rumen. *Front. Microbiol.* 6:1313.

Newbold, C. J., El Hassan, S. M., Wang, J., Ortega, M. E. and Wallace, R. J. 1997. Influence of foliage from African multipurpose trees on activity of rumen protozoa and bacteria. *Br. J. Nutr.* 78(2):237-249.

Newbold, C. J., Lassalas, B. and Jouany, J. P. 1995. The importance of methanogens associated with ciliate protozoa in ruminal methane production in vitro. *Lett. Appl. Microbiol.* 21(4):230-234.

Orpin, C. G. and Joblin, K. N. 1997. The rumen anaerobic fungi. In P. N. Hobson and C. S. Stewart (Eds.), *The Rumen Microbial Ecosystem*. London: Blackie Academic and Professional, pp. 140-195.

Osborne, J. K., Mutsvanga, T., Alzahal, O., Duffield, T. F., Bagg, R., Dick, P., Vessie, G. and McBride, B. W. 2004. Effects of monensin on ruminal forage degradability and total tract diet digestibility in lactating dairy cows during grain-induced subacute ruminal acidosis. *J. Dairy Sci.* 87(6):1840–1847.

Patra, A. K., Kamra, D. N. and Agarwal, N. 2006. Effect of plant extracts on in vitro methanogenesis, enzyme activities and fermentation of feed in rumen liquor of buffalo. *Anim. Feed Sci. Technol.* 128(3–4):276–291.

Patra, A. K. and Saxena, J. 2010. A new perspective on the use of plant secondary metabolites to inhibit methanogenesis in the rumen. *Phytochemistry* 71(11–12):1198–1222.

Patra, A. K. and Yu, Z. 2013. Effective reduction of enteric methane production by a combination of nitrate and saponin without adverse effect on feed degradability, fermentation, or bacterial and archaeal communities of the rumen. *Bioresour. Technol.* 148:352–360.

Pen, B., Sar, C., Mwenya, B., Kuwaki, K., Morikawa, R. and Takahashi, J. 2006. Effects of *Yucca schidigera* and *Quillaja saponaria* extracts on in vitro ruminal fermentation and methane emission. *Anim. Feed Sci. Technol.* 129(3–4):175–186.

Pen, B., Takaura, K., Yamaguchi, S., Asa, R. and Takahashi, J. 2007. Effects of *Yucca schidigera* and *Quillaja saponaria* with or without β 1–4 galacto-oligosaccharides on ruminal fermentation, methane production and nitrogen utilization in sheep. *Anim. Feed Sci. Technol.* 138(1):75–88.

Podolak, I., Galanty, A. and Sobolewska, D. 2010. Saponins as cytotoxic agents: A review. *Phytochem. Rev.* 9(3):425–474.

Popova, M., Guyader, J., Silberberg, M., Seradj, A. R., Saro, C., Bernard, A., Gérard, C., Martin, C. and Morgavi, D. P. 2019. Changes in the rumen microbiota of cows in response to dietary supplementation with nitrate, linseed, and saponin alone or in combination. *Appl. Environ. Microbiol.* 85(4):e02657-18.

Ramos-Morales, E., de la Fuente, G., Nash, R. J., Braganca, R., Duval, S., Bouillon, M. E., Lahmann, M. and Newbold, C. J. 2017. Improving the antiprotozoal effect of saponins in the rumen by combination with glycosidase inhibiting iminosugars or by modification of their chemical structure. *PLoS ONE* 12(9):e0184517.

Santoso, B., Mwenya, B., Sar, C., Gamo, Y., Kobayashi, T., Morikawa, R., Kimura, K., Mizukoshi, H. and Takahashi, J. 2004. Effects of supplementing galacto-oligosaccharides, *Yucca schidigera* or nisin on rumen methanogenesis, nitrogen and energy metabolism in sheep. *Livest. Prod. Sci.* 91(3):209–217.

Singer, M. D., Robinson, P. H., Salem, A. Z. M. and DePeters, E. J. 2008. Impacts of rumen fluid modified by feeding *Yucca schidigera* to lactating dairy cows on in vitro gas production of 11 common dairy feedstuffs, as well as animal performance. *Anim. Feed Sci. Technol.* 146(3–4):242–258.

Śliwiński, B. J., Kreuzer, M., Sutter, F., Machmüller, A. and Wettstein, H.-R. 2004. Performance, body nitrogen conversion and nitrogen emission from manure of dairy cows fed diets supplemented with different plant extracts. *J. Anim. Feed Sci.* 13(1):73–91.

Śliwiński, B. J., Kreuzer, M., Wettstein, H.-R. and Machmüller, A. 2002a. Rumen fermentation and nitrogen balance of lambs fed diets containing plant extracts rich in tannins and saponins, and associated emissions of nitrogen and methane. *Arch. Anim. Nutr.* 56(6):379–392.

Śliwiński, B. J., Soliva, C. R., Machmüller, A. and Kreuzer, M. 2002b. Efficacy of plant extracts rich in secondary constituents to modify rumen fermentation. *Anim. Feed Sci. Technol.* 101(1–4):101–114.

Sparg, S. G., Light, M. E. and van Staden, J. 2004. Biological activities and distribution of plant saponins. *J. Ethnopharmacol.* 94(2-3):219-243.

Szakiel, A., Pązkowski, C. and Henry, M. 2011. Influence of environmental abiotic factors on the content of saponins in plants. *Phytochem. Rev.* 10(4):471-491.

Tedeschi, L. O., Fox, D. G. and Tylutki, T. P. 2003. Potential environmental benefits of ionophores in ruminant diets. *J. Environ. Qual.* 32(5):1591-1602.

Tedeschi, L. O., Muir, J. P., Naumann, H. D., Norris, A. B., Ramírez-Restrepo, C. A. and Mertens-Talcott, S. U. 2021. Nutritional aspects of ecologically relevant phytochemicals in ruminant production. *Front. Vet. Sci.* 8:628445.

Teferedegne, B. 2000. New perspectives on the use of tropical plants to improve ruminant nutrition. *Proc. Nutr. Soc.* 59(2):209-214.

Thalib, A., Widiawati, Y., Hamid, H., Suherman, D. and Sabrani, M. 1996. The effects of saponin from *Sapindus rarak* fruit on rumen microbes and performance of sheep. *J. Ilmu Ternak Vet.* 2:17-20.

Valdez, F. R., Bush, L. J., Goetsch, A. L. and Owens, F. N. 1986. Effect of steroidal saponins on ruminal fermentation and on production of lactating dairy cows. *J. Dairy Sci.* 69(6):1568-1575.

Vinogradov, E., Egbosimba, E. E., Perry, M. B., Lam, J. S. and Forsberg, C. W. 2001. Structural analysis of the carbohydrate components of the outer membrane of the lipopolysaccharide-lacking cellulolytic ruminal bacterium *Fibrobacter succinogenes* S85. *Eur. J. Biochem.* 268(12):3566-3576.

Wallace, R. J. 2004. Antimicrobial properties of plant secondary metabolites. *Proc. Nutr. Soc.* 63(4):621-629.

Wallace, R. J., Arthaud, L. and Newbold, C. J. 1994. Influence of *Yucca shidigera* extract on ruminal ammonia concentrations and ruminal microorganisms. *Appl. Environ. Microbiol.* 60(6):1762-1767.

Wallace, R. J., McEwan, N. R., McIntosh, F. M., Teferedegne, B. and Newbold, C. J. 2002. Natural products as manipulators of rumen fermentation. *Asian. Australas J. Anim. Sci.* 15(10):1458-1468.

Wang, B., Tu, Y., Zhao, S. P., Hao, Y. H., Liu, J. X., Liu, F. H., Xiong, B. H. and Jiang, L. S. 2017. Effect of tea saponins on milk performance, milk fatty acids, and immune function in dairy cow. *J. Dairy Sci.* 100(10):8043-8052.

Wang, Y., McAllister, T. A., Newbold, C. J., Rode, L. M., Cheeke, P. R. and Cheng, K.-J. 1998. Effects of *Yucca schidigera* extract on fermentation and degradation of steroidal saponins in the rumen simulation technique (RUSITEC). *Anim. Feed Sci. Technol.* 74(2):143-153.

Wang, Y., McAllister, T. A., Yanke, L. J. and Cheeke, P. R. 2000a. Effect of steroidal saponin from *Yucca schidigera* extract on ruminal microbes. *J. Appl. Microbiol.* 88(5):887-896.

Wang, Y., McAllister, T. A., Yanke, L. J., Xu, Z. J., Cheeke, P. R. and Cheng, K.-J. 2000b. In vitro effects of steroidal saponins from *Yucca schidigera* extract on rumen microbial protein synthesis and rumen fermentation. *J. Sci. Food Agric.* 80(14):2114-2122.

Williams, A. G. and Coleman, G. S. 1992. *The Rumen Protozoa*. New York: Springer-Verlag New York Inc., p. 441.

Williams, A. G. and Withers, S. E. 1991. Effect of ciliate protozoa on the activity of polysaccharide-degrading enzymes and fibre breakdown in the rumen ecosystem. *J. Appl. Bacteriol.* 70(2):144-155.

Wilson, R. C., Overton, T. R. and Clark, J. H. 1998. Effects of *Yucca schidigera* extract and soluble protein on performance of cows and concentrations of urea nitrogen in plasma and milk. *J. Dairy Sci.* 81(4):1022–1027.

Wina, E., Meutzel, S. and Becker, K. 2005a. The impact of saponins or saponin containing plant materials on ruminant production: A review. *J. Agric. Food Chem.* 53(21):8093–8105.

Wina, E., Muetzel, S., Hoffmann, E., Makkar, H. P. S. and Becker, K. 2005b. Saponins containing methanol extract of *Sapindus rarak* affect microbial fermentation, microbial activity and microbial community structure in vitro. *Anim. Feed Sci. Technol.* 121(1–2):159–174.

Wu, Z., Sadik, M., Sleiman, F. T., Simas, J. M., Pessarakli, M. and Huber, J. T. 1994. Influence of yucca extract on ruminal metabolism in cows. *J. Anim. Sci.* 72(4):1038–1042.

Yang, C. M. J. and Russell, J. B. 1993. Effect of monensin on the specific activity of ammonia production by ruminal bacteria and disappearance of amino nitrogen from the rumen. *Appl. Environ. Microbiol.* 59(10):3250–3254.

Yoshiki,Y., Kudou, S. and Okubo, K. 1998. Relationship between chemical structures and biological activities of triterpenoid saponins from soybean (review). *Biosci. Biotechnol. Biochem.* 62(12):2291–2299.

Part 3

The use of direct-fed microbials (DFM) and other supplements to optimise nutrition

Chapter 7

The use of bacterial probiotics and direct-fed microbials as dietary supplements in dairy cow nutrition

Zhongtang Yu[1], The Ohio State University, USA; Amlan Patra[2,3], West Bengal University of Animal and Fishery Sciences, India and American Institute for Goat Research, Langston University, USA; and Ming Yan[1], The Ohio State University, USA*

1 Introduction

Over the last few decades, remarkable advances in understanding the gastrointestinal (GI) microbiome have made it clear that the GI microbiome is dynamic, and it interacts with diet and hosts, impacting host nutrition, productivity, some physiological activities, immune responses, and health and diseases (Mazmanian et al., 2005; Nakajima et al., 2018; Raabis et al., 2019). Maintaining and reconstituting a well-balanced GI microbiome can effectively preserve good health and prevent or treat GI disorders (Sekirov et al., 2010). The GI microbiome maintains interspecies balance, or eubiosis, under normal conditions, which is translated to enhanced health and production efficiency due to improved nutrient utilization, metabolic functions, and immunity

*Correspondence note: Zhongtang Yu, Department of Animal Sciences, The Ohio State University, Columbus, OH 43210. Email: yu.226 @osu.edu.
[1] Department of Animal Sciences, The Ohio State University, Columbus 42310, USA.
[2] Department of Animal Nutrition, West Bengal University of Animal and Fishery Sciences, 37 K.B. Sarani, Belgachia, Kolkata 700037, India.
[3] American Institute for Goat Research, Langston University, Langston 73050, USA.

http://dx.doi.org/10.19103/AS.2022.0117.07

and reduced GI pathogen loads and diseases. External factors including diet, maternal factors, stress, environment, and antibiotic treatment can disrupt eubiosis, leading to dysbiosis (Malmuthuge and Guan, 2017). Some live bacteria and yeast have been used as direct-fed microbials (DFMs), or probiotics, to help prevent GI microbiome dysbiosis and benefit dairy cows and calves.

Probiotics are live microorganisms, primarily bacteria, which when administered in adequate amounts, confer a health benefit to the host (Guarner and Schaafsma, 1998; FAO/WHO, 2001). The US Food and Drug Administration (FDA) and the Association of American Feed Control Officials (AAFCO) use the term DFMs to refer to products that contain a source of live (viable) naturally occurring microorganisms and are intended for use in animal feeds (FDA, 1995). The use of DFMs as feed additives is regulated by AAFCO and FDA. Both FDA and AAFCO have established a list of DFMs suitable for use in animal feeds. In this chapter, DFMs and probiotics are synonymous and used interchangeably. Yeast, primarily *Saccharomyces cerevisiae*, has been the most commonly used DFMs in ruminants, and its use in ruminants including dairy cattle has been extensively reviewed (Yoon and Stern, 1995; Seo et al., 2010; McAllister et al., 2011; Poppy et al., 2012; Amin and Mao, 2021; Ban and Guan, 2021; Zhang et al., 2022), and it will not be discussed in this chapter. Collectively, bacterial DFMs are intended to improve feed digestion and utilization, enhance animal growth and health, and prevent and reduce the risk of diseases primarily subacute rumen acidosis (SARA) and enteric diseases, particularly calf diarrhea, by maintaining or restoring GI microbiome eubiosis and improving host physiological, metabolic, and immunological activities. Species of *Lactobacillus*, *Streptococcus*, *Enterococcus*, *Bacillus*, *Clostridium*, *Bifidobacterium*, *Propionibacterium*, and *Megasphaera elsdenii* are among the most commonly evaluated or used as DFMs. In dairy calves, DFMs have been used to promote growth and prevent or treat certain diseases, particularly calf diarrhea (Signorini et al., 2012). DFMs have the potential to benefit transition cows when they experience metabolic, endocrine, and immune system perturbations, and production diseases and disorders, such as fatty liver, negative energy balance, hypocalcemia, and oxidative stress during the periparturient period (Mulligan and Doherty, 2008). In lactating cows passing the periparturient period, certain DFMs can help improve nutrient utilization and milk production and reduce the risk and severity of SARA (Ban and Guan, 2021). In this chapter, we discuss the use of bacterial DFMs (probiotics) in dairy calves and cows with respect to the modes of action and the effects on rumen microbiome and function, milk production, feed efficiency, post-ruminal (intestinal) microbiome and functions, immune response, and methane (CH_4) emissions. The challenges to effective and consistent benefits are also discussed.

2 Direct-fed microbials commonly used in dairy cattle and their modes of action

For a strain of bacteria to be registered as a probiotic, it needs to meet the following criteria: able to survive in the GI tract, non-pathogenic and non-toxigenic, taxonomically identified, safe without any undesirable side effects, beneficial to host health documented clinically, and stable and feasible for commercial production (Retta, 2016; Michalak et al., 2021). Additionally, probiotics should carry no antibiotic resistance genes (Daniali et al., 2020; Li et al., 2020). When used in animals, some DFMs are also intended to improve host nutrient utilization and productivity. The bacterial DFMs used in ruminants including dairy cattle can be broadly categorized into lactate-producing bacteria (Table 1), primarily lactic acid bacteria (LABs, primarily *Bifidobacterium*, *Lactobacillus*, *Enterococcus*, *Lactococcus*, and *Streptococcus*), lactate-utilizing bacteria (LUBs, e.g. *Megasphaera elsdenii*) (Table 1), propionate-producing bacteria (PPBs, e.g. *Propionibacterium* and *Prevotella bryantii*) (Table 2), and *Bacillus*-based species (Table 3). A few other bacteria such as *Faecalibacterium* spp., fiber-degrading bacteria (e.g. *Ruminococcus flavefaciens*), and acetogens (*Proteiniphilum acetatigenes*) have also been evaluated as DFMs in dairy cattle (Table 3). Some DFMs function in the rumen, whereas others are intended to function in the post-ruminal intestines. Some DFMs contain a single strain of bacteria, while others contain multiple strains of the same species or multiple species (Table 4). Different probiotics use different modes of action to manifest their benefits.

2.1 Modes of action in the rumen

The modes of action of DFMs in the rumen include stabilization of pH, stimulating the growth of some bacteria (e.g. fiber-degrading bacteria and LUBs), modifying fermentation characteristics, improving nutrient digestibility, and increasing post-ruminal nutrient flow (Yoon and Stern, 1995; McAllister et al., 2011; Ban and Guan, 2021) (Box 1). The specific mechanisms to achieve the above benefits can differ from those of DFMs intended to function in the post-ruminal intestine.

Table 1 Bacterial DFM of lactate-producing and -utilizing species used in dairy cattle and their effects on feed intake and efficiency, rumen fermentation characteristics, milk production traits, immune response, and some blood variables in dairy cows

Species and/or strains	Dose	Lactation stage	Main findings	References
Lactate-producing species				
Lactococcus sp.	1.6×10^9 CFU/day	Mid-lactation	No effect on milk protein or fat content; increased milk lactose content; decreased milk SCCs, TNF-α, IgG, myeloperoxidase, and *N*-acetyl-β-D-glucosaminidase activity	Gao et al. (2020)
Lactobacillus sakei KTU05-6	4×10^{11} CFU/cow	Mid-lactation	No effect on feed intake, ruminal fermentation, milk yield, or composition	Lele et al. (2019)
Pediococcus pentosaceus BaltBio02	3.2×10^{11} CFU/cow	Mid-lactation	Increased milk yield at day 66, but no effect on feed intake, milk composition, or ruminal fermentation	Lele et al. (2019)
Lactobacillus pentosus D31	3.6 or 4.6×10^{11} CFU/day	Early to mid-lactation	No effect on feed intake, feed efficiency, milk yield, milk composition, or ruminal fermentation and methane production; raised ruminal pH increased at the high dose	Jeyanathan et al. (2019)
Enterococcus faecium (1, 2, or 3 strains)	2.4×10^9 CFU/day	Early lactation (SARA challenged)	Prevented decline in milk yield; no effect on ruminal pH or VFA concentration	Chiquette et al. (2015)
Lactate-utilizing species				
Megasphaera elsdenii NCIMB 41125	1×10^{10} CFU/day	Early lactation	No effect on milk, milk protein and lactose yield, or VFA concentration; tended to decrease milk fat content; improved ruminal pH stabilization	Aikman et al. (2011)
M. elsdenii ATCC 25940	3.5×10^9 CFU/day	Mid-lactation	Tended to increase molar proportion of propionate and lowered acetate to propionate ratio and plasma total fatty acid level; no effect on feed intake, milk yield and composition, ruminal pH, or lactate concentration	Zebeli et al. (2012)
M. elsdenii NCIMB 41125	2×10^{10} CFU/day	Late lactation	Increased feed intake, milk yield and composition, mean ruminal pH; reduced duration while ruminal pH <5.6	Mazon et al. (2020)
M. elsdenii NCIMB 41125	2×10^{10} CFU/day	Mid-lactation	No effect on feed intake, milk yield, protein content, mean ruminal pH; reduced duration while ruminal pH <5.6	Mazon et al. (2020)

CFU, colony-forming unit; SCC, somatic cell count; TNF-α, tumor necrosis factor α; VFA, volatile fatty acids; IgG, immunoglobulin G, SARA, subacute ruminal acidosis; DFM, direct-fed microbial.

Table 2 Bacterial DFM of propionate-producing species and other species used in dairy cattle and their effects on feed intake and efficiency, rumen fermentation characteristics, milk production traits, immune response, and some blood variables in dairy cows

Species and/or strains	Dose	Lactation stage	Main findings	References
Propionate-producing species				
Propionibacterium freudenreichii 53-W	2.9×10^{10} CFU/day	Early to mid-lactation	No effect on feed intake, feed efficiency, milk yield, milk composition, ruminal fermentation, or methane production; increased ruminal pH	Jeyanathan et al. (2019)
Propionibacterium sp. P63	1×10^{10} CFU/day	Early to mid-lactation	No effect on feed intake, milk yield, milk composition, nutrient digestibility, ruminal fermentation, or methane production; raised ruminal pH; lowered lactate concentration	Philippeau et al. (2017)
Propionibacterium sp. P169	6 or 60×10^{10} CFU/day	14 days prepartum to mid-lactation	Increased milk production, propionate proportion; reduced acetate to propionate ratio	Stein et al. (2006)
Propionibacterium sp. P169	6×10^{11} CFU/day	14 days prepartum to early lactation	Reduced feed intake, ruminal acetate proportion; no effect on milk yield or composition; increased feed efficiency and propionate proportion (tendency)	Weiss et al. (2008)
Prevotella bryantii 25A	2×10^{11} CFU/cow	Mid-lactation (SARA model)	No effect on ruminal pH, total VFA, or proportion of individual VFA	Chiquette et al. (2012)
Prevotella bryantii 25A	2×10^{11} CFU/cow	21 days prepartum to early lactation	No effect on feed intake, milk production, or ruminal pH; tended to increase milk fat content; increased acetate and butyrate concentration post-feeding; decreased ruminal lactate concentration 2–3 h after feeding	Chiquette et al. (2008)
Other bacterial species				
Proteiniphilum acetatigenes GA03	Not reported	Mid-lactation	Increased milk production, milk protein and fat content; decreased milk SCCs and plasma total fatty acid concentration; increased blood glucose	Kim et al. (2020b)
Rhodobacter sphaeroides	Not reported	Not reported	No effect on feed intake, milk yield or composition; increased milk coenzyme Q_{10} concentration	Bae et al. (2018)

CFU, colony-forming unit; SARA, subacute ruminal acidosis; VFA, volatile fatty acids; DFM, direct-fed microbial.

Table 3 Bacterial DFM of *Bacillus* species used in dairy cattle and their effects on feed intake and efficiency, rumen fermentation characteristics, milk production traits, immune response, and some blood variables in dairy cows

Species and/ or strains	Dose of DFM	Lactation stage	Main findings	References
B. coagulans PRM101	2×10^{11} CFU/cow	Mid-lactation	No effect on milk production and composition, concentration of blood and milk amyloid A, or haptoglobin	Izadi et al. (2021)
B. megaterium 1259	5, 10, or 15 $\times 10^{5}$ CFU/cow	Not reported	Increased 4% fat-corrected milk at the highest dose; no effect on milk composition, ruminal pH, or VFA concentration	Deng et al. (2021)
B. Subtilis C-3102	3×10^{9} CFU/cow	Late lactation	No effect on feed intake, milk yield and composition, feed efficiency, nutrient digestibility, or milk SCCs	Souza et al. (2017)
B. subtilis C-3102	3×10^{9} CFU/cow	Mid-lactation	Increased milk and protein yield; decreased lactose content; no effect on milk protein and fat content, or milk SCCs	Souza et al. (2017)
B. subtilis 1.1086	2×10^{11} CFU/cow	Early lactation	No effect on feed intake, milk yield and composition, feed efficiency, ruminal digestion, microbial protein synthesis, or ruminal fermentation	Qiao et al. (2010)
B. licheniformis 1.183	2×10^{11} CFU/cow	Early lactation	No effect on feed intake; increased milk yield and fat content, feed efficiency, ruminal digestion of organic matter and fiber, microbial nitrogen flow to duodenum, VFA and acetate concentration; reduced ruminal ammonia concentration, and pH	Qiao et al. (2010)
B. subtilis	0.5, 1, and 2×0^{11} CFU/cow	Early to mid-lactation (tropical condition)	Increased milk yield linearly with no effect on milk composition and SCCs; tended to decrease serum glucose concentration	Choonkham et al. (2020)
B. subtilis	0.5×0^{11} CFU/cow	21 days prepartum to early lactation	No effect on serum total antioxidant capacity and thiobarbituric acid-reactive substances	Choonkham et al. (2021)

B. subtilis natto	5 or 10 × 10^{10} CFU/cow	Early lactation	Increased milk production and milk component yield; decreased SCCs; increased VFA concentration, molar proportion of propionate, and cellulolytic bacteria	Sun et al. (2013)
B. pumilus 8G-134	5 × 10^{9} CFU/cow	Prepartum to mid-lactation	No effect on feed intake and milk composition; increased milk yield; improved feed efficiency; increased milk IgA but lowered serum haptoglobin concentrations; lower ketosis prevalence	Luan et al. (2015)
B. subtilis natto	5 or 10 × 10^{10} CFU/cow	Early lactation	Increased milk production, milk fat yield, lactose content, feed efficiency, ruminal propionate proportion; no effect on feed intake, milk fat content, or protein yield; lowered plasma total fatty acids, ruminal acetate proportion	Peng et al. (2012)

CFU, colony-forming unit; VFA, volatile fatty acids; SCC, somatic cell count; IgA, immunoglobulin A; DFM, direct-fed microbial.

Table 4 Multispecies bacterial DFM used in dairy cattle and their effects on feed intake and efficiency, rumen fermentation characteristics, milk production traits, immune response, and some blood variables in dairy cows

Species and/or strains	Dose	Lactation stage	Main findings	References
Propionibacterium sp. P63 + *Lactobacillus plantarum* 115	1×10^{10} CFU/day each	Early to mid-lactation	No effect on feed intake, milk yield, milk composition, nutrient digestibility, ruminal VFA and lactate concentration, or methane production; raised ruminal pH	Philippeau et al. (2017)
Propionibacterium sp. P63 + *Lactobacillus rhamnosus* 32	1×10^{10} CFU/day each	Early to mid-lactation	No effect on feed intake, milk yield, milk composition, nutrient digestibility, ruminal VFA and lactate concentration, or methane production; raised ruminal pH	Philippeau et al. (2017)
Enterococcus faecium + *Lactococcus lactis* DSM 11037	2.4×10^9 CFU/day + 2.8×10^9 FU/day	Early-lactation SARA challenged	Prevented a decline in milk yield; no effect on ruminal pH or VFA concentration	Chiquette et al. (2015)
Lactobacillus acidophilus NP51 + *Propionibacterium freudenreichii* NP24	4×10^9 CFU/day (total)	Mid-lactation	No effect on feed intake, milk composition, feed efficiency; increased milk and milk protein yield, crude protein and fiber digestibility	Boyd et al. (2011)
Lactobacillus acidophilus LA747 + *Propionibacterium freudenreichii* PF24	1×10^9 CFU/day + 2×10^9 CFU/day	Early to mid-lactation	No effect on feed intake, milk yield and composition, nutrient digestibility, ruminal pH or VFA profiles	Raeth-Knight et al. (2007)
Lactobacillus acidophilus LA747 + *Propionibacterium freudenreichii* PF24 + *Lactobacillus acidophilus* LA45	1×10^9 CFU/day + 2×10^9 CFU/day + 5×10^8 CFU/day	Early to mid-lactation	No effect on feed intake, milk yield and composition, nutrient digestibility, ruminal pH, or VFA profiles	Raeth-Knight et al. (2007)

Lactobacillus animalis and *Propionibacterium freudenreichii*	1×10^9 CFU/day + 2×10^9 CFU/day	Early to mid-lactation	No effect on feed intake, milk yield and composition, digestibility, or fecal score; raised ruminal pH	Lawrence et al. (2021)
Lactobacillus casei Zhang + *Lactobacillus plantarum* P-8	3.25×10^{10} CFU/day + 3.25×10^{10} CFU/day	Early lactation	Increased milk yield and milk IgG; no effect on milk composition; decreased milk SCCs and fecal opportunistic pathogens	Xu et al. (2017)

CFU, colony-forming unit; VFA, volatile fatty acids; SCC, somatic cell count; IgG, immunoglobulin G; SARA, subacute ruminal acidosis; DFM, direct-fed microbial.

Box 1 Modes of action of different bacterial direct-fed microbials in the rumen of dairy cattle

Overall bacterial DFM effects:

- Improved ruminal fermentation;
- Increased propionate production;
- Increased feed and fiber digestibility;
- Increased microbial protein synthesis and activity; and
- Better ruminal pH stabilization with less variation.

Lactate-producing bacterial DFMs:

- Tonic supply of lactate to microbiota;
- Stimulation of LUBs;
- Adaptation of microbiota to lactate;
- Stabilization of pH; and
- Improved feed digestibility.

Lactate-utilizing bacterial DFMs:

- Rapid conversion of lactate to volatile fatty acid (VFA);
- Elevation of nadir pH;
- Greater fiber digestibility;
- Increased feed intake; and
- Greater milk production.

Other bacterial DFMs:

- Cellulolytic bacteria (e.g. Ruminococcus flavefaciens and Butyrivibrio fibrisolvens);
- Detoxification of undesirable compounds (e.g. Synergistes jonesii); and
- Acetogens (e.g. Proteiniphilum acetatigenes; hydrogen sinks).

B. fibrisolvens, *Ruminococcus flavefaciens*, and *R. albus* are common, known fiber-degrading bacteria in the rumen. Dosing the rumen with these bacteria can improve fiber digestibility. Several strains of these fiber-degrading bacteria, such as *B. fibrisolvens* YE 44 (Gregg et al., 1998; Klieve et al., 2003), *R. flavefaciens* NJ (Chiquette et al., 2007), and *R. albus* A3 (Miyagi et al., 1995), have been evaluated to improve fiber digestion in the rumen. Repeated dosing allowed the DFMs to be maintained at the effective abundance and increased fiber digestibility (Chiquette et al., 2007), but the dosed strains could not sustain the effective abundance and the positive outcome disappeared after the dosing. It is worth exploring if other isolates of fiber-degrading bacteria can maintain an effective abundance persistently.

LAB-based DFMs can elicit a positive response in the rumen. One possible mode of action is that they may produce a tonic level of lactate along with acetate. The sustained production of lactate in the rumen may sustain a basal level of LUBs, which could grow rapidly and consume the increased lactate in response to feeding highly fermentable carbohydrate diets, thereby stabilizing ruminal pH and reducing ruminal acidosis (McAllister et al., 2011). Additionally, lactate can be fermented to butyrate by some rumen LUBs, such as by *Megasphaera elsdenii*. Butyrate is an important energy source for the rumen epithelial cells in calves. Thus, *M. elsdenii* was expected to benefit the growth and development of the rumen tissue. In one study, *M. elsdenii* increased ruminal butyrate concentration and feed intake in neonatal calves and rumen development (Muya et al., 2015). However, consistent results are lacking in the literature. Lactate can also be fermented to propionate by some rumen bacteria, such as *Selenomonas ruminantium*. As the major precursor of hepatic gluconeogenesis, increased propionate benefits lactose and milk production in lactating cows and growth performance in growing dairy heifers.

In terms of energy efficiency, the fermentation of sugars (glucose) to propionate is more efficient than that to acetate. Hence, increased propionate production in the rumen can improve milk production in dairy cattle and growth in growing dairy heifers. Species of *Propionibacterium*, which are both LUBs and PPBs, have been used as DFMs in dairy cattle. *Propionibacterium* spp. can increase hepatic glucose production and positively influence energy metabolism, reducing the incidence of ketosis in early-lactation cows (Francisco et al., 2002; Weiss et al., 2008).

Bacterial DFMs of other species or guilds (functional groups) have been used to modulate rumen functions. *Bacillus*-based DFMs, including *B. subtilis*, *B. amyloliquefaciens*, *B. licheniformis*, *B. coagulans*, and *B. toyonensis*, have shown some promising benefits in both dairy calves and cows (Table 3). This group of DFMs can increase amylolytic and proteolytic bacteria in the rumen, increase the digestibility of carbohydrates and protein, promote the proliferation of lactobacilli, modify rumen fermentation, and increase total VFA production (Qiao et al., 2010; Ban and Guan, 2021). However, the modes of their function remain poorly understood. *Bacillus*-based DFMs can be produced, distributed, and stored as spores, which are highly resistant to harsh physical and chemical conditions and can survive at low pH. Therefore, *Bacillus*-based DFMs are more stable during feed processing and gastric digestion (McAllister et al., 2011). *Bacillus*-based DFMs need to be fed constantly to sustain their population because DFM spores germinate but poorly proliferate in the rumen or intestine. New species or strains of bacteria are explored for their potential to benefit rumen functions. Mechanistic research is needed to better understand their modes of action so that they can be used more effectively.

Bacteriocins are ribosomally synthesized antimicrobial peptides, and they can be secreted by different prokaryotes including DFMs. Bacteriocin can inhibit the growth of other bacteria by forming pores in the bacterial cell membrane and then resulting in dissipation of proton-motive force, leading to cell death (Kumariya et al., 2019; Benítez-Chao et al., 2021). Albusin 7 produced by *Ruminococcus albus* is inhibitory against all *R. flavefaciens* strains tested (Chen et al., 2004). A rumen isolate, *Bacillus licheniformis* 26 L-10/3RA, produces lichenin, and this bacteriocin inhibited some strains of *Streptococcus*, *Ruminococcus*, *Lactobacillus*, and *Butyrivibrio* (Pattnaik et al., 2001). It is not known if bacteriocins are an important mode of action for the DFMs that function in the rumen of dairy cattle.

2.2 Modes of action in the intestines

Bacterial DFMs used in dairy cattle, especially dairy calves, to enhance post-ruminal intestinal health and function manifest their benefits through different modes of action (Box 2). Some DFMs secrete antibacterial (bactericidal or bacteriostatic) metabolites, such as organic acids, hydrogen peroxide (H_2O_2), bacteriocins, diacetyl, and acetaldehyde (Zielińska et al., 2021), all of which can reduce enteropathogen colonization and enteric infections, especially diarrhea in calves. Also, LAB-based DFMs produce lactate as a major end product, and lactate can increase cell membrane permeability and enter pathogenic bacteria disrupting their critical cellular functions by reducing intracellular pH (Seo et al., 2010; Zhang et al., 2019). For example, *Lactobacillus acidophilus* YIT 0070 and *L. casei* Shirota can inhibit Shiga toxin-producing *Escherichia coli* by producing lactate (Ogawa et al., 2001). Some *Lactobacillus* spp. produce H_2O_2 in the presence of oxygen. As a strong oxidant producing free radicals, H_2O_2 can lead to membrane lipid peroxidation, increased membrane permeability, and DNA damage (Zielińska et al., 2021). The ability of some *Lactobacillus* species to produce H_2O_2 was suggested as a possible mode of inhibition of species of *Staphylococcus*, *Salmonella*, and *Pseudomonas* (Holzapfel et al., 1995; Pridmore et al., 2008). This mode of action may be insignificant in the anaerobic luminal environment of the intestine (McAllister et al., 2011) but may function on the surface of intestinal epithelia where low levels of oxygen are available.

Box 2 Modes of action of bacterial direct-fed microbials in the post-ruminal digestive tract of dairy calves and cows

- Production of antimicrobial compounds (e.g. bacteriocins, hydrogen peroxide, organic acids, diacetyl, and acetaldehyde);
- Stimulation of host immune responses;
- Competition with pathogens for specific binding sites in mucosa;
- Competition with pathogens for nutrients or ions;
- Maintenance of a balanced beneficial bacterial population;
- Improving gut health and barrier function;
- Improving antioxidant status;
- Reducing cell apoptosis;
- Improving intestinal epithelium microstructures; and
- Vitamin synthesis.

(Modified from Seo et al., 2010; McAllister et al., 2011).

Bacteriocins usually kill closely related strains that have a similar ecological niche, but they can also inhibit a broad range of bacteria including enteric pathogens (Chikindas et al., 2018). Some studies have investigated the direct antibacterial effects of bacterial DFMs in calves, and it is generally accepted that some bacterial DFMs, including LAB-based DFMs, may inhibit enteropathogens by producing bacteriocins as one of the mechanisms (McAllister et al., 2011).

Enterotoxigenic pathogens such as *Escherichia coli* and *Salmonella* can attach to intestinal epithelial cells and subsequently cause infection, inflammation, and diarrhea. Competitive adhesion to the epithelial surface is one mechanism by which some DFMs reduce enteropathogens. For example, *Lactobacillus rhamnosus* and *L. casei* have a stronger ability to attach to the intestinal epithelial cells than enteropathogens (e.g. enterotoxigenic *E. coli*, *Salmonella*, and *Klebsiella pneumonia*), thereby hindering the attachment of the latter to the enterocyte receptors (Forestier et al., 2001; Lee et al., 2003; Frizzo et al., 2010). The blockage of enteropathogens to epithelial cells can be pathogen dependent. For example, *Lactobacillus* strains strongly inhibit the adhesion of Gram-negative bacteria (e.g. *E. coli* and *Salmonella*) but only moderately inhibit the adhesion of Gram-positive bacteria such as *Staphylococcus aureus* (Zielińska et al., 2021) *in vitro*. Again, among the various probiotic strains, *L. brevis* O24 and *L. rhamnosus* K3 showed the most effective exclusion of pathogens and antimicrobial activities (Zielińska et al., 2021). It should be noted that dead cells of DFMs may also reduce or prevent enteropathogens from adhering to the intestinal epithelial surface by blocking their access. Indeed, heat-killed *Lactobacillus* spp. inhibited the attachment to intestinal epithelial cells by several enteric pathogens (Zielińska et al., 2021). Heat-inactivated cells of *Bifidobacterium bifidum* OLB6378 also reduced the colonization of enteropathogens on the intestinal epithelial surface. Certain bifidobacterial metabolites might upregulate the expression of immunoglobulin (Ig) receptors on the host epithelial cells (Nakamura et al., 2012).

Other modes of action can also be important for certain DFMs to benefit the host animals. Gut barrier functions are governed by several types of tight junction proteins and their regulatory molecules in the epithelial cells (Patra and Kar, 2021). Proper functioning of the gut barrier is important for the absorption of nutrients and electrolytes while preventing the entry of pathogens and antigens from the GI lumen into the blood circulation (Patra and Kar, 2021). Increased pathogen loads in the gut often impair the gut barrier regulatory mechanisms, leading to diarrhea. Some DFMs may improve barrier function by upregulating the expressions of tight junction proteins, increasing the secretion of mucin, enhancing nutrient absorption, and reducing inflammation that often results from enteric infections (Sánchez et al., 2017; Bhat et al., 2020; Al-Sadi et al., 2021; Mahesh et al., 2021). Some DFMs including lactobacilli and bifidobacteria can also enhance host antioxidant status (Luan et al., 2021; Mahesh et al., 2021). Modulation of the intestinal microbiome to restore and enhance eubiosis may be one of the mechanisms by which some DFMs produce probiotic effects. Indeed, a DFM preparation containing both *Lactobacillus plantarum* P-8 and *L. casei* Zhang increased the relative abundance of *Bacteroides*, *Ruminococcus*, *Clostridium*, *Roseburia*, *Coprococcus*, *Faecalibacterium prausnitzii*, and *Dorea* while reducing some opportunistic pathogens, such as *Bacillus cereus*, *Alkaliphilus oremlandii*, and *Cronobacter sakazakii* in the feces of early-lactation cows (Xu et al., 2017). The DFMs also maintained the α-diversity of fecal bacteria, which decreased in the control cows (Xu et al., 2017). Some bacterial DFMs can compete with pathogens for nutrients such as a carbon source and essential ions, thereby reducing pathogen colonization (Siedler et al., 2020). By these modes of action, DFMs, especially LAB-based DFMs, can promote gut health and reduce enteric infections that cause calf diarrhea by reducing enteropathogen colonization, enhancing gut barrier functions and immunity, and promoting or restoring eubiosis (Banos et al., 2013; Retta, 2016).

Most of the mechanisms of DFMs intended to function in the intestines are elucidated *in vitro* or using nonruminant animals. Some DFMs may enhance the innate and/or adaptive immune responses, thereby increasing host defense against pathogens, and regulate neurological and endocrinological systems, thereby affecting host physiological processes and behavior. These mechanisms probably also operate in the intestines of dairy cattle. It should be noted that some probiotic effects are strain specific (Srutkova et al., 2015; Sánchez et al., 2017). Such specificity may explain some of the variations and inconsistencies among DFM effects reported in different animals.

3 Effects on the rumen and intestinal microbiomes and their functions

Besides animal performance, the effects of DFMs on the rumen and/or intestinal microbiomes are often evaluated to help understand their mechanistic underpinnings. The effects of DFMs on the overall rumen and intestinal microbiome and relative abundance of individual taxa can be analyzed using metataxonomics, which entails sequencing and bioinformatic analyses of amplicons of a phylogenetic marker of rumen microbes (16S rRNA gene for bacteria and archaea, 18S rRNA gene for protozoa, and ITS1 for fungi), and metagenomics, which entails shotgun sequencing and bioinformatic analyses of all the DNA extracted from samples. The effects of DFMs on the gene expression of rumen microbes can be determined using metatranscriptomics, which entails sequencing and bioinformatic analyses of all the RNA (or enriched mRNA) isolated from samples. In most of the reported studies, select species or genera of rumen microbes or pathogens in feces are quantified using qPCR.

Unlike yeast DFMs, few bacterial DFMs have been evaluated for their effect on the rumen and/or intestinal microbiomes and functions. Most bacterial DFMs are primarily intended to alter the abundance and activities of certain microbial guilds that are known to be associated with feed digestion, rumen fermentation, and host health. Thus, DFMs are not expected to significantly alter the overall rumen/intestinal microbiome, which is rather resilient. Different DFMs containing LABs or LUBs have been evaluated for their effect on the ruminal/intestinal microbiome, and the results are mixed depending on the strains or species of DFMs used. One strain of *Lactobacillus acidophilus* (BY1386) did not affect the species richness (both observed or estimated) or diversity indexes (both Shannon and Simpson) of the rumen microbiota (both mucosal and digesta) in either pre- or post-weaned calves (Fomenky et al., 2018). Two new isolates (*L. reuteri* TP1.3B and *L. johnsonii* TP1.6) increased fecal *Bifidobacterium* and *Akkermansia* with little effect on the fecal microbiome in newborn calves (Fernandez-Ciganda et al., 2022). A commercial blend of *L. animals* and *Propionibacterium freudenreichii* also did not affect the overall rumen microbiome, even though it increased *Actinobacteria* and *Bifidobacterium* in the rumen and *Rikenellaceae* in feces of dairy cows (Lawrence et al., 2021). Another multispecies DFM product containing *Bifidobacterium animalis*, *Lactobacillus casei*, *Streptococcus faecalis*, and *Bacillus cerevisiae* did not affect the α-diversity or the overall fecal microbiome in calves, but it linearly decreased the relative abundance of *Dorea*, *Faecalibacterium*, and *Mitsuokella*, while the relative abundance of *Prevotella* tended to increase linearly as the dosage of DFMs increased (Guo et al., 2022). The commercial DFM blend containing *L. animals* and *Propionibacterium freudenreichii* had minimal effect on rumen VFA profiles even though it raised the mean and minimum ruminal pH (Lawrence et al., 2021), whereas a multispecies DFM product containing *Lactobacillus plantarum*, *Pediococcus acidilactici*, *Pediococcus pentosaceus*, and *Bacillus subtilis* altered rumen fermentation by decreasing acetate and propionate while increasing butyrate concentrations in calves without affecting the abundance of select species of rumen bacteria including *Fibrobacter succinogenes*, *B. fibrisolvens*, *Streptococcus bovis*, *Prevotella bryantii*, or *Ruminobacter amylophilus* (Wang et al., 2022). In an early study specifically quantifying the effect of DFMs on cellulolytic bacteria in the rumen of dairy calves, *Lactobacillus plantarum* GF103 alone and in combination with *Bacillus subtilis* B27 decreased *R. albus*, and the DFM mix also altered the rumen microbiota (Zhang et al., 2017). DFMs containing LABs and LUBs can affect some individual taxa but have little effect on the overall rumen or intestinal microbiome. The effects on the fermentation profile can vary depending on the DFMs used.

Megasphaera elsdenii can utilize lactate and produce butyrate, which is an important energy source of the rumen epithelia. Several strains of *M. elsdenii* have been tested to increase LUBs in the rumen to prevent or mitigate SARA in dairy cows. Indeed, supplementation with *M. elsdenii* NCIMB 41125 increased ruminal butyrate concentration even though the concentrations of total VFA, acetate, or propionate production were not affected (Muya et al., 2015). This *M. elsdenii* strain also increased starter intake, weaning body weight, weight gain (a tendency), plasma β-hydroxybutyrate concentration, reticulorumen weight, papillae width, and papillae density. No such significant effect, however, was observed in another study that also used *M. elsdenii* (Yohe et al., 2018). Age of animals, dose, and strains of *M. elsdenii* may be important factors that influence the effect of *M. elsdenii* in calves. However, supplementation with *M. elsdenii* did not increase the abundance of *M. elsdenii* in the rumen (Arik et al., 2019).

More species of *Bacillus* (*B. subtilis*, *B. amyloliquefaciens*, *B. toyonensis*, and *B. licheniformis*) and their strains than other bacteria have been evaluated as probiotics as they can promote amylolytic and proteolytic bacteria in the rumen (Sun et al., 2013; Le et al., 2017). For example, *B. subtilis* natto increased proteolytic and amylolytic bacteria in the rumen of dairy cows (Sun et al., 2013). In a comparative study (Schofield et al., 2018), ruminal microbial profiles were shifted in sheep supplemented with *B. amyloliquefaciens* H57, but the *Bacillus* strain did not affect the rumen microbiome in calves, suggesting that the probiotic effects may depend on the species (both host and DFMs) used.

Furthermore, *Lactobacillus rhamnosus* PT9 exhibited a breed-specific effect on the rumen microbiome of sheep (Mani et al., 2021). Overall, the interactions between bacterial DFMs and the rumen microbiome are not clear, which hinders the development of effective DFMs to effectively improve rumen functions. The different effects of DFMs on the rumen microbiome and specific taxa are probably attributed to the DFMs used among other factors such as diet and animal age.

In the intestine, bacterial DFMs, mostly species of *Bacillus* and LABs, are primarily used to improve gut health and reduce infections. Only some of the studies analyzed the effects of DFMs on the intestinal microbiome or bacterial populations. In a recent study (Du et al., 2018), *B. amyloliquefaciens* C-1 and *B. subtilis* individually improved weight gain and feed conversion rate in growth-retarded beef calves and increased the abundance of fecal *Proteobacteria*, *Rhodospirillaceae*, *Campylobacterales*, and *Butyricimonas* while reducing several suspected pathogens, *Anaeroplasma*, and *Acholeplasma*. *Bacillus amyloliquefaciens* C-1 also increased *Sphaerochaeta* and *Treponema*, while *Bacillus subtilis* increased *Akkermansia*, which is involved in intestinal immunity (Naito et al., 2018). However, the overall fecal microbiome was not affected. It is not known if these two *Bacillus* species would have similar effects in growth-retarded dairy calves. In dairy calves, *Lactobacillus acidophilus* BY1386 did not affect the species richness or diversity index of the intestinal microbiome (colonic and ileal, mucosal and digesta) pre- or post-weaning (Fomenky et al., 2018). This DFM, however, affected some individual genera, with different genera being affected differently in pre- vs. post-weaned calves. This DFM also significantly altered the abundance of three genera in the colonic mucosa, including a drastic decrease in *Streptococcus*, and 18 genera in the ileal mucosa, including a considerable decrease in Tyzzerella_4, Ruminococcaceae_UCG-008, *Pseudoflavonifractor*, and *Lachnoclostridium* while increasing *Fibrobacter* in pre-weaned dairy calves; in post-weaned calves, the DFMs only impacted six genera of the ileal mucosa, reducing their relative abundance. Analysis of the predicted functions suggested that *Lactobacillus acidophilus* BY1386 impacted several microbial metabolic pathways only in the rumen digesta in the pre-weaned calves and several pathways including cell cycle only in the ileal mucosa in the post-weaned calves (Fomenky et al., 2018). A commercial DFM product containing *Bacillus licheniformis*, *Bacillus subtilis*, *Lactobacillus acidophilus*, *Lactococcus lactis*, and *Bifidobacterium animalis* subsp. *lactis* did not affect total fecal coliform counts or ceftiofur-resistant and tetracycline-resistant coliform counts in pre-weaned dairy calves (Corbett et al., 2015). Collectively, DFMs have little effect on the overall intestinal microbiome, but they can alter the abundance of some individual bacteria, and such impact can vary along the intestinal tract depending on the bacterial species used.

4 Effects on feed utilization and lactation performance

Different DFMs have been tried to improve feed digestion, nutrient utilization, and lactation performance in lactating cows. These include *Bacillus*-based DFMs (Table 3), LABs (Table 2), LUBs (Table 2), PPBs, and other species (Table 3). Responses to DFMs have been inconsistent, reflecting supplementation with many different bacteria (strains and species), dosage, diet, feeding management, and animal factors. In general, however, there is evidence that DFMs can increase milk production in dairy cattle, especially high-producing cows by improving feed digestion and rumen fermentation profiles and by stabilizing ruminal pH. Many different DFMs have been evaluated or tested for their efficacy on feed utilization and lactation performance. Here, we will only discuss the commonly used DFMs.

4.1 Lactate-producing and lactate-utilizing bacterial direct-fed microbials

Lactate is an important short-chain fatty acid that is produced and utilized in the rumen, and it is implicated in feed utilization, CH_4 emissions, and SARA. Thus, LAB and LUB species have been used to improve lactation performance (Table 1). *Megasphaera elsdenii* is a common species of the rumen microbiome, and it can utilize lactate and produce propionate and butyrate besides acetate (Prabhu et al., 2012). Hypothetically, supplementation with *M. elsdenii* can increase lactate utilization and alleviate ruminal lactate accumulation and pH drop in cows fed high concentrate diets to reduce the risk and severity of SARA. In one study (Aikman et al., 2011), *M. elsdenii* NCIMB 41125 drenched in early-lactation cows shortened the duration of ruminal pH being at below 5.6; this pH-moderating effect was not accompanied by any effect on milk yield or milk composition. The same *M. elsdenii* strain, however, increased milk yield when drenched in high-yielding (> 10.000 kg milk/lactation) dairy cows fed a high-starch diet (Aikman et al., 2009). This *M. elsdenii* strain increased milk yield and milk protein content when drenched for 4 days, but not 1 day, before a SARA challenge (Mazon et al., 2020). The early drenching, not the later drenching, also raised the overall ruminal pH and reduced the duration and severity of SARA while increasing dry matter intake. Another strain of *M. elsdenii* ATCC 25940 supplemented to mid-lactation cows did not affect feed intake, milk production, or milk composition though it tended to increase the molar proportion of ruminal propionate and reduced plasma total fatty acids (Zebeli et al., 2012). The benefits of *M. elsdenii* as DFMs probably depend on the diet, lactation stage, and *M. elsdenii* strains used.

Even though lactate accumulation lowers ruminal pH, LAB-based DFMs can also elicit a positive response in the rumen because LABs may sustain the production of low concentrations of lactate and thus stimulate LUBs (see Section 2.1). Some LABs, especially species of *Lactobacillus* and *Pediococcus*, showed mixed effects in improving milk production. One DFM product containing *L. casei* Zhang and *L. plantarum* P-8 (1:1 ratio) increased milk production in early-lactation Holstein cows (Xu et al., 2017), and *Pediococcus pentosaceus* BaltBio02 increased milk production in mid-lactation cows (Lele et al., 2019), though none of these DFM strains affected feed intake, ruminal fermentation, or milk composition. However, *L. sakei* KTU05-6 had no effect in mid-lactation cows (Lele et al., 2019), while *L. pentosus* D31 did not show any benefits in early- or mid-lactation cows (Jeyanathan et al., 2019). In early-lactation cows experiencing SARA, *Enterococcus faecium* and *L. lactis* DSM 11037 prevented a decline in milk yield (Chiquette et al., 2015). The mixed effects of LAB-based DFMs on feed utilization and lactation performance in dairy cattle call for future research to define the conditions under which these DFMs may provide benefits.

4.2 Propionate-producing bacterial direct-fed microbials

Propionibacterium species can ferment lactate, in addition to sugars, to propionate. Thus, strains and species of *Propionibacterium* have been evaluated as DFMs in dairy cows, but the responses varied (Table 2). For example, *Propionibacterium* sp. P169 [at 6×10^{11} or 6×10^{10} colony-forming unit (CFU)/day] increased milk production in both primiparous and multiparous cows from early to mid-lactation (Stein et al., 2006). However, this *Propionibacterium* strain did not affect milk yield or milk composition in early-lactating cows, even though it increased feed efficiency by decreasing feed intake and ruminal acetate proportion while increasing (a tendency) the proportion of propionate (Weiss et al., 2008). Two recent studies showed that *P. freudenreichii* 53-W did not affect feed intake, feed efficiency, milk yield, milk composition, or ruminal fermentation in early- or mid-lactation cows (Jeyanathan et al., 2019), and *Propionibacterium* sp. P63 did not affect any of the above animal traits in mid-lactation cows (Philippeau et al., 2017). Strains of *Propionibacterium* alone probably result in limited improvement in lactation performance.

Combinations of LABs and species of *Propionibacterium* as LUBs have been used as multispecies DFMs to improve lactation performance. Different combinations of *Propionibacterium freudenreichii* NP24 with several LAB species, including *Lactobacillus plantarum* P63, *L. rhamnosus* 32, or *L. acidophilus* (several different strains), however, did not affect lactation performance or feed intake (Raeth-Knight et al., 2007; Philippeau et al., 2017; Lawrence et al., 2021). One study did show that a DFM mix containing *L. acidophilus* NP51 and *P. freudenreichii* NP24 increased milk yield, milk protein yield, and protein and fiber digestibility without affecting feed intake, milk composition, or feed efficiency in mid-lactation cows (Boyd et al., 2011). Lactation stage, strains and species of the combined DFMs, and diet may influence their effects. Species of *Prevotella*, mostly *Prevotella bryantii* 25A, were also tested, but no significant effect was noted on feed intake or milk production, though several ruminal fermentation characteristics were affected in early-lactating cows (Chiquette et al., 2008). Being the most predominant bacteria in the rumen, species of *Prevotella* may not be limiting therein, and the use of *Prevotella*-based DFMs may not be justified.

4.3 Bacillus-based direct-fed microbials

Bacillus-based DFMs are among the most evaluated and tested bacterial DFMs in dairy cattle. They can affect ruminal feed digestion, fermentation characteristics, and protein metabolism, which can benefit lactation performance. Table 3 summarizes the effects of *Bacillus*-based DFMs on lactation performance in dairy cattle reported in the literature. Supplementation with a solid-state fermentation product of *B. subtilis* natto (at 1×10^{10} CFU/day) to multiparous early-lactation Holstein cows improved several milk production traits, including increased feed efficiency (by 11.7%), milk yield (by 10.0%), milk fat yield (by 16.7%), and lactose content (by 3.1%) without influencing feed intake, milk fat content, or milk protein yield (Peng et al., 2012). Similar effects on lactation performance were reported in another study also using the solid-state fermentation product containing *B. subtilis* natto (at about 1×10^{11} CFU/day) but with different magnitudes of the beneficial effects (Sun et al., 2013). The responses to the same *B. subtilis* natto strain by some ruminal bacterial populations, such as proteolytic and amylolytic bacteria, and rumen fermentation characteristics could differ depending on the dose. Most studies using *B. subtilis* as DFMs are done in temperate zones, but studies also showed that *B. subtilis* (strain not reported) could also increase milk yield in tropic regions (Choonkham et al., 2020; Deng et al., 2021). Several other *Bacillus* species have also been tested as DFMs, including *B. megaterium*, *B. pumilus*, *B. licheniformis*, and *B. coagulans*. Supplementation with *B. megaterium* 1259 increased milk production (Deng et al., 2021). *B. pumilus* 8G-134 fed from 21 days before calving to mid-lactation increased milk fat and milk protein yields during the second week of lactation, but no effect was noted on milk yield or milk composition (Luan et al., 2015). The observed effects on milk production in some studies were linked to improved feed digestibility, ruminal fermentation, microbial activity, and/or protein synthesis. Indeed, *B. licheniformis* increased the ruminal digestibility of organic matter and fiber and microbial protein flow to the duodenum (Qiao et al., 2010). Inconsistency in the effects and efficacy is noted among studies using different species and/or strains of DFMs and lactation stages. In a study comparing two *Bacillus* species (*B. subtilis* 1.1086 vs. *B. licheniformis* 1.183), milk yield and milk protein were increased by *B. licheniformis* but not *B. subtilis* (Qiao et al., 2010). Also, *B. licheniformis* increased ruminal digestibility of fiber and organic matter, total VFA and acetate concentrations, and microbial protein flows into the duodenum and decreased ammonia nitrogen concentration in ruminal fluid, whereas *B. subtilis* did not affect any of the variables (Qiao et al., 2010). Different strains of the same *Bacillus* species can also result in different responses. This is exemplified by the different responses reported in lactation cows fed *B. subtilis* natto (Peng et al., 2012; Sun et al., 2013) vs. cows fed *B. subtilis* 1.1086 (Qiao et al., 2010). Cows can also respond to the same DFMs in a lactation stage-dependent manner. For example, *B. subtilis* C-3102 increased milk yield, milk protein and total solid yields, and lactose content in mid-lactation cows with high milk somatic cell counts (SCCs), but the same *B. subtilis* strain did not have any effect on milk yield, milk composition, or nutrient digestibility in late-lactation cows (Souza et al., 2017). However, the cows at the two different lactation stages were fed different diets (12% more forage was fed to the late-lactating cows). Thus, different strains or species may be needed for cows at different lactation stages or fed different diets. As previously shown (Huang et al., 2012), *B. subtilis* natto may be used in late-lactation cows.

Bacillus-based DFMs can improve energy metabolism, probably by increasing propionate production in the rumen. Cows in early lactation often experience negative energy balance, and they mobilize fat stored in fat depots as non-esterified fatty acids that are then catabolized to produce energy. This process results in elevated plasma concentrations of non-esterified fatty acids and β-hydroxybutyrate. *Bacillus*-based DFMs, such as *B. subtilis* and *B. pumilus*, may improve energy efficiency in dairy cattle. For example, *B. subtilis* (strain not specified) fed to transition dairy cows increased blood glucose concentration and tended to decrease plasma β-hydroxybutyrate concentration (Choonkham et al., 2021). *B. Subtilis* DSM15544 also decreased plasma non-esterified fatty acid concentration in non-lactating cows, even though it did not affect nutrient digestibility, nitrogen balance, or CH_4 production (Kawauchi et al., 2021). Similar effects were also reported in transition cows fed *B. pumilus* and *B. subtilis* natto (Peng et al., 2012; Luan et al., 2015). *Bacillus*-based DFMs may assist in mitigating energy balance in transition cows.

5 Effects on the immune system and health

Both calves and lactating cows suffer from diseases, especially diarrhea in calves, and SARA, immune suppression, stress, and mastitis in lactating cows. Some DFMs are beneficial to lower gut health by excluding pathogens, maintaining GI tract integrity, and establishing a eubiotic intestinal microbiome, all of which can reduce diarrhea in calves. Different DFMs have been evaluated, but the responses have been inconsistent among the numerous studies reported in the literature, owing to the difference in many factors including the DFMs, diets, and host animals used (Tables 1-4).

5.1 Enteric pathogens and calf diarrhea

Diarrhea is a costly disease among dairy calves. It is the main cause of morbidity and mortality and slows growth in dairy calves. It also has long-term adverse effects on lactation performance. Diarrhea is caused by pathogenic viruses (e.g. rotavirus and coronavirus) or bacteria (*Escherichia coli* K99, type C *Clostridium perfringens*, *Salmonella* spp.) and parasites (cryptosporidia and coccidia) (Cho and Yoon, 2014). Bacterial DFMs have the potential to prevent and reduce calf diarrhea. The effects of DFMs on pathogen load and incidence and severity of calf diarrhea (often indicated by fecal scores) are evaluated besides animal growth. Although most studies using various bacterial DFMs, including species of *Lactobacillus*, *Bacillus*, *Megasphaera*, and *Pediococcus*, showed largely no significant improvement in feed intake, digestion, utilization, or growth, many studies showed a decrease in enteric enteropathogens and diarrhea. Species of *Lactobacillus* and other LABs are the most commonly evaluated DFMs, and they are effective in reducing calf diarrhea. In one very early study, multispecies DFMs of *Lactobacillus* spp. reduced total fecal coliform and diarrhea in veal calves (Timmerman et al., 2005). *Lactobacillus plantarum* alone reduced diarrhea in addition to improving feed intake and weight gain (Casper et al., 2021). An early meta-analysis showed that supplementation with LAB-based DFMs reduced calf diarrhea (Signorini et al., 2012). Based on subanalysis, LABs reduced diarrhea only in trials that used whole milk or multi-strain LABs. DFMs containing only LABs did not improve fecal characteristics or change LAB:coliform ratio, but LAB:coliform ratio was increased in trials that used whole milk and DFMs of mono-strain LABs. It was estimated that LABs might reduce calf diarrhea and increase fecal consistency with a >70% probability, but the probability of significant improvement in diarrhea frequency and LAB:coliform ratio was higher (>85%) if whole milk was fed to calves. Whole milk feeding improved the probiotic effect of LABs on incidence of diarrhea and LAB:coliform ratio probably because milk contains nutrients that enhance the probiotic effects of LABs (Signorini et al., 2012).

Multispecies DFMs of LABs are generally more effective than single-species LABs in reducing calf diarrhea. In an early study, three LAB species (*Lactobacillus casei* DSPV 318T, *L. salivarius* DSPV 315T, and *Pediococcus acidilactici* DSPV 006T) in combination decreased the fecal consistency index and promoted starter intake; earlier development of the rumen, omasum and reticulum; and improved growth performance (Frizzo et al., 2010). One probiotic bolus containing five species of LABs (*Pediococcus acidilactici*, *Enterococcus faecium*, *Lactobacillus acidophilus*, *L. casei*, and *Bifidobacterium bifidum*), peptide extract, an enzyme blend, killed yeast extract, dried whey, and natural flavors reduced the duration of diarrhea when administered to calves at the onset of diarrhea (Renaud et al., 2019). Two new isolates of LABs, *Lactobacillus reuteri* TP1.3B and *L. johnsonii* TP1.6, decreased fecal scores and diarrhea incidences (Fernandez et al., 2020), whereas *Lactococcus lactis* subsp. *lactis*, *Pediococcus pentosaceus*, and *Lactobacillus plantarum* in combination did not affect fecal scores or weight gain in newborn calves (Liu et al., 2022). Multispecies *Lactobacillus*-based DFMs can reduce diarrhea, but the efficacy depends on the strains and species to be combined.

Other species have also been evaluated to reduce calf diarrhea. Species of *Bacillus* have attracted attention as promising DFMs to reduce calf diarrhea. *Bacillus amyloliquefaciens* CECT 5940 mixed into milk placer containing butyrate decreased fecal coliform but increased *Klebsiella oxytoca*, *Proteus vulgaris*, and fecal score in calves (Vazquez-Mendoza et al., 2018). *Bacillus megaterium* decreased diarrhea incidences in calves and increased weight gain and feed-to-gain ratio (Yao et al., 2020). *Faecalibacterium prausnitzii* lowered the incidence of severe diarrhea in dairy calves and diarrhea-associated mortality and increased weight gain (Foditsch et al., 2015). Fecal *Faecalibacterium prausnitzii* also increased in abundance, suggesting its ability to proliferate in the intestines of calves.

Multispecies DFMs containing LABs and other bacteria can be potentially more effective than single-species DFMs in preventing and reducing calf diarrhea. In these DFMs, species of LABs and *Bacillus* and/or *Saccharomyces cerevisiae* are included. One multispecies DFM blend containing LABs (*Lactobacillus plantarum*, *Pediococcus acidilactici*, and *Pediococcus pentosaceus*) and *Bacillus subtilis* decreased fecal score in Holstein calves without significantly affecting feed intake, weight gain, or feed efficiency (Wang et al., 2022). A DFM product containing *Bifidobacterium animalis*, *Lactobacillus casei*, and *Enterococcus faecalis* as LABs and *Bacillus cerevisiae* decreased the fecal consistency index and increased feed intake and weight gain in neonatal dairy calves linearly with increasing dose (Guo et al., 2022). Multispecies DFMs containing yeast also reduce diarrhea as demonstrated by decreased incidence of diarrhea in pre-weaned dairy calves fed a combination of *Lactobacillus acidophilus* S5, *Bacillus subtilis* Bzg988118, and *Saccharomyces cerevisiae* SHZ2017 during the first 4 weeks of their lives (Wu et al., 2021). The same multispecies DFMs also had a positive effect in treating calf diarrhea, improved weight gain and feed efficiency, and promoted the development of the small intestine of Holstein calves challenged with *Escherichia coli* K99 (Wu et al., 2022). However, a lack of any effects was also reported. A combination of *Bacillus subtilis* and *Lactobacillus plantarum* did not affect the duration of the first diarrhea episode or hazard of diarrhea, total *Escherichia coli*, pathogenic *E. coli*, total clostridia, or *Clostridium perfringens*, even though fecal shedding of *Cryptosporidium* oocysts was decreased (Lucey et al., 2021). Again, proper strains and species of LABs and other bacteria are needed to formulate effective multispecies DFMs.

Bacterial DFMs can be used to ferment milk or milk replacer, and the fermented liquid foods containing the DFMs (with increased quantities) are then fed to calves, eliminating the need for separate feeding of DFMs. Milk fermented with *Lactobacillus murinus* CRL1695, *L. mucosae* CRL1696, *L. johnsonii* CRL1693, and *L. salivarius* CRL1702 in combination lowered the morbidity and mortality of dairy calves and enhanced their growth (Maldonado et al., 2018). Milk replacer fermented with *Streptococcus faecalis* T-110, *Clostridium butyricum* TO-A, and *Bacillus mesentericus* TO-A in combination and the same milk replacer supplemented with *Lactobacillus plantarum* HOKKAIDO alone also helped recover suppressed feed intake in Holstein calves challenged with bovine rotavirus and decreased the severity and histopathological lesions of the intestinal mucosa due to enteritis (Kayasaki et al., 2021). The milk replacer that was fermented with probiotics, but not supplemented with the probiotics, also significantly reduced the incidence of enteritis, mortality from enteritis, the duration of a series of treatments for enteritis, number of consultations, and the cost of medical care for the disease. This discrepancy may be attributable to some of the fermentation products and/or the different species used.

5.2 Immunity

Commensal GI bacteria using their archetypal molecules (e.g. polysaccharides) promote the development and maturation of the immune systems of animals (Mazmanian et al., 2005). Some DFMs have immunobiotic properties, which can regulate phagocytic activities and different cytokine responses in immune cells, namely, dendritic cells, macrophages, and T and B lymphocytes, and alter immunological and inflammatory biomarkers (Raabis et al., 2019). Bacterial DFMs can elicit immunomodulatory responses in dairy calves and cows, such as an increase in plasma lymphocyte numbers (Adjei-Fremah et al., 2018) and milk concentrations of antibodies (Luan et al., 2015). Some bacteria can be inoculated into udders or supplemented to diets as DFMs. *Lactococcus lactis* DPC 3147 directly inoculated into the udders of healthy dairy cows eliciting considerable innate immune responses, including udder inflammation and increases in interleukin-1β (IL-1β) and IL-8 and SCCs (Beecher et al., 2009). A multi-strain DFM product containing *Lactobacillus acidophilus*, *Enterococcus faecium*, *Bacillus subtilis*, *Aspergillus oryzae*, and *Saccharomyces cerevisiae* supplemented to lactating Holstein-Frisian cows for 60 days modulated the expressions of many cytokines and chemokines and immune-related pathways such as C-X-C receptors, Toll-like receptors (TLR), and clusters of differentiation immune markers and activated the transcription of heme oxygenase 1, interferon-γ (IFN- γ) receptors, peptidoglycan recognition protein 2, nuclear factor κB1 (NF-κB1), and mitogen-activated protein kinase (Adjei-Fremah et al., 2018). Cows suffer systemic stress and inflammation during the days before and after calving. Haptoglobin is one of the most important acute-phase proteins that are associated with the severity and activity of inflammation and bacterial and viral diseases in ruminants (Joshi et al., 2018). Dietary supplementation with *Bacillus pumilus* 8G-134 increased IgA in milk (but not in blood) and tended to reduce the number of haptoglobin-positive (plasma concentration >150 mg/L) lactating cows after calving without affecting milk or blood IgM and IgG (Luan et al., 2015). In bovine endometrial epithelial cell cultures, *Lactobacillus rhamnosus* GR-1 ameliorated the disruption of cellular ultrastructure and apoptosis and attenuated the increases in mRNA expression of many immune response indicators, including TLRs, nucleotide-binding oligomerization domain-like receptors, caspase-1, myeloid differentiation antigen 88 (MyD88), NF-κB, and some proinflammatory cytokines including tumor necrosis factor α (TNF-α), IFN-β, and ILs caused by *Escherichia coli* challenge, and it was suggested that this probiotic strain might avert postpartum uterine diseases in dairy cows (Liu et al., 2016). These results indicate that DFMs can reduce inflammatory activities in postpartum cows.

Several *Bacillus-based* DFMs have been evaluated in calves for their effect on immune responses. *Bacillus subtilis* natto mixed into milk improved immunity in dairy calves as indicated by increased serum IgG and IFN-γ and tended to decrease IL-4, while not affecting IgA, IgM, IgE, IL-6, or IL-10 (Sun et al., 2010). A *Bacillus*-based DFM enhances inflammatory response for pathogen clearance in scouring calves and promotes T cell subsets including proportions of activated, mature cells (Novak et al., 2012). A recent study showed that *B. megaterium* increased serum glutathione, IgA, IgM, IgG, and IL-4 and decreased serum TNF-α and malondialdehyde in suckling calves (Yao et al., 2020). *Bacillus* species may be useful in promoting immunity and health in calves.

Bacterial DFMs have the potential to modulate the immune response in livestock animals, but research on dairy cattle is limited. A meta-analysis using 19 experiments in cattle (beef cattle and dairy cows at different ages or lactation stages) found that probiotics did not consistently affect serum IgG, amyloid A, haptoglobin, and β-hydroxybutyrate or reproductive disorders (Barreto et al., 2021). The authors concluded that DFMs did not yield a consistent beneficial immune response because of the differences in dose, dosing schedule, diet, and the physiological stages of cattle. Future mechanistic studies may help identify and select the probiotics that are effective in eliciting beneficial immune responses. A specific probiotic can be evaluated for its ability and mechanisms to elicit immune responses using bovine tissue cultures (Liu et al., 2016). As an example, in bovine intestinal epithelial cell cultures, five strains of bifidobacteria modulated the TLR2 expression (Galdeano and Perdigon, 2006; Murata et al., 2013). Several *Lactobacillus* strains across multiple species modulated TLR3-mediated innate immunity and TLR4-mediated inflammation (Mansilla et al., 2020). The currently available evidence shows that TLR-mediated immune responses to probiotics can be strain specific. Multi-strain probiotics are probably needed to help regulate different TLRs in lactating dairy cows (Adjei-Fremah et al., 2018).

5.3 Mastitis

Mastitis, both clinical and subclinical, in dairy herds is one of the most costly diseases and welfare issues due to its high prevalence resulting in reduced productivity, longevity, and profitability (Francoz et al., 2017). Antibiotics are commonly used to treat mastitis along with herbal therapy. Probiotics have been recently evaluated as alternatives because antibiotics are not always successful in treating mastitis due to antibiotic resistance of mastitis pathogens. Some bacteria have been inoculated locally to the udders to treat mastitis by improving tight junction barrier function in the epithelium and modulating the local immune system. Dietary supplementation of DFMs has also been evaluated to reduce SCCs and enhance immune response. Competitive exclusion, production of antimicrobial compounds (e.g. bacteriocins, H_2O_2, nitric oxide, and short-chain fatty acids), and enhancement of immunity are important underpinning mechanisms of such benefits.

In clinical trials with dairy cows, intramammary administration of *Lactococcus lactis* DPC3174 (Klostermann et al., 2008) and a combination of *Lactobacillus acidophilus* and *L. casei* (Greene et al., 1991) appeared to be as efficacious as common antibiotic treatments. When inoculated into the udders during the drying-off period, *Lactobacillus plantarum* CRL 1716 and *L. perolens* CRL 1724 promoted phagocytosis of *Staphylococcus* spp. in blood, reduced *Escherichia coli* and *Streptococcus uberis* in milk, and reduced *Staphylococcus aureus* and *S. haemolyticus* in both blood and milk (Pellegrino et al., 2017; Berardo et al., 2020). These two *Lactobacillus* species also enhanced lymphocyte proliferation and increased serum and milk IgG levels, all of which suggest that intramammary LAB inoculation can be a useful strategy to prevent mastitis. However, intramammary inoculation of LABs increased milk SCCs and upregulated the expression of the genes of proinflammatory cytokines, including IL-1β, IL-8, and TNF-α (Pellegrino et al., 2017), implying strong inflammatory responses to LABs in udders. Also, intramammary administration of *Bifidobacterium breve* in dairy cows increased SCCs, induced a massive influx of polymorphonuclear leukocytes into the mammary gland, and enhanced innate immune response in mammary glands. These events are likely associated with enhanced clearance of the invading minor pathogens in the mammary gland and may enhance the host defense capability in the mammary glands.

Lactobacilli administered orally can hypothetically translocate from the digestive tract to mammary glands through the entero-mammary pathway via macrophages or dendritic cells, but there is no sound scientific foundation to support this hypothesis (Martín et al., 2004; Rainard and Foucras, 2018). Dietary supplementation with *Lactococcus* (species or strain not specified), however, decreased milk SCCs, mammary gland inflammation as indicated by reduced activities of myeloperoxidase and *N*-acetyl-β-D-glucosaminidase, and milk TNF-α concentration (Gao et al., 2020). The *Lactococcus* supplementation also reduced *Streptococcus* and *Enterococcus* (some species of these two genera are pathogens of mastitis) in milk. *Saccharomyces cerevisiae* and *Lactococcus* in combination, but not individually, decreased proinflammatory cytokines IL-1β, IL-6, and TNF-α (Gao et al., 2020), suggesting a synergistic effect in attenuating inflammation of mammary glands. *Lactobacillus casei* Zhang and *L. plantarum* P-8 as DFMs decreased SCCs and enhanced immune responses as indicated by increased milk concentrations of IgG and lactoferrin, and the activities of lysozyme and lactoperoxidase in lactating primiparous Holstein cows (Xu et al., 2017). *Bacillus subtilis* natto decreased SCCs in early-lactating Holstein cows (Sun et al., 2013), but *B. subtilis* C-3102 did not decrease SCCs in mid- or late-lactation cows (Souza et al., 2017). The above studies revealed that LABs can reduce mastitis severity by regulating the microbiota and modulating the immune system in the udders.

Overall, treating mastitis in dairy cows using DFMs, through intramammary inoculation, teat spray, or dietary supplementation, has mixed results, but at the local level, DFMs can decrease SCCs, enhance immunity and eliminate mastitis pathogens, and downregulate proinflammatory cytokines (Amir et al., 2016; Rainard and Foucras, 2018). However, these benefits have only been reported in a small number of studies. Information on the use of bacterial DFMs that are administered orally is scarce. Bacterial strains or lactation stages may be factors influencing the efficacy. New probiotic strains are needed. Determination of the modes of action in enhancing immunity and pathogen clearance from mammary glands can help select efficacious strains, optimal doses, and application routes. High-quality randomized controlled trials are also needed to assess the efficacy of probiotics in preventing and treating mastitis at the herd level.

5.4 Rumen acidosis

Highly fermentable diets in large quantities are fed to early-lactation or high-yielding dairy cows to meet their energy requirements. This feeding practice, however, leads to rapid production of VFA and lactic acid and decreased saliva secretion to buffer the acids owing to decreased chewing and rumination activities, eventually resulting in ruminal pH drop for prolonged periods causing SARA (pH 5.2 to 5.6) or acute ruminal acidosis (pH < 5.2) (Humer et al., 2018; Aschenbach et al., 2019). Acute rumen acidosis is not common, but SARA is frequent in dairy herds. The low ruminal pH can damage the rumen epithelia and reduce feed intake and digestion, which results in decreased milk production. In addition, the low ruminal pH increases the release of toxic metabolites (e.g. biogenic amines) and endotoxin lipopolysaccharides (LPSs) due to increased bacterial autolysis, resulting in the disruption of the ruminal barrier integrity and systemic inflammatory responses and multiple metabolic disorders (Chiquette et al., 2015; Aschenbach et al., 2019). The rumen microbiome changes in response to SARA (Mao et al., 2013; Plaizier et al., 2017). Many dietary strategies including the use of DFMs have been tried to reduce ruminal acidosis (Plaizier et al., 2018).

Various LABs, such as *Lactobacillus plantarum*, *Enterococcus faecium*, and *Lactococcus lactis*, have been tested to reduce the severity of rumen acidosis in dairy cattle. In the rumen, lactate concentration usually remains low (<5 mM) as LUBs (e.g. *M. elsdenii* and *Selenomonas ruminantium* spp. *lactilytica*) ferment lactate to propionate (Chiquette et al., 2008). Therefore, the LUBs in the rumen remains low in population and slow in growth, making it unable to respond robustly to lactate surges when the diet is abruptly switched to highly fermentable carbohydrates (McAllister et al., 2011). Repeated supplementation with LABs may help maintain a level of lactate that supports a robust LUB population that then rapidly consumes the lactate and lessens the severity of rumen acidosis (Nocek et al., 2002; Chiquette, 2009). However, extruded rye flour fermented with *Lactobacillus sakei* KTU05-6 or *Pediococcus pentosaceus* BaltBio02 did not affect ruminal pH or lactate concentration in mid-lactating cows (Lele et al., 2019). It remains to be determined to what extent LABs alone can help maintain a robust population of LUBs and reduce lactate surge and rumen acidosis.

By consuming lactate, LUBs can reduce ruminal lactate concentration and consequently rumen acidosis. Dietary supplementation of LUBs is assumed to increase their populations so that they can prevent or reduce ruminal acidosis in cows fed rapidly fermentable diets (Mazon et al., 2020). Strains of *Propionibacterium* and *Megasphaera elsdenii* have been evaluated as LUBs to reduce rumen acidosis in dairy cattle. Species of *Propionibacterium* alone did not affect rumen pH (Stein et al., 2006; Weiss et al., 2008). Several strains of *M.*

elsdenii have been evaluated as DFMs to moderate SARA. *M. elsdenii* YE34 drenched to steers prior to switching to a grain-rich diet established itself in the rumen, and its abundance increased by 100-fold (Klieve et al., 2003). In an early study, *M. elsdenii* NCIMB *41125* administered to early-lactation cows via ruminal cannula decreased the duration of rumen pH < 5.6 and moderated day-to-day pH fluctuations (Aikman et al., 2011). The molar proportion of propionate also increased. Similar benefits from the same *M. elsdenii* strain were also recently reported in mid-lactating cows under SARA challenge (Mazon et al., 2020) and in steers and sheep (Henning et al., 2010). However, another *M. elsdenii* strain, ATCC 25940, did not reduce ruminal lactate concentration or affect ruminal pH, feed intake, milk yield, or milk composition in mid-lactation Holstein cows even though it altered some of the rumen VFA profiles (Zebeli et al., 2012). Different *M. elsdenii* strains probably differ in their ability to reduce the severity of rumen acidosis.

Combinations of LABs with LUBs and other bacteria are more effective in reducing SARA. A combination of *Lactobacillus animalis* (an LAB) and *Propionibacterium freudenreichii* (an LUB) top-dressed onto the diet fed to mid-lactation cows raised ruminal pH following an abrupt increase in dietary starch (Lawrence et al., 2021). A combination of two LABs (*Lactobacillus plantarum* and *Enterococcus faecium*) and *Saccharomyces cerevisiae* supplemented daily to dairy cows only resulted in slightly less acidic rumen conditions and higher digestion of ear corn and corn silage (Nocek et al., 2002). In Holstein dairy cows experiencing SARA, a DFM product consisting of *Enterococcus faecium* and *Saccharomyces cerevisiae* reduced the severity of SARA (Chiquette, 2009). Ruminal inoculation of *Enterococcus faecium* together with *S. cerevisiae* or *Lactococcus lactis* to lactating Holstein cows experiencing mild SARA raised mean ruminal pH numerically and reduced the decline in milk production, but not *Enterococcus faecium* alone (Chiquette et al., 2015). Different species of probiotics may enhance each other's activities and prebiotic effects, resulting in more profound effects. Generally, DFMs are more efficacious in moderating ruminal pH in cows experiencing more severe SARA (Chiquette, 2009; Chiquette et al., 2012). Collectively, although many probiotics have been evaluated, few of them have been shown to be consistently effective.

5.5 Heat stress

Dairy cows experience several types of physiological, environmental, and dietary stresses that affect their health, immunity, GI microbiome, and ultimately production performance. Heat stress is considered one of the most important stressors for dairy cattle, especially in subtropical and tropical regions, as it reduces milk production and deteriorates the health, immune responses, behaviors, and welfare of dairy cattle (Becker et al., 2020; Sammad et al., 2020). The gut has emerged as a main organ affected by heat stress because the rumen and intestinal microbiomes and their functions are impaired and pathogen colonization is increased in heat-stressed dairy cattle (Chen et al., 2018; Patra and Kar, 2021; Park et al., 2022). Therefore, supplementation with DFMs is suggested as a strategy to ameliorate heat stress-induced dysbiosis and pathophysiological alterations in the gut and the animal body. Bacterial DFMs have been evaluated in heat-stressed nonruminants, especially poultry, in many studies (Deng et al., 2012; Song et al., 2014; Patra and Kar, 2021), but only a few studies have tested bacterial DFMs in dairy cattle. In one on-farm study, supplementation with *Bacillus subtilis* increased milk production without affecting the milk composition of crossbred Holstein-Friesian cows under tropical conditions (Choonkham et al., 2020). *Bacillus subtilis* reduced negative energy balance and improved health, as indicated by keeping white blood cells and monocytes in the healthy range in crossbred Holstein-Friesian cows under tropical conditions during the transition period (Choonkham et al., 2021). Because the rumen and intestinal microbiomes can affect the nutrition, health, and productivity of dairy cows and because heat stress has profound effects on these microbiomes (Kim et al., 2020a; Park et al., 2022), some bacteria of the rumen and intestinal microbiome, especially those that are suppressed by heat stress, may be explored as DFMs to ameliorate the adverse effects of heat stress.

6 Benefit to mitigation of methane emissions

Dairy cattle like other ruminants emit CH_4, which is primarily produced by methanogens utilizing hydrogen and carbon dioxide (Janssen and Kirs, 2008). Numerous studies have evaluated different strategies to mitigate CH_4 emissions from dairy cattle (Hristov et al., 2013; Patra et al., 2017; Haque, 2018; Almeida et al., 2021). CH_4 mitigation strategies are based on one of three mechanisms: directly inhibiting methanogenesis, shifting rumen fermentation to reduce hydrogen production, and providing alternative hydrogen receptors (Patra et al., 2017). Bacterial DFMs can potentially reduce CH_4 production in the rumen using one or more of the above mechanisms (Jeyanathan et al., 2014). Different bacteria have been tested as DFMs in reducing CH_4 production *in vitro* and *in vivo*. They fall into one of three guilds: LABs, LUPs, and reductive acetogens.

LABs can reduce rumen CH_4 production by conceptually shifting rumen fermentation away from acetate and hydrogen production, directly inhibiting rumen methanogens by producing lactate and lowering pH, and inhibiting specific rumen microbes that produce the substrates of methanogenesis (hydrogen, formate, and methyl compound) (Doyle et al., 2019). Screening 45 strains of bacteria, including 26 strains of LABs (mostly species of *Lactobacillus* and *Lactococcus*), identified two strains of LABs (*Lactobacillus pentosus* D31 and *Lactobacillus bulgaricus* D1) that decreased CH_4 production *in vitro*, with *Lactobacillus pentosus* D31 also decreasing CH_4 emissions from sheep (Jeyanathan et al., 2016). Strains of *Lactobacillus pentosus*, *Leuconostoc mesenteroides*, *Lactobacillus plantarum*, *Lactococcus lactis*, *Peptostreptococcus productus*, *Streptococcus equinus*, and *Pediococcus pentosaceus* have been evaluated *in vitro* or in sheep, and the results are inconsistent (Doyle et al., 2019). No study has evaluated LABs in mitigating CH_4 emissions from dairy cattle, probably because a much larger quantity of LABs is needed. Some LAB strains are often used as silage inoculants to improve silage quality and aerobic stability, and they can survive in the rumen (Weinberg et al., 2003). A ration fermented with *Lactobacillus plantarum* Chikuso-1 as a silage inoculant reduced CH_4 emissions and improved digestibility both *in vitro* and in sheep (Cao et al., 2010b; a, 2011). Some LABs, thus, may be used as silage inoculants, rather than as DFMs, to mitigate CH_4 emissions from dairy cattle.

Most LUBs can ferment lactate to propionate, besides acetate and/or butyrate, and lactate-to-propionate fermentation consumes hydrogen, thereby reducing the availability of hydrogen for CH_4 production in the rumen (Jeyanathan et al., 2014). Therefore, LUBs can potentially mitigate CH_4 emissions from ruminants. A few strains of *Propionibacterium* have been shown to reduce CH_4 production by mixed rumen cultures *in vitro* (Alazzeh et al., 2012; Meale et al., 2014; Jeyanathan et al., 2016). However, when fed to beef heifers that were fed high- or low-forage diets, two strains of *Propionibacterium acidipropionici* (strains P5 and P169), or *Propionibacterium jensenii* P54 did not reduce CH_4 emissions (Vyas et al., 2014a; 2014b). Similarly, in dairy cattle, *Propionibacterium* sp. P63 by itself failed to reduce CH_4 emissions and together with *Lactobacillus rhamnosus* 32 only numerically reduced CH_4 yield (Philippeau et al., 2017). Another strain of *Propionibacterium freudenreichii* (strain 53-W) reduced CH_4 production *in vitro* but not in dairy cows fed either a high-starch or high-forage diet (Jeyanathan et al., 2019). LUBs alone may have limited ability to reduce CH_4 emissions from dairy cattle.

Reductive acetogens use hydrogen and carbon dioxide while producing acetate. Thus, reductive acetogenesis is an alternative hydrogen disposal pathway of methanogenesis and can help reduce CH_4 production. Indeed, adding acetogens to mixed rumen cultures *in vitro* decreased CH_4 production (Nollet et al., 1998; Lopez et al., 1999). In a recent study (Kim et al., 2020b), four strains of acetogens (*Proteiniphilum acetatigenes* DA02, GA01, GA03, and *Alkaliphilus crotonatoxidans* GA02) reduced CH_4 production by mixed rumen cultures *in vitro* when starch served as the substrate but not when casein was the substrate. In dairy cows, however, supplementation of the most promising stain, *Proteiniphilum acetatigenes* GA03, did not reduce CH_4 emissions. Because reductive acetogenesis is much less thermodynamically favorable and competitive than methanogenesis, reductive acetogens probably have little potential as DFMs to mitigate CH_4 emissions from dairy cattle.

Overall, bacterial DFMs yielded little success in mitigating CH_4 emissions from dairy cattle (Tables 1-4). They may indirectly help reduce CH_4 emissions by improving feed utilization and lactation performance because fewer animals will be needed to produce a given amount of milk. This is exemplified by *Bacillus licheniformis*, which improved dietary nitrogen and energy utilization in sheep though it did not directly reduce CH_4 emissions (Deng et al., 2018). Some DFMs, especially LABs, are often used as silage inoculants. Mitigation of CH_4 emissions by bacterial DFMs as silage inoculants may result in a spillover benefit.

7 Conclusion and future trends

Bacterial DFMs have been used primarily to modulate the functions of the microbiome in the lower gut of calves and the rumen of lactating cows and transition cows. Collectively, there is some promising evidence that bacterial DFMs can improve ruminal fermentation and milk production, reduce the severity of rumen acidosis, enhance immune responses and gut barrier function, and reduce pathogen loads and calf diarrhea. To a lesser extent, bacterial DFMs may improve feed utilization, reduce mastitis, ameliorate heat stress, and abate CH_4 emissions. However, the benefits of bacterial DFMs vary considerably among strains and species used, even within the same species or strains of DFMs among studies. The discrepancies are attributable to variations in dose, duration of supplementation, diet, animal age, or stages of lactation. More studies are needed to establish the dose and duration needed to achieve desirable benefits for a given diet, host age, and lactation stage.

DFMs containing multi-strains with complementary metabolic, immunogenic, and physiological functions can have synergistic activities and are generally more effective. Multispecies DFMs with complementary and synergistic effects could be even more effective than multi-strain DFMs of the same species because strains of the same species often have similar metabolic features (Timmerman et al., 2004; Lambo et al., 2021). This is demonstrated by combinations of some LABs, LUBs, and *Bacillus* species (discussed above). Only a few studies have tried multispecies bacterial DFMs in dairy cows, and the results are not consistent. Selections of specific strains and species to be included in multi-strain or multispecies DFMs require the understanding of the metabolic and physiological features of individual strain/species, but most of the tested bacterial DFMs are not well or poorly characterized metabolically or physiologically. Thus, future research is needed to adequately characterize the current DFMs and new strains/species and evaluate their utilities as DFMs. Furthermore, because DFMs need to be metabolically active in the rumen, intestine, or mammary gland, all of which have a microbiome, the interaction between DFMs and the indigenous microbes can affect the efficacy of DFMs. Thus, research is needed to determine the interaction between DFMs and indigenous microbes of the target sites. Additionally, *Saccharomyces cerevisiae* is a common yeast fed to dairy cattle with documented benefits. It should also be considered in formulating multi-strain or multispecies DFMs.

Few commercial DFM products are available, and their use is very limited in dairy cows. The bacteria in DFM products need to survive feed processing and storage. Spore-forming bacteria, such as those of *Bacillus*, are especially useful. When DFMs are selected, safety-associated selection criteria, such as dissemination of antibiotic resistance and enterotoxin production, are often neglected. For the safety of dairy cattle and humans, commercial DFM products should not carry any transferable antibiotic resistance genes. Probiotic effects of DFMs can be strain specific, and thus the strains included in any DFM product should be specified. Finally, the cost-effectiveness has not been determined in any of the reported studies. It can be a major barrier to the application of DFMs at the farm level.

8 Where to look for further information

Bacterial DFMs or probiotics have attracted much research interest and development efforts in the past decade. The most up-to-date scientific information regarding bacterial DFMs used in dairy cows is presented at scientific conferences and published in many scientific journals, including animal science journals and microbiology journals. A few books also have chapters reviewing bacterial DFMs.

8.1. Major scientific conferences

1. ADSA Annual Meeting, the annual meeting of the American Dairy Science Association.
2. ASAS-CSAS Annual Meeting, the joint annual meeting of the American Society of Animal Science and the Canadian Society of Animal Science.
3. Annual Meeting of the European Federation of Animal Science.
4. Asian-Australian Association of Animal Production (AAAP) Animal Science Congress.

8.2. Scientific journals

1 Animals, a journal published by MDPI, https://www.mdpi.com/journal/animals.
2 *Animal Feed Science and Technology*, a journal published by Elsevier, https://www.sciencedirect.com/journal/animal-feed-science-and-technology.
1 *Frontiers in Microbiology*, a journal published by Frontiers Media SA, https://www.frontiersin.org/journals/microbiology.
2 *Journal of Animal Science*, an official journal of the American Society of Animal Science, https://academic.oup.com/jas.
3 *Journal of Animal Sciences and Biotechnology*, a journal published by BioMed Central, https://jasbsci.biomedcentral.com/.
4 *Journal of Applied Microbiology*, an official journal of the Society for Applied Microbiology, https://ami-journals.onlinelibrary.wiley.com/journal/13652672.
5 *Journal of Dairy Science*, an official journal of the American Dairy Science Association, https://www.adsa.org/Publications/Journal-of-Dairy-Science.
6 *Livestock Science*, a journal published by Elsevier, https://www.sciencedirect.com/journal/livestock-science.
7 *Translational Animal Science*, an official journal of the American Society of Animal Science, https://academic.oup.com/tas/.

8.3. Books

1 Advances in Probiotics for Sustainable Food and Medicine, Goel, G. and Kumar, A. (eds.), Springer Singapore, 2021; ISBNs: 978-9-81-156794-0, 978-9-81-156795-7; DOI: 10.1007/978-981-15-6795-7.
2 Direct-Fed Microbials and Prebiotics for Animals. Callaway, T. and Ricke, S. (eds.), Springer New York, 2011; eBook ISBN: 978-1-4614-1311-0; DOI: 10.1007/978-1-4614-1311-0.

9 References

Adjei-Fremah, S., Ekwemalor, K., Asiamah, E. K., Ismail, H., Ibrahim, S. and Worku, M. 2018. Effect of probiotic supplementation on growth and global gene expression in dairy cows. *J. Appl. Anim. Res.* 46(1):257–263.

Aikman, P. C., Henning, P. H., Horn, C. H. and Jones, A. K. 2009. Effect of using *Megasphaera elsdenii* NCIMB 41125 as a probiotic on feed intake and milk production in early lactation dairy cows. In: Ruminant Physiology-Digestion, Metabolism and Effects of Nutrition on Reproduction and Welfare, pp. 110–111.

Aikman, P. C., Henning, P. H., Humphries, D. J. and Horn, C. H. 2011. Rumen pH and fermentation characteristics in dairy cows supplemented with *Megasphaera elsdenii* NCIMB 41125 in early lactation. *J. Dairy Sci.* 94(6):2840–2849.

Alazzeh, A. Y., Sultana, H., Beauchemin, K. A., Wang, Y., Holo, H., Harstad, O. M. and McAllister, T. A. 2012. Using strains of propionibacteria to mitigate methane emissions *in vitro*. *Acta Agric. Scand. A* 62:263–272.

Almeida, A. K., Hegarty, R. S. and Cowie, A. 2021. Meta-analysis quantifying the potential of dietary additives and rumen modifiers for methane mitigation in ruminant production systems. *Anim. Nutr.* 7(4):1219–1230.

Al-Sadi, R., Nighot, P., Nighot, M., Haque, M., Rawat, M. and Ma, T. Y. 2021. *Lactobacillus acidophilus* induces a strain-specific and toll-like receptor 2-dependent enhancement of intestinal epithelial tight junction barrier and protection against intestinal inflammation. *Am. J. Pathol.* 191(5):872–884.

Amin, A. B. and Mao, S. 2021. Influence of yeast on rumen fermentation, growth performance and quality of products in ruminants: A review. *Anim. Nutr.* 7(1):31–41. doi:10.1016/j.aninu.2020.10.005.

Amir, L. H., Griffin, L., Cullinane, M. and Garland, S. M. 2016. Probiotics and mastitis: Evidence-based marketing? *Int. Breastfeed. J.* 11:19.

Arik, H. D., Gulsen, N., Hayirli, A. and Alatas, M. S. 2019. Efficacy of *Megasphaera elsdenii* inoculation in subacute ruminal acidosis in cattle. *J. Anim. Physiol. Anim. Nutr. (Berl)* 103(2):416–426.

Aschenbach, J. R., Zebeli, Q., Patra, A. K., Greco, G., Amasheh, S. and Penner, G. B. 2019. Symposium review: The importance of the ruminal epithelial barrier for a healthy and productive cow. *J. Dairy Sci.* 102(2):1866–1882.

Bae, G. S., Choi, A., Yeo, J. M., Kim, J. N., Song, J., Kim, E. J. and Chang, M. B. 2018. Supplementing *Rhodobacter sphaeroides* in the diet of lactating Holstein cows may naturally produce coenzyme Q10-enriched milk. *Asian-Australas. J. Anim. Sci.* 31(1):40–46.

Ban, Y. and Guan, L. L. 2021. Implication and challenges of direct-fed microbial supplementation to improve ruminant production and health. *J. Anim. Sci. Biotechnol.* 12:1–22.

Banos, G., Wall, E., Coffey, M. P., Bagnall, A., Gillespie, S., Russell, G. C. and McNeilly, T. N. 2013. Identification of immune traits correlated with dairy cow health, reproduction and productivity. *PLoS ONE* 8(6):e65766.

Barreto, M. O., Soust, M., Moore, R. J., Olchowy, T. W. J. and Alawneh, J. I. 2021. Systematic review and meta-analysis of probiotic use on inflammatory biomarkers and disease prevention in cattle. *Prev. Vet. Med.* 194:105433.

Becker, C. A., Collier, R. J. and Stone, A. E. 2020. Invited review: Physiological and behavioral effects of heat stress in dairy cows. *J. Dairy Sci.* 103(8):6751–6770. doi:10.3168/jds.2019-17929.

Beecher, C., Daly, M., Berry, D. P., Klostermann, K., Flynn, J., Meaney, W., Hill, C., McCarthy, T. V., Ross, R. P. and Giblin, L. 2009. Administration of a live culture of *Lactococcus lactis* DPC 3147 into the bovine mammary gland stimulates the local host immune response, particularly IL-1β and IL-8 gene expression. *J. Dairy Res.* 76(3):340–348.

Benítez-Chao, D. F., León-Buitimea, A., Lerma-Escalera, J. A. and Morones-Ramírez, J. R. 2021. Bacteriocins: An overview of antimicrobial, toxicity, and biosafety assessment by *in vivo* models. *Front. Microbiol.* 12:630695.

Berardo, N., Bohl, L., Porporatto, C., Nader-Macias, M. E. F., Bogni, C. and Pellegrino, M. 2020. Intramammary inoculation with lactic acid bacteria at dry-off triggers an immunomodulatory response in dairy cows. *Benef. Microbes* 11(6):561–572.

Bhat, M. I., Sowmya, K., Kapila, S. and Kapila, R. 2020. Potential probiotic *Lactobacillus rhamnosus* (MTCC-5897) inhibits *Escherichia coli* impaired intestinal barrier function by modulating the host tight junction gene response. *Probiotics Antimicrob. Proteins* 12(3):1149–1160.

Boyd, J., West, J. W. and Bernard, J. K. 2011. Effects of the addition of direct-fed microbials and glycerol to the diet of lactating dairy cows on milk yield and apparent efficiency of yield. *J. Dairy Sci.* 94(9):4616–4622.

Cao, Y., Cai, Y., Takahashi, T., Yoshida, N., Tohno, M., Uegaki, R., Nonaka, K. and Terada, F. 2011. Effect of lactic acid bacteria inoculant and beet pulp addition on fermentation characteristics and in vitro ruminal digestion of vegetable residue silage. *J. Dairy Sci.* 94(8):3902–3912.

Cao, Y., Takahashi, T., Horiguchi, K. and Yoshida, N. 2010a. Effect of adding lactic acid bacteria and molasses on fermentation quality and *in vitro* ruminal digestion of total mixed ration silage prepared with whole crop rice. *Grassl. Sci.* 56(1):19–25.

Cao, Y., Takahashi, T., Horiguchi, K., Yoshida, N. and Cai, Y. 2010b. Methane emissions from sheep fed fermented or non-fermented total mixed ration containing whole-crop rice and rice bran. *Anim. Feed Sci. Technol.* 157(1–2):72–78.

Casper, D. P., Hultquist, K. M. and Acharya, I. P. 2021. *Lactobacillus plantarum* GB LP-1 as a direct-fed microbial for neonatal calves. *J. Dairy Sci.* 104(5):5557–5568.

Chen, J., Stevenson, D. M. and Weimer, P. J. 2004. Albusin B, a bacteriocin from the ruminal bacterium *Ruminococcus albus* 7 that inhibits growth of *Ruminococcus flavefaciens*. *Appl. Environ. Microbiol.* 70(5):3167–3170.

Chen, S., Wang, J., Peng, D., Li, G., Chen, J. and Gu, X. 2018. Exposure to heat-stress environment affects the physiology, circulation levels of cytokines, and microbiome in dairy cows. *Sci. Rep.* 8(1):14606. doi:10.1038/s41598-018-32886-1.

Chikindas, M. L., Weeks, R., Drider, D., Chistyakov, V. A. and Dicks, L. M. 2018. Functions and emerging applications of bacteriocins. *Curr. Opin. Biotechnol.* 49:23–28. doi:10.1016/j.copbio.2017.07.011.

Chiquette, J. 2009. Evaluation of the protective effect of probiotics fed to dairy cows during a subacute ruminal acidosis challenge. *Anim. Feed Sci. Technol.* 153(3–4):278–291.

Chiquette, J., Allison, M. J. and Rasmussen, M. 2012. Use of *Prevotella bryantii* 25A and a commercial probiotic during subacute acidosis challenge in midlactation dairy cows. *J. Dairy Sci.* 95(10):5985–5995.

Chiquette, J., Allison, M. J. and Rasmussen, M. A. 2008. *Prevotella bryantii* 25A used as a probiotic in early-lactation dairy cows: Effect on ruminal fermentation characteristics, milk production, and milk composition. *J. Dairy Sci.* 91(9):3536–3543.

Chiquette, J., Lagrost, J., Girard, C. L., Talbot, G., Li, S., Plaizier, J. C. and Hindrichsen, I. K. 2015. Efficacy of the direct-fed microbial *Enterococcus faecium* alone or in combination with *Saccharomyces cerevisiae* or *Lactococcus lactis* during induced subacute ruminal acidosis. *J. Dairy Sci.* 98(1):190–203.

Chiquette, J., Talbot, G., Markwell, F., Nili, N. and Forster, R. J. 2007. Repeated ruminal dosing of *Ruminococcus flavefaciens* NJ along with a probiotic mixture in forage or concentrate-fed dairy cows: Effect on ruminal fermentation, cellulolytic populations and in sacco digestibility. *Can. J. Anim. Sci.* 87(2):237–249.

Cho, Y. I. and Yoon, K. J. 2014. An overview of calf diarrhea - infectious etiology, diagnosis, and intervention. *J. Vet. Sci.* 15(1):1–17. doi:10.4142/jvs.2014.15.1.1.

Choonkham, W., Intanon, M., Chewonarin, T., Bernard, J. K. and Suriyasathaporn, W. 2021. Effects of supplemental *Bacillus subtilis*, injectable vitamin E plus selenium, or both on health parameters during the transition period in dairy cows in a tropical environment. *Trop. Anim. Health Prod.* 53(2):298.

Choonkham, W., Schonewille, J. T., Bernard, J. K. and Suriyasathaporn, W. 2020. Effects of on-farm supplemental feeding of probiotic *Bacillus subtilis* on milk production in lactating dairy cows under tropical conditions. *J. Anim. Feed Sci.* 29(3):199–205.

Corbett, E. M., Norby, B., Halbert, L. W., Henderson, S. T., Grooms, D. L., Manning, S. D. and Kaneene, J. B. 2015. Effect of feeding a direct-fed microbial on total and antimicrobial-resistant fecal coliform counts in preweaned dairy calves. *Am. J. Vet. Res.* 76(9):780–788. doi:10.2460/ajvr.76.9.780.

Daniali, M., Nikfar, S. and Abdollahi, M. 2020. Antibiotic resistance propagation through probiotics. *Expert Opin. Drug Metab. Toxicol.* 16(12):1207–1215.

Deng, B., Wang, L., Ma, Q., Yu, T., Liu, D., Dai, Y. and Zhao, G. 2021. Genomics analysis of *Bacillus megaterium* 1259 as a probiotic and its effects on performance in lactating dairy cows. *Animals (Basel)* 11(2):397.

Deng, K. D., Xiao, Y., Ma, T., Tu, Y., Diao, Q. Y., Chen, Y. H. and Jiang, J. J. 2018. Ruminal fermentation, nutrient metabolism, and methane emissions of sheep in response to dietary supplementation with *Bacillus licheniformis*. *Anim. Feed Sci. Technol.* 241:38–44.

Deng, W., Dong, X. F., Tong, J. M. and Zhang, Q. 2012. The probiotic *Bacillus licheniformis* ameliorates heat stress-induced impairment of egg production, gut morphology, and intestinal mucosal immunity in laying hens. *Poult. Sci.* 91(3):575–582.

Doyle, N., Mbandlwa, P., Kelly, W. J., Attwood, G., Li, Y., Ross, R. P., Stanton, C. and Leahy, S. 2019. Use of lactic acid bacteria to reduce methane production in ruminants, a critical review. *Front. Microbiol.* 10:2207. doi:10.3389/fmicb.2019.02207.

Du, R., Jiao, S., Dai, Y., An, J., Lv, J., Yan, X., Wang, J. and Han, B. 2018. Probiotic *Bacillus amyloliquefaciens* C-1 improves growth performance, stimulates GH/IGF-1, and regulates the gut microbiota of growth-retarded beef calves. *Front. Microbiol.* 9:2006.

FAO/WHO. 2001. *Health and Nutritional Properties of Probiotics in Food Including Powder Milk With Live Lactic Acid Bacteria. Rep. Jt. FAOWHO Expert Consult*, pp. 1–34.

FDA 1995. *CPG: Direct-Fed Microbial Products*. Accessed March 15, 2022. Available at: https://www.fda.gov/regulatory-information/search-fda-guidance-documents/cpg-sec-689100-direct-fed-microbial-products Sec. 689.100.

Fernandez, S., Fraga, M., Castells, M., Colina, R. and Zunino, P. 2020. Effect of the administration of *Lactobacillus* spp. strains on neonatal diarrhoea, immune parameters and pathogen abundance in pre-weaned calves. *Benef. Microbes* 11(5):477–488. doi:10.3920/BM2019.0167.

Fernandez-Ciganda, S., Fraga, M. and Zunino, P. 2022. Probiotic *Lactobacilli* administration induces changes in the fecal microbiota of preweaned dairy calves. *Probiotics Antimicrob. Proteins* 14(5):804–815. doi:10.1007/s12602-021-09834-z.

Foditsch, C., Pereira, R. V., Ganda, E. K., Gomez, M. S., Marques, E. C., Santin, T. and Bicalho, R. C. 2015. Oral administration of *Faecalibacterium prausnitzii* decreased the incidence of severe diarrhea and related mortality rate and increased weight gain in preweaned dairy heifers. *PLoS ONE* 10(12):e0145485. doi:10.1371/journal.pone.0145485.

Fomenky, B. E., Do, D. N., Talbot, G., Chiquette, J., Bissonnette, N., Chouinard, Y. P., Lessard, M. and Ibeagha-Awemu, E. M. 2018. Direct-fed microbial supplementation influences the bacteria community composition of the gastrointestinal tract of pre- and post-weaned calves. *Sci. Rep.* 8(1):14147. doi:10.1038/s41598-018-32375-5.

Forestier, C., De Champs, C., Vatoux, C. and Joly, B. 2001. Probiotic activities of *Lactobacillus casei rhamnosus*: *In vitro* adherence to intestinal cells and antimicrobial properties. *Res. Microbiol.* 152(2):167–173.

Francisco, C. C., Chamberlain, C. S., Waldner, D. N., Wettemann, R. P. and Spicer, L. J. 2002. Propionibacteria fed to dairy cows: Effects on energy balance, plasma metabolites and hormones, and reproduction. *J. Dairy Sci.* 85(7):1738–1751.

Francoz, D., Wellemans, V., Dupré, J. P., Roy, J. P., Labelle, F., Lacasse, P. and Dufour, S. 2017. Invited review: A systematic review and qualitative analysis of treatments other than conventional antimicrobials for clinical mastitis in dairy cows. *J. Dairy Sci.* 100(10):7751–7770.

Frizzo, L. S., Soto, L. P., Zbrun, M. V., Bertozzi, E., Sequeira, G., Armesto, R. R. and Rosmini, M. R. 2010. Lactic acid bacteria to improve growth performance in young calves fed milk replacer and spray-dried whey powder. *Anim. Feed Sci. Technol.* 157(3–4):159–167.

Galdeano, C. M. and Perdigon, G. 2006. The probiotic bacterium *Lactobacillus casei* induces activation of the gut mucosal immune system through innate immunity. *Clin. Vaccine Immunol.* 13(2):219–226.

Gao, J., Liu, Y. C., Wang, Y., Li, H., Wang, X. M., Wu, Y., Zhang, D. R., Gao, S. and Qi, Z. L. 2020. Impact of yeast and lactic acid bacteria on mastitis and milk microbiota composition of dairy cows. *AMB Express* 10(1):22.

Greene, W. A., Gano, A. M., Smith, K. L., Hogan, J. S. and Todhunter, D. A. 1991. Comparison of probiotic and antibiotic intramammary therapy of cattle with elevated somatic cell counts. *J. Dairy Sci.* 74(9):2976–2981.

Gregg, K., Hamdorf, B., Henderson, K., Kopecny, J. and Wong, C. 1998. Genetically modified ruminal bacteria protect sheep from fluoroacetate poisoning. *Appl. Environ. Microbiol.* 64(9):3496–3498.

Guarner, F. and Schaafsma, G. J. 1998. Probiotics. *Int. J. Food Microbiol.* 39(3):237–238.

Guo, Y., Li, Z., Deng, M., Li, Y., Liu, G., Liu, D., Liu, Q., Liu, Q. and Sun, B. 2022. Effects of a multi-strain probiotic on growth, health, and fecal bacterial flora of neonatal dairy calves. *Anim. Biosci.* 35(2):204–216. doi:10.5713/ab.21.0084.

Haque, M. N. 2018. Dietary manipulation: A sustainable way to mitigate methane emissions from ruminants. *J. Anim. Sci. Technol.* 60:15.

Henning, P. H., Horn, C. H., Leeuw, K.-J., Meissner, H. H. and Hagg, F. M. 2010. Effect of ruminal administration of the lactate-utilizing strain *Megasphaera elsdenii* (Me) NCIMB 41125 on abrupt or gradual transition from forage to concentrate diets. *Anim. Feed Sci. Technol.* 157(1–2):20–29.

Holzapfel, W. H., Geisen, R. and Schillinger, U. 1995. Biological preservation of foods with reference to protective cultures, bacteriocins and food-grade enzymes. *Int. J. Food Microbiol.* 24(3):343–362.

Hristov, A. N., Oh, J., Firkins, J. L., Dijkstra, J., Kebreab, E., Waghorn, G., Makkar, H. P., Adesogan, A. T., Yang, W., Lee, C., Gerber, P. J., Henderson, B. and Tricarico, J. M. 2013. Special topics - Mitigation of methane and nitrous oxide emissions from animal operations: I. A review of enteric methane mitigation options. *J. Anim. Sci.* 91(11):5045–5069.

Huang, L., Zhou, L., Bu, D., Cheng, J. and Ji, Y. 2012. Effects of probiotics on performance of late lactation dairy cows. *Acta Agric. Boreali-Occidentalis Sin.* 27:406–409.

Humer, E., Petri, R. M., Aschenbach, J. R., Bradford, B. J., Penner, G. B., Tafaj, M., Südekum, K. H. and Zebeli, Q. 2018. Invited review: Practical feeding management recommendations to mitigate the risk of subacute ruminal acidosis in dairy cattle. *J. Dairy Sci.* 101(2):872–888.

Izadi, B., Mohebbi-Fani, M., Hosseinzadeh, S., Shekarforoush, S. S., Nazifi, S. and Rasooli, A. 2021. Alteration of fatty acid profile of milk in Holstein cows fed *Bacillus coagulans* as probiotic: A field study. *Iran. J. Vet. Res.* 22(2):100–106.

Janssen, P. H. and Kirs, M. 2008. Structure of the archaeal community of the rumen. *Appl. Environ. Microbiol.* 74(12):3619–3625.

Jeyanathan, J., Martin, C., Eugène, M., Ferlay, A., Popova, M. and Morgavi, D. P. 2019. Bacterial direct-fed microbials fail to reduce methane emissions in primiparous lactating dairy cows. *J. Anim. Sci. Biotechnol.* 10:41.

Jeyanathan, J., Martin, C. and Morgavi, D. P. 2014. The use of direct-fed microbials for mitigation of ruminant methane emissions: A review. *Animal* 8(2):250–261.

Jeyanathan, J., Martin, C. and Morgavi, D. P. 2016. Screening of bacterial direct-fed microbials for their antimethanogenic potential in vitro and assessment of their effect on ruminal fermentation and microbial profiles in sheep. *J. Anim. Sci.* 94(2):739–750.

Joshi, V., Gupta, V. K., Bhanuprakash, A. G., Mandal, R. S. K., Dimri, U. and Ajith, Y. 2018. Haptoglobin and serum amyloid A as putative biomarker candidates of naturally occurring bovine respiratory disease in dairy calves. *Microb. Pathog.* 116:33–37.

Kawauchi, D., Angthong, W., Keaokliang, O., Ishida, T., Takahashi, T. and Kawashima, T. 2021. Effect of feeding *Bacillus subtilis* on rumen fermentation, blood metabolites, nutrient digestibility, and energy and nitrogen balances in non-lactating crossbred cows. *Anim. Sci. J.* 92(1):e13531.

Kayasaki, F., Okagawa, T., Konnai, S., Kohara, J., Sajiki, Y., Watari, K., Ganbaatar, O., Goto, S., Nakamura, H., Shimakura, H., Minato, E., Kobayashi, A., Kubota, M., Terasaki, N., Takeda, A., Noda, H., Honma, M., Maekawa, N., Murata, S. and Ohashi, K. 2021. Direct evidence of the preventive effect of milk replacer-based probiotic feeding in calves against severe diarrhea. *Vet. Microbiol.* 254:108976. doi:10.1016/j.vetmic.2020.108976.

Kim, D. H., Kim, M. H., Kim, S. B., Son, J. K., Lee, J. H., Joo, S. S., Gu, B. H., Park, T., Park, B. Y. and Kim, E. T. 2020a. Differential dynamics of the ruminal microbiome of Jersey Cows in a heat stress environment. *Animals (Basel)* 10(7):1127.

Kim, S. H., Mamuad, L. L., Islam, M. and Lee, S. S. 2020b. Reductive acetogens isolated from ruminants and their effect on in vitro methane mitigation and milk performance in Holstein cows. *J. Anim. Sci. Technol.* 62(1):1–13.

Klieve, A. V., Hennessy, D., Ouwerkerk, D., Forster, R. J., Mackie, R. I. and Attwood, G. T. 2003. Establishing populations of *Megasphaera elsdenii* YE 34 and *Butyrivibrio fibrisolvens* YE 44 in the rumen of cattle fed high grain diets. *J. Appl. Microbiol.* 95(3):621–630.

Klostermann, K., Crispie, F., Flynn, J., Ross, R. P., Hill, C. and Meaney, W. 2008. Intramammary infusion of a live culture of *Lactococcus lactis* for treatment of bovine mastitis: Comparison with antibiotic treatment in field trials. *J. Dairy Res.* 75(3):365–373.

Kumariya, R., Garsa, A. K., Rajput, Y. S., Sood, S. K., Akhtar, N. and Patel, S. 2019. Bacteriocins: Classification, synthesis, mechanism of action and resistance development in food spoilage causing bacteria. *Microb. Pathog.* 128:171–177.

Lambo, M. T., Chang, X. and Liu, D. 2021. The recent trend in the use of multistrain probiotics in livestock production: An overview. *Animals (Basel)* 11(10):2805.

Lawrence, M., Polukis, S., Barnard, A. M., Miller, M. A., Kung, L. Jr. and Gressley, T. F. 2021. Evaluating the effects of *Lactobacillus animalis* and *Propionibacterium freudenreichii* on performance and rumen and fecal measures in lactating dairy cows. *J. Dairy Sci.* 104(4):4119–4133.

Le, O. T., Dart, P. J., Harper, K., Zhang, D., Schofield, B., Callaghan, M. J., Lisle, A. T., Klieve, A. V. and McNeill, D. M. 2017. Effect of probiotic *Bacillus amyloliquefaciens* strain H57 on productivity and the incidence of diarrhoea in dairy calves. *Anim. Prod. Sci.* 57(5):912–919.

Lee, Y. K., Puong, K. Y., Ouwehand, A. C. and Salminen, S. 2003. Displacement of bacterial pathogens from mucus and Caco-2 cell surface by lactobacilli. *J. Med. Microbiol.* 52(10):925–930.

Lele, V., Zelvyte, R., Monkeviciene, I., Kantautaite, J., Stankevicius, R., Ruzauskas, M., Sederevicius, A., Antanaitis, R. and Bartkiene, E. 2019. Milk production and ruminal parameters of dairy cows fed diets containing *Lactobacillus sakei* KTU05-6 and *Pediococcus pentosaceus* BaltBio02. *Pol. J. Vet. Sci.* 22(2):327–335.

Li, T., Teng, D., Mao, R., Hao, Y., Wang, X. and Wang, J. 2020. A critical review of antibiotic resistance in probiotic bacteria. *Food Res. Int.* 136:109571.

Liu, B., Wang, C., Huasai, S., Han, A., Zhang, J., He, L. and Aorigele, C. 2022. Compound probiotics improve the diarrhea rate and intestinal microbiota of newborn calves. *Animals (Basel)* 12(3):322. doi:10.3390/ani12030322.

Liu, M., Wu, Q., Wang, M., Fu, Y. and Wang, J. 2016. *Lactobacillus rhamnosus* GR-1 limits *Escherichia coli*-induced inflammatory responses via attenuating MyD88-dependent and MyD88-independent pathway activation in bovine endometrial epithelial cells. *Inflammation* 39(4):1483–1494.

Lopez, S., McIntosh, F. M., Wallace, R. J. and Newbold, C. J. 1999. Effect of adding acetogenic bacteria on methane production by mixed rumen microorganisms. *Anim. Feed Sci. Technol.* 78(1–2):1–9.

Luan, S., Duersteler, M., Galbraith, E. A. and Cardoso, F. C. 2015. Effects of direct-fed *Bacillus pumilus* 8G-134 on feed intake, milk yield, milk composition, feed conversion, and health condition of pre-and postpartum Holstein cows. *J. Dairy Sci.* 98(9):6423–6432.

Luan, X., Feng, M. and Sun, J. 2021. Effect of *Lactobacillus plantarum* on antioxidant activity in fermented sausage. *Food Res. Int.* 144:110351.

Lucey, P. M., Lean, I. J., Aly, S. S., Golder, H. M., Block, E., Thompson, J. S. and Rossow, H. A. 2021. Effects of mannan-oligosaccharide and *Bacillus subtilis* supplementation to preweaning Holstein dairy heifers on body weight gain, diarrhea, and shedding of fecal pathogens. *J. Dairy Sci.* 104(4):4290–4302. doi:10.3168/jds.2020-19425.

Mahesh, M. S., Mohanta, R. K. and Patra, A. K. 2021 Probiotics in livestock and poultry nutrition and health. In: Goel, G. and Kumar, A. (Eds.), Advances in Probiotics for Sustainable Food and Medicine. Springer, Singapore, pp. 149–179.

Maldonado, N. C., Chiaraviglio, J., Bru, E., De Chazal, L., Santos, V. and Nader-Macias, M. E. F. 2018. Effect of milk fermented with lactic acid bacteria on diarrheal incidence, growth performance and microbiological and blood profiles of newborn dairy calves. *Probiotics Antimicrob. Proteins* 10(4):668–676. doi:10.1007/s12602-017-9308-4.

Malmuthuge, N. and Guan, L. L. 2017. Understanding host-microbial interactions in rumen: Searching the best opportunity for microbiota manipulation. *J. Anim. Sci. Biotechnol.* 8:8.

Mani, S., Aiyegoro, O. A. and Adeleke, M. A. 2021. Characterization of rumen microbiota of two sheep breeds supplemented with direct-fed lactic acid bacteria. *Front. Vet. Sci.* 7:1199.

Mansilla, F., Takagi, M., Garcia-Castillo, V., Aso, H., Nader-Macias, M. E., Vignolo, G., Kitazawa, H. and Villena, J. 2020. Modulation of toll-like receptor-mediated innate immunity in bovine intestinal epithelial cells by lactic acid bacteria isolated from feedlot cattle. *Benef. Microbes* 11(3):269–282.

Mao, S. Y., Zhang, R. Y., Wang, D. S. and Zhu, W. Y. 2013. Impact of subacute ruminal acidosis (SARA) adaptation on rumen microbiota in dairy cattle using Pyrosequencing. *Anaerobe* 24:12–19.

Martín, R., Langa, S., Reviriego, C., Jiménez, E., Marín, M. L., Olivares, M., Boza, J., Jiménez, J., Fernández, L., Xaus, J. and Rodríguez, J. M. 2004. The commensal microflora of human milk: New perspectives for food bacteriotherapy and probiotics. *Trends Food Sci. Technol.* 15(3–4):121–127.

Mazmanian, S. K., Liu, C. H., Tzianabos, A. O. and Kasper, D. L. 2005. An immunomodulatory molecule of symbiotic bacteria directs maturation of the host immune system. *Cell* 122(1):107–118.

Mazon, G., Campler, M. R., Holcomb, C., Bewley, J. M. and Costa, J. H. C. 2020. Effects of a *Megasphaera elsdenii* oral drench on reticulorumen pH dynamics in lactating dairy cows under subacute ruminal acidosis challenge. *Anim. Feed Sci. Technol.* 261:114404.

McAllister, T. A., Beauchemin, K. A., Alazzeh, A. Y., Baah, J., Teather, R. M. and Stanford, K. 2011. Review: The use of direct fed microbials to mitigate pathogens and enhance production in cattle. *Can. J. Anim. Sci.* 91(2):193–211.

Meale, S. J., Ding, S., He, M. L., Dugan, M. E. R., Ribeiro, G. O. Jr., Alazzeh, A. Y., Holo, H., Harstad, O. M., McAllister, T. A. and Chaves, A. V. 2014. Effect of *Propionibacterium freudenreichii* on ruminal fermentation patterns, methane production and lipid biohydrogenation of beef finishing diets containing flaxseed oil in a rumen simulation technique. *Can. J. Anim. Sci.* 94(4):685–695.

Michalak, M., Wojnarowski, K., Cholewińska, P., Szeligowska, N., Bawej, M. and Pacoń, J. 2021. Selected alternative feed additives used to manipulate the rumen microbiome. *Animals (Basel)* 11(6):1542.

Miyagi, T., Kaneichi, K., Aminov, R. I., Kobayashi, Y., Sakka, K., Hoshino, S. and Ohmiya, K. 1995. Enumeration of transconjugated *Ruminococcus albus* and its survival in the goat rumen microcosm. *Appl. Environ. Microbiol.* 61(5):2030–2032.

Mulligan, F. J. and Doherty, M. L. 2008. Production diseases of the transition cow. *Vet. J.* 176(1):3–9. doi:10.1016/j.tvjl.2007.12.018.

Murata, K., Villena, J. C., Tomosada, Y., Risa, H., Chiba, E., Shimazu, T., Aso, H., Suda, Y., Iwabuchi, N. and Xiao, J.-Z. 2013. Bifidobacteria upregulate expression of toll-like receptor negative regulators counteracting enterotoxigenic *Escherichia coli* mediated inflammation in bovine intestinal epitheliocytes. *Open J. Vet. Med.* 3:143–155.

Muya, M. C., Nherera, F. V., Miller, K. A., Aperce, C. C., Moshidi, P. M. and Erasmus, L. J. 2015. Effect of *Megasphaera elsdenii* NCIMB 41125 dosing on rumen development, volatile fatty acid production and blood β-hydroxybutyrate in neonatal dairy calves. *J. Anim. Physiol. Anim. Nutr. (Berl)* 99(5):913–918.

Nakajima, A., Vogelzang, A., Maruya, M., Miyajima, M., Murata, M., Son, A., Kuwahara, T., Tsuruyama, T., Yamada, S., Matsuura, M., Nakase, H., Peterson, D. A., Fagarasan, S. and Suzuki, K. 2018. IgA regulates the composition and metabolic function of gut microbiota by promoting symbiosis between bacteria. *J. Exp. Med.* 215(8):2019–2034.

Nakamura, Y., Terahara, M., Iwamoto, T., Yamada, K., Asano, M., Kakuta, S., Iwakura, Y. and Totsuka, M. 2012. Upregulation of polymeric immunoglobulin receptor expression by the heat-inactivated potential probiotic *Bifidobacterium bifidum* OLB6378 in a mouse intestinal explant model. *Scand. J. Immunol.* 75(2):176–183.

Naito, Y., Uchiyama, K., and Takagi. 2018. A next-generation beneficial microbes. *Akkermansia muciniphila. J. Clin. Biochem. Nutr.* 63(1):33–35

Nocek, J. E., Kautz, W. P., Leedle, J. A. Z. and Allman, J. G. 2002. Ruminal supplementation of direct-fed microbials on diurnal pH variation and in situ digestion in dairy cattle. *J. Dairy Sci.* 85(2):429–433.

Nollet, L., Mbanzamihigo, L., Demeyer, D. and Verstraete, W. 1998. Effect of the addition of *Peptostreptococcus productus* ATCC 35244 on reductive acetogenesis in the ruminal ecosystem after inhibition of methanogenesis by cell-free supernatant of *Lactobacillus plantarum* 80. *Anim. Feed Sci. Technol.* 71(1-2):49–66.

Novak, K. N., Davis, E., Wehnes, C. A., Shields, D. R., Coalson, J. A., Smith, A. H. and Rehberger, T. G. 2012. Effect of supplementation with an electrolyte containing a *Bacillus*-based direct-fed microbial on immune development in dairy calves. *Res. Vet. Sci.* 92(3):427–434. doi:10.1016/j.rvsc.2011.04.008.

Ogawa, M., Shimizu, K., Nomoto, K., Tanaka, R., Hamabata, T., Yamasaki, S., Takeda, T. and Takeda, Y. 2001. Inhibition of in vitro growth of Shiga toxin-producing *Escherichia coli* O157: H7 by probiotic *Lactobacillus* strains due to production of lactic acid. *Int. J. Food Microbiol.* 68(1-2):135–140.

Park, T., Ma, L., Gao, S. and Bu, D. 2022. Heat stress impacts the multi-domain rumen microbiota and some of the functional features independent of its effect on feed intake in lactating dairy cows. *J. Anim. Sci. Biotechnol* 13:1–15.

Patra, A., Park, T., Kim, M. and Yu, Z. 2017. Rumen methanogens and mitigation of methane emission by anti-methanogenic compounds and substances. *J. Anim. Sci. Biotechnol.* 8:1–18.

Patra, A. K. and Kar, I. 2021. Heat stress on microbiota composition, barrier integrity, and nutrient transport in gut, production performance, and its amelioration in farm animals. *J. Anim. Sci. Technol.* 63(2):211–247.

Pattnaik, P., Kaushik, J. K., Grover, S. and Batish, V. K. 2001. Purification and characterization of a bacteriocin-like compound (Lichenin) produced anaerobically by *Bacillus licheniformis* isolated from water buffalo. *J. Appl. Microbiol.* 91(4):636–645.

Pellegrino, M., Berardo, N., Giraudo, J., Nader-Macías, M. E. F. and Bogni, C. 2017. Bovine mastitis prevention: Humoral and cellular response of dairy cows inoculated with lactic acid bacteria at the dry-off period. *Benef. Microbes* 8(4):589–596.

Peng, H., Wang, J. Q., Kang, H. Y., Dong, S. H., Sun, P., Bu, D. P. and Zhou, L. Y. 2012. Effect of feeding *Bacillus subtillis* natto on milk production and composition, blood metabolites and rumen fermentation in early lactation dairy cows. *J. Anim. Physiol. Anim. Nutr.* 96(3):506–512.

Philippeau, C., Lettat, A., Martin, C., Silberberg, M., Morgavi, D. P., Ferlay, A., Berger, C. and Nozière, P. 2017. Effects of bacterial direct-fed microbials on ruminal characteristics, methane emission, and milk fatty acid composition in cows fed high-or low-starch diets. *J. Dairy Sci.* 100(4):2637–2650.

Plaizier, J. C., Danesh Mesgaran, M., Derakhshani, H., Golder, H., Khafipour, E., Kleen, J. L., Lean, I., Loor, J., Penner, G. and Zebeli, Q. 2018. Review: Enhancing gastrointestinal health in dairy cows. *Animal* 12(s2):s399–s418. doi:10.1017/S1751731118001921.

Plaizier, J. C., Li, S., Danscher, A. M., Derakshani, H., Andersen, P. H. and Khafipour, E. 2017. Changes in microbiota in rumen digesta and feces due to a grain-based subacute ruminal acidosis (SARA) challenge. *Microb. Ecol.* 74(2):485–495.

Poppy, G. D., Rabiee, A. R., Lean, I. J., Sanchez, W. K., Dorton, K. L. and Morley, P. S. 2012. A meta-analysis of the effects of feeding yeast culture produced by anaerobic fermentation of *Saccharomyces cerevisiae* on milk production of lactating dairy cows. *J. Dairy Sci.* 95(10):6027–6041. doi:10.3168/jds.2012-5577.

Prabhu, R., Altman, E. and Eiteman, M. A. 2012. Lactate and acrylate metabolism by *Megasphaera elsdenii* under batch and steady-state conditions. *Appl. Environ. Microbiol.* 78(24):8564–8570. doi:10.1128/AEM.02443-12.

Pridmore, R. D., Pittet, A. C., Praplan, F. and Cavadini, C. 2008. Hydrogen peroxide production by *Lactobacillus johnsonii* NCC 533 and its role in anti-*Salmonella* activity. *FEMS Microbiol. Lett.* 283(2):210–215.

Qiao, G. H., Shan, A. S., Ma, N., Ma, Q. Q. and Sun, Z. W. 2010. Effect of supplemental *Bacillus* cultures on rumen fermentation and milk yield in Chinese Holstein cows. *J. Anim. Physiol. Anim. Nutr. (Berl)* 94(4):429–436.

Raabis, S., Li, W. and Cersosimo, L. 2019. Effects and immune responses of probiotic treatment in ruminants. *Vet. Immunol. Immunopathol.* 208:58–66.

Raeth-Knight, M. L., Linn, J. G. and Jung, H. G. 2007. Effect of direct-fed microbials on performance, diet digestibility, and rumen characteristics of Holstein dairy cows. *J. Dairy Sci.* 90(4):1802–1809.

Rainard, P. and Foucras, G. 2018. A critical appraisal of probiotics for mastitis control. *Front. Vet. Sci.* 5:251.

Renaud, D. L., Kelton, D. F., Weese, J. S., Noble, C. and Duffield, T. F. 2019. Evaluation of a multispecies probiotic as a supportive treatment for diarrhea in dairy calves: A randomized clinical trial. *J. Dairy Sci.* 102(5):4498–4505.

Retta, K. 2016. Role of probiotics in rumen fermentation and animal performance: A review. *Int. J. Livest. Prod.* 7(5):24–32.

Sammad, A., Wang, Y. J., Umer, S., Lirong, H., Khan, I., Khan, A., Ahmad, B. and Wang, Y. 2020. Nutritional physiology and biochemistry of dairy cattle under the influence of heat stress: Consequences and opportunities. *Animals (Basel)* 10(5):793. doi:10.3390/ani10050793.

Sánchez, B., Delgado, S., Blanco-Míguez, A., Lourenço, A., Gueimonde, M. and Margolles, A. 2017. Probiotics, gut microbiota, and their influence on host health and disease. *Mol. Nutr. Food Res.* 61(1):1600240.

Schofield, B. J., Lachner, N., Le, O. T., McNeill, D. M., Dart, P., Ouwerkerk, D., Hugenholtz, P. and Klieve, A. V. 2018. Beneficial changes in rumen bacterial community profile in sheep and dairy calves as a result of feeding the probiotic *Bacillus amyloliquefaciens* H57. *J. Appl. Microbiol.* 124(3):855–866.

Sekirov, I., Russell, S. L., Antunes, L. C. M. and Finlay, B. B. 2010. Gut microbiota in health and disease. *Physiol. Rev.* 90(3):859–904.

Seo, J. K., Kim, S.-W., Kim, M. H., Upadhaya, S. D., Kam, D. K. and Ha, J. K. 2010. Direct-fed microbials for ruminant animals. *Asian-Australas. J. Anim. Sci.* 23(12):1657–1667.

Siedler, S., Rau, M. H., Bidstrup, S., Vento, J. M., Aunsbjerg, S. D., Bosma, E. F., McNair, L. M., Beisel, C. L. and Neves, A. R. 2020. Competitive exclusion is a major bioprotective mechanism of lactobacilli against fungal spoilage in fermented milk products. *Appl. Environ. Microbiol.* 86(7):e02312-19.

Signorini, M. L., Soto, L. P., Zbrun, M. V., Sequeira, G. J., Rosmini, M. R. and Frizzo, L. S. 2012. Impact of probiotic administration on the health and fecal microbiota of young calves: A meta-analysis of randomized controlled trials of lactic acid bacteria. *Res. Vet. Sci.* 93(1):250–258. doi:10.1016/j.rvsc.2011.05.001.

Song, J., Xiao, K., Ke, Y. L., Jiao, L. F., Hu, C. H., Diao, Q. Y., Shi, B. and Zou, X. T. 2014. Effect of a probiotic mixture on intestinal microflora, morphology, and barrier integrity of broilers subjected to heat stress. *Poult. Sci.* 93(3):581–588. doi:10.3382/ps.2013-03455.

Souza, V. L., Lopes, N. M., Zacaroni, O. F., Silveira, V. A., Pereira, R. A. N., Freitas, J. A., Almeida, R., Salvati, G. G. S. and Pereira, M. N. 2017. Lactation performance and diet digestibility of dairy cows in response to the supplementation of *Bacillus subtilis* spores. *Livest. Sci.* 200:35–39.

Srutkova, D., Schwarzer, M., Hudcovic, T., Zakostelska, Z., Drab, V., Spanova, A., Rittich, B., Kozakova, H. and Schabussova, I. 2015. *Bifidobacterium longum* CCM 7952 promotes epithelial barrier function and prevents acute DSS-induced colitis in strictly strain-specific manner. *PLoS ONE* 10(7):e0134050.

Stein, D. R., Allen, D. T., Perry, E. B., Bruner, J. C., Gates, K. W., Rehberger, T. G., Mertz, K., Jones, D. and Spicer, L. J. 2006. Effects of feeding propionibacteria to dairy cows on milk yield, milk components, and reproduction. *J. Dairy Sci.* 89(1):111–125.

Sun, P., Wang, J. Q. and Deng, L. F. 2013. Effects of *Bacillus subtilis* natto on milk production, rumen fermentation and ruminal microbiome of dairy cows. *Animal* 7(2):216–222.

Sun, P., Wang, J. Q. and Zhang, H. T. 2010. Effects of *Bacillus subtilis* natto on performance and immune function of preweaning calves. *J. Dairy Sci.* 93(12):5851-5855. doi:10.3168/jds.2010-3263.

Timmerman, H. M., Koning, C. J. M., Mulder, L., Rombouts, F. M. and Beynen, A. C. 2004. Monostrain, multistrain and multispecies probiotics - a comparison of functionality and efficacy. *Int. J. Food Microbiol.* 96(3):219–233.

Timmerman, H. M., Mulder, L., Everts, H., Van Espen, D. C., Van Der Wal, E., Klaassen, G., Rouwers, S. M. G., Hartemink, R., Rombouts, F. M. and Beynen, A. C. 2005. Health and growth of veal calves fed milk replacers with or without probiotics. *J. Dairy Sci.* 88(6):2154–2165.

Vazquez-Mendoza, P., Elghandour, M. M. M., Alaba, P. A., Sanchez-Aparicio, P., Alonso-Fresan, M. U., Barbabosa-Pliego, A. and Salem, A. Z. M. 2018. Antimicrobial and bactericidal impacts of *Bacillus amyloliquefaciens* CECT 5940 on fecal shedding of pathogenic bacteria in dairy calves and adult dogs. *Microb. Pathog.* 114:458–463. doi:10.1016/j.micpath.2017.11.040.

Vyas, D., McGeough, E. J., McGinn, S. M., McAllister, T. A. and Beauchemin, K. A. 2014a. Effect of *Propionibacterium* spp. on ruminal fermentation, nutrient digestibility, and methane emissions in beef heifers fed a high-forage diet. *J. Anim. Sci.* 92(5):2192–2201.

Vyas, D., McGeough, E. J., Mohammed, R., McGinn, S. M., McAllister, T. A. and Beauchemin, K. A. 2014b. Effects of *Propionibacterium* strains on ruminal fermentation, nutrient digestibility and methane emissions in beef cattle fed a corn grain finishing diet. *Animal* 8(11):1807–1815.

Wang, H., Yu, Z., Gao, Z., Li, Q., Qiu, X., Wu, F., Guan, T., Cao, B. and Su, H. 2022. Effects of compound probiotics on growth performance, rumen fermentation, blood parameters, and health status of neonatal Holstein calves. *J. Dairy Sci.* 105(3):2190–2200. doi:10.3168/jds.2021-20721.

Weinberg, Z. G., Muck, R. E. and Weimer, P. J. 2003. The survival of silage inoculant lactic acid bacteria in rumen fluid. *J. Appl. Microbiol.* 94(6):1066–1071.

Weiss, W. P., Wyatt, D. J. and McKelvey, T. R. 2008. Effect of feeding propionibacteria on milk production by early lactation dairy cows. *J. Dairy Sci.* 91(2):646–652.

Wu, Y., Wang, L., Luo, R., Chen, H., Nie, C., Niu, J., Chen, C., Xu, Y., Li, X. and Zhang, W. 2021. Effect of a multispecies probiotic mixture on the growth and incidence of diarrhea, immune function, and fecal microbiota of pre-weaning dairy calves. *Front. Microbiol.* 12:681014. doi:10.3389/fmicb.2021.681014.

Wu, Y. Y., Nie, C. X., Xu, C., Luo, R. Q., Chen, H. L., Niu, J. L., Bai, X. and Zhang, W. 2022. Effects of dietary supplementation with multispecies probiotics on intestinal epithelial development and growth performance of neonatal calves challenged with Escherichia coli K99. *J. Sci. Food Agric.* 102(10):4373–4383. doi:10.1002/jsfa.11791.

Xu, H., Huang, W., Hou, Q., Kwok, L., Sun, Z., Ma, H., Zhao, F., Lee, Y.-K. and Zhang, H. 2017. The effects of probiotics administration on the milk production, milk components and fecal bacteria microbiota of dairy cows. *Sci. Bull.* 62(11):767–774.

Yao, J., Wang, L., Zhang, W., Liu, M. and Niu, J. 2020. Effects of *Bacillus megaterium* on growth performance, serum biochemical parameters, antioxidant capacity, and immune function in suckling calves. *Open Life Sci.* 15(1):1033–1041. doi:10.1515/biol-2020-0106.

Yohe, T. T., Enger, B. D., Wang, L., Tucker, H. L. M., Ceh, C. A., Parsons, C. L. M., Yu, Z. and Daniels, K. M. 2018. Short communication: Does early-life administration of a *Megasphaera elsdenii* probiotic affect long-term establishment of the organism in the rumen and alter rumen metabolism in the dairy calf? *J. Dairy Sci.* 101(2):1747–1751. doi:10.3168/jds.2017-12551.

Yoon, I. K. and Stern, M. D. 1995. Influence of direct-fed microbials on ruminal microbial fermentation and performance of ruminants: A review. *Asian-Australas. J. Anim. Sci.* 8(6):533–555.

Zebeli, Q., Terrill, S. J., Mazzolari, A., Dunn, S. M., Yang, W. Z. and Ametaj, B. N. 2012. Intraruminal administration of *Megasphaera elsdenii* modulated rumen fermentation profile in mid-lactation dairy cows. *J. Dairy Res.* 79(1):16–25.

Zhang, C., Zhang, J., Yu, Z., Zhou, G. and Yao, J. 2022. Effects of supplementation with *Saccharomyces cerevisiae* products on dairy calves: A meta-analysis. *J. Dairy Sci.* 105(9):7386–7398.

Zhang, C., Zhang, S., Liu, W., Guo, T., Gu, R. and Kong, J. 2019. Potential application and bactericidal mechanism of lactic acid-hydrogen peroxide consortium. *Appl. Biochem. Biotechnol.* 189(3):822–833.

Zhang, R., Dong, X., Zhou, M., Tu, Y., Zhang, N., Deng, K. and Diao, Q. 2017. Oral administration of *Lactobacillus plantarum* and *Bacillus subtilis* on rumen fermentation and the bacterial community in calves. *Anim. Sci. J.* 88(5):755–762.

Zielińska, D., Lepecka, A., Oldak, A., Dlugosz, E. and Kolożyn-Krajewska, D. 2021. Growth and adhesion inhibition of pathogenic bacteria by live and heat-killed food-origin *Lactobacillus strains* or their supernatants. *FEMS Microbiol. Lett.* 368(5):fnab024.

Chapter 8

The use of exogenous enzymes as dietary supplements in dairy cow nutrition

Diwakar Vyas, Kathy G. Arriola, Felipe X. Amaro, Halima Sultana, Mohammad Malekkhahi, Samia Farooq, Adeoye Oyebade and Adegbola T. Adesogan, Institute of Food and Agricultural Sciences, University of Florida-Gainesville, USA

1 Introduction

The demand for animal source foods including eggs, fish, meat, and milk is projected to increase, especially in economies where livestock systems are inefficient (Herrero et al., 2013). For the dairy production systems, the challenge is to increase the efficiency of nutrient utilization by lactating dairy cows without increasing its environmental impact (Asner and Archer, 2010). Substitution of dietary structural carbohydrates with non-structural carbohydrates in diets results in greater feed efficiency (milk yield per unit of feed consumed) (Lascano et al., 2011). However, increasing the proportion of non-structural carbohydrates in diets beyond a certain threshold may adversely affect rumen health. Another way of increasing feed efficiency is to increase the bioavailability of nutrients from dietary feed ingredients. Exogenous enzymes can be a cost-effective way to increase feed efficiency by increasing dietary nutrient bioavailability but results on their efficacy of enzymes have been equivocal.

Exogenous enzymes used in ruminant diets can be classified into three major categories including fibrolytic, amylolytic, and proteolytic enzymes. The objective of this chapter is to recap fundamental understanding on using

http://dx.doi.org/10.19103/AS.2022.0117.08

exogenous enzymes to improve utilization of ruminant diets and to propose future research focus areas to improve their efficacy at enhancing forage and byproduct digestibility and subsequently animal performance.

2 Exogenous fibrolytic enzymes

Forages represent the single most important feed resource for ruminant animal production even under the intensive concentrate feeding systems in the United States; however, their cell-wall structure controls the quality and determine the digestibility (Adesogan et al., 2019). Since improvements in forage quality through breeding are only realized in the long term, several short-term alternatives have been developed with varying levels of success. Physical treatments such as grinding, chopping, pelleting, and steaming have led to small, though significant improvements in the quality of straw and poor-quality forages. Chemical treatment of straw has given the most improvements in digestibility and reductions in cell wall content. Alkalis, particularly sodium hydroxide (NaOH) have been effectively used to increase forage digestibility; however, the hazardous nature of NaOH and the potential toxicity from residual Na in the forage has limited the widespread adoption of this method. Exogenous fibrolytic enzymes (EFE) have also been used as strategy to improve fiber degradability and increase forage quality. As the use of EFE to increase dietary nutrient utilization has escalated during the past years, various manuscripts reported the effects of EFE on dairy and beef cattle performance (Beauchemin et al., 2004; Beauchemin and Holtshausen, 2011; Adesogan et al., 2014; Meale et al., 2014). Some meta-analyses revised and condensed data from different studies to evaluate the outcome of EFE on dairy cattle performance and ruminant diets (Arriola et al., 2017; Tirado-Gonzalez et al., 2017).

2.1 Enzyme activities

Exogenous fibrolytic enzymes used in ruminant diets have two principal enzyme activities: cellulase and xylanase. Previous reviews reported some of the major fungal source of commercial enzymes such as *Trichoderma reesei*, *Aspergillus oryzae*, or *Saccharomyces cerevisiae* or bacteria such as *Bacillus subtilis* and *Streptococcus faecium* (Pendleton, 2000, Beauchemin et al., 2004). These commercial enzymes included in their preparation cellulases or xylanases; however, because most are crude extracts, they also contain additional enzyme activities such as amylases, proteases, esterases, or pectinases (McAllister et al., 2001). Regardless of the existence of additional enzyme activities, most of the commercial enzyme preparations mainly describe cellulase or xylanase activities. This imprecise designation of enzymes (Adesogan et al., 2014) is partly responsible for the considerable variability in literature on the efficacy of exogenous enzymes at improving dietary nutrient utilization in ruminants.

Cellulase and xylanase activities can be measured by several assays (Wood and Bhat, 1988). Cellulose hydrolysis is accomplished primarily by three major enzymes: (1) endoglucanase (EC 3.2.1.4), hydrolyzing cellulose chains at random to produce cellulose oligomers of varying degrees of polymerization; (2) exoglucanase (EC 3.2.1.9.1), hydrolyzing the cellulose chain from the non-reducing end producing cellobiose; and (3) β-glucosidase (EC 3.2.1.2.1) releases glucose from cellobiose and hydrolyzes short chain cello-oligosaccharides from both reducing and non-reducing ends (Beauchemin et al., 2003; Bhat and Hazlewood, 2001; Figure 1). The most common substrate for measuring cellulase activity is carboxymethyl cellulose, which measures endo-β-1,4-glucanase activity (Wood and Bhat, 1988). Alternative substrates like avicel and cellobiose measure exoglucanase and β-glucosidase activities, respectively (Bhat and Hazlewood, 2001).

Xylanase is degraded primarily by two principal enzymes: xylanase (EC 3.2.1.8) and β-1,4 xylosidase (EC 3.2.1.37) (Bhat and Hazlewood, 2001). Authors also reported the use of oat spelt or birchwood xylan as substrates for the estimation of xylanase activity. Xylanases are regularly described as

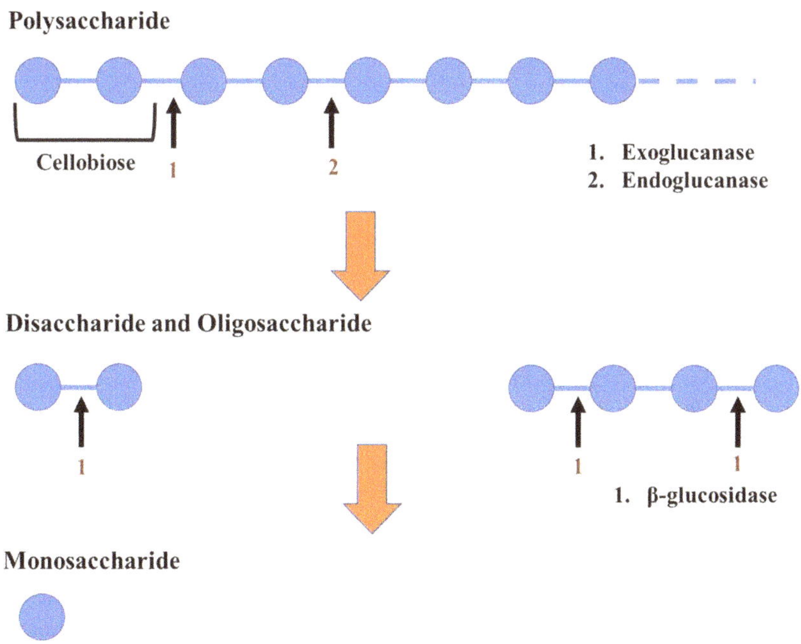

Figure 1 Schematic representation of the major enzymes involved in cellulose hydrolysis. Source: Adapted from Beauchemin et al. (2004).

endoxylanases and they permit production of xylo-oligomers when they break down xylan, whereas, β-1,4 xylosidases yields xylose (White et al., 1993).

Cellulases and xylanases may not completely effective at degrading plant cell-wall because of the cell-wall arrangement and the distribution of lignin. Ferulic acid cross-links polysaccharides and lignin (Hatfield et al., 1999) (Figure 2) by making ether bonds with lignin and ester bonds with arabinoxylans (Iiyama et al., 1990) contributing to the cell-wall recalcitrance. It was suggested that adding exogenous enzymes with esterase activity could increase the efficacy of fiber degradation in the rumen (Nsereko et al., 2000; Dhiman et al., 2002). Eun and Beauchemin (2007) observed that the application of a mixture of xylanase, cellulase and esterase has potential to enhance *in vitro* dry matter (DM) and neutral detergent fiber (NDF) degradability of alfalfa hay and corn silage by 25%. Krueger et al. (2008) applied an esterase enzyme using tropical grass as substrate and reported greater NDF hydrolysis and 96 h *in vitro* DM degradability (IVDMD). Animal studies have also shown efficacy of esterase inclusion in fibrolytic enzyme preparations (Arriola et al., 2011a).

For measuring esterase activity, aryl and carboxyl esterase activity is typically measured using *p*-nitrophenyl acetate and α-naphthyl acetate as substrates (Gonzalez et al., 2006) while ethyl ferulate is used as substrate for ferulic acid esterase activity (Mastihuba et al., 2002). However, esterase enzymes cannot hydrolyze the most potent plant cell-wall cross linkages including those involving dimers of ferulic acid and ether linkages. Rather, specific enzymes of fungal origin including laccase, lignin peroxidase, manganese peroxidase, and versatile peroxidase have lignin-degrading ability (Asemoloye et al., 2021).

Some authors have demonstrated synergism between endo and exoglucanases when crystalline cellulose is being dissolved (Wood et al.,

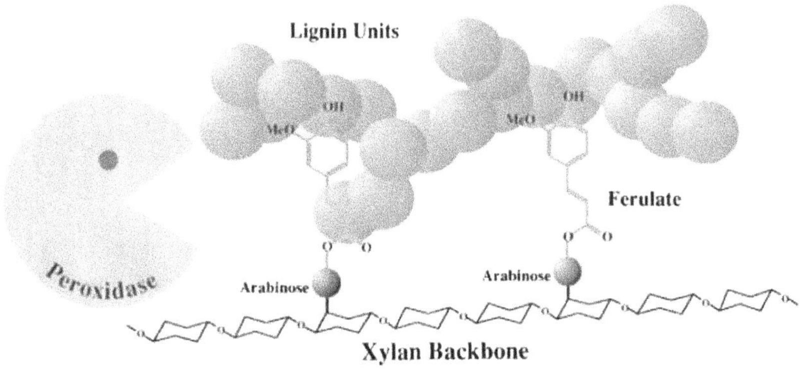

Figure 2 Schematic diagram of ferulate radical coupling to lignin, both coniferyl and sinapyl alcohol residues become linked to accessible ferulates. Source: Hatfield et al. (1999).

1989; Klyosov, 1990). One of the most evaluated type of synergy is between endoglucanase and exoglucanase or different types of exoglucanases or endoglucanases. Coughlan et al. (1993) recognized that absolute and systematic hydrolysis of xylan needed the synergism between the principal and the side-chain-cleaving enzymes with unlike precision. Some studies evaluated how the synergy between cellulases and xylanases could affect the degradation on different type of forages such as bahiagrass hay (Krueger and Adesogan, 2008), oat hulls (Yu et al., 2005), corn stover, and rice straw (Song et al., 2016).

2.2 Mechanism of action

Fibrolytic enzymes are catalytic proteins with high cellulase and or xylanase activity for hydrolyzing beta glucans and xylans in plant cell walls exerting effects through a combination of several pre- or postingestive processes (McAllister et al., 2001).

2.2.1 Preingestive effects

In the preingestive stage, enzyme action is largely affected by direct hydrolysis, which entails cleavage of the glycosidic linkages of the cell wall polysaccharides such that their constituent monomeric units are released. The process removes structural barriers restricting accessibility to the more digestible cell contents within the forages resulting in greater extent of degradation (Nsereko et al., 2000) and greater release of sugars due to partial solubilization of NDF and acid detergent fiber (ADF) (Hristov et al., 1996; Krause et al., 1998; Krueger and Adesogan, 2008; Arriola and Adesogan., 2013; Romero et al., 2015). However, we need more studies validating this potential of EFE. The preingestive hydrolysis allows the attachment of the enzyme to forage particle and may provide greater protection from proteolysis in the rumen (Fontes et al., 1995).

Several studies (Hristov et al., 1996; Krause et al., 1998; Krueger and Adesogan, 2008; Arriola and Adesogan., 2013; Romero et al., 2015) documented the evidence that ruminant feeds might be pretreated with EFE to exploit preingestive fiber degradation and increase release of fermentable sugars. We evaluated enzyme activities (xylanase, endoglucanase, exoglucanase, and β-glucosidase) from 18 EFE at different pH and temperature. Xylanase activity was optimal at pH 5.0 and maximum endoglucanase activity was observed at pH 4.0 (Figure 3), while maximum xylanase and endoglucanase activities were observed at 50°C (Figure 4) (Arriola et al., 2011b). Xylanase activity at rumen pH and temperature was lowered by 89% and 93%, respectively, compared with maximum activity observed at pH 5.0 and temperature 50°C. Similarly, xylanase activity was lowered by 87% at ambient temperature compared

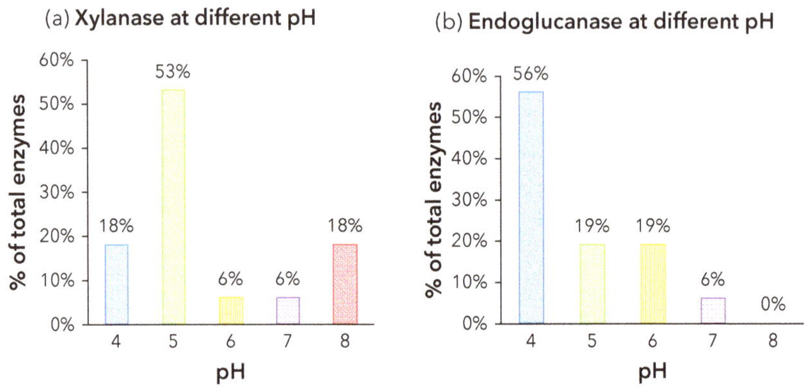

Figure 3 (a) Xylanase and (b) endoglucanase activities from 18 exogenous fibrolytic enzymes analyzed at different pH levels. Source: Arriola et al. (2011b).

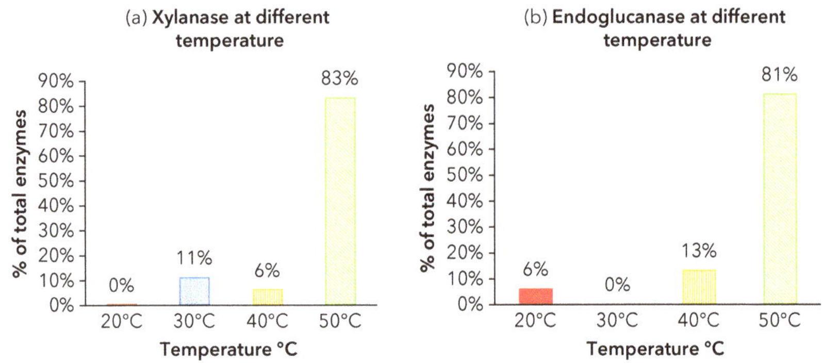

Figure 4 (a) Xylanase and (b) endoglucanase activities from 18 exogenous fibrolytic enzymes analyzed at different temperature levels. Source: Arriola et al. (2011b).

with optimal activity at 50°C. Endoglucanase activity was lowered by 66% and 84% at rumen pH and temperature, respectively, when compared with optimal activity observed at pH 4.0 and 50°C. The activity was completely lost at room temperature. These data indicate that many of the existing EFE exhibit maximum key activities at pH and temperatures that exceed those in the rumen, which suggests that though such EFE are promising when activity is tested in the laboratory, their actual effects in diets are far less. Therefore, more strategic selection of EFE that exhibit optimal activity under ruminal conditions is needed. New EFE preparations have to be screened *in vitro* using several application rates and a variety of feed sources to determine effects on NDF degradability prior to inclusion in the animal diets.

Romero et al. (2015) evaluated the effects of 12 EFE on *in vitro* NDF degradability and preingestive hydrolysis using 4-week regrowth bermudagrass haylage as substrate. The authors observed an improvement of NDFD with 9 out of 12 EFE, 5 enzymes showed greater molar proportion of acetate whereas 6 enzymes showed improvement in molar proportion of propionate and 4 enzymes showed lower acetate-to-propionate ratio compared with untreated bermudagrass haylage. During preingestive hydrolysis, 3 enzymes decreased NDF concentration, even though 10 enzymes increased the release of water-soluble carbohydrates. This illustrates the variability in EFE response attributed primarily to their activities on different substrates. Since pure substrates were used in this study, some have advocated that EFE activity testing should be done using the target forage or diet as the substrate.

2.2.2 Postingestive effects

The enzyme action is thought to be mediated through direct hydrolysis, improved palatability probably due to the increased production of monosaccharides, reduced viscosity, increased proximity between substrates and microbes, synergy between ruminal fibrolytic and exogenous enzymes (Morgavi et al., 2000), and stimulation of bacterial activity by exogenous enzymes and other substances introduced with the enzyme complex.

Hristov et al. (1998) observed that xylanases are particularly stable in the rumen and the effect of direct enzyme supplementation in lower digestive tract is possible. Similar responses were observed in other studies (Morgavi et al., 2000, 2001). Addition of EFE to ruminant diets increased the rate of ruminal particle passage in some (Romero et al., 2013); however, other studies did not observe this trend (Beauchemin et al., 1999). This is because in addition to preingestive substrate hydrolysis, EFE increase the capacity of hydrolysis in the rumen (Beauchemin et al., 2004; Meale et al., 2014). Previously, McAllister et al. (2001) described that EFE represent just a fragment of the enzyme activity in the rumen taking in consideration the capacity of ruminal microbiota to digest fiber; hence, a rise in fiber degradation might not be accredited only to direct EFE hydrolysis. Morgavi et al. (2000) suggested that greater understanding of EFE supplementation on outcome in the rumen is possible by better insights on interaction between EFE and rumen microbes, which enhances ruminal cellulose, xylan, and corn silage digestion.

An early study revealed that addition of EFE to the diet of heifers augmented xylanase and endoglucanase activity in the duodenum by 30% and 5%, respectively (Histrov et al., 1998). Their subsequent work showed increment of xylanase activity in the large intestine and feces of heifers fed treated diets with xylanase enzyme (Hristov et al., 2000). Morgavi et al. (2001) observed that certain EFE were able to get through fermentation in the rumen and abomasum and this possibly impact the activity for specific time in the small intestine. Beauchemin et al. (1999) reported that application of EFE to barley grain diets for dairy cows increased total tract digestion, which was partially accredited to an enhancement in digestibility of fiber and starch in the lower tract. However, postruminal effects may only account for a minor component of any positive responses to EFE addition and their practical implications are not fully understood (Meale et al., 2014).

2.3 Effects of exogenous fibrolytic enzymes application on dairy cow performance

Exogenous fibrolytic enzymes have been applied to different parts of diets and at different time points (e.g. at ensiling or just before feeding). Enzyme treatment at ensiling reduced NDF and ADF concentrations and improved NDF digestibility (Kung et al., 1991; Dean et al., 2005); however, no effects were observed on milk yield (Sheperd and Kung, 1996; Dean et al., 2013). Milk yield was improved in some studies when enzyme was applied to the forage portion of the diet (Lewis et al., 1999; Kung et al., 2000); however, other studies did not observe the same trend (Dhiman et al., 2002; Vicini et al., 2003). Milk yield (Gado et al., 2009; Mohamed et al., 2013) and feed efficiency (Arriola et al., 2011a; Holtshausen et al., 2011; Mohamed et al., 2013) was improved when enzymes were applied directly to the total mixed ration (TMR); however, other studies found no improvement in milk yield (Bernard et al., 2010; Peters et al., 2015). Based on a meta-analysis (Arriola et al., 2017), application of enzymes to the TMR instead of forage or concentrate may increase milk yield, milk protein content, and protein yield. However, the effects of EFE supplementation on performance of lactating dairy cows have been very inconsistent and we need further validation and less variability before EFE can be used consistently for lactation diets (Meale et al., 2014).

The variable effects of EFE supplementation on performance of lactating dairy cows may be attributed to various factors including physiological stage of lactation, insufficient number of animals resulting in low statistical power, experimental design with short periods, enzymes with variable enzyme activities, etc. These factors have been described in detail in a review paper (Adesogan et al., 2014). Arriola et al. (2017) conducted a meta-analysis using 15 studies with 17 experiments and 36 observations. The selection criteria included papers published in English, used control treatment, used confined or housed lactating dairy cows in experiments employing continuous experimental designs, and supplemented with at least one fibrolytic enzyme as a dietary treatment. The meta-analysis revealed that EFE applied to lactation diets increased yield of milk (0.82 kg/day), milk protein (0.03 kg/day), and milk lactose (0.05 kg/day) (Table 1).

Table 1 Meta-analysis of exogenous fibrolytic enzymes applied to dairy diets

		RMD (95% CI)[b]			Heterogeneity			SMD (95% CI)[c]	
Item	*n*[a]	Controlmean	Random effect	*P* value	Chi-squared (*Q*)	*P* value	I^2 (%)	Random effect	*P* value
DMI, kg/day	36	23.2 (2.18)	0.22 (0.07, 0.49)	0.132	58.4	0.008	40.1	0.09 (0.07, 0.24)	0.276
Milk yield, kg/day	36	36.0 (3.26)	0.83 (0.36, 1.30)	0.001	51.4	0.037	31.8	0.25 (0.10, 0.39)	0.001
FCM, kg/day	26	35.7 (3.43)	0.55 (-0.005, 1.11)	0.052	38.3	0.044	34.6	0.16 (0.003, 0.33)	0.001
Milk fat, %	36	3.59 (0.41)	0.05 (0.09, 0.01)	0.008	30.2	0.697	0.0	0.16 (0.27, 0.04)	0.006
Milk protein yield, kg/day	27	1.13 (0.14)	0.03 (0.007, 0.04)	0.007	32.8	0.169	20.6	0.20 (0.04, 0.37)	0.014
Milk lactose yield, kg/day	13	1.77 (0.19)	0.05 (0.02, 0.08)	0.001	7.43	0.828	0.0	0.31 (0.13, 0.48)	0.001
Dry matter digestibility, %	9	68.4 (4.44)	1.36 (0.14, 2.86)	0.077	16.7	0.033	52.2	0.31 (0.04, 0.66)	0.080
NDF digestibility, %	9	50.1 (7.45)	2.30 (0.42, 0.503)	0.098	17.5	0.025	54.4	0.31 (0.05, 0.67)	0.093

Source: Adapted from Arriola et al. (2017).[a]Number of comparisons.
[b]RMD = raw mean difference between enzyme and control treatments.
[c]SMD = standardized mean difference between enzyme and control treatments.

Table 2 Meta-analysis of exogenous fibrolytic enzymes (EFE) applied to dairy diets with high and low forage.

Variable		F:C ratio <50%			F:C ratio ≥ 50%	
	n	Control	EFE		Control	EFE
Dairy cows						
DMI	52	20.3	20.1		21.8	21.2
Milk yield (kg/day)	52	28.4	28.1		33	34.9
Milk protein (g/day)	49	924.6	881.3		1007.5	1106.9
Milk fat (g/day)	49	1136.6	1101.2		1179.6	1262.6
	P value					
	EFE	F	F:C	EFE*F	EFE*F:C	F*F:C
DMI	0.003	0.0001	0.0001	0.98	0.8	0.0001
Milk yield	0.0001	0.0001	0.0001	0.15	0.06	0.0001
Milk protein	0.0001	0.0005	0.0001	0.31	0.06	0.0001
Milk fat	0.0004	0.015	0.023	0.33	0.18	0.0001

Adapted from Tirado-Gonzalez et al. (2017). *n*, number of experiments; F, forage; F:C, forage-to-concentrate ratio source.

Another meta-analysis examined the effect of EFE applied to diets of livestock, forage portion of these diets were based on legumes and grasses at different levels (Tirado-Gonzalez et al., 2017). This study evaluated data from 52 experiments using dairy cows and concluded that addition of EFE to high-forage diets increased milk and milk protein yield; while addition of EFE to low-forage-based diets (forage:concentrate < 50%) showed no improvement of milk yield or milk components (Table 2).

3 Exogenous amylases

Exogenous amylases have long been used in biochemical processes such as production of ethanol, fructose and glucose syrup, detergents, and coated paper and it is estimated that such processes account for approximately 25% of world's enzyme market (Souza, 2010). Studies on the efficacy of amylases at improving starch digestibility or utilization in non-ruminants started more than half a century ago in piglets (Cunningham and Brisson, 1957).

Amylolytic enzymes were not routinely tested for or used in ruminant rations until exogenous enzymes were shown to be resistant to ruminal protease degradation (Hristov et al., 1998), which implied that could improve diet digestibility and consequently animal performance (Eun and Beauchemin, 2005; Klingerman et al., 2009).

Amylolytic enzymes are the enzymes that catalyze the hydrolysis of starch into sugars. These enzymes are classified as hydrolases and belong to the glycoside hydrolase (GH) family (Cantarel et al., 2008). α-Amylases (EC 3.2.1.1) are possibly the most important members of the GH family; they are endo-specific enzymes that can hydrolyze internal 1,4-linkages in amylose and amylopectin in starch (Bedford and Partridge, 2010). Glucoamylases (EC 3.2.1.3) are amylolytic enzymes that differ from α-amylases because they breakdown starch into mostly free glucose instead of a combination of glucose, maltose, and maltooligosaccharides (Pazur et al., 1977). Glucoamylases are also called exo-amylases because they hydrolyze 1,4-alpha-glycosidic bonds from the non-reducing ends and the 1,6-alpha-glycosidic linkages of polysaccharides.

3.1 Amylolytic enzyme activity assay

Previous studies described methods for quantitative determination of catalytic activity of amylolytic enzymes by estimating the formation of reducing sugar (glucose or maltose) with Somogyi copper reagent (Nelson, 1944) and dinitro-salicylic acid (DNS) (Bernfield, 1955; Miller, 1959). The iodine assays for amylolytic enzyme activity were also developed for quick determination of starch-converting enzymes and based on the formation of a colored complex between iodine and linear chains in starch (Boyer and Preiss, 1978; Xiao et al., 2006). Jung and Vogel (2008) proposed a spectrophotometric method that has been widely used in ruminant nutrition studies to assay the α-amylase concentration.

3.2 Exogenous amylases for ruminants

Starch digestion is critical for increasing the performance of animals fed grains (Huntington, 1997). Amylolytic enzymes have potential to increase the rate of starch digestion and consequently improve animal performance when added to ruminant diets (Rojo-Rubio et al., 2001). Due to the complexity of the rumen environment and the multiple interactions among several microbial species, the mechanisms by which amylase enzymes improve animal production remain unclear and need to be further explored. Some authors suggest that shifting the site of starch digestion from the rumen to the small intestine, modulating cross-feeding interactions among ruminal microorganisms, and changes in the

VFA profile (Beauchemin and Holtshausen, 2010; Bedford and Partridge, 2010) for exogenous amylases.

Morgavi et al. (2000) observed that amylase activity from *Aspergillus niger* was stable after 6 h of ruminal incubation. Similarly, Rojo et al. (2005) observed that amylases were slowly degraded in the rumen; probably due to glycosylation. These studies confirm that certain exogenous amylases are stable in the rumen, which may explain their efficacy at increasing starch digestion in several studies. Ruminal *in vitro* starch digestion of sorghum was greater after it was treated with α-amylase from *Bacillus licheniformis* (82.85% vs. 75.13%, respectively; Mora-Jaimes et al., 2002). Similarly, several *in vitro* and *in situ* studies showed greater dry matter (DM) and starch digestibility of grains treated with amylase from *B. licheniformis* (Rojo-Rubio et al., 2001). Amaro et al. (2021) tested ten exogenous enzymes with amylolytic activity sourced from several bacteria and fungi and reported that seven increased *in vitro* DM and starch digestibility of ground dent corn grain after 7 h of incubation. Clearly, exogenous enzymes have shown promise for increasing *in vitro* starch digestibility. When such commercial and experimental enzyme preparations containing amylolytic activity have been evaluated in lactation trial, increases in dry matter intake (DMI), ruminal and apparent total tract starch digestibility, feed efficiency, and milk production have been reported in some studies (Tricarico et al., 2005; Harrison and Tricarico, 2007; Klingerman et al., 2009) but less so in others (Ferraretto et al., 2011; Weiss et al., 2011). Amylase addition to diets increased average daily gain (ADG) in beef cattle and sheep (Crosby et al., 2006; Tricarico et al., 2007; Salem et al., 2013) and glucoamylase addition to finishing feedlot diets increased apparent total tract starch digestibility and feed conversion in beef steers (Devant et al., 2020; Pitalluga et al., 2021). However, effects of amylase supplementation on animal performance have not been consistent (DiLorenzo et al., 2011), which poses a big challenge to adoption of this technology by the dairy and beef industry.

Recently, Cueva et al. (2021) evaluated the effects of feeding amylase-enabled corn silage on performance of lactating dairy cows. Due to the presence of a bacterial transgene, these corn silage hybrids express high levels of thermotolerant α-amylase in the grain endosperm. This enzyme has the ability to cleave α-1,4-glycosidic bonds resulting in production of linear and branched oligosaccharides. Feeding amylase-enabled corn silage increased milk yield, milk true protein, feed efficiency, and lactose yield while it tended to increase energy-corrected milk yield. Furthermore, the intensity of methane emissions was lower for amylase-enabled corn silage treatment.

4 Exogenous proteases

Proteolytic enzymes (EC 3.4.) catalyze the hydrolysis of peptide bonds between amino acid residues of proteins and are categorized into two main groups of protease families; exopeptidases or endopeptidases (Barrett et al., 2012). Endopeptidases cleave within the protein chains, releasing and solubilizing larger protein fragments whereas exopeptidases act at or near the ends of the peptide chains, and they are designated as aminopeptidases and carboxypeptidases to indicate their action at the N- or C-terminals of the peptide chain. Furthermore, proteolytic enzymes are classified into four groups according to their mechanism of action based on the catalytic amino acid residue (Rao et al., 1998): (1) serine proteases, (2) aspartate proteases, (3) cysteine proteases, and (4) metalloproteases.

Previous studies have shown that most proteolytic enzymes used in animal feeds exist as alkaline and serin-type proteases, which have high potential to increase feed digestibility in ruminants (Eun and Beauchemin, 2007; Colombatto and Beauchemin, 2009). This may be due to their ability to bypass rumen degradation and contribute to protein digestion in the small intestine. The serine proteases are characterized by a precisely oriented catalytic triad of a carboxylate group of aspartic acid (Asp102), an imidazole ring of histidine (His-57), and a hydroxyl group of serine (Ser-195) (Blow et al., 1969). Most commercial alkaline proteases are isolated from *Bacillus* species (Maurer, 2004), while a considerable number of fungal species including *Aspergillus* are the main sources of alkaline serine proteases (Hajji et al., 2007).

4.1 Mode of action

Although the specific modes of action of exogenous proteases in ruminant diets remain unknown, it is speculated that proteolytic enzymes disrupt cell-wall integrity and enhance fiber degradation by attacking some of the cell-wall N with components that act as physical barriers to degradation (Colombatto et al., 2003). In addition, considering the cell-wall effects of alkaline serine proteases, proteolytic enzymes may remove structural barriers including lignified middle lamella/primary walls, allowing ruminal microorganisms to access digestible nutrients (Jung et al., 2000). These structural barriers may be composed of structural proteins from the cell-wall, which might act as barriers to microbial digestion. Proteolytic enzymes may also boost enzyme activities of the ruminal fluid (specifically xylanase and endoglucanase) by stimulating ruminal microbial populations or may have synergistic effects with hydrolases from rumen microorganisms (Eun and Beauchemin, 2005). By hydrolyzing the protein matrix surrounding starch granules in grains (McAllister et al., 1993), proteolytic enzymes can also increase starch accessibility resulting in faster rate of starch degradation in rumen (Ferraretto et al., 2015). This protein matrix typically impedes microbial attachment and fermentation in the rumen or hydrolytic and enzymatic digestion in the abomasum and the small intestine (Giuberti et al., 2014).

4.2 Exogenous proteases for ruminants

Exogenous proteases are routinely supplemented in swine and poultry diets to degrade proteins stored in diverse plant feeds and proteinaceous antinutrients in the seeds of legumes (Philipps, 2018). Proteins, like prolamin in corn, form a matrix with starch in kernels thereby restricting access to starch by rumen microbes (Owens et al., 1996). In ruminants, supplementation with exogenous proteolytic enzymes (EPE) is not as common as in non-ruminants due to the abundance of endogenous proteolytic enzymes produced by ruminal microorganisms. However, EPE treatment has been proposed as a way of improving silage quality and animal performance (Colombatto and Beauchemin, 2009; Young et al., 2012). This is because they can target N cross-linkages in the cell-wall of forages thereby increasing digestibility of low-quality forages. For instance, Colombatto and Beauchemin (2009) reported greater DM and NDF degradability of alfalfa hay in response to exogenous proteases without any measurable cellulase or xylanase activity. Similarly, enzymes having mainly protease activity increased fiber degradability of alfalfa hay in lactating dairy cows and the response was maximized with a high concentrate diet (Eun and Beauchemin, 2005). Further, Sucu et al. (2014) reported greater feed efficiency and milk lactose concentration in dairy cows fed an EPE-treated TMR. In feedlot cattle, EPE addition to the diet improved ADG and final body weight, and increased the digestibility of DM, CP, NDF, and ADF (Vera et al., 2012). Eun and Beauchemin (2008) conducted a meta-analysis of *in vitro* experiments using EPE as feed additives and reported a positive linear relationship between *in vitro* gas production and elevated protease activity; however, no effects were observed on feed degradability. The inconsistency of protease efficacy among studies was attributed to the method used for measuring its enzyme activity. In addition, the type of protease used may also result in equivocal responses.

Besides the inclusion of EPE in the ruminant diets, there has been growing interest in using proteases to hydrolyze the protein matrix within the corn endosperm for improving the accessibility and ruminal degradability of starch in the corn silage. Ensiling initiates the proteolytic process for degrading prolamin-starch matrix (Hoffman et al., 2011); however, several months are required for the protein matrix to be sufficiently hydrolyzed during ensiling to optimize starch availability in corn silage. Inclusion of EPE in fresh forage before ensiling may accelerate hydrolysis of the prolamin-starch matrix (Young et al., 2012; Windle et al., 2014; Ferraretto et al., 2015). For instance, Young et al. (2012) reported that the inclusion of EPE (2000 mg/kg, AB Vista, Wiltshire, UK) in whole-corn plant accelerated the proteolytic processes during ensiling and increased *in vitro* starch degradation after 45 days of ensiling. Similarly, Windle et al. (2014) noted that treating corn plants with a high dose of experimental protease (2000 mg/kg /kg of wet forage, AB Vista, Wiltshire, UK) at harvest time accelerated proteolysis during ensiling, leading to corn silages with high levels of *in vitro* starch degradability (IVSD) after 45 days of ensiling. Most recently, Der Bedrosian and Kung (2019) treated chopped whole-plant corn (37% DM) with 0, 20, 200, 1000, or 2000 mg of protease followed by ensiling for 45 days. *In vitro* degradability of starch was improved with the 200 mg/kg dose of protease while concentrations of soluble protein and NH_3-N were increased by 1000 mg/kg and 2000 mg/kg doses of protease compared to the untreated silage.

Few studies have investigated the effect of EPE on starch availability in ensiled high-moisture corn (HMC) (Kung et al., 2014; Ferraretto et al., 2015). Kung et al. (2014) examined the effect of exogenous protease on the fermentation of ground HMC and observed greater concentrations of soluble protein and NH_3-N after 70 days and 140 days of ensiling, respectively, with protease treatment. They noted that the latter result suggested that added proteases do not solely target prolamin proteins rather, they also degrade albumin, globulin, and glutelin protein fractions (Landry et al., 2000). In addition, Kung et al. (2014) revealed that an exogenous protease accelerated IVSD hydrolysis of fermented HMC (70 days and 140 days vs. 0 day), which may be important because of the long time it takes (8 mo) for HMC to reach maximal starch degradation (Ferraretto et al., 2013). Similarly, Ferraretto et al. (2015) reported higher *in vitro* starch degradability, soluble protein, and NH_3-N in HMC treated with exogenous protease compared to the control group after 30 days of ensiling. Overall, the literature indicates that adding a commercial protease enzyme to ruminant diets or ensiled forages improves starch and protein degradability. However, further research is warranted on effects of exogenous proteases on efficiency of nitrogen use in ruminants.

4.3 Proteolytic enzyme assay

The methods used for measuring proteolytic enzymatic activities include spectrophotometry, fluorescence, and radiolabeling techniques. Proteolytic enzyme catalyzes the hydrolysis of proteins and various hydrolysis products of proteins into peptides and amino acids. An early attempt to quantify proteolytic enzyme activity by a caseinolytic spectrophotometric assay was reported by Kunitz (1947). Twenty years later, a more sensitive spectrophotometric method was described (Lin et al., 1969) involving *N,N*-dimethyl protein as a substrate and a direct measurement of the bond's splits and terminal amino groups with trinitrobenzene sulfonic acid. Subsequently, Brock et al. (1982) described the proteolytic activity of rumen microorganisms with azocasein as a substrate, which has been adopted widely for ruminant nutrition studies.

5 Fungal enzymes

Fungi are some of the most commonly used direct-fed microbes in animal nutrition. They are known to produce fibrolytic enzymes including cellulolytic, xylanolytic, and ligninolytic enzymes, and non-fibrolytic enzymes including amylolytic, lipolytic, proteolytic enzymes, and phytases (Green and Beezhold, 2011; Abrão et al., 2017). In addition, fungi such as *Aspergillus* spp. have been reported to be one of the generally recognized as safe microbes used as enzymes producers (Headon and Walsh, 1994).

Fungi are major contributors to natural degradation of lignocellulose via the lignocellulolytic enzymes they produce. Abrão et al. (2017) isolated fungi from the digestive tract of grazing cattle and reported it contained fibrolytic enzymes including cellulases, xylanases, and ligninolytic enzymes from several fungi, especially *Aspergillus* spp. Some of the lignin-degrading enzymes and enzyme complexes of fungal origin include laccase, lignin peroxidase, manganese peroxidase, and versatile peroxidase (Asemoloye et al., 2021). In addition to these ligninolytic enzymes, certain white-rot fungi (such as *Trametes versicolor, Heterobasidium annosum*, and *Irpex lacteus*) are known to produce cellulose-degrading enzymes (β-glucosidase, cellobiohydrolase, and β-xylosidase) (Vrsanska et al., 2016), resulting in simultaneous degradation of lignin and cellulose components. Abrão et al. (2017) evaluated enzyme production by *Aspergillus* spp. isolates and reported high levels of cellulase and phenoloxidases (critical for lignin degradation). These complexes of enzymes from fungi can significantly improve digestibility and nutrient utilization of low-quality forages such as wheat straw and bermudagrass (Akin et al., 1993).

Fungi are also known to produce three main types of amylolytic enzymes: α-amylase (EC 3.2.1.1), glucoamylase (EC 3.2.1.3), and α-glucosidase (EC 3.2.1.20) (Nakamura et al., 2006; Yuan et al., 2008; Chen et al., 2012). In the classification of GH in the Carbohydrate-Active enZymes (CAZy: http://www.cazy.org) system, fungal α-amylases are classified into two subfamilies; GH13_1 and GH13_5 (Stam et al., 2006). Enzymes in the subfamily GH13_1 are extracellular and fungal-specific α-amylases, whereas those belonging to GH13_5 are intracellular, with higher sequence similarities to α-amylases from bacteria (Van der Kaaij et al., 2007). A major commercial application of exogenous α-amylase from fungi (such as *Aspergillus niger* and *Aspergillus oryzae*) is in thinning of starch for industrial production of sugar, beer, and alcohol. The glucoamylases (γ-amylases) are employed in commercial solubilization of starch and other polysaccharides for the production of glucose. Industrial production of glucoamylases mainly use filamentous fungi from the genus *Aspergillus* and *Rhizopus* (e.g. *Aspergillus niger* and *Rhizopus oryzae*) (Coutinho and Reilly, 1997). Commercial production of fungal α-glucosidases are mainly from *Aspergillus spp.* and *Mucor spp* (Papagianni, 2004).

Howson and Davis (1983) reported 10–80% reduction in phytate concentration in rapeseed meal inoculated with fungi (such as *Aspergillus ficuum* and *Aspergillus niger*) due to extracellular phytase production. Insoluble phytates are components of cereals and oilseed, which closely associate with protein component of these feedstuffs, and limit the bioavailability of their constituent amino acids (Fontaine et al., 1946; Barré and van Huot, 1965). Peak extracellular phytase activity for *A. ficumm* at 25°C was reported to occur 10 days post-incubation (Howson and Davis, 1983).

5.1 Modes of action

As opposed to conventional pretreatment methods such as acid and alkali pretreatments of lignocellulosic biomass, ligninolytic enzymes might be a safer and likewise efficient means of decreasing the lignin network and its negative effect on enzymatic hydrolysis of lignocellulosic substrates. The ligninolytic enzymes, laccase, and manganese peroxidase from fungal species act primarily on phenols, although they can also metabolize the non-phenolic moieties under certain conditions (Rouches et al., 2016). Lignin peroxidase acts on the non-phenolic lignin components (which could amount to 90% of the polymer), whereas the versatile peroxidase acts both on phenolic and non-phenolic components (Martinez et al., 2005; Rouches et al., 2016).

Starch is composed of amylose, a linear polymer of glucose molecules linked by α-1,4-glycosidic bonds, and amylopectin, a highly branched polymer composed of α-1,4-linked glucose molecules with varying proportions of α-1,6-linked side chains that are composed of 15–45 glucose units (van der Maarel et al., 2002). Fungal α-amylase like other α-amylases catalyzes the hydrolysis of internal α-1,4-glycosidic bonds within starch into shorter maltooligosaccharides and α-limit dextrins (Chen et al., 2012). However, fungal glucoamylases (γ-amylases) catalyze hydrolysis of both the α-1,4- and α-1,6-glycosidic bonds, releasing β-D-glucose only from the non-reducing end of the polysaccharide (Sauer et al., 2000; Chen et al., 2012). The α-glucosidases could similarly hydrolyze both the α-1,4- and α-1,6-glycosidic bonds, resulting in release of α-D-glucose also from the non-reducing end (Vihinen and Mantsiila, 1989; Ernst et al., 2006).

There are only a few studies validating the potential of fungal enzymes at improving feed digestibility. Fazaeli et al. (2004) treated wheat straw with a lignin-selective strain, *Pleurotus ostreatus* (P-41) and observed greater *in vivo* DMI (12.2 kg/day vs. 10.6 kg/day), total-tract DM digestibility (58.8% vs 52.3%), NDF digestibility (42.3% vs. 34.3%), and milk yield (9 kg/day vs. 7.5 kg/day), in lactating dairy cows in late-lactation when treated and untreated straw were fed as 30% of a TMR. Similarly, Shrivastava et al. (2014) reported greater DMI (per kg of metabolic body weight), DM digestibility (57.82% vs. 52.07%), NDF digestibility (53.3% vs. 45.8%), and 50 g/day higher average BW gain, when buffalo calves were fed wheat straw treated with *Crinipellis* species, RCK-1, instead of untreated wheat straw. Shahzad et al. (2016) also reported greater DMI and digestibility when white-rot fungi-treated wheat straw was used to replace 33% of the untreated wheat straw in a TMR fed to buffalo calves. White-rot fungi are still not widely included in ruminant diets partly due to extended pretreatment time required (van Kuijk et al., 2015).

6 Future trends in research

6.1 Role of metagenomics

Metagenomics is the direct analysis of DNA from environmental samples. This methodology was implemented to understand how dietary manipulations affect the rumen microbiome and environment in recent years. Krause et al. (2003) evaluated data focused on enhancing degradation of the fiber in the rumen using treatments such as inoculants, fungal treatment, and exogenous enzymes. The authors outlined that though genomes of *Fibrobacter succinogenes*, *Ruminococcus albus*, and *Ruminococcus flavefaciens* have been sequenced, the enzymatic mechanism by which they hydrolyze cellulose is still unclear. Equivocal responses observed in response to dietary inclusion of fibrolytic enzymes may be attributed to our lack of proper understanding of the complex ruminal environment. In addition, research focus has always been on ruminal availability of exogenous enzymes, however, the actual requirement for enzymes and the functional genomic framework within which the rumen operates has not been understood well. A study that examined the ruminal enzymes in dairy cows using a metagenomics library reported that 9 endo-β-1,4-glucanases, 12 esterases, and 1 cyclodextrinase (Ferrer et al., 2005) were present. However, when Hess et al. (2011) assembled 15 uncultured microbial genomes, they identified 27 755 predicted ruminal carbohydrate-active enzymes expressing 90 candidate proteins with 57% of proteins being enzymatically active against cellulosic substrates. The latter reveals the vast number of active fibrolytic enzymes that have not been properly characterized from uncultured ruminal microbes. Recently, Ribeiro et al. (2016) examined ongoing methods and strategies to recognize rumen fibrolytic enzymes and originate beneficial ruminant fibrolytic feed enzyme cocktails. They concluded that metagenomics and metatranscriptomics are helping to characterize rumen enzymes and microbes involved in lignocellulose degradation, but this work is in the early stages and much progress is needed to enable development of enzyme cocktails for optimizing feed degradation

Badhan et al. (2014) observed synergistic release of glucose and xylose from barley straw and alfalfa hay when rumen enzyme mixtures were supplemented with aerobic enzymes, indicating that combining enzymes from aerobic and anaerobic organisms may increase efficacy relative to the normal practice of using enzymes from anaerobic organisms alone in ruminant nutrition. Future research should determine the rate-limiting enzymes in fiber deconstruction and improve knowledge of synergies between rumen microbes and enzymes involved in plant cell-wall deconstruction. Cultivation-independent metagenomics and metatranscriptomics that focus on the latter will enable the development of more effective ruminant fibrolytic feed enzyme cocktails.

6.2 Expansin-like proteins

Expansins and expansin-like proteins (BsEXLX1) are a group of nonhydrolytic proteins with the unique ability to induce cell-wall relaxation or loosening (Cosgrove, 2000; Georgelis et al., 2011). The most remarkable characteristic of expansins and expansin-like proteins is their ability to synergize with exogenous fibrolytic enzymes to increases hydrolysis of cellulose and hemicellulose (Kim et al., 2009; Bunterngsook et al., 2015; Liu et al., 2015). Wang et al. (2011) observed synergistic increase in cellulase activity in the presence of fungal swollenin, an expansin-like protein, despite the lack of cellulolytic activity from swollenins. Preliminary studies showed that synergistic effects between exogenous fibrolytic enzymes and BsEXLX1 increased hydrolysis of cellulose-based but not hemicellulose-based substrates under ruminal conditions as well as pre- and pos-ingestive hydrolysis and *in vitro* digestibility of NDF when bermudagrass haylage (Pech-Cervantes et al., 2017). These preliminary results suggest that addition of BsEXLX1 could increase the efficacy of EFE at improving fiber digestion of certain forages by ruminants. More efficient methods of producing BsEXLX1 are required to supply sufficient quantities for testing in cattle diets. Yoon et al. (2016) successfully expressed 1000% more expansin, cEXPA1, from cucumber gene in transgenic corn plants compared with traditional methods. Synergies between the cellulase and plant expansin from cucumber (cEXPA1) increased (by 2.5 folds) corn stover and tobacco leaf hydrolysis. Hence, transgenics may offer less expensive methods to produce expansins at large scale, even though expression of cEXPA1 in corn plants has not always been successful. Future research should also examine the fiber degradation potential of expansins from live microbes that secrete them.

7 Conclusion

This chapter summarized the basic knowledge about the main fibrolytic enzymes involved in plant cell-wall digestibility, their modes of action, and their effects on animal performance. The chapter emphasized the importance of *in vitro* screening enzymes to confirm efficacy and noted that EFE activity testing should involve similar to those the EFE is expected to degrade and should be conducted at ruminal pH and temperature. Preparations of EFE that are promising after the latter screening approach can then be tested in animals. This approach will reduce costs of enzyme testing and accelerate identification of the most promising candidates. The number of studies using exogenous fibrolytic enzymes in ruminant diets have been increased in the last years; however, results are still variable because the strategies applied were unsuccessful.

Despite considerable advances in understanding how to optimize EFE efficacy in ruminant diets, most existing preparations cannot degrade the main cross-links between lignin and cell-wall polysaccharides that reduce intake and digestibility of feeds. More research is needed on more potent EFE preparations, particularly those containing fungal EFE that can degrade these recalcitrant linkages. Publications on these new enzyme preparations should fully characterize other enzyme activities present in cocktails. Future research should also elucidate the mechanism of interaction between EFE and ruminal rumen using effective and comprehensive approaches such as metagenomics and metatranscriptomics. More research is also required to mainstream using expansin-like proteins to optimize efficacy of fibrolytic enzymes.

8 References

Abrão, F. O., Duarte, E. R., Pessoa, M. S., Santos, V. L. D., Freitas Júnior, L. F. D., Barros, K. D. O., Hughes, A. F. D. S., Silva, T. D. and Rodrigues, N. M. (2017). Notable fibrolytic enzyme production by *Aspergillus spp.* isolates from the gastrointestinal tract of beef cattle fed in lignified pastures. *PLoS ONE* 12(8):e0183628. https://doi.org/10.1371/journal.pone.0183628.

Adesogan, A. T., Arriola, K. G., Jiang, Y., Oyebade, A., Paula, E. M., Pech-Cervantes, A. A., Romero, J. J., Ferraretto, L. F. and Vyas, D. (2019). Symposium review: Technologies for improving fiber utilization. *J. Dairy Sci.* 102(6):5726–5755.

Adesogan, A. T., Ma, Z. X., Romero, J. J. and Arriola, K. G. (2014). Ruminant nutrition symposium: Improving cell wall digestion and animal performance with fibrolytic enzymes. *J. Anim. Sci. Rumin. Nutr.* 92(4):1317–1330.

Akin, D. E., Sethuraman, A., Morrison, W. H., Martin, S. A. and Eriksson, K. E. (1993). Microbial delignification with white rot fungi improves forage digestibility. *Appl. Environ. Microbiol.* 59(12):4274–4282.

Amaro, F. X., Kim, D., Agarussi, M. C., Silva, V. P., Fernandes, T., Arriola, K. G., Jiang, Y., Cervantes, A. P., Adesogan, A. T., Ferraretto, L. F., Yu, S., Li, W. And Vyas, D. (2021). Effects of exogenous α-amylases, glucoamylases, and proteases on ruminal in vitro dry matter and starch digestibility, gas production, and volatile fatty acids of mature dent corn grain. *Tran. Anim. Sci.* 5(1):222.

Arriola, K. G. and Adesogan, A. T. (2013). Effect of fibrolytic enzyme application on the digestibility of corn silage, alfalfa hay, concentrates, and complete diets under simulated ruminal and preruminal conditions. *Anim. Nutr. Feed Technol.* 13:537-550.

Arriola, K. G., Kim, S. C., Staples, C. R. and Adesogan, A. T. (2011a). Effect of fibrolytic EFE application to low- and high-concentrate diets on the performance of lactating dairy cattle. *J. Dairy Sci.* 94(2):832-841.

Arriola, K. G., Romero, J. R. and Adesogan, A. T. (2011b). Effects of pH and temperature on fibrolytic enzyme activities of various commercial exogenous enzyme preparations. *J. Dairy Sci.* 94:554.

Arriola, K. G., Oliveira, A. S., Ma, Z. X., Lean, I. J., Giurcanu, M. C. and Adesogan, A. T. (2017). A meta-analysis on the effect of dietary application of exogenous fibrolytic enzymes on the performance of dairy cows. *J. Dairy Sci.* 100(6):4513-4527.

Asemoloye, M. D., Marchisio, M. A., Gupta, V. K. and Pecoraro, L. (2021). Genome-based engineering of ligninolytic enzymes in fungi. *Microb. Cell Fact.* 20(1):20.

Asner, G. P. and Archer, S. R. (2010). Livestock and the global carbon cycle. In Steinfeld, H., Mooney, H. A., Schneider, F. and Neville, L. E. (Eds.), *Livestock in a Changing Landscape: Drivers, Consequences, and Responses*. Washington, DC: Island Press, 69-82.

Badhan, A., Wang, Y., Gruninger, R., Patton, D., Powlowski, J., Tsang, A. and McAllister, T. (2014). Formulation of enzyme blends to maximize the hydrolysis of alkaline peroxide pretreated alfalfa hay and barley straw by rumen enzymes and commercial cellulases. *BMC Biotechnol.* 14:31.

Barré, R. and van Huot, N. (1965). Study of the combination of phytic acid with native, acetylated and deaminated human serum albumin. *Bull. Soc. Chim. Biol. (Paris)* 47(7):1399-1417.

Barrett, A. J., Rawlings, N. D. and Woessner, J. F. (2012). *Handbook of Proteolytic Enzymes, Edition*. Waltham, MA: Academic Press.

Beauchemin, K. and Holtshausen, L. (2010). Developments in enzyme usage in ruminants. *Enzymes Farm Anim. Nutr.* 206-230.

Beauchemin, K. A., Colombatto, D. and Morgavi, D. P. (2004). A rationale for the development of feed enzyme products for ruminants. *Can. J. Anim. Sci.* 84(1):23-36.

Beauchemin, K. A., Colombatto, D., Morgavi, D. P. and Yang, W. Z. (2003). Use of exogenous fibrolytic enzymes to improve feed utilization by ruminants. *J. Anim. Sci.* 81:E37-E47.

Beauchemin, K. A. and Holtshausen, L. (2011). Development in enzyme usage in ruminants. In Bedford, M. R. and Partridge, G. C. (Eds.), *Enzymes in Farm Animal Nutrition* (2nd edn.). Wallingford, UK: CABI Publishing, 206-230.

Beauchemin, K. A., Rode, L. M. and Sewalt, V. J. H. (1995). Fibrolytic enzymes increase fiber digestibility and growth rate of steers fed dry forages. *Can. J. Anim. Sci.* 75(4):641-644.

Beauchemin, K. A., Yang, W. Z. and Rode, L. M. (1999). Effects of grain source and enzyme additive onsite and extent of nutrient digestion in dairy cows. *J. Dairy Sci.* 82(2):378-390.

Bedford, M. R. and Partridge, G. G. (2010). *Enzymes in Farm Animal Nutrition*. Wallingford, UK: CABI Publishing, 15-16.

Bernard, J. K., Castro, J. J., Mullis, N. A., Adesogan, A. T., West, J. W. and Morantes, G. (2010). Effect of adding alfalfa hay or Tifton 85 Bermudagrass haylage with or without a cellulase EFE on performance of Holstein cows. *J. Dairy Sci.* 93(11):5280-5285.

Bernfeld, P. (1955). Amylase α and β. *Methods Enzymol.* 1:149-158.

Bhat, K. M. and Hazlewood, G. P. (2001). Enzymology and other characteristics of cellulases and xylanases. In Bedford, M. R. and Partridge, G. C. (Eds.), *Enzymes in Farm Animal Nutrition*. Wallingford, UK: CABI Publishing, 11-60.
Blow, D. M., Birktoft, J. J. and Hartley, B. S. (1969). Role of a buried acid group in the mechanism of action of chymotrypsin. *Nature* 221(5178):337-340.
Boyer, C. D. and Preiss, J. (1978). Multiple forms of (1→4)-α-D-glucan, (1→4)-α-D-glucan-6-glycosyl transferase from developing *Zea mays* L. Kernels. *Carb. res. Carbohydr. Res.* 61(1):321-334.
Brock, F. M., Forsberg, C. W. and Buchanan-Smith, J. G. (1982). Proteolytic activity of rumen microorganisms and effects of proteinase inhibitors. *Appl. Environ. Microbiol.* 44(3):561-569.
Bunterngsook, B., Mhuantong, W., Champreda,V., Thamchaipenet, A. and Eurwilaichitr, L. (2014). Identification of novel bacterial expansins and their synergistic actions on cellulose degradation. *Bioresour. Technol.* 159:64-71.
Cantarel, B. L., Coutinho, P. M., Rancurel, C., Bernard, T., Lombard, V. and Henrissat, B. (2009). The Carbohydrate-Active EnZymes database (CAZy): an expert resource for Glycogenomics. *Nucleic Acids Research*. 37:D233-D238.
Chen, W., Xie, T., Shao, Y. and Chen, F. (2012). Phylogenomic relationships between amylolytic enzymes from 85 strains of fungi. *PLoS ONE* 7(11):e49679.
Colombatto, D. and Beauchemin, K. A. (2009). A protease additive increases fermentation of alfalfa diets by mixed ruminal microorganisms in vitro. *J. Anim. Sci.* 87(3):1097-1105.
Colombatto, D., Morgavi, D. P., Furtado, A. F. and Beauchemin, K. A. (2003). Screening of exogenous enzymes for ruminant diet: Relationship between biochemical characteristics and in vitro ruminal degradation. *J. Anim. Sci.* 81(10):2628-2638.
Cosgrove, D. J. (2000). Loosening of plant cell-walls by expansins. *Nature* 407(6802):321-326.
Coughlan, M. P., Tuohy, M. G., Filho, E. X. F., Puls, J., Clayessens, M., Vrsanská, M. and Hughes, M. M. (1993). Enzymological aspects of microbial hemicellulases with emphasis on fungal systems. In Coughlan, M. P. and Hazlewood, G. P. (Eds.), *Hemicellulose and Hemicellulases.* London: Portland Press, 53-84.
Coutinho, P. M. and Reilly, P. J. (1997). Glucoamylase structural, functional, and evolutionary relationships. *Proteins* 29(3):334-347.
Crosby, M. M., Mendoza, G. D., Melgoza, L. M., Bárcena, R., Plata, F. X. and Aranda, E. M. (2006). Effects of *Bacillus licheniformis* amylase on starch digestibility and sheep performance. *J. Appl. Anim. Res.* 30(2):133-136.
Cueva, S. F., Stefenoni, H., Melgar, A., Raisanen, S. E., Lage, C. F. A., Wasson, D. E., Fetter, M. E., Pelaez, A. M., Roth, G. W. and Hristov, A. N. (2021). Lactational performance, rumen fermentation, and enteric methane emission of dairy cows fed an amylase-enabled corn silage. *J. Dairy Sci.* 104(9):9827-9841.
Cunningham, H. M. and Brisson, G. J. (1957). The effect of amylases on the digestibility of starch by baby pigs. *J. Anim. Sci.* 16(2):370-376.
Dean, D. B., Adesogan, A. T., Krueger, N. and Littell, R. C. (2005). Effect of fibrolytic enzymes on the fermentation characteristics, aerobic stability, and digestibility of Bermudagrass silage. *J. Dairy Sci.* 88(3):994-1003.
Dean, D. B., Staples, C. R., Littell, R. C., Kim, S. C. and Adesogan, A. T. (2013). Effect of method of adding a fibrolytic enzyme to dairy cow diets on feed intake digestibility, milk production, ruminal fermentation, and blood metabolites. *Anim. Nutr. Feed Technol.* 13:337-353.

Der Bedrosian, M. C. and Kung, L. (2019). The effect of various doses of an exogenous acid protease on the fermentation and nutritive value of corn silage. *J. Dairy Sci.* 102(12):10925–10933.

Devant, M., Yu, S., Genís, S., Larsen, T. and Wenting, L. (2020). Effects of exogenous glucoamylase enzymes alone or in combination with a neutral protease on apparent total tract digestibility and feces D-lactate in crossbred Angus bulls fed a ration rich in rolled corn. *Animals (Basel)* 10(6):1077.

Dhiman, T. R., Zaman, M. S., Gimenez, R. R., Walters, J. L. and Treacher, R. J. (2002). Performance of dairy cows fed forage treated fibrolytic enzymes prior to feeding. *Anim. Feed Sci. Technol.* 101(1–4):115–125.

DiLorenzo, N., Smith, D. R., Quinn, M. J., May, M. L., Ponce, C. H., Steinberg, W., Engstrom, M. A. and Galyean, M. L. (2011). Effects of grain processing and supplementation with exogenous amylase on nutrient digestibility in feedlot diets. *Livest. Sci.* 137(1–3):178–184.

Ernst, H. A., Leggio, L. L., Willemoës, M., Leonard, G., Blum, P. and Larsen, S. (2006). Structure of the sulfolobus solfataricus α-glucosidase: Implications for domain conservation and substrate recognition in GH31. *J. Mol. Biol.* 358(4):1106–1124.

Eun, J.-S. and Beauchemin, K. A. (2005). Effects of a proteolytic feed enzyme on intake, digestion, ruminal fermentation, and milk production. *J. Dairy Sci.* 88(6):2140–2153.

Eun, J.-S. and Beauchemin, K. A. (2007). Enhancing in vitro degradation of alfalfa hay and corn silage using feed enzymes. *J. Dairy Sci.* 90(6):2839–2851.

Eun, J.-S. and Beauchemin, K. A. (2008). Relationship between enzymic activities and in vitro degradation of alfalfa hay and corn silage. *Anim. Feed Sci. Technol.* 145(1–4):53–67.

Eun, J.-S., Beauchemin, K. A. and Schulze, H. (2007). Use of an in vitro fermentation bioassay to evaluate improvements in degradation of alfalfa hay due to exogenous feed enzymes. *Anim. Feed Sci. Technol.* 135(3–4):315–328.

Fazaeli, H., Mahmodzadeh, H., Jelan, Z. A., Rouzbehan, Y., Liang, J. B. and Azizi, A. (2004). Utilization of fungal treated wheat straw in the diet of late lactating cow. *Asian-Australas. J. Anim. Sci.* 17(4):467–472.

Ferraretto, L. F., Crump, P. M. and Shaver, R. D. (2013). Effect of cereal grain type and corn grain harvesting and processing methods on intake, digestion, and milk production by dairy cows through a meta-analysis. *J. Dairy Sci.* 96(1):533–550.

Ferraretto, L. F., Crump, P. M. and Shaver, R. D. (2015). Effect of ensiling time and exogenous protease addition to whole-plant corn silage of various hybrids, maturities, and chop lengths on nitrogen fractions and ruminal in vitro starch digestibility. *J. Dairy Sci.* 98(12):8869–8881.

Ferraretto, L. F., Shaver, R. D., Espineira, M., Gencoglu, H. and Bertics, S. J. (2011). Influence of a reduced-starch diet with or without exogenous amylase on lactation performance by dairy cows. *J. Dairy Sci.* 94(3):1490–1499.

Ferrer, M., Golyshina, O. V., Cherniknova, T. N., Khachane, A. N., Reyes-Duarte, D., Santos, V. A. P. M. D., Strompl, C., Elborough, K., Javis, G., Neef, A., Yakimov, M. M., Timmis, K. N. and Golyshin, P. N. (2005). Novel hydrolase diversity retrieved from a metagenome library of bovine rumen microflora. *Environ. Microbiol.* 7(12):1996–2010.

Fontaine, T. D., Pons, W. A. and Irving, G. W. (1946). Protein-phytic acid relationship in peanuts and cottonseed. *J. Biol. Chem.* 164(2):487–507.

Fontes, C. M. G. A., Hall, J., Hirst, B. H., Hazelwood, G. P. and Gilbert, H. J. (1995). The resistance of cellulases and xylanases to proteolysis inactivation. *Appl. Microbiol. Biotechnol.* 43(1):52–57.

Gado, H. M., Salem, A. Z. M., Robinson, P. H. and Hassan, M. (2009). Influence of exogenous enzymes on nutrient digestibility, extent of ruminal fermentation as well as milk production and composition in dairy cows. *Anim. Feed Sci. Technol.* 154(1-2):36-46.

Georgelis, N., Tabuchi, A., NIkolaidis, N. and Cosgrove, D. J. (2011). Structure-function analysis of the bacterial expansin EXLX1. *J. Biol. Chem.* 286(19):16814-16823.

Giuberti, G., Gallo, A., Masoero, F., Ferraretto, L. F., Hoffman, P. C. and Shaver, R. D. (2014). Factors affecting starch utilization in large animal food production system: A review. *Starch* 66(1-2):72-90.

Gonzalez, C. F., Proudfoot, M., Brown, G., Korniyenko, Y., Mori, H., Savchenko, A. V. and Yakunin, A. F. (2006). Molecular basis of formaldehyde detoxification: Characterization of two S-formylglutathione hydrolases from *Escherichia coli*, FrmB and YeiG. *J. Biol. Chem.* 281(20):14514-14522.

Green, B. J. and Beezhold, D. H. (2011). Industrial fungal enzymes: An occupational allergen perspective. *J. Allerg.* 2011:682574.

Hajji, M., Kanoun, S., Nasri, M. and Gharsallah, N. (2007). Purification and characterization of an alkaline serine-protease produced by a new isolated *Aspergillus clavatus* ES1. *Process Biochem.* 42(5):791-797.

Harrison, G. A. and Tricarico, J. M. (2007). Effects of an *Aspergillus oryzae* extract containing s amylase activity on lactational performance in commercial dairy herds. *Prof. Anim. Scient* 23(3):291-294.

Hatfield, R. D., Ralph, J. and Grabber, J. H. (1999). Review: Cell-wall cross-linking by ferulates and diferulates in grasses. *J. Sci. Food Agric.* 79(3):403-407.

Headon, D. R. and Walsh, G. (1994) The industrial production of enzymes. *Biotechnol. Adv.* 12(4):635-646.

Herrero, M., Grace, D., Njuki, J., Johnson, N., Nahoro, D. E., Silvestri, S. and Rufino, M. C. (2013). The roles of livestock in developing countries. *Animal* 7(Suppl. 1):3-18.

Hess, M., Sczyrba, A., Egan, R., Kim, T. W., Chokhawala, H., Schroth, G., Luo, S., Clark, D. S., Chen, F., Zhang, T., Mackie, R. I., Pennacchio, L. A., Tringe, S. G., Visel, A., Woyke, T., Wang, Z. and Rubin, E. M. (2011). Metagenomic discovery of biomass-degrading genes and genomes from cow rumen. *Science* 331(6016):463-467.

Hoffman, P. C., Esser, N. M., Shaver, R. D., Coblentz, W. K., Scott, M. P., Bodnar, A. L., Schmidt, R. J. and Charley, R. C. (2011). Influence of ensiling time and inoculation on alteration of the starch-protein matrix in high-moisture corn. *J. Dairy Sci.* 94(5):2465-2474.

Holtshausen, L., Chung, Y. H., Gerardo-Cuervo, H., Oba, M. and Beauchemin, K. A. (2011). Improved milk production efficiency in early lactation dairy cattle with dietary addition of a developmental fibrolytic EFE additive. *J. Dairy Sci.* 94(2):899-907.

Howson, S. J. and Davis, R. P. (1983). Production of phytate-hydrolysing enzyme by some fungi. *Enzyme Microb. Technol.* 5(5):377-382.

Hristov, A. N., McAllister, T. A. and Cheng, K. J. (1998). Effect of dietary or abomasal supplementation of exogenous polysaccharide-degrading enzymes on rumen fermentation and nutrient digestibility. *J. Anim. Sci.* 76(12):3146-3156.

Hristov, A. N., McAllister, T. A. and Cheng, K. J. (2000). Intraruminal supplementation with increasing levels of exogenous polysaccharide-degrading enzymes: Effects on nutrient digestion in cattle fed and barley grain diet. *J. Anim. Sci.* 78(2):477-487.

Hristov, A. N., Rode, L. M., Beauchemin, K. A. and Wuerfel, R. L. (1996). Effect of a commercial enzyme preparation on barley silage in vitro and in sacco dry matter degradability. *Proc. Western Sect. Meet. Am. Soc. Anim. Sci.* 47:282-284.

Huntington, G. B. J. (1997). Starch utilization by ruminants: From basics to the bunk. *J. Anim. Sci.* 75(3):852–867.

Iiyama, K., Lam, T. B. T. and Stone, B. A. (1994). Covalent cross-links in the cell-wall. *Plant Physiol.* 104(2):315–320.

Jung, H.-J. G., Jorgensen, M. A., Linn, J. G. and Engels, F. M. (2000). Impact of accessibility and chemical composition on cell-wall polysaccharide degradability of maize and lucerne stems. *J. Sci. Food Agric.* 80(3):419–427.

Jung, S. and Vogel, K. (2008). *Determination of Ronozyme Rumi-Star Alpha-Amylase Activity in Feed and Per Se Samples.* Regulatory Report No. 2500706. Basel, Switzerland: DSM Nutritional Products Ltd.

Kim, E. S., Lee, H. J., Bang, W. G., Choi, I. G. and Kim, K. H. (2009). Functional characterization of a bacterial expansin from *Bacillus subtilis* for enhanced enzymatic hydrolysis of cellulose. *Biotechnol. Bioeng.* 102(5):1342–1353.

Klingerman, C. M., Hu, W., McDonell, E. E., DerBedrosian, M. C. and Kung, L., Jr. (2009). An evaluation of exogenous enzymes with amylolytic activity for dairy cows. *J. Dairy Sci.* 92(3):1050–1059.

Klyosov, A. A. (1990). Trends in biochemistry and enzymology of cellulose degradation. *Biochemistry* 29(47):10577–10585.

Krause, D. O., Denman, S. E., Mackie, R. I., Morrison, M., Rae, A. L., Attwood, G. T. and McSweeney, C. S. (2003). Opportunities to improve fiber digestion in the rumen: Microbiology, ecology, and genomics. *FEMS Microbiol. Rev.* 27(5):663–693.

Krause, M., Beauchemin, K. A., Rode, L. M., Farr, B. I. and Norgaard, P. (1998). Fibrolytic enzyme treatment of barley grain and source of forage in high-grain diets fed to growing cattle. *J. Anim. Sci.* 76(11):2912–2920.

Krueger, N. A. and Adesogan, A. T. (2008). Effects of different mixtures of fibrolytic enzymes on digestion and fermentation of bahiagrass hay. *Anim. Feed Sci. Technol.* 145(1–4):84–94.

Krueger, N. A., Adesogan, A. T., Staples, C. R., Krueger, W. K., Dean, D. B. and Littell, R. C. (2008). The potential to increase digestibility of tropical grasses with a fungal, ferulic acid esterase enzyme preparation. *Anim. Feed Sci. Technol.* 145(1–4):95–108.

Kung, L., Jr., Treacher, R. J., Nauman, G. A., Smagala, A. M., Endres, K. M. and Cohen, M. A. (2000). The effect of treating forages with fibrolytic EFEs on its nutritive value and lactation performance of dairy cows. *J. Dairy Sci.* 83(1):115–122.

Kung, L., Jr., Tung, R. S., Maciorowski, K. G., Buffum, K., Knutsen, K. and Aimutis, W. R. (1991). Effects of plant cell-wall degrading enzymes and lactic acid bacteria on silage fermentation and composition. *J. Dairy Sci.* 74(12):4284–4296.

Kung, L., Jr., Windle, M. C. and Walker, N. (2014). The effect of an exogenous protease on the fermentation and nutritive value of high moisture corn. *J. Dairy Sci.* 97(3):1707–1712.

Kunitz, M. (1947). Crystalline soybean trypsin inhibitor: II. General properties. *J. Gen. Physiol.* 30(4):291–310.

Landry, J., Delhaye, S. and Damerval, C. (2000). Improved method for isolating and quantitating α-amino nitrogen as nonprotein, true protein, salt-soluble proteins, zeins, and true glutelins in maize endosperm. *Cereal Chem.* 77:620–626.

Lascano, C. E., Carulla, J. E. and Vargas, J. J. (2011). Strategies for reducing methane emissions from ruminants. *Rev. Bras. Geog. Fis.* 6:1315–1335.

Lewis, G. E., Sanchez, W. K., Hunt, C. W., Guy, M. A., Pritchard, G. T., Swanson, B. I. and Treacher, R. J. (1999). Effect of direct-fed fibrolytic EFEs on the lactational performance of dairy cows. *J. Dairy Sci.* 82(3):611–617.

Lin, Y., Means, G. E. and Feeney, R. E. (1969). Action of proteolytic enzymes on N, N-dimethyl proteins: Basis for a micro-assay for proteolytic enzymes. *J. Biol. Chem.* 244(3):789 –793.

Liu, X., Ma, Y. and Zhang, M. (2015). Research advances in expansins and expansion-like proteins involved in lignocellulose degradation. *Biotechnol. Lett.* 37(8):1541–1551.

Martinez, A. T., Speranza, M., Ruiz-Duenas, F. J., Ferreira, P., Camarero, S., Guillen, F., Martinez, M. J., Gutierrez, A. and del Rio, J. C. (2005). Biodegradation of lignocellulosics: Microbial, chemical, and enzymatic aspects of the fungal attack of lignin. *Int. Microbiol.* 8(3):195–204.

Mastihuba, V., Kremnický, L., Mastihubová, M., Willett, J. L. and Côté, G. L. (2002). A spectrophotometric assay for feruloyl esterases. *Anal. Biochem.* 309(1):96–101.

Maurer, K. H. (2004). Detergent proteases. *Curr. Opin. Biotechnol.* 15(4):330–334.

McAllister, T. A., Hristov, A. N., Beauchemin, K. A., Rode, L. M. and Cheng, K. J. (2001). Enzymes in ruminant diets. In Bedford, M. R. and Partridge, G. G. (Ed.), *Enzymes in Farm Animal Nutrition*. Oxford, UK: CABI Publishing, 273–298.

McAllister, T. A., Phillippe, R. C., Rode, L. M. and Cheng, K. J. (1993). Effect of the protein matrix on the digestion of cereal grains by ruminal microorganisms. *J. Anim. Sci.* 71(1):205–212.

Meale, S. J., Beauchemin, K. A., Hristov, A. N., Chaves, A. V. and McAllister, T. A. (2014). Board-invited review: Opportunities and challenges in using exogenous enzymes to improve ruminant production. *J. Anim. Sci.* 92(2):427–442.

Millar, G. L. (1959). Use of dinitrosalicylic acid reagent for determination of reducing sugar. *Anal. Chem.* 31(3):426–428.

Mohamed, D. E., Borhami, B. E., El-Shazly, K. A. and Sallam, S. M. A. (2013). Effect of dietary supplementation with fibrolytic enzymes on the productive performance of early lactation dairy cows. *J. Agric. Sci.* 6:146–155.

Mora-Jaimes, G., Barcena-Gama, R., Mendoza-Martínez, G., González-Muñoz, S. and Herrera-Haro, J. (2002). Performance and ruminal fermentation in lambs fed sorghum grain treated with amylases. *Agrociencia (Montecillo)* 36(1):31–39.

Morgavi, D. P., Beauchemin, K. A., Nsereko, V. L., Rode, L. M., Iwaasa, A. D., Yang, W. Z., McAllister, T. A. and Wang, Y. (2000). Synergy between ruminal fibrolytic enzymes and enzymes from *Trichoderma longibrachiatum*. *J. Dairy Sci.* 83(6):1310–1321.

Morgavi, D. P., Beauchemin, K. A., Nsereko, V. L., Rode, L. M., McAllister, T. A., Iwaasa, A. D., Wang, Y. and Yang, W. Z. (2001). Resistance of feed enzymes to proteolytic inactivation by rumen microorganisms and gastrointestinal proteases. *J. Anim. Sci.* 79(6):1621–1630.

Nakamura, T., Maeda, Y., Tanoue, N., Makita, T., Kato, M. and Kobayashi, T. (2006). Expression profile of amylolytic genes in *Aspergillus nidulans*. *Biosci. Biotechnol. Biochem.* 70(10):2363–2370.

Nelson, N. (1944). A photometric adaption of the Somogyi method for the determination of glucose. *J. Biol. Chem.* 153(2):375–380.

Nsereko, V. L., Morgavi, D. P., Rode, L. M., Beauchemin, K. A. and McAllister, T. A. (2000). Effect of fungal enzyme preparations on hydrolysis and subsequent degradation of alfalfa hay fiber by mixed rumen microorganisms in vitro. *Anim. Feed Sci. Technol.* 88(3–4):153–170.

Owens, F. N., Zinn, R. A. and Kim, Y. K. (1986). Limits to starch digestion in the ruminant small intestine. *J. Anim. Sci.* 63(5):1634–1648.

Papagianni, M. (2004). Fungal morphology and metabolite production in submerged mycelial processes. *Biotechnol. Adv.* 22(3):189–259.

Pazur, J. H., Cepure, A., Okada, S. and Forsberg, L. S. (1977). Comparison of the action of glucoamylase and glucosyltransferase on D-glucose, maltose, and malto-ligosaccharides. *Carbohydr. Res.* 58(1):193–202.

Pech-Cervantes, A. A., Muhammad, I., Ogunade, I. M., Jiang, Y., Kim, D. H., Gonzalez, C. F., Hackmann, T. J., Oliveira, A. S., Vyas, D. and Adesogan, A. T. (2019). Exogenous fibrolytic enzymes and recombinant bacterial expansins synergistically improve hydrolysis and in vitro digestibility of Bermudagrass haylage. *J. Dairy Sci.* 102(9):8059–8073.

Pendleton, B. (2000). The regulatory environment. In Muirhead, S. and Minnetonka, M. N. (Eds.), *Direct-Fed Microbial, Enzyme and Forage Additive Compendium*. The Miller Publishing Company, 49–55.

Peters, A., Meyer, U. and Danicke, S. (2015). Effect of exogenous fibrolytic enzymes on performance and blood profile in early and mid-lactation Holstein cows. *Anim. Nutr.* 1(3):229–238.

Philipps, W. P. (2018). Proteases-animal feed. In Simões, N. C. and Kumar, V. (Eds.), *Enzymes in Human and Animal Nutrition*. Cambridge, MA: Academic Press, 279–297.

Pittaluga, A. M., Yu, S., Li, W. and McCann, J. C. (2021). Effect of exogenous glucoamylase inclusion on in vitro fermentation and growth performance of feedlot steers fed a dry-rolled corn-based diet. *Transl. Anim. Sci.* 5(2):txab082.

Rao, M. B., Tanksale, A. M., Ghatge, M. S. and Deshpande, V. V. (1998). Molecular and biotechnological aspects of microbial proteases. *Microbiol. Mol. Biol. Rev.* 62(3):597–635.

Ribeiro, G. O., Gruninger, R. J., Badhan, A. and McAllister, T. A. (2016). Mining the rumen for fibrolytic feed enzymes. *Anim. Front.* 6(2):20–26.

Rojo, R., Mendoza, G. D., González, S. S., Landois, L., Bárcena, R. and Crosby, M. M. (2005). Effects of exogenous amylases from Bacillus licheniformis and *Aspergillus niger* on ruminal starch digestion and lamb performance. *Anim. Feed Sci. Technol.* 123–124:655–665.

Rojo-Rubio, R., Mendoza-Martinez, G. and Crosby-Galvan, M. (2001). Use of thermostable amylase from Bacillus licheniformis on in vitro starch digestion of sorghum and corn. *Agrociencia* 35:423–427.

Romero, J. J., Zarate, M. A., Arriola, K. G., Gonzalez, C. F., Silva-Sanchez, C., Staples, C. R. and Adesogan, A. T. (2015). Screening exogenous fibrolytic enzyme preparations for improved in vitro digestibility of Bermudagrass haylage. *J. Dairy Sci.* 98(4):2555–2567.

Romero, J. J., Zarate, M. A., Queiroz, O. C. M., Han, J. H., Shin, J. H., Staples, C. R., Brown, W. F. and Adesogan, A. T. (2013). Fibrolytic enzyme and ammonia application effects on the nutritive value, intake, and digestion kinetics of Bermudagrass hay in beef cattle. *J. Anim. Sci.* 91(9):4345–4356.

Rouches, E., Herpoël-Gimbert, I., Steyer, J. P. and Carrere, H. (2016). Improvement of anaerobic degradation by white-rot fungi pretreatment of lignocellulosic biomass: A review. *Renew. Sustain. Energy Rev.* 59:179–198.

Salem, A. Z. M., Gado, H. M., Colombatto, D. and Elghandour, M. M. Y. (2013). Effects of exogenous enzymes on nutrient digestibility, ruminal fermentation and growth performance in beef steers. *Livest. Sci.* 154(1–3):69–73.

Sauer, J., Sigurskjold, B. W., Christensen, U., Frandsen, T. P., Mirgorodskaya, E., Harrison, M., Roepstorff, P. and Svensson, B. (2000). Glucoamylase: Structure/function relationships, and protein engineering. *Biochim. Biophys. Acta* 1543(2):275–293.

Shahzad, F., Abdullah, M., Chaudhry, A. S., Bhatti, J. A., Jabbar, M. A., Ahmed, F., Mehmood, T., Asim, M., Ahmed, S., Kamran, Z., Irshad, I. and Tahir, M. N. (2016). Effects of varying levels of fungal (*Arachniotus* sp.) treated wheat straw as an ingredient of total mixed ration on growth performance and nutrient digestibility in Nili Ravi buffalo calves. *Asian-Australas. J. Anim. Sci.* 29(3):359–364.

Sheperd, A. C. and Kung, L., Jr. (1996). An enzyme additive for corn silage: Effects on silage composition and animal performance. *J. Dairy Sci.* 79(10):1760–1766.

Shrivastava, B., Jain, K. K., Kalra, A. and Kuhad, R. C. (2014). Bioprocessing of wheat straw into nutritionally rich and digested cattle feed. *Sci. Rep.* 4:6360.

Song, H. T., Gao, Y., Yang, Y. M., Xiao, W. J., Liu, S. H., Xia, W. C., Liu, Z. L., Yi, L. and Jiang, Z. B. (2016). Synergistic effect of cellulase and xylanase during hydrolysis of natural lignocellulosic substrates. *Bioresour. Technol.* 219:710–715.

Souza, P. M. and de Oliveira Magalhães, P. (2010). Application of microbial α-amylase in industry - A review. *Braz. J. Microbiol.* 41(4):850–861.

Stam, M. R., Danchin, E. G., Rancurel, C., Coutinho, P. M. and Henrissat, B. (2006). Dividing the large glycoside hydrolase family 13 into subfamilies: Towards improved functional annotations of α-amylase-related proteins. *Protein Eng. Des. Sel.* 19(12):555–562.

Sucu, E., Nayeri, A., Sanz-Fernandez, M. V., Upah, N. C. and Baumgard, L. H. (2014). The effects of supplemental protease enzymes on production variables in lactating Holstein cows. *Ital. J. Anim. Sci.* 13(2):348–351.

Tirado-Gonzalez, D. N., Miranda-Romero, L. A., Ruiz-Flores, A., Medina-Cuellar, S. E., Ramirez-Velarde, R. and Tirado-Estrada, G. (2018). Meta-analysis: Effect of exogenous fibrolytic enzymes in ruminant diets. *J. Appl. Anim. Res.* 46(1):771–783.

Tricarico, J. M., Abney, M. D., Galyean, M. L., Rivera, J. D., Hanson, K. C., McLeod, K. R. and Harmon, D. L. (2007). Effects of a dietary Aspergillus oryzae extract containing α-amylase activity on performance and carcass characteristics of finishing beef cattle. *J. Anim. Sci.* 85(3):802–811.

Tricarico, J. M., Johnston, J. D., Dawson, K. A., Hanson, K. C., McLeod, K. R. and Harmon, D. L. (2005). The effects of an Aspergillus oryzae extract containing alpha-amylase activity on ruminal fermentation and milk production in lactating Holstein cows. *Anim. Sci.* 81(3):365–374.

Van der Kaaij, R. M., Janeček, Š., van der Maarel, M. J. E. C. and Dijkhuizen, L. (2007). Phylogenetic and biochemical characterization of a novel cluster of intracellular fungal α-amylase enzymes. *Microbiology (Reading)* 153(12):4003–4015.

van der Maarel, M. J., van der Veen, B., Uitdehaag, J. C., Leemhuis, H. and Dijkhuizen, L. (2002). Properties and applications of starch-converting enzymes of the alpha-amylase family. *J. Biotechnol.* 94(2):137–155.

van Kuijk, S. J. A., Sonnenberg, A. S. M., Baars, J. J. P., Hendriks, W. H. and Cone, J. W. (2015). Fungal treated lignocellulosic biomass as ruminant feed ingredient: A review. *Biotechnol. Adv.* 33(1):191–202.

Vera, J. M., Smith, A. H., ZoBell, D. R., Young, A. J. and Eun, J.-S. (2012). Effects of an exogenous proteolytic enzyme on growth performance of beef steers and in vitro ruminal fermentation in continuous culture. *Prof. Anim. Sci.* 28(4):452–463.

Vicini, J. L., Bateman, H. G., Bhat, M. K., Clark, J. H., Erdman, R. A., Phipps, R. H., Van Amburgh, M. E., Hartnell, G. F., Hintz, R. L. and Hard, D. L. (2003). Effect of feeding supplemental fibrolytic enzymes or soluble sugars with malic acid on milk production. *J. Dairy Sci.* 86(2):576–585.

Vihinen, M. and Mantsiila, P. (1989). Microbial amylolytic enzymes. *Crit. Rev. Biochem. Mol. Biol.* 24(4):329–418.

Vrsanska, M., Voberkova, S., Langer, V., Palovcikova, D., Moulick, A., Adam, V. and Kopel, P. (2016). Induction of laccase, lignin peroxidase and manganese peroxidase activities in white-rot fungi using copper complexes. *Molecules* 21(11):1553.

Wang, Y., Tang, R., Tao, J., Gao, G., Wang, X., Mu, Y. and Feng, Y. (2011). Quantitative investigation of non-hydrolytic disruptive activity on crystalline cellulose and application to recombinant swollenin. *Appl. Microbiol. Biotechnol.* 91(5):1353–1363.

Weiss, W. P., Steinberg, W. and Engstrom, M. A. (2011). Milk production and nutrient digestibility by dairy cows when fed exogenous amylase with coarsely ground dry corn. *J. Dairy Sci.* 94(5):2492–2499.

White, B. A., Mackie, R. I. and Doerner, K. C. (1993). Enzymatic hydrolysis of forage cell-walls. In Jung, H. G., Buxton, D. R., Hatfield, R. D. and Ralph, J. (Eds.), *Forage Cell-Wall Structure and Digestibility*. Madison, WI: CSSA, ASA. SSSA, 455–484.

Windle, M. C., Walker, N. and Kung, L., Jr. (2014). Effects of an exogenous protease on the fermentation and nutritive value of corn silage harvested at different dry matter contents and ensiled for various lengths of time. *J. Dairy Sci.* 97(5):3053–3060.

Wood, T. M. and Bhat, M. K. (1988). Methods for measuring cellulase activities. In Wood, W. A. and Kellogg, S. T. (Eds.), *Methods Enzymol.* London, UK: Academic Press Inc. 160:87–112.

Wood, T. M., McCrae, S. I. and Bhat, K. M. (1989). The mechanism of fungal cellulose action. Synergism between enzyme components of *Penicillium pinophilum* cellulase in solubilizing hydrogen bond ordered cellulose. *Biochem. J.* 260(1):37–43.

Xiao, Z., Storms, R. and Tsang, A. (2006). A quantitative starch-iodine method for measuring alpha-amylase and glucoamylase activities. *Anal. Biochem.* 351(1):146–148.

Yoon, S., Devaiah, S. P., Choi, S. E., Bray, J., Love, R., Lane, J., Drees, C., Howard, J. H. and Hood, E. E. (2016). Over-expression of the cucumber expansin gene (Cs-EXPA1) in transgenic maize seed for cellulose deconstruction. *Transgenic Res* 25:173–186.

Young, K. M., Lim, J. M., Der Bedrosian, M. C. and Kung, L. (2012). Effect of exogenous protease enzymes on the fermentation and nutritive value of corn silage. *J. Dairy Sci.* 95(11):6687–6694.

Yu, P., McKinnon, J. J. and Christensen, D. A. (2005). Improving the nutritional value of oat hulls for ruminant animals with pretreatment of a multienzyme cocktail: In vitro studies. *J. Anim. Sci.* 83(5):1133–1141.

Yuan, X. L., van der Kaaij, R. M., van den Hondel, C. A., Punt, P. J., van der Maarel, M. J., Dijkhuizen, L. and Ram, A. F. (2008). Aspergillus niger genome-wide analysis reveals a large number of novel alpha-glucan acting enzymes with unexpected expression profiles. *Mol. Genet. Genomics* 279(6):545–561.

Chapter 9

Amino acids in dairy nutrition: enhancing milk protein synthesis and beyond

Juan J. Loor, University of Illinois, USA

1 Introduction

Recognition that 'crude protein' (CP) was an important nutrient in diet formulation for lactating dairy cows dates back more than 60 years (Schwab and Broderick, 2017). Although not routinely done, due to cost, the application of amino acid (AA) analysis in common feedstuffs has allowed for a better appreciation of the profiles of these nutrients reaching the ruminal environment (Table 1). An advantage of having AA profiles of feedstuffs is the ability to 'complement' feed sources to achieve the desired amount of CP and profile of AA in a given diet. Unless feedstuffs contain a high proportion of 'rumen undegradable protein' (RUP), provided there is sufficient energy from carbohydrate fermentation, the ruminal microflora can utilize the free AA produced during proteolysis (Bach et al., 2005) to synthesize 'microbial crude protein' (MCP) (Table 2). Although estimates vary across published studies, a range of 50–67% of total intestinal CP is contributed by the MCP fraction (Clark et al., 1992; Shabi et al., 2000). Clearly, the MCP fraction can have a substantial impact on the amount and profiles of AA that enter the lower gastrointestinal tract of the animal (Loosli et al., 1949; Thomas et al., 1951; Virtanen, 1966). Along with the RUP and undigested feed protein, these three fractions form the basis of the 'metabolizable protein'

http://dx.doi.org/10.19103/AS.2022.0117.09

Table 1 Amino acid composition in some common feedstuffs used in dairy cow diet formulations

Item	Corn silage	Grass silage	Alfalfa hay	Corn, ground	Soybean meal, solvent extracted	Soyhulls	Soybean meal, expeller	Raw soybeans, cracked	Blood meal, ring dried
DM, %	33.0	32.7	90.1	87.8	91.4	89.0	94.0	89.0	92.0
CP, % of DM	9.8	16.8	17.8	9.5	52.2	13.1	48.2	40.2	95.5
NDF, % of DM	43.7	54.6	44.9	8.6	12.1	62.0	30.0	27.8	-
Amino acid, % of CP									
Ala	7.96	6.43	4.10	7.68	4.02	4.03	3.92	4.38	6.95
Asp	4.80	7.26	11.46	7.47	10.57	8.75	10.25	11.32	8.77
Cys	1.53	1.55	0.28	2.21	1.15	1.73	1.12	1.12	0.02
Glu	9.69	7.02	10.11	18.63	16.59	10.88	15.73	17.51	8.07
Gly	3.27	4.05	5.06	4.00	3.85	7.71	3.78	4.18	3.92
Ser	2.86	3.10	4.89	4.74	4.41	4.51	4.40	4.80	4.59
Tyr	2.14	2.44	3.82	3.89	3.05	3.73	2.95	3.41	2.73
Arg	1.84	2.62	4.61	4.74	6.67	4.55	6.06	7.19	3.70
His	1.33	1.37	2.25	2.74	2.36	2.43	2.24	2.56	4.73
Ile	2.65	3.21	3.88	3.16	3.75	3.51	3.36	4.18	0.94
Leu	7.55	6.37	8.20	12.32	6.88	6.33	6.64	7.56	10.44
Lys	2.14	3.63	4.83	3.16	5.65	6.07	4.75	6.19	7.70
Met	1.02	0.95	0.84	1.79	0.92	1.13	0.89	1.09	1.48
Phe	2.86	3.63	5.06	4.84	4.35	3.64	4.13	4.80	5.87
Thr	2.65	3.27	4.55	3.68	3.58	3.42	3.57	3.88	4.41
Val	3.98	4.76	6.80	4.84	4.21	4.33	3.88	4.65	7.35

Source: Data from Shabi et al. (2000), Socha et al. (2005) and Ordway et al. (2009).

(MP) concept used in protein evaluation systems for dairy cows (Schwab and Broderick, 2017). It is clearly a more biologically relevant metric because dietary CP content does not provide information on the fate of the various protein fractions in a given diet.

The mixed bacteria and protozoa contribute the majority of 'microbial-derived' AA (Table 2) in the small intestine and along with RUP constitute the main sources of AA available for absorption. Through a combination of feeding studies and analysis of plasma and abomasal AA profiles in cattle and sheep, early researchers also were keen to recognize that the composite protein reaching the small intestine of ruminants was 'limited' in AA such as methionine (Met) and lysine (Lys) (Lofgreen et al., 1947). Early studies established that AA nitrogen ('true protein') as a proportion of total nitrogen in bacterial protein is not constant across diets (Weller et al., 1958); it averages ~70–80% of CP in bacteria and protozoa from cattle (Table 2), and in some cases dietary characteristics (e.g. CP level, starch level and neutral detergent fiber level) can affect some, but not all, AA profiles of microbes (Hvelplund, 1986; Mabjeesh et al., 1997). Diet evaluation models such as the NRC (2001) and the Cornell Net Carbohydrate and Protein System (CNCPS) (Fox et al., 2004) in the United States and INRAtion (Rulquin et al., 1998) in France use different approaches to predict the duodenal flows of AA. Although the main goal of this chapter is not to discuss the adequacy of available models in the context of AA nutrition of dairy cows, our current protein systems still do not predict accurately duodenal AA or MP flow (Sok et al., 2017), which constitutes a serious problem.

2 Reducing dietary crude protein in dairy cow diets

Increasing costs of feed protein and concerns about excess nitrogen excretion into the environment are two key issues driving the ongoing efforts to reduce CP levels in dairy diets. Readers are directed to the excellent reviews from Sinclair et al. (2014) and Broderick (2018) discussing, in detail, various aspects related to optimizing dietary CP in lactating cow diets and also approaches to optimize the conversion of feed CP to a human food protein. The fact that lactating cows can only capture ~25–35% of dietary CP and secrete it in milk has been known for decades (Broderick, 2003). A meta-analysis of the effects of dietary CP concentration and degradability on milk protein yield, and efficiency of utilization of dietary nitrogen (N) for milk protein synthesis, concluded that the CP concentration of the diet is the most important dietary factor influencing milk N efficiency (Huhtanen and Hristov, 2009). Reducing dietary CP is the most significant means by which to increase the efficiency of dietary protein utilization (Sinclair et al., 2014; Broderick, 2018). Some of the recommendations

Table 2 Means and standard errors of amino acid (AA) composition (g AA/100 g total AA) in 22 ruminal bacteria and two protozoal species cultured in vitro (Purser and Buechler, 1966), mixed ruminal bacteria from lactating cows (Shabi et al., 2000), or cattle (Sok et al., 2017; Gresner et al., 2021)[a]

	Bacteria				Protozoa				
		Rumen			Cultures		Rumen		
Amino acid (AA)	Cultures	Lactating cows[b]	Cattle/sheep[c]	Cattle/sheep[d]	Entodinium	Isotricha	Lactating cows[b]	Cattle/sheep[c]	Cattle/sheep[d]
Ala	6.5 ± 0.7	6.6 ± 0.2	7.2 ± 0.1	6.7 ± 0.1	4.6	4.1	3.9 ± 0.1	4.2 ± 0.1	4.2 ± 0.1
Arg	5.4 ± 0.6	4.6 ± 0.1	4.6 ± 0.1	5.0 ± 0.1	7.6	4.3	4.2 ± 0.1	4.5 ± 0.1	4.5 ± 0.1
Asp	11.1 ± 1.3	10.9 ± 0.6	11.9 ± 0.2	11.9 ± 0.2	12.4	16.1	13.3 ± 0.2	13.4 ± 0.2	12.6 ± 0.2
Cys	1.0 ± 0.7	-	1.4 ± 0.3	1.8 ± 0.3	1.5	2.4	-	1.9 ± 0.3	1.7 ± 0.3
Glu	11.9 ± 2.4	13.5 ± 1.0	12.8 ± 0.2	13.3 ± 0.2	15.7	18.0	15.9 ± 0.5	14.4 ± 0.2	14.3 ± 0.2
Gly	6.1 ± 0.7	5.5 ± 0.2	5.6 ± 0.1	5.0 ± 0.1	4.4	4.1	4.2 ± 0.1	4.3 ± 0.1	4.3 ± 0.1
Ser	3.8 ± 0.9	4.5 ± 0.1	4.5 ± 0.1	4.6 ± 0.1	6.2	5.4	4.4 ± 0.1	4.1 ± 0.1	4.2 ± 0.1
Tyr	4.2 ± 0.2	5.0 ± 0.2	5.2 ± 0.2	5.3 ± 0.2	4.9	4.1	4.5 ± 0.1	4.9 ± 0.2	4.8 ± 0.2
His	2.3 ± 0.3	0.5 ± 0.3	1.9 ± 0.1	2.0 ± 0.1	3.2	4.3	1.4 ± 0.3	1.8 ± 0.1	1.8 ± 0.1
Ile	6.4 ± 0.7	6.1 ± 0.2	5.5 ± 0.1	5.4 ± 0.1	5.8	5.7	6.7 ± 0.2	6.5 ± 0.1	6.3 ± 0.1
Leu	7.3 ± 1.2	6.9 ± 0.3	7.6 ± 0.1	7.7 ± 0.1	8.1	7.3	7.3 ± 0.1	7.9 ± 0.1	7.6 ± 0.1
Lys	9.3 ± 1.7	9.0 ± 0.3	7.6 ± 0.2	7.6 ± 0.2	13.1	9.2	11.9 ± 0.2	10.9 ± 0.2	10.4 ± 0.2
Met	2.6 ± 0.5	2.7 ± 0.1	2.4 ± 0.1	2.3 ± 0.1	1.4	2.0	2.4 ± 0.1	2.2 ± 0.1	2.0 ± 0.1
Phe	5.1 ± 0.4	7.7 ± 0.9	5.2 ± 0.1	5.4 ± 0.1	6.1	3.8	6.3 ± 0.2	5.5 ± 0.1	5.8 ± 0.1
Pro	-	3.6 ± 0.1	3.6 ± 0.1	3.6 ± 0.1	-	-	3.4 ± 0.1	3.3 ± 0.1	3.5 ± 0.1
Thr	5.5 ± 0.8	5.5 ± 0.1	5.6 ± 0.1	5.4 ± 0.1	5.8	5.2	4.7 ± 0.1	5.0 ± 0.1	5.0 ± 0.1
Trp	-	-	1.3 ± 0.1	1.4 ± 0.1	-	-	-	0.8 ± 0.1	1.9 ± 0.1
Val	6.6 ± 1.7	7.1 ± 0.2	5.9 ± 0.1	5.7 ± 0.1	4.6	4.0	5.5 ± 0.2	4.9 ± 0.1	5.2 ± 0.1

	Bacteria				Protozoa				
		Rumen			Cultures		Rumen		
Amino acid (AA)	Cultures	Lactating cows[b]	Cattle/sheep[c]	Cattle/sheep[d]	Entodinium	Isotricha	Lactating cows[b]	Cattle/sheep[c]	Cattle/sheep[d]
Mean AA nitrogen, % of total	–	–	–	79.0 ± 1.7	–	–	–	–	70.6 ± 3.2
Total AA abomasal flow, kg/d	–	0.60 ± 0.20	–	–	–	–	0.28 ± 0.12	–	–
Total CP abomasal flow, kg/d	–	0.92 ± 0.29	–	–	–	–	0.31 ± 0.15	–	–

[a]Database from Sok et al. (2017) encompassing 22 papers with data for fluid-associated bacteria, 7 papers with data for particle-associated bacteria and 8 papers with data for protozoa. Database form Gresner et al. (2021) focused on published data dealing with Central European diets and a maximum proportion of concentrate of 60% of total diet dry matter.
[b]Days in milk: 174 ± 29 days. Diet was based on corn silage (36% of diet DM), cracked corn or expanded corn (36% of diet DM) and soybean meal (10% of diet DM). Dry matter intake averaged 16.2 ± 0.6 kg/d. There were no treatment effects on production variables, AA profiles or AA flows to the duodenum (Shabi et al., 2000).
[c]Data for fluid-associated bacteria only reported in Sok et al. (2017). Differences in individual AA profiles were minor when comparing fluid and particle-associated bacteria.
[d]Data for total bacteria from Gresner et al. (2021).

by Broderick (2018) to improve the yield of milk and milk protein by dairy cows include the following:

- Group-feeding dairy cows according to production, and feeding diets higher in RUP;
- Supplementing rumen-protected (RP) essential amino acids (EAA);
- Employing nutritional models to more accurately meet animal requirements; and
- Developing nutritional strategies to maximize milk component yield rather than milk volume.

Although a thorough discussion of each of these recommendations is beyond the scope of this chapter, it should be noted that feeding higher RUP as a way to increase MP supply even in early lactation when EAA needs are the greatest decreases milk N efficiency (Amanlou et al., 2017; Tebbe and Weiss, 2021; Zang et al., 2021a). From a production and health standpoint, however, the first 3-4 weeks postpartum might be a period during which higher RUP (e.g. 5-9% of diet dry matter [DM]) could help optimize milk yield and DMI, potentially reducing the risk of cows to develop a metabolic disorder (Amanlou et al., 2017; Tebbe and Weiss, 2021; Zang et al., 2021a). In the context of dietary RUP, even when diets are formulated to provide adequate levels of MP, the choice of feedstuffs to increase RUP also could impact production outcomes and milk N efficiency (Tebbe and Weiss, 2021). For example, using a blend of RUP and RPEAA to increase MP supply to adequate levels for cows during the first 25 days postpartum resulted in greater DMI and greater energy-corrected milk (ECM) yield during the ~10 weeks following the end of the treatment period (Tebbe and Weiss, 2021).

The focus on the incorporation of RP AA (RPAA) or 'analogues' in dairy cow diets is particularly important because of the recognition long ago (1960s through 1970s) (Polan et al., 1970; Broderick et al., 1970) that certain EAA such as Met and Lys are often deficient in most diets fed to high-producing dairy cows (e.g. Holstein) in the Americas, Europe and China (Schwab and Broderick, 2017). Work from Australian groups in sheep demonstrating improvements in wool production in response to post-ruminal supply of sulfur AA was a factor driving the interest in dairy cows (Jacobson et al., 1967). The study of Polan et al. (1970) was among the first to demonstrate a positive effect of enhancing post-ruminal supply of Met by feeding Met-hydroxy-analogue (MOH). That study is especially noteworthy because supplementation of MOH began 2 weeks prior to parturition and extended through the first 90 days of lactation; i.e. it encompassed the peripartal ('transition') period and the first stages of lactation when milk production is the highest. Broderick et al. (1970) reported the first use of RP Met (RPMet) in lactating dairy cows. Subsequent work from Schwab et al. (1976) confirmed that Met and Lys were first-limiting AA for cows fed corn-based diets. Besides Lys and Met, research in the late 1990s also identified histidine (His) as the first-limiting AA when cows were fed non-corn diets without protein supplements (Vanhatalo et al., 1999; Varvikko et al., 1999).

Although most of the work on EAA supply has been performed in lactating cows, i.e. with a focus on milk protein yield, the past 20–25 years have seen a gradual increase in research focused on examining the benefits of AA during the transition period (discussed in more depth in subsequent sections). The use of AA by the cow clearly is important for protein deposition, be it in muscle or milk, but it is evident that AAs are used by cells or tissues for a myriad of functions some of which have important implications in terms of health (Figs 1 and 2). Focus on the transition period stems largely from the recognition that EAA such as Met and Lys could impact metabolic processes beneficial to the cow in tissues such as the liver (Piepenbrink et al., 2004; Fehlberg et al., 2020). Knowledge that glutamine (Gln) is an essential AA during the immune response also led to in vivo research (Jafari et al., 2006) using intravenous infusion during the transition period and established that some liver-derived immune-related proteins (i.e. acute-phase proteins [APP]) responded to the increased supply of this AA. Subsequent research with isolated blood neutrophils indicated that, although Gln may be important during the immune response, availability of other AA is essential to furnish sources of carbon for energy metabolism, especially after parturition (Garcia et al., 2016). Clearly, the area of AA nutrition in dairy cows during the transition period is particularly important because, not only is the fetus still growing in the last weeks of pregnancy (Bell et al., 1995) but after parturition the need of AA for physiological processes other than milk protein synthesis is also increased (Coleman et al., 2020).

3 Protein feeding and amino acid balancing

The review from Patton et al. (2014) is an excellent source for detailed information on the various aspects related to balancing diets for AA in lactating dairy cows, and will not be covered in this chapter. Rather, it should be emphasized that the recent application of meta-analytical methods has allowed a clearer understanding of the associations among the post-ruminal supply of AA and milk protein yield, milk protein percentage and milk yield in lactating dairy cows (Vyas and Erdman, 2009; Sinclair et al., 2014; Lean et al., 2018). Such analyses also have been performed in studies where CP of the diet was considered 'low', i.e. ≤15% CP of DM, and the RPAA was a combination of RPMet and RPLys (Sinclair et al., 2014).

Among the EAA, a substantial body of published data supports the view that Met and Lys are the most limiting AA in dairy cow diets. In the 2001 version of *Nutrient Requirements of Dairy Cattle* (NRC, 2001), these are recommended in amounts of 7.2% of MP for Lys and 2.4% of MP for Met (NRC, 2001), i.e.

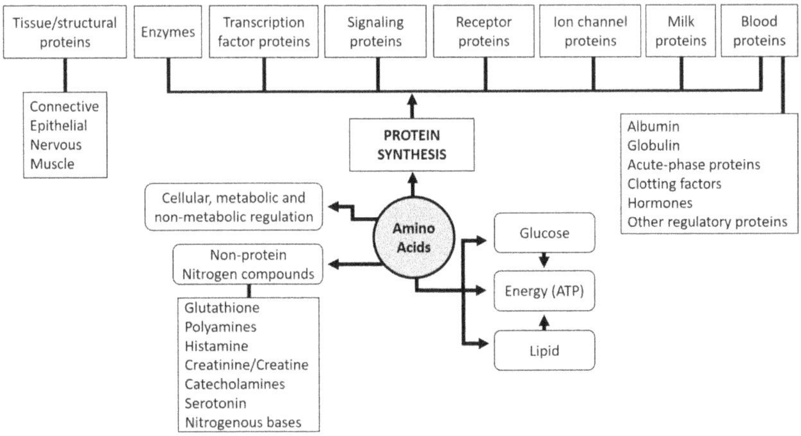

Figure 1 Schematic summary of the major fates of amino acids in animal cells and tissues.

a Lys:Met ratio in the MP should be achieved to ensure the greatest benefits in terms of milk protein yield. From an energy and AA nutrition standpoint, however, it should be noted that animals have the metabolic flexibility to utilize C-containing substrates, e.g. the C backbone of AA, when energy intake is low that could result in an inefficient use of AA as energy (Coleman et al., 2020). For this reason, it has been suggested to supply 3.03 g of Lys/Mcal of metabolizable energy (ME) and 1.14 g of Met/Mcal of ME to allow for adequate utilization of these AAs by dairy cows (Van Amburgh et al., 2015). Although this chapter does not attempt to discuss this topic in any depth, it is clear that deficiencies in Met and Lys can arise depending on limited and variable concentrations in common feedstuffs (Table 1). For instance, Lys concentrations are adequate in blood meal, less in soybean meal and the least in corn gluten meal. Similarly, Met concentrations are low in blood meal and soybean meal (Erasmus et al., 1994). In addition to providing high levels of Lys, blood meal is an adequate source of RUP for dairy cows; however, digestibility of Lys in blood meal is dependent on processing methodology. When subjected to heating, digestibility of Lys in blood meal decreases (Stein et al., 2007).

Because of the known variabilities across typical feedstuffs used to formulate diets for dairy cows, ensuring an adequate delivery of limiting AA to the small intestine was a major driver for the development of 'rumen protection' techniques, or techniques that provide some protection and still deliver the AA to the mammary gland - i.e. MOH: 2-hydroxy-4-(methylthio) butanoic acid (HMB or HMBTa) and the isopropyl ester of 2-hydroxy-4-(methylthio) butanoic acid (HMBi) (Fowler et al., 2015). Meta-analyses have confirmed that RPMet, HMB/HMBTa and HMBi fed to lactating cows are effective at increasing milk protein percentage, milk protein yield and often milk yield (Zanton et al., 2014; Patton,

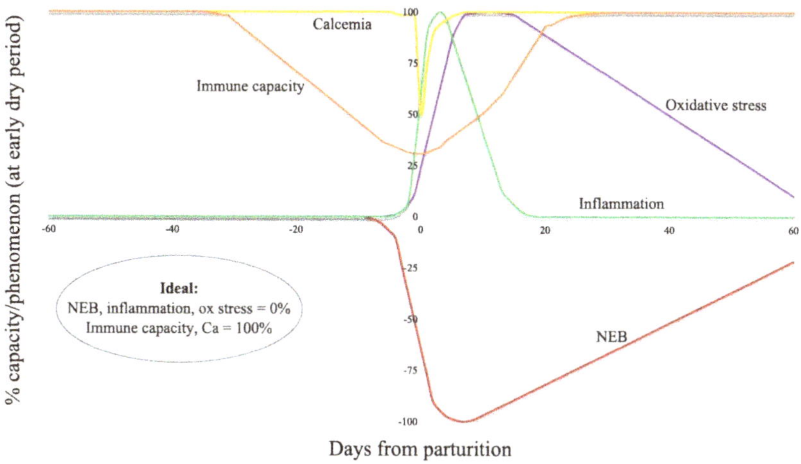

Figure 2 Theoretical profiles of major physiological events experienced by dairy cows around parturition. An ideal 'state' in terms of negative energy balance (NEB), inflammation and oxidative stress would in theory mean a value of zero and would signify the absence of a given phenomenon. In contrast, an ideal state for immune capacity and calcemia would be achieved close to 100% of the optimal level. Source: Lopreiato et al. (2020).

2010). A compound like HMB/HMBTa or HMBi is thought to be transported into the ruminal epithelium where enzymatic reactions generate free Met from the analogues (Phipps et al., 2008), but such mechanism has not been proven experimentally (Breves et al., 2010). It is speculated that a concentration gradient between mucosal and serosal sides of ruminal epithelium drives the transport of these compounds via passive diffusion (similar to VFA) or through the action of monocarboxylate transporters (e.g. MCT-1).

Regardless of dietary CP, using data from 23 published studies involving post-ruminal supplementation of Met and Lys, it was determined that milk protein yield responses to supplemental Met decreased from 16 g to 4 g of milk protein per gram of metabolizable Met intake in cows when Met intake increased from 25 g to 70 g of digestible Met per day. Similarly, milk protein responses to supplemental Lys decreased from 5.0 g to 3.2 g of milk protein per gram of metabolizable Lys in cows when Lys intake varied from 80 g to 203 g per cow per day (Vyas and Erdman, 2009). These data not only emphasized the need to fine-tune Met and Lys supply to the cow, which can be achieved through feeding RPMet or RPLys and knowledge of the degree of ruminal protection and intestinal bioavailability, but also that efficiency of Met and Ly use for milk protein yield is not constant, as some diet formulation models assume (Pacheco et al., 2012). The point about efficiency has been highlighted recently in a study utilizing jugular infusions of mixtures of Lys, Met, plus His (MKH); Ile plus Leu

(IL); or MKH plus IL (MKH+IL). This approach allowed enhancing the intestinal supply of these EAA in lactating cows fed a 15% CP diet with adequate RUP, but 15% deficient in estimated MP (Yoder et al., 2020). Among the key outcomes was the recognition that the mix of EAA reaching the small intestine determined the uptake and utilization of individual AA by the mammary gland. Further, the mix of EAA changed the efficiency of utilization of individual AA. Thus, this experiment offers further confirmation that using fixed and constant efficiencies across all AA is not appropriate when attempting to represent mammary utilization.

The recent meta-analysis of Lean et al. (2018) (without considering diet CP content) not only emphasized the positive association between metabolizable Met and milk protein yield and percentage, but also positive associations between metabolizable His, Leu, Trp and Thr with milk yield. In the context of these AAs, the potential for His to be a limiting or co-limiting AA has been recognized and studied since the late 1990s (Vanhatalo et al., 1999; Varvikko et al., 1999; Kim et al., 1999). Histidine was first identified as first-limiting AA in cows fed grass silage-based diets with low inclusion of plant protein supplements. Subsequent studies indicated that when ruminally derived microbial protein provides most of the MP to the cow, His may limit milk protein yield when corn silage- and alfalfa hay-based diets are fed (Lee et al., 2012b). Since concentrations of His in microbial protein are among the lowest relative to other EAA (Table 2), potential limitations on its supply to the small intestine are easy to appreciate. Subsequent studies confirmed that, in conditions where MP is deficient, enhanced supply of post-ruminal His in addition to Lys and Met can alleviate the reductions in milk and milk protein yield observed with low MP diets (Lee et al., 2012a; Zang et al., 2021b; Raisanen et al., 2021).

Overall, available data on EAA nutrition of lactating cows not only confirm that Met and Lys are first-limiting but also that His is potentially co-limiting across a wide range of dietary scenarios (Lean et al., 2018). As long as the balance of Met and Lys is close to predicted requirements, the available data indicate that responses to post-ruminal His supply are most pronounced when MP is deficient or when high-bypass proteins fed to cows are low in His (Zang et al., 2021b; Morris and Kononoff, 2020; Raisanen et al., 2021).

4 Physiological roles of essential amino acids: beyond milk protein

4.1 Synthesis

Historically, emphasis on EAA supply to dairy cows (i.e. Met and Lys) has centered on the premise that a shortfall compromises milk protein synthesis, which a number of studies have confirmed, particularly for cows in confinement and fed typical total-mixed rations (TMR) (Schwab and Broderick, 2017). The suggestion that an EAA such as Met plays a role in lipid metabolism was first proposed by McCarthy et al. (1968) in studies focused on the development of ketosis postpartum. The period around parturition (Fig. 2), 'periparturient' or 'transition' period', clearly is uniquely different than other stages of lactation, and in the context of AA metabolism, the mobilization of body protein in dairy cows close to calving compensates in part for the shortfall in the post-ruminal availability of EAA and non-essential AA (NEAA) as a result of decreased voluntary dry matter intake (DMI) (Komaragiri and Erdman, 1997). Thus, the supplementation of rumen-protected EAA has gradually been also deployed to the transition period (see review by Coleman et al., 2020). Because of the unique physiological adaptations that dairy cows undergo around parturition, the availability of EAA to tissues and cells other than the mammary gland is currently an important focus in the field (Coleman et al., 2020). The sections that follow attempt to discuss briefly the published data presenting evidence of a functional role for some of the EAA, including Met, Lys, branched-chain AA (BCAA), Trp and the NEAA Glu during the transition period (Tables 3 and 4).

During the transition from late-pregnancy into lactation, cows will normally experience an increase in adipose tissue lipolysis due to changes in hormones such as insulin (decrease) and growth hormone (increase) along with a decrease in voluntary DMI (Herbein et al., 1985) that renders the cow in a state of negative energy and protein balance (Fig. 2). Major physiological responses to this catabolic state include high concentrations of fatty acids and hydroxybutyrate in blood, both of which are interrelated through fatty acid oxidation and ketogenesis in the liver (Drackley, 1999). A major organelle within hepatocytes where fatty acid oxidation takes place is the mitochondria, and carnitine is essential for the transport of fatty acids from cytosol into mitochondria for subsequent β-oxidation (Drackley, 1999).

A transient state of inflammation, oxidative stress and immune dysfunction also afflicts cows during the transition period (Fig. 2) (Lopreiato et al., 2020). The inflammatory response is characterized by an increase in the hepatic production of positive APP, such as haptoglobin and serum amyloid A (SAA), and a concomitant decrease in the production of negative APP such as albumin (Lopreiato et al., 2020). At the level of the liver, the well-established triggers of these responses are the pro-inflammatory cytokines IL-6, IL-1β and TNF-α (Bertoni et al., 2009). In contrast, oxidative stress is driven by the imbalance between the production of reactive oxygen metabolites (ROM) and the neutralizing capacity of antioxidant mechanisms in tissues and in blood. Fatty acids and circulating oxylipids can trigger pro-inflammatory cytokines and subsequently cause oxidative stress, suggesting interrelationships among various factors (Sordillo and Mavangira, 2014).

The transient inflammatory status around parturition appears to be a 'normal' aspect of the adaptations to lactation (Bradford et al., 2015), with its positive or negative impact depending on the degree and length of the response (Fig. 2). Cows that approach parturition with a greater (but still subclinical) level of circulating cytokines experience greater inflammation and oxidative stress, and lower liver function often through the first month of lactation together with lower milk yield and lower postpartum DMI (Trevisi et al., 2015). In addition to their fundamental function in immunity, cytokines (ILs), interferons (IFNs) and TNF-α also elicit pathophysiological effects. This leads to what is commonly known as the 'sickness behavior', whose primary manifestation is satiety. Similar to how cows react during an inflammatory state around parturition, the reduction in DMI around calving is an example of this behavior. In mice, these cytokines can reduce meal size and duration, as well as decrease meal frequency and prolong inter-meal intervals (Plata-Salaman, 1995). Furthermore, cytokines directly affect the hypothalamus; IL-1β and IFN act directly and specifically on the glucose-sensitive neurons in the brain's 'satiety' and 'hunger' sites (Plata-Salaman, 1995).

The fact that both inflammation and oxidative stress reduce liver function in periparturient dairy cows has been recognized for some time (Bionaz et al., 2007; Trevisi et al., 2015). Thus, one important physiological target of AA nutrition in transition cows is to fine-tune the post-ruminal supply of those AAs that can serve as precursors of molecules such as carnitine, phosphatidylcholine (PC), antioxidants such as glutathione (GSH) and taurine, or AA such as Glu and Gln that can serve as fuels for the gut and other tissues (Fig. 3) (Coleman et al., 2020, 2021). Although most of the published data on the feeding of RPAA during the transition period pertains to RPMet (Table 3), studies with RPLys, RPBCAA, RPTrp and RPGlu are beginning to emerge. The sections that follow provide a brief discussion of the available data from transition cows organized by type of AA fed in RP form or as MOH.

5 Methionine, lysine and other essential and non-essential amino acids during the periparturient period

5.1 Methionine

Prior to the study of Overton et al. (1996), there were few published studies reporting the feeding of RPMet alone or in combination with RPLys during the transition period. Polan et al. (1970) were the first to feed MOH (at incremental rates) during the transition period and into mid-lactation, with positive effects reported for milk yield. The work of Piepenbrink et al. (2004) was the first that focused specifically on the link between the post-ruminal supply of Met via feeding MOH on various aspects of hepatic metabolism, including triacylglycerol (TAG) accumulation and fatty acid oxidation (Table 4).

Table 3 Summary of production performance results (relative changes from control cows: + positive effect, = no change, (−) negative effect; listed chronologically) in which the dietary supply of lysine alone, lysine plus methionine or branched-chain amino acids (BCAA) was enhanced by feeding supplements that conferred some degree of protection from ruminal degradation

Study	DMI[a]	Milk yield or ECM	Protein %	Fat %	Diet CP %	Lys and Met in the MP	Amount fed	Source[b]	Stage of lactation
Lysine									
Girma et al. (2019)	+[c]	+	=	+	15.1% prepartum 18% postpartum	7.02% Lys and 2.29% Met prepartum; 6.76% Lys and 2.15% Met postpartum	40 g	RPL*	−21 to 21
Fehlberg et al. (2020)	+[d]	+[d]	=	=	14.2% prepartum 16.8% postpartum	8.24% Lys and 2.94% Met prepartum	0.54% of DM prepartum with or without 0.40% of DM postpartum	AjiPro-L	−28 to 28
Lysine plus methionine									
Xu et al. (1998)	+[e]	+	+	+	18.6% prepartum and postpartum	7.5% Lys and 1.9% Met prepartum; 6.6% Lys and 1.9% Met postpartum	40 g Lys plus 13 g Met	RPL# and RPMet#	−21 to 56
Socha et al. (2005)	=	+	=	=	15.6% prepartum 18.5% or 16% postpartum	7.5% Lys and 2.4% Met	16 g Lys + 10.2 g Met	SM and SML	−14 to 105

(Continued)

Table 3 *(Continued)*

Study	DMI[a]	Milk yield or ECM	Protein %	Fat %	Diet CP %	Lys and Met in the MP	Amount fed	Source[b]	Stage of lactation
Lee et al. (2019)	=	=	=	=	12.2% prepartum 16% postpartum	7.48% Lys and 2.46% Met prepartum; 7.07% Lys and 2.34% Met postpartum	0.14% of DM RPMet plus 0.24% of DM RPL prepartum, with or without 0.22% of DM RPL plus 0.36% of DM RPL Postpartum	MetiPEARL USA Lysine	−21 to 22
Branched-chain amino acids									
Leal Yepes et al. (2019)	=	=	=	=	13.6% prepartum 15.5% postpartum	9.8% Lys 2.5% Met	550 g	RPBCAA	1 to 35
Tryptophan									
Jo et al. (2021)	+	=	=	=	n.a.	n.a.	15 g	ACT	−42 to 0
Glutamate									
Hisadomi et al. (2022)	=	(−)	=	=	17.1% prepartum 18.8% postpartum	n.a.	4% of DM prepartum 3% of DM postpartum	RPG	−21 to 21

Emphasis was placed on published studies where supplementation began in the prepartal period and extended through end of the transition period or beyond.
[a]DMI, dry matter intake; CP, crude protein; ECM, energy-corrected milk yield.
[b]*RPL, rumen-protected lysine (Ascor Chimici Srl, Beijing, China); AjiPro-L (AjiPro-L Generation 3, Ajinomoto Heartland Inc., Chicago, IL, USA); #RPL and RPMet, rumen-protected lysine and rumen-protected methionine (Ajinomoto Co. Inc., Tokyo, Japan); SML, Smartamine ML, Adisseo USA, Inc., Alpharetta, GA, USA); MetiPearl, 48% methionine (Kemin Industries Inc., USA); USA Lysine, 55% lysine (Kemin Industries Inc., USA); RPBCAA, 375 g L-leucine + 91 g L-valine + 85 g L-isoleucine with 27%, 67% and 48% rumen protection, respectively (Balchem Corporation, New Hampton, NY, USA); ACT, *N*-acetyl-L-tryptophan (42% L-tryptophan) (CJ CheilJedang, Suwon, Korea); RPG, rumen-protected monosodium Glu monohydrate prototype supplement (Ajinomoto Co. Inc., Tokyo, Japan).
[c]Positive response especially when the prepartum diet contained higher energy density (1.53 Mcal/kg DM versus 1.37 Mcal/kg DM).
[d]Positive response only when RPL was fed in the prepartum period.
[e]Positive response in the prepartum and postpartum.

Table 4 Summary of production performance results (relative changes from control cows: + positive effect, = no change; listed chronologically) in which the dietary supply of methionine was enhanced by feeding supplements that conferred some degree of protection from ruminal degradation

Reference	DMI[a]	Milk yield or ECM	Protein %	Fat %	Diet CP %	Lys and Met in the MP	Amount fed	Source[b]	Days in milk
Overton et al. (1996)	=	+	=	=	19% postpartum*	n.a.	17 g	RPMet*	−7 to 126
Piepenbrink et al. (2004)	=	+	=	=	14% prepartum 16.5% postpartum	2.34% Met prepartum and 2.36% postpartum§	0.10% of DM prepartum; 0.15% of DM postpartum	Alimet®	−21 to 84
Socha et al. (2005)	=	=	+[c]	=	15.6% prepartum 18.5% or 16% postpartum	7.5% Lys and 2.4% Met	10.5 g	SM	−14 to 105
Ordway et al. (2009)	+[d]	=	+	=	13.7% prepartum 16.3% postpartum	7.26% Lys and 2.41% Met prepartum; 6.92% Lys and 2.31% Met postpartum	0.35% pre and 0.54% of DM for MS; 0.06% prepartum and 0.10% postpartum for SM	MS and SM	−21 to 140
Preynat et al. (2009)	=	=	+	=	12.8% prepartum 17.3% postpartum	6.18% Lys and 1.85% Met prepartum; 6.3% Lys and 1.83% Met postpartum	15.3 g	Mepron-85	−21 to 112
Ardalan et al. (2010)	n.a.	+	=	=	16.5% prepartum 18% postpartum	n.a.	14.4 g	SM	−28 to 70
Osorio et al. (2013)	+	+	+	=	15.1% prepartum 17.5% postpartum	6.61% Lys and 2.36% Met prepartum; 6.08% Lys and 2.15% Met postpartum	0.19% MS or 0.07 % SM of DM	MS and SM	−21 to 30

(Continued)

Table 4 *(Continued)*

Reference	DMI[a]	Milk yield or ECM	Protein %	Fat %	Diet CP %	Lys and Met in the MP	Amount fed	Source[b]	Days in milk
Zhou et al. (2016b)	+	+	+	=	14.5% prepartum 17.3% postpartum	6.66% Lys and 2.37% Met prepartum; 6.24% Lys and 2.30% Met postpartum	0.08 % of DM	SM	−21 to 30
Sun et al. (2016)	+[d]	+	+	+	11.9% prepartum 15% postpartum	3.13:1 Lys to Met prepartum; 3.15:1 Lys to Met postpartum	15 g	Mepron-85	−21 to 21
Batistel et al. (2017)	+	+	+	=	15.6% prepartum 17.5% postpartum	6.51% Lys and 2.30% Met prepartum; 6.38% Lys and 2.24% Met postpartum	0.09% of DM	Mepron-85	−28 to 60
Potts et al. (2020)	=	+[e]	+[e]	+[e]	14.3% prepartum 16.7% postpartum	6.71% Lys and 1.77% Met prepartum; 6.74% Lys and 1.82% Met postpartum	12 g prepartum, 18 g postpartum	SM	−21 to 35
Toledo et al. (2021)	=	=	+	+	14.6% prepartum 16% postpartum	7.0–7.46% Lys and 2.22–2.85% Met prepartum; 6.75–7.07% Lys and 2.58–2.65% Met postpartum	0.095% of DM	SM	−20 to 112
Cardoso et al. (2021)	+[f]	=	=	+	15.8% prepartum 18.6% postpartum	6.85% Lys and 2.36% Met prepartum; 6.26% Lys and 2.98% Met postpartum	0.09% of DM prepartum, 0.13% of DM postpartum	Mepron-85	−18 to 45

Reference	DMI[a]	Milk yield or ECM	Protein %	Fat %	Diet CP %	Lys and Met in the MP	Amount fed	Source[b]	Days in milk
Abreu Michelotti et al. (2021)	n.a.	=	+	+	15.9% prepartum 17.2% postpartum	7.31% Lys and 2.01% Met prepartum; 6.37% Lys and 2.20% Met postpartum	0.05% of DM	SM	1–29

Emphasis was placed on published studies where supplementation began in the prepartal period and extended through end of the transition period or beyond. Concentrations of lysine and methionine in the metabolizable protein (MP) were only available for some studies.

[a]DMI, dry matter intake; CP, crude protein; ECM, energy-corrected milk yield.

[b]*RPMet, rumen-protected methionine (Degussa Corp.); Alimet®, methionine hydroxy analogue (Novus International Inc., St. Charles, MO, USA); SM, Smartamine-M (75% DL-methionine, Adisseo USA, Inc., Alpharetta, GA); MS, Metasmart, isopropyl ester of methionine hydroxy analogue (Adisseo USA, Inc., Alpharetta, GA); Mepron-85 (85% DL-methionine, Evonik Operations GmbH, Hanau, Germany)

[c]Positive response with 18.5% CP in diet.

[d]Positive response in the postpartum with MS or Mepron-85.

[e]Positive response only with multiparous cows.

[f]Positive response in the prepartum only.

*No information on the CP content of the prepartum diet or estimated concentrations of methionine and lysine in the metabolizable protein was reported.

[§]No information on Lys in the MP was available.

n.a., information not available in the paper.

As a lipotropic agent, Met is directly involved in very-low-density lipoprotein (VLDL) synthesis via the generation of *S*-adenosylmethionine (SAM), the most important methyl donor (Coleman et al., 2021). In turn, SAM can be used to methylate phosphatidylethanolamine (PE) to generate PC, which is essential for VLDL synthesis (Fig. 4). In the context of VLDL synthesis and liver lipid metabolism, PC is indispensable for the synthesis and release of chylomicrons and VLDL. Although no changes in measures of hepatic metabolism were detected by Piepenbrink et al. (2004), feeding MOH led to greater milk yield. Subsequent studies from Socha et al. (2005) through Ardalan et al. (2010) on feeding RPMet around parturition reported consistently greater milk protein percentage despite inconsistent responses in DMI, milk, or energy-corrected milk yield (ECM) (Table 4). A direct comparison across published studies (even from the same research groups) clearly is difficult partly because the source of RPMet or MOH was not the same. Dietary CP percentage in the prepartum and postpartum also differed across studies and could confound the interpretation of the data. Despite these limitations, the consistent increase in milk protein percentage suggests that Met was the most limiting EAA.

The work of Osorio et al. (2013) was the first, to our knowledge, where the emphasis on the post-ruminal supply of Met (via RPMet or MOH) specifically attempted to address responses in the synthesis of antioxidants (GSH, taurine) in the liver. Both in vitro (Hartman et al., 2002) and in vivo (Tabachnick and Tarver, 1955) studies using radioactive-labeled Met demonstrated hepatic incorporation of [^{35}S] into GSH. Because the utilization of Met in the 1-carbon metabolism pathway is known to generate a number of important intermediate molecules or metabolites (Fig. 2), the notion that the supply of Met to the liver could alter those pathways was appealing. This and subsequent work with transition cows fed different sources of RPMet or MOH (Zhou et al., 2016b; Sun et al., 2016; Batistel et al., 2017) demonstrated beneficial effects on GSH synthesis and a number of markers associated with inflammation and oxidative stress (Table 5). Perhaps the most unexpected response to feeding RPMet or MOH starting in the dry period has been the consistent increase in DMI (Table 4). That response has been accompanied by greater milk yield and concentrations of protein and fat (in most cases). Although direct mechanisms that account for the greater DMI have not been assessed directly, it is speculated that the reduction in inflammation due to the greater synthesis of antioxidants (Table 5) may be one of the key causes (Coleman et al., 2021). It is well-known in non-ruminants that pro-inflammatory cytokines (e.g. IL-1β) directly affect the hypothalamus and specifically the glucose-sensitive neurons in the brain 'satiety' and 'hunger' sites (Plata-Salaman, 1995).

Using changes in plasma concentrations of albumin and bilirubin (examples of hepatic-derived blood proteins, Fig. 1) along with cholesterol, Bertoni and Trevisi (2013) developed the liver functionality index (LFI), which characterizes

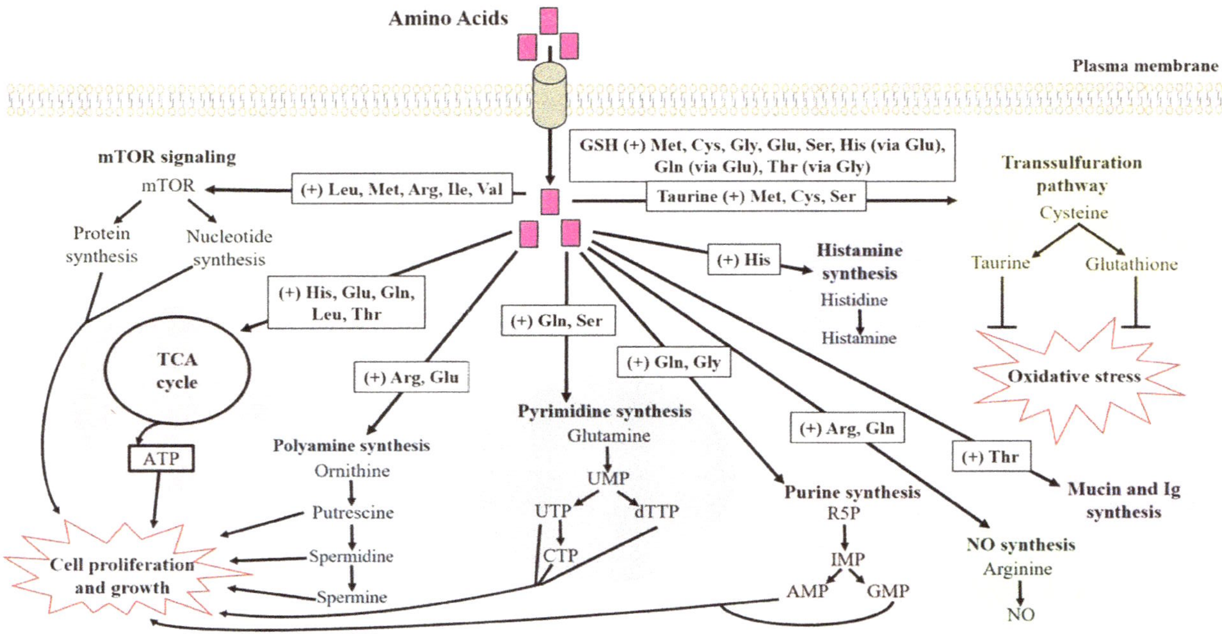

Figure 3 Theoretical model of cellular AA utilization. AMP, adenosine monophosphate; ATP, adenosine triphosphate; CTP, cytidine triphosphate; dTTP, deoxythymidine triphosphate; GMP, guanosine monophosphate; IMP, inosine monophosphate; NO, nitric oxide; R5P, ribose 5-phosphate; TCA cycle, tricarboxylic acid cycle; UMP, uridine monophosphate; UTP, uridine-5′-triphosphate. Source: Coleman et al. (2020).

the extent of the inflammatory response and helps predict consequences on health and well-being of the cow. For instance, a low LFI value is indicative of a pronounced inflammatory response, suggestive of a more difficult transition from gestation to lactation, whereas a high LFI value is suggestive of a smooth transition. In some of the experiments feeding RPMet or MOH, these treatments consistently increased blood albumin (Osorio et al., 2014; Zhou et al., 2016a; Batistel et al., 2018) well above the concentration range in cows with adequate postpartum liver function (Bionaz et al., 2007). Furthermore, feeding RPMet decreased the concentrations of inflammation-related biomarkers such as ceruloplasmin, haptoglobin, IL-1β and SAA (Osorio et al., 2014; Zhou et al., 2016a). Using the cow data from the Zhou et al. (2016a) experiment, compared with 35% of the cows not receiving RPMet, which ended up in the low LFI group, only 10% of the RPM-fed cows ended up in the low LFI group (Zhou et al., 2017), hence supporting the existence of a positive effect of RPMet on liver function. Because Met metabolism in the Met cycle generates SAM (Fig. 4), endogenous production of carnitine in a tissue such as the liver (Carlson et al., 2007) would be expected to be responsive to RPMet or MOH. Indeed, cows fed RPMet or MOH had greater hepatic concentrations of carnitine (82.1 versus 37.5 nmol/g of tissue) (Osorio et al., 2014). Thus, the data seem to suggest that a greater hepatic availability of Met can methylate Lys via reactions in the Met cycle (Fig. 4).

5.2 Lysine

In addition to its role in protein synthesis, methylated Lys residues in structural proteins (tri-methyl-Lys) and free Lys from proteolysis can provide the C backbone for the synthesis of carnitine (Ringseis et al., 2018). The addition of methyl groups from the synthesis of SAM via the Met cycle allows free Lys to serve as a substrate for carnitine synthesis (Bremer, 1983). As indicated in a previous section, carnitine is essential for mitochondrial β-oxidation of long-chain FA, assisting in the shuttling of long-chain FA into the mitochondria via the carnitine-acylcarnitine system through carnitine palmitoyltransferase-1 (CPT1). This initiates β-oxidation of FA and leads to the production of energy (ATP) (Drackley, 1999). This system is activated after calving and assists with the oxidation of FA in the liver to prevent excessive TAG accumulation and the risk of metabolic disorders (Carlson et al., 2007).

Although combinations of RPMet and RPL have been fed to transition cows and published in 1998 (Xu et al., 1998), only recently has the focus shifted to feeding RPL alone, the main hypothesis being that it could aid hepatic lipid metabolism through its linkage with carnitine synthesis (Girma et al., 2019; Fehlberg et al., 2020). In both studies published, and in spite of being different RPL sources, feeding RPL starting on the last month prepartum led to greater DMI and milk or ECM yield during the first 28 days in milk (Table 3). Although

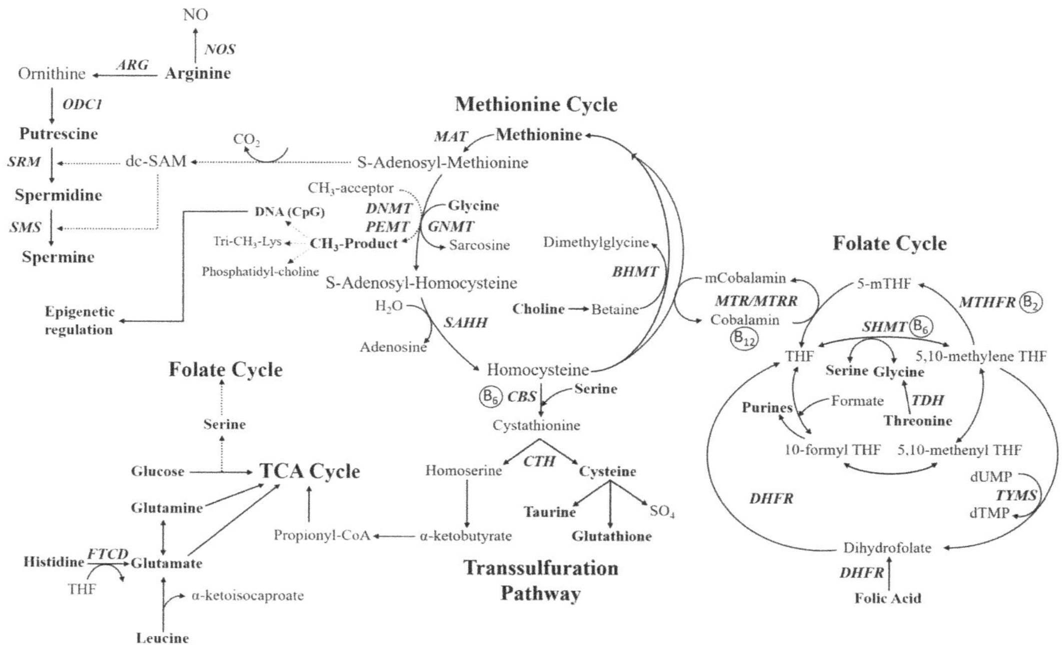

Figure 4 Interrelationships among components of the one-carbon metabolism pathways. 5-mTHF = 5-methyl-tetrahydrofolate; B2 = riboflavin; B6 = pyridoxal 5′-phosphate; B12 = cobalamin; BHMT = betaine homocysteine methyltransferase; CBS = cystathionine beta synthase; CTH = cystathionine gama-lyase; DHFR = dihydrofolate reductase; DNMT = DNA methyltransferase; dUMP = deoxyuridine monophosphate; dTMP = thymidine monophosphate; GNMT = glycine *N*-methyltransferase; MAT = methionine adenosyltransferase; MTHFR = methylenetetrahydrofolate reductase; MTR = 5-methyltetrahydrofolate-homocysteine methyltransferase; MTRR = 5-methyltetrahydro folate-homocysteine methyltransferase reductase; PEMT = phosphatidylethanolamine *N*-methyltransferase; SAHH = *S*-adenosyl-homocysteine hydrolase; SHMT = serine hydroxymethyltransferase; THF = tetrahydrofolate; TYMS = thymidylate synthetase; VLDL = very-low-density lipoprotein. Source: Revised from Coleman et al. (2021).

Table 5 Summary of additional beneficial effects of feeding rumen-protected methionine during the transition period and early lactation

Biomarker	Response[a]	Site	Biological function
Metabolism			
Carnitine	↑	Liver	β-Oxidation of fatty Acids
Cholesterol	↑	Plasma	Lipoprotein metabolism
One-carbon metabolism			
Cystathionine beta-synthase activity	↑	Liver	Antioxidant synthesis
Cystathionine	↑	Plasma	H_2S biosynthesis, redox status
Cystine	↑	Plasma	Redox status
Homocystine	↑	Plasma	Methylation reactions
Inflammation			
IL-1β	↓	Plasma	Pro-inflammatory cytokine
Haptoglobin	↓	Plasma	Inflammation signal
Albumin	↑	Plasma	Acute-phase response
Oxidative stress			
Reactive oxygen metabolites (ROM)	↔/↓	Plasma	Peroxides, superoxide, OH-radicals
Glutathione	↑	Liver, blood	Antioxidant
Taurine	↔/↑	Plasma	Antioxidant
Antioxidant capacity	↔/↑	Plasma	Total antioxidants in blood
Paraoxonase	↑	Plasma	Antioxidant enzyme

↑ = beneficial increase; ↓ = beneficial decrease; ↔ = no change in concentration relative to a control treatment.
[a]Relative to a control or a rumen-protected choline supplemented diet (Osorio et al., 2014; Zhou et al., 2016a; Sun et al., 2016; Batistel et al., 2018).

the ratios of Lys:Met in the MP estimated for the prepartum diet was 3.06 in Girma et al. (2019) and 2.80 in Fehlberg et al. (2020), supplemental RPL was effective. The positive effect was particularly evident when RPL was included in the prepartum diet and continued through 28 days in milk (Fehlberg et al., 2020). Cows that received RPL from calving until 28 days did not seem to benefit (at least from a statistical standpoint). Although in the study of Girma et al. (2019) the prepartum diet may have still benefitted from supplemental RPMet (Osorio et al., 2013; Zhou et al., 2016b; Batistel et al., 2017), the study demonstrated that when the dietary energy density is high (1.53 Mcal/kg of DM), the addition of RPL could be beneficial. Cows in the high-energy diet not only had greater DMI but also lower concentrations of BHB in the plasma, which was indicative of better hepatic capacity to handle exogenous fatty acids. Whether such effect was due to the synthesis of carnitine from Lys in the RPL was not evaluated.

Responses to feeding RPL along with RPMet during the transition period have been variable. Whereas the study of Xu et al. (1998) reported beneficial responses in terms of DMI, milk and ECM yield, milk protein percentage and milk fat percentage, only milk and ECM yield were affected in a subsequent study (Socha et al., 2005), and the most recent study did not report production differences (Lee et al., 2019). At first glance, a clear difference among these experiments is the level of dietary CP, i.e. 18.6%, 15.6% or 12.2% prepartum in Xu et al. (1998), Socha et al. (2005) and Lee et al. (2019), respectively. The reported ratios of Lys:Met in the MP also varied widely, i.e. 3.94, 3.12 or 3.04, respectively. Another aspect that complicates the comparison across these studies is the fact that different sources of RPMet and RPL were fed. Since these experiments did not report, or in Lee et al. (2019) there were no differences, plasma biomarkers of energy metabolism or immune function, it is not possible to determine the nature of the physiological changes that cows experienced.

5.3 Other essential and non-essential amino acids

The availability of various commercial sources of RPMet and RPL clearly has facilitated the study of these EAA. Undoubtedly, advances from these studies have been an incentive to develop other RP products, and recent experiments report feeding not only EAA such as BCAA or Trp, but also NEAA such as Glu (Table 3). The case for feeding RPBCAA during the transition period stems largely from the fact that the BCAA (Group II AA) (Lapierre et al., 2012) bypass the liver and travel directly to extrahepatic tissues such as mammary gland, skeletal muscle and adipose tissue for utilization. At the mammary gland level, such a feature is especially relevant due to the well-known ability of BCAA to regulate milk protein synthesis through its effect on the kinase mechanistic target of rapamycin (mTOR) (Cant et al., 2018). The few data available for the hepatic metabolism of BCAA suggest that around parturition, there is net uptake of these AAs (~5% of total flux through the liver), but with little change between the prepartum and postpartum periods (Larsen and Kristensen, 2013). Net uptake of Trp by the liver is almost negligible and does not change around parturition. Since these AAs are glucogenic, their contribution to that process around parturition clearly is minor relative to AA such as Ala (Larsen and Kristensen, 2013). Furthermore, their net uptake by the liver is markedly lower relative to those AAs considered in Group I. For instance, the net uptake of Phe, Met and Thr averages 7% (Thr) to 10% (Met) and 15% (Phe) of their total influx in the prepartum, and for Phe and Met it remains above 5% of their total influx after parturition (Larsen and Kristensen, 2013).

Although production performance was not affected by feeding RPBCAA (Leal Yepes et al., 2021), trends for lower plasma FA concentrations and greater DMI before parturition suggested these AAs could have altered some physiological mechanisms, e.g. insulin sensitivity in adipose tissue (Ji et al., 2012). Coupled with numerical increases in milk and ECM when RPBCAA was fed, the physiological data indicate that additional studies are needed to clarify the benefits of BCAA during the transition period beyond the synthesis of milk protein.

The sole published study where RPTrp was fed exclusively during the dry period indicated positive effects in terms of DMI (Jo et al., 2021). The tendency for greater milk yield during the first 8 weeks postpartum provided some evidence for a beneficial, carryover effect of RPTrp feeding during pregnancy. However, the plasma data reported indicated greater plasma FA in those cows throughout the prepartum and postpartum periods, which seems inconsistent with the DMI data. Because only four cows per treatment were used, these data should be evaluated with caution. What seems evident is that given the myriad of important roles of Trp in the body, e.g. neurotransmitter and niacin synthesis (antilipolytic molecule), additional studies could shed light on its potential benefits to the transition cow.

Although seldom discussed in the dairy nutrition literature, glutamate metabolism by the liver (at least in non-ruminants) is at the center of AA metabolism by this organ (Brosnan and Brosnan, 2009). The myriad of liver enzymes in which glutamate plays important roles, e.g. some produce glutamate and some use glutamate, likely contributes to the paucity of data in dairy cattle. An extensive discussion of glutamate metabolism by the liver is beyond the scope of this chapter. However, it is important to highlight that (1) glutamate is essential during the process of AA deamination catalyzed by 'aminotransferases' (reversible reactions) for which glutamate and alpha-ketoglutarate are reaction substrates, and (2) several AAs (e.g. Gln, His, Arg, Orn and Pro) are converted to glutamate during their catabolism (Brosnan and Brosnan, 2009).

From the standpoint of glutamate supply during the transition period, a recent study with RPglutamate is helpful in illustrating how some physiological adaptations might benefit from this non-essential AA (Hisadomi et al., 2022). This work was based on the premise that the growth of the gastrointestinal tract of the cow especially after calving would benefit from enhanced glutamate supply because this AA is well-known as the primary energy source in gut tissue (Reeds et al., 2000; El-Kadi et al., 2009). It was hypothesized that supplemental glutamate could enhance intestinal villi height, absorptive capacity and overall intestinal barrier function (Hisadomi et al., 2022). Although there were no effects of RPglutamate in terms of performance or plasma markers of energy metabolism, there were trends for greater total tract digestibility of DM, OM and NDF in the first week postpartum. The lower plasma concentration of His in those cows suggested potential alterations in the metabolism of this EAA (Fig. 5), which can be used to synthesize glutamate through a series of reactions the last of which requires folic acid, an important metabolite in the 1-carbon metabolism pathway (Fig. 4). Similarly, the lower plasma concentration of Thr in response to RPglutamate was suggestive of a reduction in utilization of glutamate carbon. It is tempting to speculate that the greater post-ruminal supply of glutamate 'channeled' this AA toward energy production within gut tissue, which may help explain the positive effects in total tract digestibility.

6 Conclusion and future trends

The field of AA nutrition in dairy cows remains very active, driven in part by worldwide interest among nutritionists in feeding lower protein diets, and discoveries regarding functional aspects of EAA and NEAA that were previously unrecognized. With the advent of commercial RPAA besides Met and Lys, we are now able to generate new information on the potential roles, beyond protein synthesis, that other AA may have on the physiology of the cow. Recognition that the AA needs of the dairy cow differ during the transition from

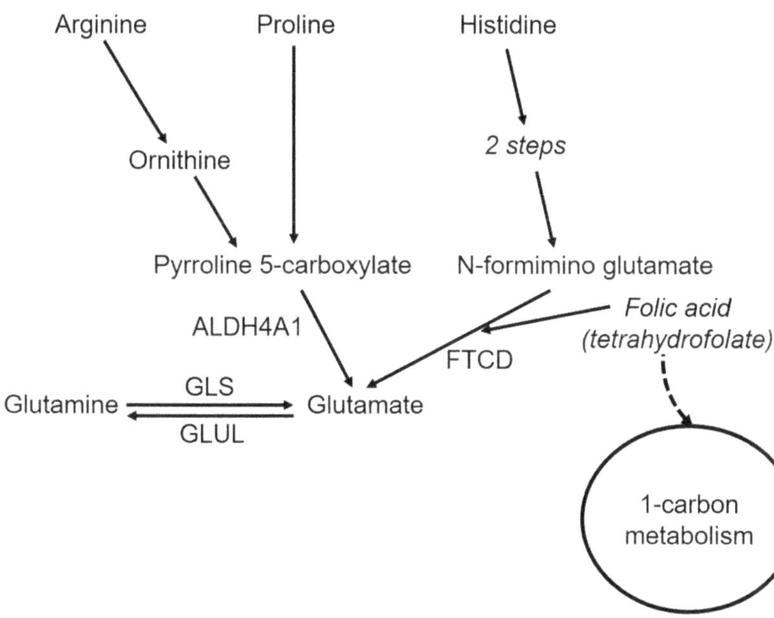

Figure 5 Simplified summary of the key catabolic reactions of the glutamate family of amino acids occurring in tissues such as the liver. ALDH4A1, aldehyde dehydrogenase 4 family member A1 (formerly known as 1-pyrroline-5-carboxylate dehydrogenase (P5CD); FTCD, formimidoyltransferase cyclodeaminase; GLS, glutaminase; GLUL, glutamate-ammonia ligase. Source: Revised from Brosnan and Brosnan (2009).

late-pregnancy to lactation compared with established lactation has sparked interest in developing a greater understanding of the role of these molecules in important organs such as the liver and adipose tissue, and also the immune system. Moving forward, it will be important to continue research efforts on the efficacy of EAA and NEAA supply during periods of environmental stress, e.g. heat stress. A greater understanding of the efficiency of AA use by tissues such as the mammary gland will enhance the predictive ability of diet formulation models. Along with these animal-level types of information, a more routine and reliable characterization of AA profiles in feedstuffs will, together, continue supporting the development of efficient feeding management strategies.

7 Where to look for further information

The National Academies of Sciences, Engineering, and Medicine has recently published the Eight revised edition of the Nutrient Requirements of Dairy Cattle with a comprehensive update on the postabsorptive use of amino acids by dairy cattle:

- https://nap.nationalacademies.org/catalog/25806/nutrient-requirements-of-dairy-cattle-eighth-revised-edition.

8 References

Amanlou, H., Farahani, T. A. and Farsuni, N. E. (2017), Effects of rumen undegradable protein supplementation on productive performance and indicators of protein and energy metabolism in Holstein fresh cows, *J. Dairy Sci.*, 100(5), 3628–3640.

Ardalan, M., Rezayazdi, K. and Dehghan-Banadaky, M. (2010), Effect of rumen-protected choline and methionine on physiological and metabolic disorders and reproductive indices of dairy cows, *J. Anim. Physiol. Anim. Nutr. (Berl)*, 94(6), e259–e265.

Bach, A., Calsamiglia, S. and Stern, M. D. (2005), Nitrogen metabolism in the rumen, *J. Dairy Sci.*, 88(Suppl. 1), E9–21.

Batistel, F., Arroyo, J. M., Bellingeri, A., Wang, L., Saremi, B., Parys, C., Trevisi, E., Cardoso, F. C. and Loor, J. J. (2017), Ethyl-cellulose rumen-protected methionine enhances performance during the periparturient period and early lactation in Holstein dairy cows, *J. Dairy Sci.*, 100(9), 7455–7467.

Batistel, F., Arroyo, J. M., Garces, C. I. M., Trevisi, E., Parys, C., Ballou, M. A., Cardoso, F. C. and Loor, J. J. (2018), Ethyl-cellulose rumen-protected methionine alleviates inflammation and oxidative stress and improves neutrophil function during the periparturient period and early lactation in Holstein dairy cows, *J. Dairy Sci.*, 101(1), 480–490.

Bell, A. W., Slepetis, R. and Ehrhardt, R. A. (1995), Growth and accretion of energy and protein in the gravid uterus during late pregnancy in Holstein cows, *J. Dairy Sci.*, 78(9), 1954–1961.

Bertoni, G. and Trevisi, E. (2013), Use of the liver activity index and other metabolic variables in the assessment of metabolic health in dairy herds, *Vet. Clin. North Am. Food Anim. Pract.*, 29(2), 413–431.

Bertoni, G., Trevisi, E. and Lombardelli, R. (2009), Some new aspects of nutrition, health conditions and fertility of intensively reared dairy cows, *Ital. J. Anim. Sci.*, 8(4), 491–518.

Bionaz, M., Trevisi, E., Calamari, L., Librandi, F., Ferrari, A. and Bertoni, G. (2007), Plasma paraoxonase, health, inflammatory conditions, and liver function in transition dairy cows, *J. Dairy Sci.*, 90(4), 1740–1750.

Bradford, B. J., Yuan, K., Farney, J. K., Mamedova, L. K. and Carpenter, A. J. (2015), Invited review: inflammation during the transition to lactation: new adventures with an old flame, *J. Dairy Sci.*, 98(10), 6631–6650.

Bremer, J. (1983), Carnitine--metabolism and functions, *Physiol. Rev.*, 63(4), 1420–1480.

Breves, G., Schroder, B., Heimbeck, W. and Patton, R. A. (2010), Short communication: transport of 2-hydroxy-4-methyl-thio-butanoic isopropyl ester by rumen epithelium in vitro, *J. Dairy Sci.*, 93(1), 260–264.

Broderick, G. A. (2003), Effects of varying dietary protein and energy levels on the production of lactating dairy cows, *J. Dairy Sci.*, 86(4), 1370–1381.

Broderick, G. A. (2018), Review: optimizing ruminant conversion of feed protein to human food protein, *Animal*, 12(8), 1722–1734.

Broderick, G. A., Kowalczyk, T. and Satter, L. D. (1970), Milk production response to supplementation with encapsulated methionine per os or casein per abomasum, *J. Dairy Sci.*, 53(12), 1714–1721.

Brosnan, M. E. and Brosnan, J. T. (2009), Hepatic glutamate metabolism: a tale of 2 hepatocytes, *Am. J. Clin. Nutr.*, 90(3), 857S–861S.

Cant, J. P., Kim, J. J. M., Cieslar, S. R. L. and Doelman, J. (2018), Symposium review: Amino acid uptake by the mammary glands: where does the control. lie? *J. Dairy Sci.*, 101(6), 5655–5666.

Cardoso, F. F., Donkin, S. S., Pereira, M. N., Pereira, R. A. N., Peconick, A. P., Santos, J. P., Silva, R. B., Caproni, V. R., Parys, C. and Danes, M. A. C. (2021), Effect of protein level and methionine supplementation on dairy cows during the transition period, *J. Dairy Sci.*, 104(5), 5467–5478. DOI: 10.3168/jds.2020-19181.

Carlson, D. B., McFadden, J. W., D'Angelo, A., Woodworth, J. C. and Drackley, J. K. (2007), Dietary L-carnitine affects periparturient nutrient metabolism and lactation in multiparous cows, *J. Dairy Sci.*, 90(7), 3422–3441.

Clark, J. H., Klusmeyer, T. H. and Cameron, M. R. (1992), Microbial protein synthesis and flows of nitrogen fractions to the duodenum of dairy cows, *J. Dairy Sci.*, 75(8), 2304–2323.

Coleman, D. N., Alharthi, A. S., Liang, Y., Lopes, M. G., Lopreiato, V., Vailati-Riboni, M. and Loor, J. J. (2021), Multifaceted role of one-carbon metabolism on immunometabolic control and growth during pregnancy, lactation and the neonatal period in dairy cattle, *J. Anim. Sci. Biotechnol.*, 12(1), 27.

Coleman, D. N., Lopreiato, V., Alharthi, A. and Loor, J. J. (2020), Amino acids and the regulation of oxidative stress and immune function in dairy cattle, *J. Anim. Sci.* 98(Suppl. 1), S175–S193.

Drackley, J. K. (1999), ADSA foundation scholar award. Biology of dairy cows during the transition period: the final frontier? *J. Dairy Sci.*, 82(11), 2259–2273.

El-Kadi, S. W., Baldwin, R. L., McLeod, K. R., Sunny, N. E. and Bequette, B. J. (2009), Glutamate is the major anaplerotic substrate in the tricarboxylic acid cycle of isolated rumen epithelial and duodenal mucosal cells from beef cattle, *J. Nutr.*, 139(5), 869–875.

Erasmus, L. J., Botha, P. M., Cruywagen, C. W. and Meissner, H. H. (1994), Amino acid profile and intestinal digestibility in dairy cows of rumen-undegradable protein from various feedstuffs, *J. Dairy Sci.*, 77(2), 541–551.

Fehlberg, L. K., Guadagnin, A. R., Thomas, B. L., Sugimoto, Y., Shinzato, I. and Cardoso, F. C. (2020), Feeding rumen-protected lysine prepartum increases energy-corrected milk and milk component yields in Holstein cows during early lactation, *J. Dairy Sci.*, 103(12), 11386–11400.

Fowler, C. M., Plank, J. E., Devillard, E., Bequette, B. J. and Firkins, J. L. (2015), Assessing the ruminal action of the isopropyl ester of 2-hydroxy-4-(methylthio) butanoic acid in continuous and batch cultures of mixed ruminal microbes, *J. Dairy Sci.*, 98(2), 1167–1177.

Fox, D. G., Tedeschi, L. O., Tylutki, T. P., Russell, J. B., Van Amburgh, M. E., Chase, L. E., Pell, A. N. and Overton, T. R. (2004), The Cornell Net Carbohydrate and Protein System model for evaluating herd nutrition and nutrient excretion, *Anim. Feed Sci. Technol.*, 112(1–4), 29–78.

Garcia, M., Elsasser, T. H., Juengst, L., Qu, Y., Bequette, B. J. and Moyes, K. M. (2016), Short communication: amino acid supplementation and stage of lactation alter apparent utilization of nutrients by blood neutrophils from lactating dairy cows in vitro, *J. Dairy Sci.*, 99(5), 3777–3783.

Girma, D. D., Ma, L., Wang, F., Jiang, Q. R., Callaway, T. R., Drackley, J. K. and Bu, D. P. (2019), Effects of close-up dietary energy level and supplementing rumen-protected lysine on energy metabolites and milk production in transition cows, *J. Dairy Sci.*, 102(8), 7059–7072.

Gresner, N., Rodehutscord, M. and Südekum, K. H. (2021), Amino acid pattern of rumen microorganisms in cattle fed mixed diets-an update, *J. Anim. Physiol. Anim. Nutr. (Berl)*, 106(4), 752–771. DOI: 10.1111/jpn.13676.

Hartman, N. R., Cysyk, R. L., Bruneau-Wack, C., Thénot, J. P., Parker, R. J. and Strong, J. M. (2002), Production of intracellular 35S-glutathione by rat and human hepatocytes for the quantification of xenobiotic reactive intermediates, *Chem. Biol. Interact.*, 142(1–2), 43–55.

Herbein, J. H., Aiello, R. J., Eckler, L. I., Pearson, R. E. and Akers, R. M. (1985), Glucagon, insulin, growth hormone, and glucose concentrations in blood plasma of lactating dairy cows, *J. Dairy Sci.*, 68(2), 320–325.

Hisadomi, S., Haruno, A., Fujieda, T., Sugino, T. and Oba, M. (2022), Effects of rumen-protected glutamate supplementation during the periparturient period on digestibility, inflammation, metabolic responses, and performance in dairy cows, *J. Dairy Sci.*, 105(4), 3129–3141. DOI: 10.3168/jds.2021-21357.

Huhtanen, P. and Hristov, A. N. (2009), A meta-analysis of the effects of dietary protein concentration and degradability on milk protein yield and milk N efficiency in dairy cows, *J. Dairy Sci.*, 92(7), 3222–3232.

Hvelplund, T. (1986), The influence of diet on nitrogen and amino-acid content of mixed rumen bacteria, *Acta Agric. Scand.*, 36(3), 325–331.

Jacobson, D. R., Soewardi, B., Hatton, R. H. and Carr, S. B. (1967), Sulfur nitrogen and amino acid balance and digestibility from lactating cows on low-sulfur and sulfur-supplemented diets, *J. Dairy Sci.*, 50(6), 980.

Jafari, A., Emmanuel, D. G., Christopherson, R. J., Thompson, J. R., Murdoch, G. K., Woodward, J., Field, C. J. and Ametaj, B. N. (2006), Parenteral administration of glutamine modulates acute phase response in postparturient dairy cows, *J. Dairy Sci.*, 89(12), 4660–4668.

Ji, P., Osorio, J. S., Drackley, J. K. and Loor, J. J. (2012), Overfeeding a moderate energy diet prepartum does not impair bovine subcutaneous adipose tissue insulin signal transduction and induces marked changes in peripartal gene network expression, *J. Dairy Sci.*, 95(8), 4333–4351.

Jo, J. H., Lee, J. S., Ghassemi Nejad, J., Kim, W. S., Moon, J. O. and Lee, H. G. (2021), Effects of dietary supplementation of acetate and L-tryptophan conjugated bypass amino acid on productivity of Pre- and post-partum dairy cows and their offspring, *Animals (Basel)*, 11(6), 1726.

Kim, C. H., Choung, J. J. and Chamberlain, D. G. (1999), Determination of the first-limiting amino acid for milk production in dairy cows consuming a diet of grass silage and a cereal-based supplement containing feather meal, *J. Sci. Food Agric.*, 79(12), 1703–1708.

Komaragiri, M. V. and Erdman, R. A. (1997), Factors affecting body tissue mobilization in early lactation dairy cows. 1. Effect of dietary protein on mobilization of body fat and protein, *J. Dairy Sci.*, 80(5), 929–937.

Lapierre, H., Lobley, G. E., Doepel, L., Raggio, G., Rulquin, H. and Lemosquet, S. (2012), Triennial Lactation symposium: mammary metabolism of amino acids in dairy cows, *J. Anim. Sci.*, 90(5), 1708–1721.

Larsen, M. and Kristensen, N. B. (2013), Precursors for liver gluconeogenesis in periparturient dairy cows, *Animal*, 7(10), 1640–1650.

Leal Yepes, F. A., Mann, S., Overton, T. R., Behling-Kelly, E., Nydam, D. V. and Wakshlag, J. J. (2021), Hepatic effects of rumen-protected branched-chain amino acids with or without propylene glycol supplementation in dairy cows during early lactation, *J. Dairy Sci.*, 104(9), 10324–10337.

Leal Yepes, F. A., Mann, S., Overton, T. R., Ryan, C. M., Bristol, L. S., Granados, G. E., Nydam, D. V. and Wakshlag, J. J. (2019), Effect of rumen-protected branched-chain amino acid supplementation on production- and energy-related metabolites during the first 35 days in milk in Holstein dairy cows, *J. Dairy Sci.*, 102(6), 5657–5672. DOI: 10.3168/jds.2018-15508.

Lean, I. J., de Ondarza, M. B., Sniffen, C. J., Santos, J. E. P. and Griswold, K. E. (2018), Meta-analysis to predict the effects of metabolizable amino acids on dairy cattle performance, *J. Dairy Sci.*, 101(1), 340–364.

Lee, C., Hristov, A. N., Cassidy, T. W., Heyler, K. S., Lapierre, H., Varga, G. A., de Veth, M. J., Patton, R. A. and Parys, C. (2012a), Rumen-protected lysine, methionine, and histidine increase milk protein yield in dairy cows fed a metabolizable protein-deficient diet, *J. Dairy Sci.*, 95(10), 6042–6056.

Lee, C., Hristov, A. N., Heyler, K. S., Cassidy, T. W., Lapierre, H., Varga, G. A. and Parys, C. (2012b), Effects of metabolizable protein supply and amino acid supplementation on nitrogen utilization, milk production, and ammonia emissions from manure in dairy cows, *J. Dairy Sci.*, 95(9), 5253–5268.

Lee, C., Lobos, N. E. and Weiss, W. P. (2019), Effects of supplementing rumen-protected lysine and methionine during prepartum and postpartum periods on performance of dairy cows, *J. Dairy Sci.*, 102(12), 11026-11039.

Lofgreen, G. P., Loosli, J. K. and Maynard, L. A. (1947), The influence of protein source upon nitrogen retention by sheep, *J. Anim. Sci.*, 6(3), 343-347.

Loosli, J. K., Williams, H. H., Thomas, W. E., Ferris, F. H. and Maynard, L. A. (1949), Synthesis of amino acids in the rumen, *Science*, 110(2849), 144-145.

Lopreiato, V., Mezzetti, M., Cattaneo, L., Ferronato, G., Minuti, A. and Trevisi, E. (2020), Role of nutraceuticals during the transition period of dairy cows: a review, *J. Anim. Sci. Biotechnol.*, 11: 96.

Mabjeesh, S. J., Arieli, A., Bruckental, I., Zamwell, S. and Tagari, H. (1997), Effect of ruminal degradability of crude protein and nonstructural carbohydrates on the efficiency of bacterial crude protein synthesis and amino acid flow to the abomasum of dairy cows, *J. Dairy Sci.*, 80(11), 2939-2949.

McCarthy, R. D., Porter, G. A. and Griel, L. C. (1968), Bovine ketosis and depressed fat test in milk: a problem of methionine metabolism and serum lipoprotein aberration, *J. Dairy Sci.*, 51(3), 459-462.

Michelotti, T. CdA., Pacheco, H. A., Leão, G. F. M., Lopes, F., Osorio, J. S. and Almeida, Rd (2021), Effects of rumen-protected methionine supplementation on dairy cows during early postpartum, *J. Appl. Anim. Res.*, 49(1), 257-263. DOI: 10.1080/09712119.2021.1942882.

Morris, D. L. and Kononoff, P. J. (2020), Effects of rumen-protected lysine and histidine on milk production and energy and nitrogen utilization in diets containing hydrolyzed feather meal fed to lactating Jersey cows, *J. Dairy Sci.*, 103(8), 7110-7123.

National Research Council (U.S.). Subcommittee on Dairy Cattle Nutrition. (2001), *Nutrient Requirements of Dairy Cattle*. Washington, DC: National Academy Press.

Ordway, R. S., Boucher, S. E., Whitehouse, N. L., Schwab, C. G. and Sloan, B. K. (2009), Effects of providing two forms of supplemental methionine to periparturient Holstein dairy cows on feed intake and lactational performance, *J. Dairy Sci.*, 92(10), 5154-5166.

Osorio, J. S., Ji, P., Drackley, J. K., Luchini, D. and Loor, J. J. (2013), Supplemental Smartamine M or MetaSmart during the transition period benefits postpartal cow performance and blood neutrophil function, *J. Dairy Sci.*, 96(10), 6248-6263.

Osorio, J. S., Trevisi, E., Ji, P., Drackley, J. K., Luchini, D., Bertoni, G. and Loor, J. J. (2014), Biomarkers of inflammation, metabolism, and oxidative stress in blood, liver, and milk reveal a better immunometabolic status in peripartal cows supplemented with Smartamine M or MetaSmart, *J. Dairy Sci.*, 97(12), 7437-7450.

Overton, T. R., LaCount, D. W., Cicela, T. M. and Clark, J. H. (1996), Evaluation of a ruminally protected methionine product for lactating dairy cows, *J. Dairy Sci.*, 79(4), 631-638. DOI: 10.3168/jds.S0022-0302(96)76408-1.

Pacheco, D., Patton, R. A., Parys, C. and Lapierre, H. (2012), Ability of commercially available dairy ration programs to predict duodenal flows of protein and essential amino acids in dairy cows, *J. Dairy Sci.*, 95(2), 937-963.

Patton, R. A. (2010), Effect of rumen-protected methionine on feed intake, milk production, true milk protein concentration, and true milk protein yield, and the factors that influence these effects: a meta-analysis, *J. Dairy Sci.*, 93(5), 2105-2118.

Patton, R. A., Hristov, A. N. and Lapierre, H. (2014), Protein feeding and balancing for amino acids in lactating dairy cattle, *Vet. Clin. North Am. Food Anim. Pract.*, 30(3), 599-621.

Phipps, R. H., Reynolds, C. K., Givens, D. I., Jones, A. K., Geraert, P. A., Devillard, E. and Bennett, R. (2008), Short communication: effects of 2-hydroxy-4-(methylthio) butanoic acid isopropyl ester on milk production and composition of lactating Holstein dairy cows, *J. Dairy Sci.*, 91(10), 4002–4005.

Piepenbrink, M. S., Marr, A. L., Waldron, M. R., Butler, W. R., Overton, T. R., Vázquez-Añón, M. and Holt, M. D. (2004), Feeding 2-hydroxy-4-(methylthio)-butanoic acid to periparturient dairy cows improves milk production but not hepatic metabolism, *J. Dairy Sci.*, 87(4), 1071–1084.

Plata-Salaman, C. R. (1995), Cytokines and feeding suppression: an integrative view from neurologic to molecular levels, *Nutrition*, 11(5), 674–677.

Polan, C. E., Chandler, P. T. and Miller, C. N. (1970), Methionine hydroxy analog: varying levels for lactating cows, *J. Dairy Sci.*, 53(5), 607–610.

Potts, S. B., Scholte, C. M., Moyes, K. M. and Erdman, R. A. (2020), Production responses to rumen-protected choline and methionine supplemented during the periparturient period differ for primi- and multiparous cows, *J. Dairy Sci.*, 103(7), 6070–6086. DOI: 10.3168/jds.2019-17591.

Preynat, A., Lapierre, H., Thivierge, M. C., Palin, M. F., Matte, J. J., Desrochers, A. and Girard, C. L. (2009), Influence of methionine supply on the response of lactational performance of dairy cows to supplementary folic acid and vitamin B12, *J. Dairy Sci.*, 92(4), 1685–1695. DOI: 10.3168/jds.2008-1572.

Purser, D. B. and Buechler, S. M. (1966), Amino acid composition of rumen organisms, *J. Dairy Sci.*, 49(1), 81–84.

Raisanen, S. E., Lage, C. F. A., Oh, J., Melgar, A., Nedelkov, K., Chen, X., Miura, M. and Hristov, A. N. (2021), Histidine dose-response effects on lactational performance and plasma amino acid concentrations in lactating dairy cows: 1. Metabolizable protein-adequate diet, *J. Dairy Sci.*, 104(9), 9902–9916.

Reeds, P. J., Burrin, D. G., Stoll, B. and Jahoor, F. (2000), Intestinal glutamate metabolism, *J. Nutr.* 130(4S Suppl), 978S–982S.

Ringseis, R., Keller, J. and Eder, K. (2018), Regulation of carnitine status in ruminants and efficacy of carnitine supplementation on performance and health aspects of ruminant livestock: a review, *Arch. Anim. Nutr.*, 72(1), 1–30.

Rulquin, H., Guinard, J. and Verite, R. (1998), Variation of amino acid content in the small intestine digesta of cattle: development of a prediction model, *Livest. Prod. Sci.*, 53(1), 1–13.

Schwab, C. G. and Broderick, G. A. (2017), A 100-year review: protein and amino acid nutrition in dairy cows, *J. Dairy Sci.*, 100(12), 10094–10112.

Schwab, C. G., Satter, L. D. and Clay, B. (1976), Response to lactating dairy cows to abomasal infusion of amino acids, *J. Dairy Sci.*, 59(7), 1254–1270.

Shabi, Z., Tagari, H., Murphy, M. R., Bruckental, I., Mabjeesh, S. J., Zamwel, S., Celik, K. and Arieli, A. (2000), Partitioning of amino acids flowing to the abomasum into feed, bacterial, protozoal, and endogenous fractions, *J. Dairy Sci.*, 83(10), 2326–2334.

Sinclair, K. D., Garnsworthy, P. C., Mann, G. E. and Sinclair, L. A. (2014), Reducing dietary protein in dairy cow diets: implications for nitrogen utilization, milk production, welfare and fertility, *Animal*, 8(2), 262–274.

Socha, M. T., Putnam, D. E., Garthwaite, B. D., Whitehouse, N. L., Kierstead, N. A., Schwab, C. G., Ducharme, G. A. and Robert, J. C. (2005), Improving intestinal amino acid supply of pre- and postpartum dairy cows with rumen-protected methionine and lysine, *J. Dairy Sci.*, 88(3), 1113–1126.

Sok, M., Ouellet, D. R., Firkins, J. L., Pellerin, D. and Lapierre, H. (2017), Amino acid composition of rumen bacteria and protozoa in cattle, *J. Dairy Sci.*, 100(7), 5241–5249.

Sordillo, L. M. and Mavangira, V. (2014), The nexus between nutrient metabolism, oxidative stress and inflammation in transition cows, *Anim. Prod. Sci.*, 54(9), 1204–1214.

Stein, H. H., Seve, B., Fuller, M. F., Moughan, P. J., de Lange, C. F. and Committee on Terminology to Report AA Bioavailability and Digestibility (2007), Invited review: amino acid bioavailability and digestibility in pig feed ingredients: terminology and application, *J. Anim. Sci.*, 85(1), 172–180.

Sun, F., Cao, Y., Cai, C., Li, S., Yu, C. and Yao, J. (2016), Regulation of nutritional metabolism in transition dairy cows: energy homeostasis and health in response to post-ruminal choline and methionine, *PLoS ONE*, 11(8), e0160659.

Tabachnick, M. and Tarver, H. (1955), The conversion of methionine-S35 to cystathionine-S35 and taurine-S35 in the rat, *Arch. Biochem. Biophys.*, 56(1), 115–122.

Tebbe, A. W. and Weiss, W. P. (2021), Concurrent and carryover effects of feeding blends of protein and amino acids in high-protein diets with different concentrations of forage fiber to fresh cows. 2. Protein balance and body composition, *J. Dairy Sci.*, 104(5), 5601–5616.

Thomas, W. E., Loosli, J. K., Williams, H. H. and Maynard, L. A. (1951), The utilization of inorganic sulfates and urea nitrogen by lambs, *J. Nutr.*, 43(4), 515–523.

Toledo, M. Z., Stangaferro, M. L., Gennari, R. S., Barletta, R. V., Perez, M. M., Wijma, R., Sitko, E. M., Granados, G., Masello, M., Van Amburgh, M. E., Luchini, D., Giordano, J. O., Shaver, R. D. and Wiltbank, M. C. (2021), Effects of feeding rumen-protected methionine pre- and postpartum in multiparous Holstein cows: lactation performance and plasma amino acid concentrations, *J. Dairy Sci.*, 104(7), 7583–7603. DOI: 10.3168/jds.2020-19021.

Trevisi, E., Jahan, N., Bertoni, G., Ferrari, A. and Minuti, A. (2015), Pro-inflammatory cytokine profile in dairy cows: consequences for new lactation, *Ital. J. Anim. Sci.*, 14(3).

Van Amburgh, M. E., Collao-Saenz, E. A., Higgs, R. J., Ross, D. A., Recktenwald, E. B., Raffrenato, E., Chase, L. E., Overton, T. R., Mills, J. K. and Foskolos, A. (2015), The Cornell Net carbohydrate and Protein System: updates to the model and evaluation of version 6.5, *J. Dairy Sci.*, 98(9), 6361–6380.

Vanhatalo, A., Huhtanen, P., Toivonen, V. and Varvikko, T. (1999), Response of dairy cows fed grass silage diets to abomasal infusions of histidine alone or in combinations with methionine and lysine, *J. Dairy Sci.*, 82(12), 2674–2685.

Varvikko, T., Vanhatalo, A., Jalava, T. and Huhtanen, P. (1999), Lactation and metabolic responses to graded abomasal doses of methionine and lysine in cows fed grass silage diets, *J. Dairy Sci.*, 82(12), 2659–2673.

Virtanen, A. I. (1966), Milk production of cows on protein-free feed, *Science*, 153(3744), 1603–1614.

Vyas, D. and Erdman, R. A. (2009), Meta-analysis of milk protein yield responses to lysine and methionine supplementation, *J. Dairy Sci.*, 92(10), 5011–5018.

Weller, R. A., Gray, F. V. and Pilgrim, A. F. (1958), The conversion of plant nitrogen to microbial nitrogen in the rumen of the sheep, *Br. J. Nutr.*, 12(4), 421–429.

Xu, S., Harrison, J. H., Chalupa, W., Sniffen, C., Julien, W., Sato, H., Fujieda, T., Watanabe, K., Ueda, T. and Suzuki, H. (1998), The effect of ruminal bypass lysine and methionine on milk yield and composition of lactating cows, *J. Dairy Sci.*, 81(4), 1062–1077.

Yoder, P. S., Huang, X., Teixeira, I. A., Cant, J. P. and Hanigan, M. D. (2020), Effects of jugular infused methionine, lysine, and histidine as a group or leucine and isoleucine as a group on production and metabolism in lactating dairy cows, *J. Dairy Sci.*, 103(3), 2387–2404.

Zang, Y., Ji, P., Morrison, S. Y., Koba, Y., Grant, R. J. and Dann, H. M. (2021a), Reducing metabolizable protein supply: effects on milk production, blood metabolites, and health in early-lactation dairy cows, *J. Dairy Sci.*, 104(12), 12443–12458.

Zang, Y., Silva, L. H. P., Geng, Y. C., Ghelichkhan, M., Whitehouse, N. L., Miura, M. and Brito, A. F. (2021b), Dietary starch level and rumen-protected methionine, lysine, and histidine: effects on milk yield, nitrogen, and energy utilization in dairy cows fed diets low in metabolizable protein, *J. Dairy Sci.*, 104(9), 9784–9800.

Zanton, G. I., Bowman, G. R., Vazquez-Anon, M. and Rode, L. M. (2014), Meta-analysis of lactation performance in dairy cows receiving supplemental dietary methionine sources or postruminal infusion of methionine, *J. Dairy Sci.*, 97(11), 7085–7101.

Zhou, Z., Bulgari, O., Vailati-Riboni, M., Trevisi, E., Ballou, M. A., Cardoso, F. C., Luchini, D. N. and Loor, J. J. (2016a), Rumen-protected methionine compared with rumen-protected choline improves immunometabolic status in dairy cows during the peripartal period, *J. Dairy Sci.*, 99(11), 8956–8969.

Zhou, Z., Trevisi, E., Luchini, D. N. and Loor, J. J. (2017), Differences in liver functionality indexes in peripartal dairy cows fed rumen-protected methionine or choline are associated with performance, oxidative stress status, and plasma amino acid profiles, *J. Dairy Sci.*, 100(8), 6720–6732.

Zhou, Z., Vailati-Riboni, M., Trevisi, E., Drackley, J. K., Luchini, D. N. and Loor, J. J. (2016b), Better postpartal performance in dairy cows supplemented with rumen-protected methionine compared with choline during the peripartal period, *J. Dairy Sci.*, 99(11), 8716–8732.

Chapter 10

The use of supplements to mitigate enteric methane emission in dairy cattle

Alexander N. Hristov, The Pennsylvania State University, USA; Claudia Arndt, International Livestock Research Institute, Kenya; Audino Melgar, Instituto de Innovation Agropecuaria de Panama, City of Knowledge, Panama and The Pennsylvania State University, USA; and Derek Wasson, The Pennsylvania State University, USA

1 Introduction

Livestock enteric CH_4 mitigation is an old feed energy efficiency problem with new environmental dimensions. Government and public are interested in finding solutions to climate change, and it is believed that mitigation of agricultural GHG emissions is part of the solution. In the USA, agricultural activities are responsible for the generation of GHG such as CO_2, CH_4, and N_2O, with the last two gases being of primary interest as agriculture contributed 39% (CH_4) and 80% (N_2O) of their total emissions in 2019, on a CO_2-equivalent basis (USEPA, 2021). Within agriculture, livestock is responsible for 94% of all CH_4 emissions in the USA (USEPA, 2021).

Methane mitigation has been identified as important for addressing climate change, and a recent analysis suggested that strategies exist to decrease global CH_4 emissions from human activities by 50% within the next decade

http://dx.doi.org/10.19103/AS.2022.0117.10

(Ocko et al., 2021). These authors estimated that immediate implementation of available mitigation measures could slow global warming by about 30% in the next decade and argued that efforts aimed at benefiting climate in the near-term, such as CH_4 mitigation, should be pursued simultaneously with long-term goals of reducing global CO_2 emissions.

A Food and Agriculture Organization (FAO)-commissioned analysis of livestock non-CO_2 GHG mitigation strategies identified a number of viable options that can decrease CH_4 and N_2O emissions from enteric fermentation and manure management (Hristov et al., 2013a). That report and other recent reviews (i.e. Beauchemin et al., 2020; Hegarty et al., 2021; Arndt et al., 2022) outlined opportunities and limitations of practices aimed at mitigating enteric CH_4. Developments in the field, such as recent research with the CH_4 inhibitor 3-NOP (Melgar et al., 2020a,b, 2021; Schilde et al., 2021) and the macroalga *Asparagopsis taxiformis* (Li et al., 2018; Roque et al., 2019a; Kinley et al., 2020; Stefenoni et al., 2021), call for an updated analysis of available enteric CH_4 mitigation options. Therefore, this chapter will highlight the most promising nutritional strategies for enteric CH_4 mitigation, will summarize findings of a recent comprehensive meta-analysis of enteric CH_4 mitigation strategies conducted under the GLOBAL NETWORK project, and will present the results of a meta-analysis of production and CH_4 emission data from 3-NOP experiments conducted at The Pennsylvania State University.

This chapter is focused on enteric CH_4 emissions; however, it is important to point out that agricultural GHG mitigation strategies should not be limited to direct livestock emissions (from enteric fermentation or manure management) but should also include indirect emissions from crop production. Although beyond the scope of this chapter, GHG emissions and sinks from growing livestock feed are an important part of the carbon cycle and balance in an agricultural production system and must be considered in the context of sustainable global food production (see Hristov et al., 2018).

2 Dietary interventions

For an overview of nutritional interventions for mitigating enteric CH_4 emission in ruminant systems, the reader is referred to several comprehensive reviews published during the last decade (Hristov et al., 2013b; Knapp et al., 2014; Cobellis et al., 2016; Beauchemin et al., 2020; Hegarty et al., 2021; Arndt et al., 2022). This section will only offer a brief summary of two dietary approaches to decrease enteric CH_4 emission: improving forage quality and concentrate inclusion. Increasing forage digestibility and digestible forage intake has the potential for decreasing CH_4 emission intensity (CH_4 emitted per kilogram of milk, energy-corrected milk (ECM), or average daily gain in growing animals) and was one of the major recommended mitigation practices in the analysis by

Hristov et al. (2013b). Although responses vary, enteric CH_4 emissions can be reduced when corn silage replaces grass silage in the diet (Hassanat et al., 2013; Benchaar et al., 2015; Ramin et al., 2021; van Gastelen et al., 2022). Feeding legume silages could also lower CH_4 emissions compared to grass silage due to their lower fiber concentration (Waghorn et al., 2002; Archimède et al., 2011). However, the mitigation effect of legumes (vs. grasses) is less consistent than that of corn silage and with some species, specifically red clover, slight increases in CH_4 yield (i.e. gram CH_4 per kilogram of DMI) have been reported (Gidlund et al., 2017). Maturity at harvest is an important factor determining CH_4 emissions from ruminants fed predominantly forage diets (Eugène et al., 2021). These authors also concluded that legumes rich in tannins can decrease overall GHG emissions by mitigating both enteric CH_4 and manure N_2O emission by shifting N excretion from urine to feces. Additionally, to properly compare the carbon footprint of different forages in dairy cow nutrition, one has to account for GHG from forage and crop production and soil carbon sequestration (Beauchemin et al., 2010; Thoma et al., 2013), which is beyond the scope of the current text.

Another mitigation practice that has been consistently found to decrease daily CH_4 emissions, yield and intensity is increasing the inclusion of concentrate feeds in the diet in lieu of forages. However, implementation of this practice can be limited by feed availability, cost, and the risk of rumen function disturbances. Further, marked improvements in the Y_m factor (i.e. emitted CH_4 energy as percent of feed gross energy intake) can be expected beyond 35–40% concentrate inclusion in the diet, but the effect is also dependent on the overall level of feed intake (Sauvant and Giger-Reverdin, 2009). Arndt et al. (2022) reported that decreasing dietary forage-to-concentrate ratio decreased CH_4 yield by an average of 13% (over control treatments within studies), increased milk yield (17%), and decreased CH_4 intensity (9%). These effects were not associated with decreased fiber digestibility, which is typically reported for higher concentrate inclusion rates (Nousiainen et al., 2009; Agle et al., 2010).

3 Lipids

The recent meta-analysis of published treatment means data by Arndt et al. (2022) identified several CH_4 mitigation practices (Table 1) that substantially decrease daily CH_4 emissions and do not negatively affect DMI and animal production to the extent that CH_4 yield and intensity would be increased.

Dietary lipids and lipid supplements are well known to decrease enteric CH_4 emission and, along with oilseeds, were one of the recommended mitigation strategies identified by Arndt et al. (2022). In that analysis, lipids and oilseeds decreased daily CH_4 emissions by an average of 19.5% and

Table 1 Effective nutritional strategies for mitigating daily enteric CH_4 emission and emission yield and intensity in ruminants[a]

Mitigation strategy	*n*[b]	Mean effect[c]	95% CI[d]	*P*-value[e]	I^2
Daily CH_4, g/day					
Inhibitors	23	−35.2	(−40.4; −29.5)	<0.001	76.9
Electron sinks	54	−17.1	(−20.1; −14.0)	<0.001	70.6
Oils and fats[f]	63	−19.5	(−23.6; −15.2)	<0.001	96.0
Tanniferous forages	42	−11.6	(−16.1; −6.8)	<0.001	86.0
CH_4 yield, g/kg DMI					
Inhibitors	23	−33.6	(−39.9; −26.6)	<0.001	77.3
Electron sinks	51	−15.4	(−18.1; −12.6)	<0.001	72.0
Oils and fats[g]	52	−14.8	(−18.0; −11.4)	<0.001	84.2
Tanniferous forages	39	−10.0	(−14.0; −5.9)	<0.001	84.0
CH_4 intensity, g/kg milk yield (dairy animals)					
Inhibitors	3	−31.8	(−40.0; −22.4)	<0.001	0.0
Electron sinks	12	−12.7	(−16.3; −9.0)	<0.001	17.3
Oils and fats[h]	24	−11.9	(−17.6; −5.7)	0.001	77.1
Tanniferous forages	7	−17.8	(−26.4; −8.1)	0.003	68.2

I = Heterogeneity statistic (%).
[a]Adapted from Arndt et al. (2022).
[b]*n* = number of treatment comparisons.
[c]Decrease from control, %.
[d]Lower and upper 95% confidence interval, %.
[e]*P*-value for the mitigation effect.
[f]Similar effect was observed for oilseeds (n = 26; mean effect = −19.5%; 95% CI = −24.0% and −14.8%; $P < 0.001$).
[g]Similar effect was observed for oilseeds (n = 18; mean effect = −14.3%; 95% CI = −19.9% and −8.2%; $P < 0.001$).
[h]Similar effect was observed for oilseeds (n = 6; mean effect = −11.6%; 95% CI = −18.9% and −3.6%; $P = 0.02$).

decreased CH_4 yield and intensity [by 15% and 12–22% (milk or meat production), respectively]. However, feed intake and fiber digestibility were also decreased (by 5.7%, $P < 0.001$ and 4.2%, $P = 0.005$, respectively), although this did not result in decreased animal production. Lipids decrease CH_4 intensity by providing a highly digestible source of energy to the host animal that can boost milk production (Rabiee et al., 2012). Arndt et al. (2022) indicated that within the oil and fat category, the CH_4 mitigation effect was significant for vegetable oils but not for animal fat, which confirmed previous reports (Grainger and Beauchemin, 2011; Hristov et al., 2013b). Care must be taken when feeding vegetable oils to ruminants because of its detrimental effect on rumen fermentation, specifically that of polyunsaturated fatty acids, often leading to milk fat depression (Palmquist and Jenkins, 2017).

The cost of lipid supplementation, particularly in developing countries, may also hinder the adoption of this mitigation strategy. As concluded by Hristov et al. (2013a), '*...lipids are effective in reducing enteric CH_4 emission, but the feasibility of this mitigation practice depends on its cost-effectiveness and potential effects on feed intake (negative), productivity (negative) and milk fat content in lactating animals (positive or negative)*'. The Arndt et al. (2022) analysis indicated that oilseeds have a CH_4 mitigation effect similar to that of oils, with perhaps the advantage that the oil may be released at a slower rate in the rumen, thus having a less harmful effect on the microbial ecosystem. Some by-product feeds routinely fed to dairy cows, such as extruded oilseed meals or high-oil distiller grains, could also decrease CH_4 emission (Hristov et al., 2013b). The effect of these feeds on CH_4 emission and animal performance, however, appears to be inconsistent. As an example, substitution of solvent-extracted soybean meal (1.3% oil) with extruded soybean meal (8.6% oil) had no effect on daily CH_4 emission and CH_4 yield and intensity in dairy cows, despite a 25% increase in dietary ether extract concentration (Harper et al., 2019).

4 Electron sinks

Alternative electron sinks (fumarate, data from 11 studies, and nitrates, data from 24 studies) were another effective CH_4 mitigation strategy identified by Arndt et al. (2022). These rumen modifiers decreased daily CH_4 emissions and both CH_4 yield and intensity (Table 1) and appeared to have increased slightly milk yield (by 3%; $P = 0.001$), despite decreased DMI (1.5%; $P = 0.008$). Similar mitigation effect for nitrates was reported by Feng et al. (2020) in a recent meta-analysis but that study pointed to a differential response in beef compared with dairy cattle, attributing the lower mitigation effect in beef to inclusion of slow-release nitrate studies in the database. Although effective, the nitrate strategy has been criticized as unsafe for the animal due to toxicity, particularly if applied in commercial settings. Another potential issue with nitrates could be increased N excretion in urine and, consequently, increased NH_3 and N_2O emissions from manure during storage or following soil application, although this has not been systematically studied yet. A recent report by Villar et al. (2021) found 49% of the ^{15}N from labeled nitrate in urine 141 h after being fed to sheep, approximately 41% of which was recovered as urinary urea. This led the authors to conclude that plasma nitrate is reduced to NH_3 and partially converted to urea in the liver and excreted in urine suggesting the possibility of increased manure N emissions when nitrates are fed to ruminants. The latter could be particularly problematic when diets are high in overall crude protein and rumen-degraded protein, causing an increased excretion of nitrate-N in urine. There are also indications that the practice can increase, although not substantially, enteric

N_2O emission (Petersen et al., 2015) and was determined to be economically unfeasible by European studies (Doreau et al., 2014; van Middelaar et al., 2014).

5 Plant-based bioactive compounds

Botanicals and specifically essential oils (EO) have also been extensively studied as CH_4 mitigants (Calsamiglia et al., 2008). However, the effect of this category of compounds on CH_4 emissions in controlled studies has been inconsistent, and the strategy has not been recommended in the analysis of Hristov et al. (2013b). Many studies have reported mitigation effects of botanicals or EOs *in vitro*; but it is widely accepted that *in vitro* systems, although well suited for screening of large number of treatments, are not representative of live animal responses, and it is strongly recommended that *in vitro* data are confirmed in *in vivo* experiments (Hristov et al., 2012; Flachowsky and Lebzien, 2012; Yáñez-Ruiz et al., 2016). Overall, variability in animal responses to EO-based additives has been extremely high (Hegarty et al., 2021). These compounds generally have a broad spectrum of activities and in some cases negatively affect overall rumen fermentation. It has been suggested that the field may benefit from a more targeted compound approach (Cobellis et al., 2016). Recent results for some commercial products have been moderately encouraging (Roque et al., 2019b; Belanche et al., 2020; Silvestre et al., 2021) and need to be confirmed. In their analysis, Hegarty et al. (2021) rated the CH_4 reduction potential of EO as low (≤ 5% reduction) and the confidence in their efficiency at 2 [low agreement and medium evidence, based on IPCC methodology (Smith et al., 2014)].

Tannins as feed supplements or as tanniferous plants have often, but not always (Beauchemin et al., 2007a), shown a potential for reducing CH_4 emission by up to 20% (Woodward et al., 2001; Sliwinski et al., 2002; Waghorn et al., 2002; Zhou et al., 2011; Staerfl et al., 2012). Screening experiments at The Pennsylvania State University (A. N. Hristov, unpublished data) have shown up to 54% reduction in CH_4 production when condensed tannin products were included at up to 150 mg/L in the incubation medium *in vitro* (Fig. 1). Tannins, however, would inevitably have antinutritional effects, particularly in protein-limited diets, because they tend to reduce amino acid absorption (Waghorn, 2008). Thus, it is important that benefits of reduced CH_4 yields do not overshadow detrimental effects of tannins on digestion and production. A meta-analysis of *in vivo* experiments with tannins by Jayanegara et al. (2012) reported a relatively close relationship between dietary tannin concentration and CH_4 production per unit of digestible organic matter. These authors, however, reported a trend for decreased feed intake and a statistically significant decrease in nutrient digestibility, particularly of crude protein, with increasing dietary tannin concentration. The Arndt et al. (2022) meta-analysis concluded that supplementation of the diet with condensed tannins did not

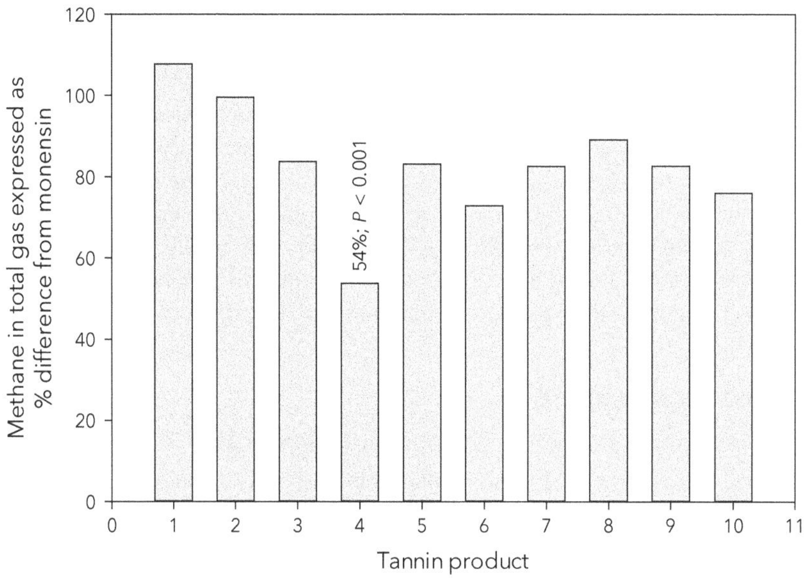

Figure 1 Methane mitigation effect of tannin products *in vitro*. Methane (percent of gas produced) is expressed as percent difference over monensin (5 mg/L medium, used as a positive control). Tannin products were tested at three concentrations (15 mg/L, 75 mg/L, and 150 mg/L medium). Overall treatment effect, $P < 0.001$; inclusion rate effect, $P \geq 0.006$; treatment × inclusion rate, $P \geq 0.01$. Total gas production was not affected ($P \geq 0.06$) by treatment.

affect daily CH_4 emissions but decreased CH_4 yield by 12% (seven publications; $P = 0.01$) and fiber digestibility by 12% (five publications; $P = 0.003$). Overall, the risk of impaired rumen function and animal productivity with tannins is greater than with saponins (Goel and Makkar, 2012). The anti-methanogenic effect of saponins is less pronounced than that of tannins (Goel and Makkar, 2012). Reports of a strong CH_4 mitigation effect of tea saponins (Wang et al., 2012) have not been confirmed.

Inclusion of tanniferous forages in the diet was a mitigation strategy recommended by the Arndt et al. (2022) meta-analysis. The mitigation effect was lower relative to the other recommended strategies (Table 1), and the authors indicated that fiber digestibility was also decreased (6.9%; $P = 0.02$), which could potentially increase manure CH_4 emissions. Among the forages included in the database, *Sericea lespedeza* (*Lespedeza cuneata*) reduced daily CH_4 emissions by 32% (three publications; $P < 0.001$). Other tanniferous forages that had some effect on CH_4 emissions were *Leucaena* and *Lotus* (*corniculatus and pedunculatus*) with 8% average reduction (seven and two studies at $P = 0.10$ and $P = 0.01$, respectively). Studies with these forages did not report data on animal performance, and, therefore, effects on CH_4 intensity could not be elucidated.

As mentioned above, tannins (both from grazing tanniferous forages and through dietary supplementation of condensed tannins) have the ability to shift N excretion from urine to feces, which would decrease N_2O (and NH_3) emissions from manure. Feeding of 'or grazing' tanniferous forages is particularly attractive for extensive and pasture-based production systems, but more long-term research with high-producing animals is needed to conclude with confidence that tannins and tanniferous forages are a viable and cost-effective mitigation strategy for decreasing livestock GHG emissions. Due to these uncertainties, Hegarty et al. (2021) rated the CH_4 reduction potential of tannins as low and the confidence in their efficacy at 2 (low agreement and medium evidence).

6 Chemical inhibitors

Methane inhibitors (3-NOP and bromochloromethane) had the largest CH_4 mitigation effect and had no effect on DMI, fiber digestibility, and milk production or average daily gain in the Arndt et al. (2022) meta-analysis. Since 2018 [the last year data were included in the Arndt et al. (2022) database], new data from 3-NOP experiments with beef and dairy cattle were published. For example, an extensive study on a commercial feedlot in Alberta, Canada, reported a 26% decrease in CH_4 emissions and a slight numerical decrease in DMI, which resulted in a trend for increased feed efficiency in beef cattle fed a backgrounding diet (Alemu et al., 2021).

An earlier meta-analysis of 3-NOP data (11 experiments) in dairy and beef cattle concluded that the CH_4 mitigation effect of the inhibitor was less pronounced in beef (22% reduction in daily CH_4 emissions) than in dairy (39% reduction) cattle and that neutral-detergent fiber content of the diet impairs the effect of the additive (Dijkstra et al., 2018). Another meta-analysis (18 experiments) reported a trend for decreased DMI in beef (but not dairy) cattle and trends for decreased milk yield and increased concentration of milk fat and protein in dairy cows (Kim et al., 2020).

A meta-analysis of Penn State's 3-NOP data from five experiments with lactating dairy cows was conducted to elucidate the effects of the inhibitor on CH_4 emissions and animal production variables. Cows involved in these studies were both primi- and multiparous and in mid-lactation (with the exception of Melgar et al., 2020a, where cows entered the study within 3 days after calving). The average milk production was above 40 kg/day, with milk components typical for Penn State's dairy herd. Diets contained on average 31% amylase-treated neutral-detergent fiber, 44% non-fiber carbohydrates, and 5.3% ether extract.

The analysis showed no effect of 3-NOP on DMI, milk yield, ECM yield, feed efficiency, and body weight and body weight change of the cows (Table 2). Milk true protein and lactose concentrations and yields were also not affected by 3-NOP. Concentration of milk fat, however, was increased ($P = 0.04$) by 0.19%

units and fat yield tended to be increased (P = 0.06) by 90 g/day by 3-NOP compared with the control. Across studies, milk urea nitrogen concentration was also increased (P = 0.03) by 3-NOP compared with the control. Likely explanation for the milk fat effect is the observed increase in rumen butyrate concentration (one of the main precursors for *de novo* synthesized fatty acids in milk) with 3-NOP and consequent increase in short-chain fatty acids, accompanied by a decrease in *trans* fatty acids, in milk fat (Melgar et al., 2021). The hypothesis for explaining the milk urea nitrogen effect of 3-NOP is based on the observed consistent increase in ruminal butyrate concentration by the inhibitor (Lopes et al., 2016; Guyader et al., 2017; Melgar et al., 2020a). Butyrate is known to stimulate blood flow and, consequently, NH_3 absorption in the rumen (Engelhardt et al., 1978; Rémond et al., 1993), which is supported by the observed decrease in rumen NH_3 concentration by 3-NOP in some experiments (Lopes et al., 2016; Melgar et al., 2020a). Thus, increased absorption of NH_3 would likely explain the increase in milk urea nitrogen in several studies with 3-NOP included in the current meta-analysis (Melgar et al., 2020a,b, 2021).

The average reduction (P < 0.001) in daily CH_4 emissions by 3-NOP was 123 g/day (or 28%; 310 g/day vs. 433 g/day for 3-NOP and control, respectively; Table 2). The reductions (P < 0.001) in CH_4 yield and intensity were 4.91 g/kg DMI (or 28%; 12.3 g/kg and 17.2 g/kg, respectively) and 3.74 g/kg ECM (or 32%; 8.0 g/kg and 11.8 g/kg, respectively). Emissions of H_2 were largely increased (P < 0.001; 1.41 g/day vs. 0.13 g/day, respectively) by 3-NOP, but daily CO_2 emission and emission yield were not affected by the inhibitor. The intensity of CO_2 emission, however, was slightly decreased (by 18 g/kg ECM, or 5%; P = 0.03) by 3-NOP compared with the control.

The relationship between 3-NOP intake and daily CH_4 emissions was exponential (Fig. 2; P = 0.002), with the mitigation effect diminishing beyond 2 to 2.4 g 3-NOP/cow/day. As discussed in Melgar et al. (2020b), although numerically maximum mitigation effect was achieved at 150 mg 3-NOP/kg DMI (corresponding to an average 3-NOP intake of 3.7 g/cow/day), there was no difference in CH_4 emissions between 100 mg/kg, 150 mg/kg, and 200 mg/kg DMI. Similarly, CH_4 yield and intensity were exponentially decreased by 3-NOP intake (Fig. 2; P = 0.004 for both metrics), and there was no further improvement in the mitigation effect beyond 100 mg 3-NOP/kg DMI.

The mitigation effect of 3-NOP is reversible, as evident from Fig. 3 (data from Melgar et al., 2020a). At the end of that experiment, cows receiving 3-NOP in Phase 2 of the main experiment were switched to the control treatment, and cows that were on the control treatment were fed 3-NOP for another 3 weeks (experimental Phase 3) and CH_4 emissions were measured during week 3. As Fig. 3 shows, control cows in Phase 3 (that were on 3-NOP in Phase 2) had CH_4 yield similar to control cows in Phase 2 (14.3 g/kg vs. 15.1 g/kg DMI, respectively). Likewise, cows that were on the control treatment in Phase 2

Table 2 A meta-analysis of the effect of 3-nitrooxypropanol (3-NOP) on production variables and enteric gas emissions in Penn State's experiments with dairy cows[a]

Variable[b]	Difference in means[c]	Standardized difference in means[d]	*SE*	Variance	Lower 95% CL for mean	Upper 95% CL for mean	*P*-value[e]
DMI, kg/day	0.40	0.08	0.161	0.026	−0.238	0.391	0.63
Milk yield, kg/day	0.22	0.03	0.159	0.025	−0.278	0.346	0.83
Milk yield/DMI, kg/kg	−0.01	−0.02	0.160	0.026	−0.335	0.293	0.89
ECM, kg/day	1.25[f]	0.20	0.160	0.026	−0.110	0.519	0.20
ECM/DMI, kg/kg	−0.03	0.18	0.160	0.026	−0.137	0.492	0.27
Milk components, %							
Fat	0.19	0.34	0.161	0.026	0.023	0.654	0.04
True protein	0.06	0.23	0.160	0.026	−0.081	0.545	0.15
Lactose	−0.03	−0.02	0.160	0.026	−0.517	0.110	0.20
MUN, mg/dL	0.35	0.34	0.161	0.026	0.026	0.657	0.03
Milk components, kg/day							
Fat	0.09	0.30	0.161	0.026	−0.014	0.618	0.06
True protein	0.03	0.16	0.160	0.026	−0.154	0.473	0.32
Lactose	0.006	0.011	0.160	0.025	−0.301	0.324	0.95
BW, kg	6.84	0.014	0.160	0.026	−0.299	0.327	0.93
BW change, g/day	168	−0.036	0.165	0.027	−0.360	0.288	0.83

Daily enteric gas emissions, g/day							
CH_4	−123	−2.03	0.191	0.036	−2.408	−1.659	<0.001
CO_2	11.6	−1.23	0.160	0.026	−0.437	0.190	0.44
H_2	1.27	2.30	0.207	0.043	1.900	2.706	<0.001
Emission yield, g/kg DMI							
CH_4	−4.91	−2.01	0.194	0.038	−2.390	−1.651	<0.001
CO_2	−1.84	−0.11	0.160	0.026	−0.042	0.208	0.51
Emission intensity, g/kg ECM							
CH_4	−3.74	−1.73	0.183	0.034	−2.080	−1.366	<0.001
CO_2	−18.0	−0.35	0.160	0.026	−0.666	−0.038	0.03

[a]Data from: Hristov et al. (2015); Lopes et al. (2016); Melgar et al. (2020a,b); and Melgar et al. (2021).
[b]DMI, dry matter intake; ECM, energy-corrected milk; MUN, milk urea nitrogen; BW, body weight.
[c]3-NOP treatments - control.
[d]Fixed effects.
[e]*P*-value for the standardized difference in means.
[f]Due to a +2.9 kg/day difference in one experiment (Lopes et al., 2016).

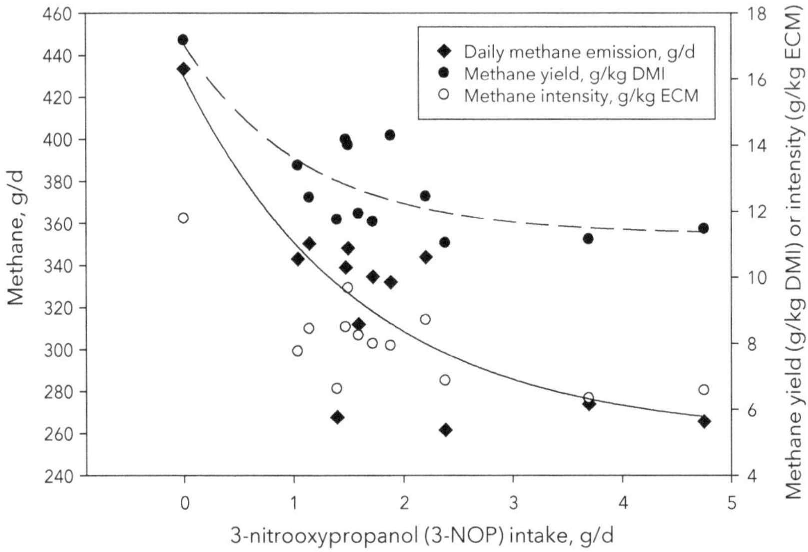

Methane, g/cow/d = 430.1 - 170.1 × [1 - exp(-0.62 × 3-NOP intake, g/cow/d)]
Adj. R^2 = 0.63, P = 0.002

Methane, g/kg DMI = 11.30 - 5.77 × [1 - exp(-0.907 × 3-NOP intake, g/cow/d)]
Adj. R^2 = 0.61, P = 0.004

Methane, g/kg ECM = 6.53 - 5.02 × [1 - exp(-0.766 × 3-NOP intake, g/cow/d)]
Adj. R^2 = 0.61, P = 0.004

Figure 2 Relationship of 3-nitrooxypropanol (3-NOP) intake and methane emissions [as g/day, g/kg dry matter intake (DMI), or g/kg energy-corrected milk (ECM)] in dairy cows. Data are treatment means from the following experiments: Hristov et al. (2015); Lopes et al. (2016); Melgar et al. (2020a,b); and Melgar et al. (2021).

decreased their CH_4 yield to levels similar to those of 3-NOP cows in Phase 2 (12.4 g/kg vs. 10.3 g/kg DMI, respectively) when they received 3-NOP in Phase 3. These data suggest that the mitigation effect of 3-NOP disappears rapidly after the compound is removed from the feed and also that the effect is taking place within 2 weeks of treatment. Indeed, the first published dairy cow study with 3-NOP (Reynolds et al., 2014) reported a sharp decline in CH_4 emissions immediately after the inhibitor was administered through the rumen cannula, but the effect disappeared rapidly (within 2–3 h) after dosing. As pointed out in Hristov and Melgar (2020), due to its chemical structure and solubility, the residence time of 3-NOP in the rumen is likely short. These authors observed marked differences in the mitigation effect of 3-NOP during a 24-h feeding cycle; CH_4 emission was similar between control and 3-NOP 2 h before feeding, but 3-NOP cows had 45% lower CH_4 emission, compared with the control, immediately after feeding, with the mitigation effect decreasing to 13% 4 h

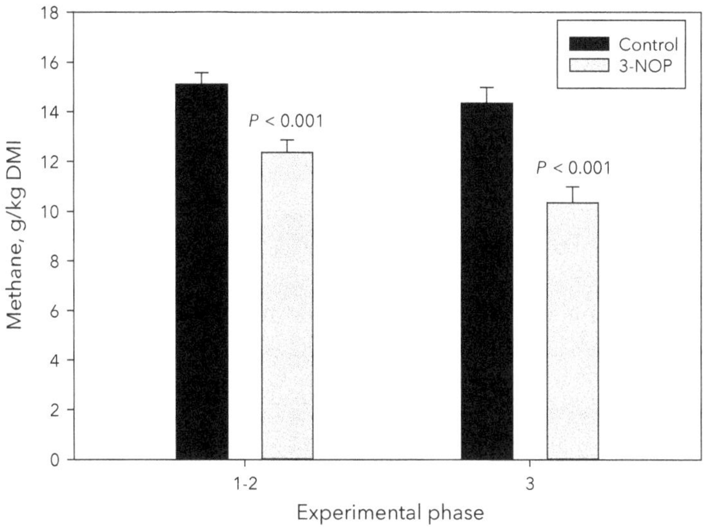

Figure 3 Enteric methane yield [g/kg dry matter intake (DMI)] of dairy cows receiving 3-nitrooxypropanol (3-NOP). Data are from Melgar et al. (2020a) and represent treatment means and associated standard error of the mean. In experimental Phase 1–2, treatment cows received 60 mg/kg DMI 3-NOP for 15 weeks, and data shown in graph are from experimental week 15. In Phase 3, control cows from Phase 1–2 received 60 mg/kg DMI 3-NOP for 3 weeks and methane emissions were measured during week 3. Cows receiving 3-NOP in Phase 1–2 were control cows in Phase 3.

before feeding. These trends were clearly related to DMI (i.e. 3-NOP intake) by the cows. Thus, Hristov and Melgar (2020) concluded that the CH_4 mitigation effect of 3-NOP is highest immediately after feeding (highest 3-NOP intake) and lowest, or non-existent, before feeding (lowest 3-NOP intake).

The mitigation effect of 3-NOP was proven in the long-term experiments with dairy cows discussed here and in trials with beef cattle (Vyas et al., 2018; and more recently, Alemu et al., 2021). An interesting observation, however, was recently reported by Schilde et al. (2021). In that study with lactating dairy cows, the mitigation effect of 3-NOP (applied at around 48 mg/kg DMI) disappeared after 8 weeks on treatment, when cows were fed a high-forage/low-concentrate diet (40% neutral-detergent fiber and 25% starch, feed DM basis) but not a low-forage/high-concentrate diet (34% neutral-detergent fiber and 31% starch). This observation was in line with the Dijkstra et al. (2018) meta-analysis where dietary neutral-detergent fiber was identified as a factor that may be decreasing the mitigation effect of 3-NOP. What causes this interaction is unclear. Schilde et al. (2021) hypothesized that lower DMI and increased rumen retention time with the high-forage/low-concentrate diet resulted in enhanced H_2 accumulation, which may have played a role in the diminishing mitigation effect of 3-NOP over time. Similarly, McGinn et al.

(2019) reported a 7.5–26.1% (depending on the measurement method used) decrease over 90 days in the enteric CH_4 emission reduction by 3-NOP (fed at 125 mg/kg feed DM) in beef cattle on a commercial feedlot. Other studies from the Lethbridge group, however, did not report signs of adaptation to 3-NOP over periods of 112 days (Romero-Perez et al., 2015) or 108 days (Alemu et al., 2021).

We examined the dynamics of CH_4 emissions in two long-term studies with 3-NOP from our laboratory (Melgar et al., 2020a, 2021). The averaged mitigation effect of 3-NOP (expressed as percent reduction in CH_4 yield compared with the control within experiment) was regressed over the duration of the experiments, as shown in Fig. 4. There was an apparent linear decrease [adjusted $R^2 = 0.58$ and 0.81; $P = 0.15$ and 0.02 for Melgar et al. (2020a) and Melgar et al. (2021), respectively] in the mitigation effect of 3-NOP over time. In the case of Melgar et al. (2020a), the mitigation effect decreased from 24% during experiment week 2 to 17% during week 15. Similarly, the mitigation effect of 3-NOP in the Melgar et al. (2021) study decreased from 31% during week 3 to 24% during week 15. In the first experiment, the rate of decrease was 0.61 g/kg DMI per week, and in the second experiment, the decrease was 0.56 g/kg per week. These observations suggest a possible transient, or inconsistent, effect of 3-NOP over time in diets that are around 28–31% (or higher, as implied by Schilde et al.,

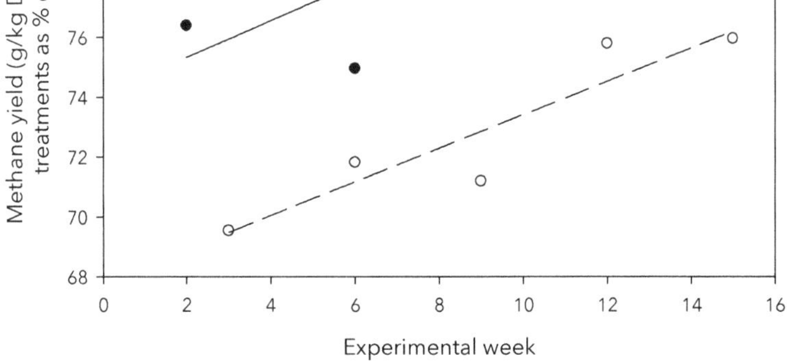

Figure 4 Enteric methane yield of dairy cows receiving 3-nitrooxypropanol (3-NOP) treatments expressed as percentage of the control within experiment (DMI = dry matter intake). Data are from Melgar et al. (2020a) and Melgar et al. (2021) and represent differences treatment means during each methane measurement period within experiment.

2021) neutral-detergent fiber and require further investigation in long-term, full -lactation, or multiple-lactations studies, as proposed by Tricarico et al. (2022).

7 Macroalgae

The interest in macroalgae for mitigation of enteric CH_4 emissions in ruminants has dramatically increased in recent years and since Li et al. (2018) documented a strong anti-methanogenic effect of the red alga *Asparagopsis taxiformis* in sheep. Research groups around the globe have screened red, brown, and green macroalgae for anti-methanogenic effects (Dubois et al., 2013; Machado et al., 2014; Maia et al., 2016; Wasson et al., 2021), and while some species have shown promising results, *Asparagopsis* spp. (*taxiformis* and *armata*) appear to be the only ones with confirmed mitigating effect in *in vivo* experiments with dairy and beef cattle (Li et al., 2018; Roque et al., 2019a; Kinley et al., 2020; Stefenoni et al., 2021). The current understanding is that the anti-methanogenic activity of *Asparagopsis* spp. is based on its content of low-molecular-weight halogenated compounds, of which the brominated halomethane bromoform is dominant (Genovese et al., 2012). *Asparagopsis* spp. do cause a dramatic decrease in CH_4 emissions *in vivo*, but DMI may also decrease (Stefenoni et al., 2021) and there are concerns with the environmental impact of bromoform (ozone layer

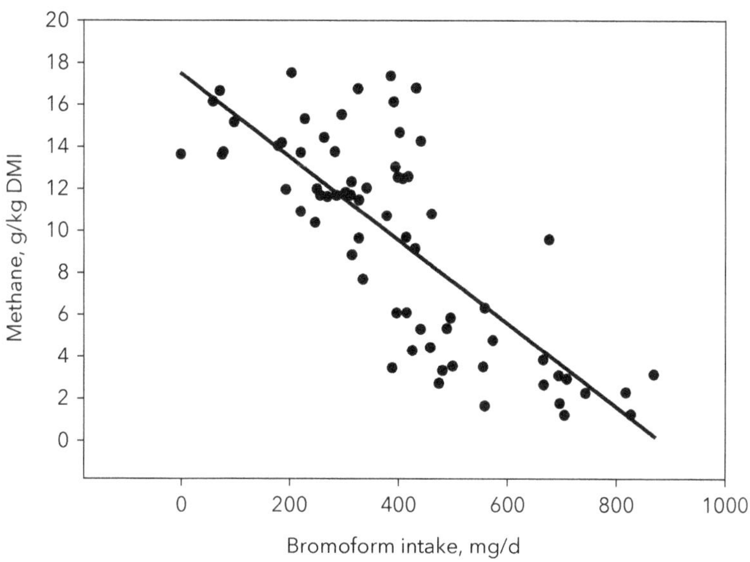

Figure 5 Relationship of bromoform intake from *Asparagopsis taxiformis* and enteric methane emission yield in dairy cows [data from Stefenoni et al. (2021) and unpublished data from A. N. Hristov, The Pennsylvania State University]. Data are individual animal observations. Overall relationship: CH_4 yield, g/kg dry matter intake (DMI) = 17.4733 (*SE* = 0.8772) - 0.0198 (*SE* = 0.0020) × bromoform intake, mg/day (R^2 = 0.59, $P < 0.001$).

depletion) and effects on animal health and milk quality (Stefenoni et al., 2021; Muizelaar et al., 2021; Hegarty et al., 2021). Bromoforms are volatile and activity may decrease over prolonged storage, or if the seaweed is exposed to light or heat (Stefenoni et al., 2021). As our data clearly show, decreasing bromoform concentration and its intake will linearly diminish the mitigation potential of *A. taxiformis* (Fig. 5; data from Stefenoni et al., 2021 and unpublished data by A. N. Hristov, The Pennsylvania State University). Based on these data, CH_4 yield will decrease by 1.5 g/kg to 2.0 g/kg DMI for every 100 mg/day increase in bromoform intake. However, data from a recent study with lactating dairy cows indicated a decrease in the mitigation effect of *A. taxiformis* beyond 10 weeks of treatment, which may suggest rumen adaptation (unpublished data by D. Wasson and A. N. Hristov, The Pennsylvania State University). Thus, long-term effects of *A. taxiformis* (and all seaweeds) on animal productivity, health, reproduction, and milk quality need to be studied, and the economics of mass application in the global dairy and beef industries are unclear. As a result of these uncertainties, Hegarty et al. (2021) rated the confidence in *Asparagopsis* spp. efficacy in their meta-analysis at 1 (low agreement and limited evidence). Other seaweed species have also shown promising preliminary results *in vitro* (Fig. 6; unpublished data by D. Wasson and A. N. Hristov, The Pennsylvania State University) that need to be confirmed in animal experiments. Research in this novel and exciting field will certainly continue in the near future, but its long-term impact on livestock GHG emissions is difficult to predict.

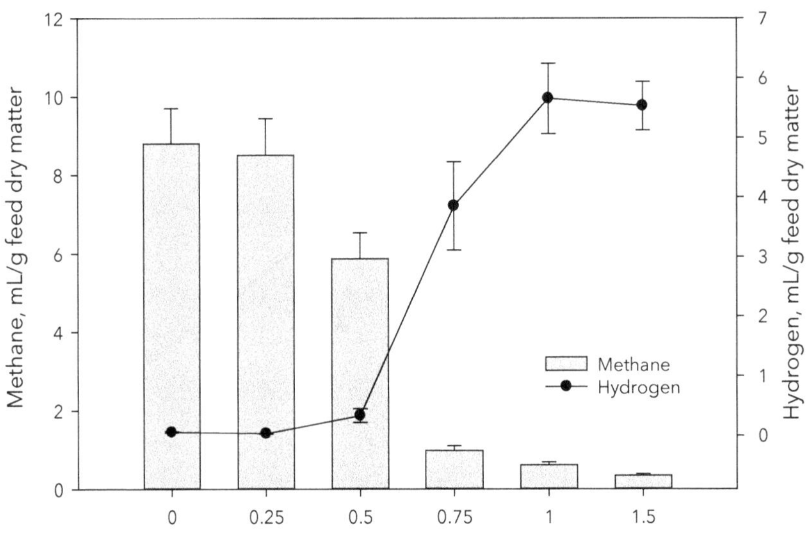

Figure 6 Effect of seaweed X on methane and hydrogen emissions *in vitro*. Data are means and associated *standard error* (unpublished data by D. Wasson and A. N. Hristov, The Pennsylvania State University).

8 Additive effects of feed supplements

An important question when it comes to dietary feed additives as enteric CH_4 mitigants is that of additivity of their effects. Depending on the cost (see next section), it may be practical to use more than one proven mitigation practice, which may enhance the overall mitigation effect. In theory, it is more plausible to expect additive effects from supplements that have contrasting modes of action. As an example, Zhang et al. (2021) investigated the effect of simultaneous addition of 3-NOP and canola oil on CH_4 emission in beef cattle. The inhibitor 3-NOP decreased CH_4 yield by 32% and the canola oil decreased it by 27%. Interestingly, simultaneous addition of the mitigants resulted in a 51% reduction in CH_4 yield. It is noted, however, that: (1) application rates for both additives were high - 200 mg/kg and 50 g/kg feed DM for 3-NOP and canola oil, respectively and (2) this was a short-term (28-day experimental periods), crossover experiment and the persistency of the additive effect is unknown. Not surprisingly, at this high inclusion rate, canola oil decreased fiber and overall dietary DM digestibility. A similar additive mitigation effect on CH_4 emission was reported when lipids were combined with nitrates (Guyader et al., 2015) or tannins (Williams et al., 2020). Thus, additivity has to be considered when effects on GHG emission and the wider impact on climate change of enteric CH_4 mitigation strategies are estimated.

9 Economics

One element that is often missing in enteric CH_4 discussions is the cost of implementing mitigation strategies with proven effectiveness. Our intention here is not to comprehensively discuss the literature on this topic but to bring the attention of the reader to this very important detail that may be a limiting factor in many national GHG mitigation programs. The reader is referred to several analyses where the economics of mitigating livestock GHG emissions are discussed in detail (Doreau et al., 2014; van Middelaar et al., 2014; Herrero et al., 2016). As stated in Herrero et al. (2016), although the technical mitigation potential for the livestock sector may be large, the share that can be achieved at a reasonable economic cost is likely to be much smaller. As an example, the current price of wild-harvest, freeze-dried *A. taxiformis* (discussed above) from one source (seaExpert, Feteira, Ilha Do Faial, Portugal) was €167.2 per kg, excluding transportation. At a feeding rate of around 0.5% of dietary DM (Stefenoni et al., 2021), inclusion of this seaweed in a dairy cow diet will cost the producer around $24/day, which is obviously entirely unrealistic. It should be noted, however, that this price is expected to rapidly decrease if/when technologies for aquaculture production of *Asparagopsis* are developed and global supplies increase. Thus, it is safe to assume that without revolutionary improvements in technology or government and market interventions (i.e. carbon credits/market and/or tax), the efficacy of a mitigation strategy alone will not be enough to ensure its adoption by the livestock industries. In addition, a number of key factors on the global scale need to be considered in the various mitigation scenarios, including emission leakage (increasing livestock emissions in one region as a result of decreased animal production in another region), rebound effects (i.e. increased demand for animal products as production efficiency increases and production costs decrease), emission pricing and demand-side effects (including dietary changes), and others (Herrero et al., 2016). Therefore, efficacy, although critically important, is only one piece of the complex puzzle for adoption of mitigation strategies for livestock GHG emissions.

10 Conclusion and future trends in research

A recent meta-analysis of a large database by Arndt et al. (2022) identified several enteric CH_4 mitigation strategies that were proven effective in *in vivo* studies with cattle and small ruminants. The recommended strategies included CH_4 inhibitors, alternative electron sinks, vegetable oils and oilseeds, and tanniferous forages. These strategies were effective in decreasing CH_4 emission yield and intensity as well, despite potential negative effects on DMI and fiber digestibility (e.g. a higher inclusion rate of oils may negatively affect feed intake, rumen function, and animal performance, specifically milk components in dairy cows). In the case of nitrates (electron sinks), concerns with animal health may be impeding their adoption in practice. Tannins and tanniferous forages may also have a negative impact on nutrient digestibility, and much research is needed to confirm their effects on overall animal performance in long-term experiments with high-producing animals. A meta-analysis of studies with lactating dairy cows fed the CH_4 inhibitor 3-NOP at The Pennsylvania State University showed a consistent 28–32% decrease in daily CH_4 emission or emission yield and intensity. The inhibitor had no effect on DMI, milk production, and body weight and body weight change of the cows but slightly increased milk fat concentration and yield (0.19%-units and 90 g/day, respectively).

The red macroalga *Asparagopsis taxiformis* has a strong CH_4 mitigation effect, but studies are needed to determine its feasibility, long-term efficacy, and effects on animal production and health. In general, long-term studies are lacking for all mitigation strategies that have shown efficacy in short-term experiments. Such studies must be conducted with a wide variety of diets and animal breeds and categories for these strategies to be adopted by the livestock industries. Several recommended practices have shown additivity in their CH_4 mitigation effect. However, more studies are needed to quantitatively determine the size of emission reductions (and possible co-benefits and emission trade-offs) when two or more enteric CH_4 mitigation practices are simultaneously used.

11 References

Agle, M., Hristov, A. N., Zaman, S., Schneider, C., Ndegwa, P. M. and Vaddella, V. K. (2010). Effect of dietary concentrate on rumen fermentation, digestibility, and nitrogen losses in dairy cows, *J. Dairy Sci.* 93(9), 4211–4222.

Alemu, A. W., Pekrul, L. K. D., Shreck, A. L., Booker, C. W., McGinn, S. M., Kindermann, M. and Beauchemin, K. A. (2021). 3-Nitrooxypropanol decreased enteric methane production from growing beef cattle in a commercial feedlot: implications for sustainable beef cattle production, *Front. Anim. Sci.* 2, 641590. doi:10.3389/fanim.2021.641590.

Anon. (2021). *Inventory of U.S. Greenhouse Gas Emissions and Sinks: 1990–2019, USEPA (United States Environmental Protection Agency).* US EPA, Washington, DC.

Archimède, H., Eugène, M., Magdeleine, C. M., Boval, M., Martin, C., Morgavi, D. P., Lecomte, P. and Doreau., M. (2011). Comparison of methane production between C3 and C4 grasses and legumes, *Anim. Feed Sci. Technol.* 166–167, 59–64.

Arndt, C., Hristov, A. N., Price, W. J., McClelland, S. C., Pelaez, A. M., Cueva, S. F., Oh, J., Bannink, A., Bayat, A. R., Crompton, L. A., Dijkstra, J., Eugène, M. A., Enahoro, D., Kebreab, E., Kreuzer, M., McGee, M., Martin, C., Newbold, C. J., Reynolds, C. K., Schwarm, A., Shingfield, K. J., Veneman, J. B., Yáñez-Ruiz, D. R. and Yu, Z. (2022). Full adoption of strategies to mitigate enteric methane emissions by ruminants and how they can help to meet the 1.5°C climate target by 2030 but not 2050, *Proc. Natl. Acad. Sci. U. S. A.* 119, e2111294119. doi:10.1073/pnas.2111294119.

Beauchemin, K. A., Janzen, H. H., Little, S. M., McAllister, T. A. and McGinn, S. M. (2010). Life cycle assessment of greenhouse gas emissions from beef production in western Canada: a case study, *Agric. Syst.* 103(6), 371–379. doi:10.1016/j.agsy.2010.03.008.

Beauchemin, K. A., McGinn, S. M., Martinez, T. F. and McAllister, T. A. (2007). Use of condensed tannin extract from quebracho trees to reduce methane emissions from cattle, *J. Anim. Sci.* 85(8), 1990–1996.

Beauchemin, K. A., Ungerfeld, E. M., Eckard, R. J. and Wang, M. (2020). Review: fifty years of research on rumen methanogenesis: lessons learned and future challenges for mitigation, *Animal* 14(S1) (Suppl.1), s2–s16. doi:10.1017/S1751731119003100.

Belanche, A., Newbold, C. J., Morgavi, D. P., Bach, A., Zweifel, B. and Yáñez-Ruiz, D. R. (2020). A meta-analysis describing the effects of the essential oils blend agolin ruminant on performance, rumen fermentation and methane emissions in dairy cows, *Animals (Basel)* 10(4), 620. doi:10.3390/ani10040620.

Benchaar, C., Hassanat, F., Martineau, R. and Gervais, R. (2015). Linseed oil supplementation to dairy cows fed diets based on red clover silage or corn silage: effects on methane production, rumen fermentation, nutrient digestibility, N balance, and milk production, *J. Dairy Sci.* 98(11), 7993–8008. doi:10.3168/jds.2015-9398.

Calsamiglia, S., Busquet, M., Cardozo, P. W., Castillejos, L. and Ferret, A. (2008). Invited review: essential oils as modifiers of rumen microbial fermentation, *J. Dairy Sci.* 90(6), 2580–2595. doi:10.3168/jds.2006-644.

Cobellis, G., Trabalza-Marinucci, M. and Yu, Z. (2016). Critical evaluation of essential oils as rumen modifiers in ruminant nutrition: a review, *Sci. Total Environ.* 545–546, 556–568. doi:10.1016/j.scitotenv.2015.12.103.

Dijkstra, J., Bannink, A., France, J., Kebreab, E. and van Gastelen, S. (2018). Short communication: antimethanogenic effects of 3-nitrooxypropanol depend on supplementation dose, dietary fiber content, and cattle type, *J. Dairy Sci.* 101(10), 9041-9047. doi:10.3168/jds.2018-14456.

Doreau, M., Bamière, L., Pellerin, S., Lherm, M. and Benoit, M. (2014). Mitigation of enteric methane for French cattle: potential extent and cost of selected actions, *Anim. Prod. Sci.* 54(9), 1417-1422. doi:10.1071/AN14207.

Dubois, B., Tomkins, N. W., Kinley, R. D., Bai, M., Scott, S., Paul, N. A. and de Nys, R. (2013). Effect of tropical algae as additives on rumen in vitro gas production and fermentation characteristics, *Am. J. Plant Sci.* 4, 34-43. doi:10.4236/ajps.2013.412A2005.

Engelhardt, W., Hinderer, S. and Wipper, E. (1978). Factors influencing the endogenous urea-N secretion and utilization in the gastrointestinal tract. In: Osboum, D. F., Beever, D. E. and Thomson, D. J. (Eds), *Ruminants Digestion and Feed Evaluation*. Agricultural Research Council, London, pp. 4.1-4.12.

Eugène, M., Klumpp, K. and Sauvant, D. (2021). Methane mitigating options with forages fed to ruminants, *Grass Forage Sci.* 76(2), 196-204. doi:10.1111/gfs.12540.

Feng, X. Y., Dijkstra, J., Bannink, A., van Gastelen, S., France, J. and Kebreab, E. (2020). Antimethanogenic effects of nitrate supplementation in cattle: a meta-analysis, *J. Dairy Sci.* 103(12), 11375-11385. doi:10.3168/jds.2020-18541.

Flachowsky, G. and Lebzien, P. (2012). Effects of phytogenic substances on rumen fermentation and methane emissions: a proposal for a research process, *Anim. Feed Sci. Technol.* 176(1-4), 70-77. doi:10.1016/j.anifeedsci.2012.07.009.

Genovese, G., Faggio, C., Gugliandolo, C., Torre, A., Spanò, A., Morabito, M. and Maugeri, T. L. (2012). In vitro evaluation of antibacterial activity of *Asparagopsis taxiformis* from the Straits of Messina against pathogens relevant in aquaculture, *Mar. Environ. Res.* 73, 1-6. doi:10.1016/j.marenvres.2011.10.002.

Gidlund, H., Hetta, M. and Huhtanen, P. (2017). Milk production and methane emissions from dairy cows fed a low or high proportion of red clover silage and an incremental level of rapeseed expeller, *Livest. Sci.* 197, 73-81. doi:10.1016/j.livsci.2017.01.009.

Goel, G. and Makkar, H. P. S. (2012). Methane mitigation from ruminants using tannins and saponins, *Trop. Anim. Health Prod.* 44(4), 729-739.

Grainger, C. and Beauchemin, K. A. (2011). Can enteric methane emissions from ruminants be lowered without lowering their production?, *Anim. Feed Sci. Technol.* 166-167, 308-320. doi:10.1016/j.anifeedsci.2011.04.021.

Guyader, J., Eugène, M., Meunier, B., Doreau, M., Morgavi, D. P., Silberberg, M., Rochette, Y., Gerard, C., Loncke, C. and Martin, C. (2015). Additive methane-mitigating effect between linseed oil and nitrate fed to cattle, *J. Anim. Sci.* 93(7), 3564-3577.

Guyader, J., Ungerfeld, E. M. and Beauchemin, K. A. (2017). Redirection of metabolic hydrogen by inhibiting methanogenesis in the rumen simulation technique (RUSITEC), *Front. Microbiol.* 8, 393. doi:10.3389/fmicb.2017.00393.

Harper, M. T., Oh, J., Melgar, A., Nedelkov, K., Räisänen, S., Chen, X., Martins, C. M. M. R., Young, M., Ott, T. L., Kniffen, D. M., Fabin, R. A. and Hristov, A. N. (2019). Production effects of feeding extruded soybean meal to early-lactation dairy cows, *J. Dairy Sci.* 102(10), 8999-9016. doi:10.3168/jds.2019-16551.

Hassanat, F., Gervais, R., Julien, C., Massé, D. I., Lettat, A., Chouinard, P. Y., Petit, H. V. and Benchaar, C. (2013). Replacing alfalfa silage with corn silage in dairy cow diets: effects on enteric methane production, ruminal fermentation, digestion, N balance, and milk production, *J. Dairy Sci.* 96(7), 4553-4567. doi:10.3168/jds.2012-6480.

Hegarty, R. S., Cortez Passetti, R. A., Dittmer, K. M., Wang, Y., Shelton, S., Emmet-Booth, J., Wollenberg, E., McAllister, T., Leahy, S., Beauchemin, K. and Gurwick, N. (2021). *An Evaluation of Emerging Feed Additives to Reduce Methane Emissions from Livestock* (1st edn.). A report coordinated by Climate Change, Agriculture and Food Security (CCAFS) and the New Zealand Agricultural Greenhouse Gas Research Centre (NZAGRC) initiative of the Global Research Alliance (GRA).

Herrero, M., Henderson, B.,Havlík, P., Thornton, P. K., Conant, R. T., Smith, P., Wirsenius, S., Hristov, A. N., Gerber, P., Gill, M., Butterbach-Bahl, K., Valin, H., Garnett, T. and Stehfest, E. (2016) Greenhouse gas mitigation potentials in the livestock sector, *Nat. Clim. Change* 6(5), 452–461. doi:10.1038/nclimate2925.

Hristov, A. N., Lee, C., Hristova, R. A., Huhtanen, P. and Firkins, J. L. (2012). A meta-analysis of the variability in continuous culture rumen fermentation and digestibility data, *J. Dairy Sci.* 95(9), 5299–5307. doi:10.3168/jds.2012-5533.

Hristov, A. N., Johnson, J. M. F., Rice, C. W., Brown, M. E., Conant, R. T., Del Grosso, S. J., Gurwick, N. P., Rotz, C. A., Sainju, U. M., Skinner, R. H., West, T. O., Runkle, B. R. K., Janzen, H., Reed, S. C., Cavallaro, N. and Shrestha, G. (2018). Chapter 5. Agriculture. In: Cavallaro, N., Shrestha, G., Birdsey, R., Mayes, M. A., Najjar, R. G., Reed, S. C., Romero-Lankao, P. and Zhu, Z. (Eds). *Second State of the Carbon Cycle Report (SOCCR2): A Sustained Assessment Report*. U.S. Global Change Research Program, Washington, DC, pp. 229–263. doi:10.7930/SOCCR2.2018.Ch5.

Hristov, A. N. and Melgar, A. (2020). Short communication: relationship of dry matter intake with enteric methane emission measured with the GreenFeed system in dairy cows receiving a diet without or with 3-nitrooxypropanol, *Animal* 14(S3), s484–s490. doi:10.1017/S1751731120001731.

Hristov, A. N., Oh, J., Giallongo, F., Frederick, T. W., Harper, M. T., Weeks, H. L., Branco, A. F., Moate, P. J., Deighton, M. H., Williams, S. R. O., Kindermann, M. and Duval, S. (2015). An inhibitor persistently decreased enteric methane emission from dairy cows with no negative effect on milk production, *Proc. Natl. Acad. Sci. U. S. A.* 112(34), 10663-10668. doi:10.1073/pnas.1504124112.

Hristov, A. N., Oh, J., Lee, C., Meinen, R., Montes, F., Ott, T., Firkins, J., Rotz, A., Dell, C., Adesogan, A., Yang, W., Tricarico, J., Kebreab, E., Waghorn, G., Dijkstra, J. and Oosting, S. (2013a). Mitigation of greenhouse gas emissions in livestock production - a review of technical options for non-CO_2 emissions, *FAO Anim. Prod. Health Pap.* Gerber, P. J., Henderson, B. and Harinder, P. S. (Eds) Makkar. Food and Agriculture Organization, Rome, Italy. Available at: http://www.fao.org/3/i3288e/i3288e.pdf 177.

Hristov, A. N., Oh, J., Firkins, J. L., Dijkstra, J., Kebreab, E., Waghorn, G., Makkar, H. P. S., Adesogan, A. T., Yang, W., Lee, C., Gerber, P. J., Henderson, B. and Tricarico, J. M. (2013b). Special topics-mitigation of methane and nitrous oxide emissions from animal operations: I. A review of enteric methane mitigation options, *J. Anim. Sci.* 91(11), 5045–5069. doi:10.2527/jas.2013-6583.

Jayanegara, A., Leiber, F. and Kreuzer, M. (2012). Meta-analysis of the relationship between dietary tannin level and methane formation in ruminants from in vivo and in vitro experiments, *J. Anim. Physiol. Anim. Nutr. (Berl)* 96(3), 365-375. doi:10.1111/j.1439-0396.2011.01172.x.

Kim, H., Lee, H. G., Baek, Y. C., Lee, S. and Seo, J. (2020). The effects of dietary supplementation with 3-nitrooxypropanol on enteric methane emissions, rumen fermentation, and production performance in ruminants: a meta-analysis, *J. Anim. Sci. Technol.* 62(1), 31–42. doi:10.5187/jast.2020.62.1.31.

Kinley, R. D., Martinez-Fernandez, G., Matthews, M. K., de Nys, R., Magnusson, M. and Tomkins, N. W. (2020). Mitigating the carbon footprint and improving productivity of ruminant livestock agriculture using a red seaweed, *J. Cleaner Prod.* 259, 120836. doi:10.1016/j.jclepro.2020.120836.

Knapp, J. R., Laur, G. L., Vadas, P. A., Weiss, W. P. and Tricarico, J. M. (2014). Invited Review: enteric methane in dairy cattle production: quantifying the opportunities and impact of reducing emissions, *J. Dairy Sci.* 97(6), 3231–3261:/doi:10.3168/jds.2013-7234.

Li, X., Norman, H. C., Kinley, R. D., Laurence, M., Wilmot, M., Bender, H., de Nys, R. and Tomkins, N. (2018). *Asparagopsis taxiformis* decreases enteric methane production from sheep, *Anim. Prod. Sci.* 58(4), 681–688. doi:10.1071/AN15883.

Lopes, J. C., de Matos, L. F., Harper, M. T., Giallongo, F., Oh, J., Gruen, D., Ono, S., Kindermann, M., Durval, S. and Hristov, A. N. (2016). Effect of 3-nitrooxipropanol on ruminal fermentation, methane and hydrogen emissions, and methane isotopic signature in dairy cows, *J. Dairy Sci.* 99(7), 5335–5344. doi:10.3168/jds.2015-10832.

Machado, L., Magnusson, M., Paul, N. A., de Nys, R. and Tomkins, N. E. (2014). Effects of marine and freshwater macroalgae on in-vitro total gas and methane production, *PLoS ONE* 9(1), e85289. doi:10.1371/journal.pone.0085289.

Maia, M. R. G., Fonseca, A. J. M., Oliveria, H. M., Medonca, C. and Cabrita, A. R. J. (2016). The potential role of seaweeds in the natural manipulation of rumen fermentation and methane production, *Sci. Rep.* 6, 32321. doi:10.1038/srep32321.

McGinn, S. M., Flesch, T. K., Beauchemin, K. A., Shreck, A. and Kindermann, M. (2019). Micrometeorological methods for measuring methane emission reduction at beef cattle feedlots: evaluation of 3-nitrooxypropanol feed additive, *J. Environ. Qual.* 48(5), 1454–1461. doi:10.2134/jeq2018.11.0412.

Melgar, A., Lage, C. F. A., Nedelkov, K., Räisänen, S. E., Stefenoni, H., Fetter, M. E., Chen, X., Oh, J., Duval, S., Kindermann, M., Walker, N. D. and Hristov, A. N. (2021). Enteric methane emission, milk production and composition of dairy cows fed 3-nitrooxypropanol, *J. Dairy Sci.* 104(1), 357–366. doi:10.3168/jds.2020-18908.

Melgar, A., Harper, M. T., Oh, J., Giallongo, F., Young, M. E., Ott, T. L., Duval, S. and Hristov, A. N. (2020a). Effects of 3-nitrooxypropanol on rumen fermentation, lactational performance, and the resumption of ovarian cyclicity in dairy cows, *J. Dairy Sci.* 103(1), 410–432. doi:10.3168/jds.2019-17085.

Melgar, A., Welter, K. C., Nedelkov, K., Martins, C. M. M. R., Harper, M. T., Oh, J., Räisänen, S. E., Chen, X., Cueva, S. F., Duval, S. and Hristov, A. N. (2020b). Dose-response effect of 3-nitrooxypropanol on enteric methane emission in dairy cows, *J. Dairy Sci.* 103(7), 6145–6156. doi:10.3168/jds.2019-17840.

Muizelaar, W., Groot, M., van Duinkerken, G., Peters, R. and Dijkstra, J. (2021). Safety and transfer study: transfer of bromoform present in *Asparagopsis taxiformis* to milk and urine of lactating dairy cows, *Foods* 10(3), 584. doi:10.3390/foods10030584.

Nousiainen, J., Rinne, M. and Huhtanen, P. (2009). A meta-analysis of feed digestion in dairy cows. 1. The effects of forage and concentrate factors on total diet digestibility, *J. Dairy Sci.* 92(10), 5019–5030. doi:10.3168/jds.2008-1833.

Ocko, I. B., Sun, T., Shindell, D., Oppenheimer, M., Hristov, A. N., Pacala, S. W., Mauzerall, D. L., Xu, Y. and Hamburg, S. P. (2021). Acting rapidly to deploy readily available methane mitigation measures by sector can immediately slow global warming, *Environ. Res. Lett.* 16(5), 054042. doi:10.1088/1748-9326/abf9c8.

Palmquist, D. L. and Jenkins, T. C. (2017). A 100-year review: fat feeding of dairy cows, *J. Dairy Sci.* 100(12), 10061–10077. doi:10.3168/jds.2017-12924.

Petersen, S. O., Hellwing, A. L. F., Brask, M., Højberg, O., Poulsen, M., Zhu, Z., Baral, K. R. and Lund, P. (2015). Dietary nitrate for methane mitigation leads to nitrous oxide emissions from dairy cows, *J. Environ. Qual.* 44(4), 1063–1070. doi:10.2134/jeq2015.02.0107.

Rabiee, A. R., Breinhild, K., Scott, W., Golder, H. M., Block, E. and Lean, I. J. (2012). Effect of fat additions to diets of dairy cattle on milk production and components: a meta-analysis and metaregression, *J. Dairy Sci.* 95(6), 3225–3247. doi:10.3168/jds.2011-4895.

Ramin, M., Chagas, J. C., Smidt, H., Exposito, R. G. and Krizsan, S. J. (2021). Enteric and fecal methane emissions from dairy cows fed grass or corn silage diets supplemented with rapeseed oil, *Animals (Basel)* 11(5), 1322. doi:10.3390/ani11051322.

Rémond, D., Chaise, J. P., Delval, E. and Poncet, C. (1993). Net transfer of urea and ammonia across the ruminal wall of sheep, *J. Anim. Sci.* 71(10), 2785–2792. doi:10.2527/1993.71102785x.

Reynolds, C. K., Humphries, D. J., Kirton, P., Kindermann, M., Duval, S. and Steinberg, W. (2014). Effects of 3-nitrooxypropanol on methane emission, digestion, and energy and nitrogen balance of lactating dairy cows, *J. Dairy Sci.* 97(6), 3777–3789. doi:10.3168/jds.2013-7397.

Romero-Perez, A., Okine, E. K., McGinn, S. M., Guan, L. L., Oba, M., Duval, S. M., Kindermann, M. and Beauchemin, K. A. (2015). Sustained reduction in methane production from long-term addition of 3-nitrooxypropanol to a beef cattle diet, *J. Anim. Sci.* 93(4), 1780–1791. doi:10.2527/jas.2014-8726.

Roque, B. M., Salwen, J. K., Kinley, R. and Kebreab, E. (2019a). Inclusion of *Asparagopsis armata* in lactating dairy cows' diet reduces enteric methane emission by over 50 percent, *J. Clean. Prod.* 234, 132–138. doi:10.1016/j.jclepro.2019.06.193.

Roque, B. M., van Lingen, H. J., Vrancken, H. and Kebreab, E. (2019b). Effect of Mootral–a garlic- and citrus-extract-based feed additive–on enteric methane emissions in feedlot cattle, *Transl. Anim. Sci.* 3(4), 1383–1388. doi:10.1093/tas/txz133.

Sauvant, D. and Giger-Reverdin, S. (2009). Modélisation des interactions digestives et de la production de méthane chez les ruminants, (In French.), *INRA Prod. Anim.* 22(5), 375–384 INRA (Prods). Anim 22, 375–384.

Schilde, M., von Soosten, D., Hüther, L., Meyer, U., Zeyner, A. and Dänicke, S. (2021). Effects of 3-nitrooxypropanol and varying concentrate feed proportions in the ration on methane emission, rumen fermentation and performance of periparturient dairy cows, *Arch. Anim. Nutr.* 75(2), 79–104. doi:10.1080/1745039X.2021.1877986.

Silvestre, T., Räisänen, S. E., Wasson, D. E., Cueva, S. F., Lage, C. F. A., Wall, E. H. and Hristov, A. N. (2021). Effects of rumen protected botanicals on lactational performance, methane emission, and blood metabolites of dairy cows, *J. Dairy Sci.* 104 (Suppl. 1), 318.

Sliwinski, B. J., Kreuzer, M., Wettstein, H. R. and Machmuller, A. (2002). Rumen fermentation and nitrogen balance of lambs fed diets containing plant extracts rich in tannins and saponins and associated emissions of nitrogen and methane, *Arch. Anim. Nutr.* 56, 379–392.

Smith, P. C., Heinrich, G., Suklitsch, M., Gobiet, A., Stoffel, M. and Fuhrer, J. (2014). Station-scale bias correction and uncertainty analysis for the estimation of irrigation water

requirements in the Swiss Rhone catchment under climate change, *Clim. Change* 127(3–4), 521–534.
Staerfl, S. M., Amelchanka, S. L., Kälber, T., Soliva, C. R., Kreuzer, M. and Zeitz, J. O. (2012). Effect of feeding dried high-sugar ryegrass ('AberMagic') on methane and urinary nitrogen emissions of primiparous cows, *Livest. Sci.* 150(1–3), 293–301.
Stefenoni, H. A., Räisänen, S. E., Welchez, S. F., Wasson, D. E., Lage, C. F. A., Melgar, A., Fetter, M. E., Smith, P., Hennessy, M., Vecchiarelli, B., Bender, J., Pitta, D., Cantrell, C. L., Yarish, C. and Hristov, A. N. (2021). Effects of the macroalga *Asparagopsis taxiformis* and oregano leaves on methane emission, rumen fermentation, and lactational performance of dairy cows, *J. Dairy Sci.* 104(4), 4157–4173. doi:10.3168/jds.2020-19686.
Thoma, G., Popp, J., Nutter, D., Shonnard, D., Ulrich, R., Matlock, M., Kim, D. S., Neiderman, Z., Kemper, N., East, C. and Adom, F. (2013). Greenhouse gas emissions from milk production and consumption in the United States: a cradle-to-grave life cycle assessment circa 2008, *Int. Dairy J.* 31(Suppl. 1), S3–S14. doi:10.1016/j.idairyj.2012.08.013.
Tricarico, J. M., de Haas, Y., Hristov, A. N., Kebreab, E., Kurt, T., Mitloehner, F. and Pitta, D. (2022), Symposium review: the Greener Cattle Initiative is a program developed to support research on enteric methane mitigation from ruminants, *J. Dairy Sci.* 105(10), 8535–8542. doi:10.3168/jds.2021-21397.
van Gastelen, S., Dijkstra, J., Heck, J. M. L., Kindermann, M., Klop, A., de Mol, R., Rijnders, D., Walker, N. and Bannink, A. (2022). Methane mitigation potential of 3-nitrooxypropanol in lactating cows is influenced by basal diet composition, *J. Dairy Sci.* 105(5), 4064–4082. doi:10.3168/jds.2021-20782.
van Middelaar, C. E., Dijkstra, J., Berentsen, P. B. M. and De Boer, I. J. M. (2014). Cost-effectiveness of feeding strategies to reduce greenhouse gas emissions from dairy farming, *J. Dairy Sci.* 97(4), 2427–2439. doi:10.3168/jds.2013-7648.
Villar, M. L., Godwin, I. R., Hegarty, R. S., Erler, D. V., Farid, H. T. and Nolan, J. V. (2021). Nitrate and nitrite absorption, recycling and retention in tissues of sheep, *Small Rumin. Res.* 200, 106392. doi:10.1016/j.smallrumres.2021.106392.
Vyas, D., Alemu, A. W., McGinn, S. M., Duval, S. M., Kindermann, M. and Beauchemin, K. A. (2018). The combined effects of supplementing monensin and 3-nitrooxypropanol on methane emissions, growth rate, and feed conversion efficiency in beef cattle fed high forage and high grain diets, *J. Anim. Sci.* 96(7), 2923–2938. doi:10.1093/jas/sky174.
Waghorn, G. C. (2008). Beneficial and detrimental effects of dietary condensed tannins for sustainable sheep and goat production – progress and challenges, *Anim. Feed Sci. Technol.* 147(1–3), 116–139.
Waghorn, G. C., Tavendale, M. H. and Woodfield, D. R. (2002). Methanogenesis from forages fed to sheep, *ProNZG*. 64, 167–171.
Wang, J. K., Ye, J. A. and Liu, J. X. (2012). Effects of tea saponins on rumen microbiota, rumen fermentation, methane production and growth performance–a review, *Trop. Anim. Health Prod.* 44(4), 697–706.
Wasson, D. E., Stefenoni, H., Welchez, S., Lage, C., Räisänen, S., Melgar, A., Fetter, M., Yarish, C. and Hristov, A. N. (2021). Screening of macroalgae species for enteric methane mitigation effect in vitro, *J. Dairy Sci.* 104 (Suppl. 1), 271.

Williams, S. R. O., Hannah, M. C., Eckard, R. J., Wales, W. J. and Moate, P. J. (2020). Supplementing the diet of dairy cows with fat or tannin reduces methane yield, and additively when fed in combination, *Animal* 14(S3), s464–s472.

Woodward, S. L., Waghorn, G. C., Ulyatt, M. J. and Lassey, K. R. (2001). Early indications that feeding Lotus will reduce methane emissions from ruminants, *Proc. N. Z. Soc. Anim. Prod.* 61, 23–26.

Yáñez-Ruiz, D. R., Bannink, A., Dijkstra, J., Kebreab, E., Morgavi, D. P., O'Kiely, P., Reynolds, C. K., Schwarm, A., Shingfield, K. J., Yu, Z. T. and Hristov, A. N. (2016). Design, implementation and interpretation of in vitro batch culture experiments to assess methane mitigation in ruminants – a review, *Anim. Feed Sci. Technol.* 216, 1–18. doi:10.1016/j.anifeedsci.2016.03.016.

Zhang, X. M., Smith, M. L., Gruninger, R. J., Kung, L., Vyas, D., McGinn, S. M., Kindermann, M., Wang, M., Tan, Z. L. and Beauchemin, K. A. (2021). Combined effects of 3-nitrooxypropanol and canola oil supplementation on methane emissions, rumen fermentation and biohydrogenation, and total tract digestibility in beef cattle, *J. Anim. Sci.* 99(4), 1–10.

Zhou, Y. Y., Mao, H. L., Jiang, F., Wang, J. K., Liu, J. X. and McSweeney, C. S. (2011). Inhibition of rumen methanogenesis by tea saponins with reference to fermentation pattern and microbial communities in Hu sheep, *Anim. Feed Sci. Technol.* 166–167, 93–100.

Part 4

Assessing alternative feed sources from agricultural co-products

Chapter 11

Assessing alternative fiber sources from by-products

Maurice L. Eastridge[1], The Ohio State University, USA

1 Introduction

The usefulness of by-products has long been identified, but how to define them has been debated for many years. Some of the products were once regarded as 'waste products', but it was quickly realized that most of these products have nutritional value and need not be wasted. The coining of 'by-product' has been accepted by many with the understanding that they are 'secondary products produced in addition to the principal product' (AAFCO, 2016). Yet, many people think this is a negative identity and have referred to them as co-products (two or more products originating from a given process with each having value) or alternative feeds. In this article, these feeds will be referred to as by-products with intrinsic nutritional value. Although by-products can originate from the processing of plants or animals, plant-based by-products will be discussed in this article since it focuses on the fiber from such feedstuffs. Animal by-products can serve as valuable sources of starch, protein, fat, and minerals, which is addressed in other articles in this publication. Plant by-products can serve as important sources of energy, contributed by the starch, fiber, protein, and fat they contain; however, this article also will focus on the value of the fiber relative to rumen function and for dilution of starch from diets for dairy cattle.

[1] Department of Animal Sciences, The Ohio State University, Columbus, USA, eastridge.1@osu.edu.

http://dx.doi.org/10.19103/AS.2022.0117.12

Although by-products are defined as secondary products, they provide valuable attributes to diets for dairy cattle. On the flip side, the ability of ruminants to effectively utilize by-products adds tremendously in minimizing the carbon footprint of food production systems and thus contributes to the environmental sustainability of agriculture. Livestock has been estimated to recycle more than 4.32×10^{10} kg of human-inedible food and fiber processing by-products, converting them into human-edible food, pet food, and industrial products (Fig. 1; White and Hall, 2017). It has been estimated that about 35% of the diets for lactating dairy cows in the USA consists of by-products and human non-edible grains (Knapp, 2015). About 25% of the wheat and 44% of the corn processed and 60–80% of the oilseeds for human food result in animal by-product feeds (Knapp, 2015). Thus, the use of by-products, human non-edible grains, and high fiber plant sources (e.g. forages) must be taken into account for life cycle assessments of livestock production systems.

Fiber from by-product feeds contributes physically effective neutral detergent fiber (peNDF; stimulates chewing) for maintaining ruminal pH (NASEM, 2021), but both peNDF and non-effective NDF (ne-NDF) contribute energy to the diet and serve as diluters of starch which adds to the modulation of ruminal pH. These attributes of by-products, the economic value of peNDF

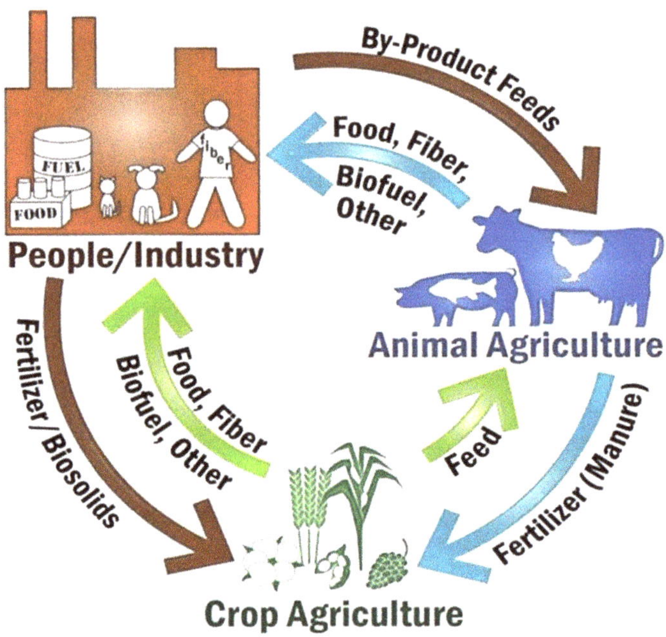

Figure 1 Illustration of the food production system, the connectivity of each component, and the importance of livestock in the utilization of by-product feeds. Source: White and Hall (2017).

and ne-NDF, and some limitations to feeding by-products will be presented in the following section.

2 By-products as alternative fiber sources

Many different plant-based by-product feeds are available from the processing of cereal grains, oilseeds, nuts, fruits, and vegetables (NASEM, 2021; Yang et al., 2021). However, this article will focus on the ones that are typically available widely in the USA and are commonly chosen as non-forage fiber sources (NFFS) for feeding to dairy cattle (Table 1). Although all of these by-product feeds are available in dry form, some of them are often available in local markets in wet form, especially brewers' grains and distillers' grains (DGs). Albeit the dry form increases the market potential, even for exporting, the drying process adds considerably to their costs. In addition, transporting wet commodities is costly per unit of dry matter (DM), and stability of the feedstuff is a concern. Some of these feeds have become more highly processed to meet different market demands and needs of animals in different productive stages. For example, pelleting increases the density of the feed per unit of volume, and thus a pelleted feed is more conducive to shipping long distances. On the other hand, pelleting reduces the particle size of the feed and thus can alter the level of pe-NDF. Lintless whole cottonseed (WCS; e.g. Pima cottonseed) has lower digestibility compared to linted WCS due to its more rapid exit from the rumen (Sullivan et al., 1993). Pelleting WCS reduces its particle size and results in the oil being in a free form, potentially having negative effects on ruminal fiber digestibility (Reveneau et al., 2005). Coating WCS with starch to increase its density for shipping and reduce lodging in storage has been demonstrated to result in similar performance of lactating cows in comparison to unprocessed WCS, but a reduction in fiber digestibility may occur and the processing increases the cost of the commodity (Bernard et al., 1999). DGs have become more refined as they can be purchased with and without solubles, some of the fat is extracted to provide a low-fat DGs (Morris et al, 2018), and the feed can be processed into different particle fractions for providing a high-protein feedstuff and commodities with different fiber fractions. It is important to have a clear understanding of each product when making choices in purchasing and in formulations of rations for livestock.

Many reviews of feeding by-products to ruminants have been written (NAFS, 1991, 1995, 2003; Firkins, 2003; NASEM, 2021; Yang et al., 2021), providing a cumulation of data from many studies with specific by-products and their influence on animal production, health, and efficiency. Much of the recent research has focused on optimal feeding levels in relation to the type, quality, and level of forage in the diet and any effects of additives to diets containing appreciable levels of by-products, such as urea and monensin. Considerable focus has occurred on the impacts of the by-products on yields of milk fat and protein. There has been focus on milk fat both because of the value of milk fat in the market and because of the use of milk fat as an indicator of adequate peNDF in diets (NASEM, 2021). Some of the research has focused on new products resulting from changes in processing of the by-products. It is known that variability in chemical composition of some by-products can be high (Table 1). Thus, to address these concerns, feeds need to be analyzed routinely from the same supplier, purchase should be done from suppliers that focus on process control for uniformity of the by-products, and for by-products with high variably, low inclusion level in diets will reduce the risk of negatively affecting animal performance or health.

With peNDF focusing primarily on chewing response for maintaining ruminal conditions, forages have a much higher peNDF than by-product feeds. Although particle size is only one of the factors affecting peNDF (chemical composition and fragility of particles are also important), it has been one of the major considerations for maintaining rumen pH given that larger particles require more chewing, provide for the formation of a greater particulate mat in the dorsal rumen that slows down particle passage to increase ruminal digestibility, and thus are less likely to pass quickly from the rumen. However, using particle size separation of feeds as a basis for determination of peNDF has its challenges, including the type of screen and screening on a dry versus wet basis. For example, wet particles are more likely to cling together, and wet particles are heavier and may change conformation to fit through smaller screens. peNDF is commonly measured as particles >8 mm, but this is typically used for forages and explains why forage NDF (fNDF) was concluded as the primary contributor to peNDF in diets for dairy cattle (NASEM, 2021). Given that by-products are not forages and there is a lack of data with by-products using particle size as a direct measure of peNDF, it is difficult to select among the by-product feeds relative to the expected results on rumen health and animal performance. Mertens (1997) suggested a physically effective fiber (pef) value of 30–40% for most NFFS in contrast to most forages having a pef of 100%. Mertens (1997) also suggested using particle sizes >1.18 mm as another reference point for effective fiber for feeds. With this latter approach, corn gluten feed and wheat bran fall within the 30–40% pef suggested. On the other hand, DGs with solubles, wheat middlings, and soybean hulls only have 2–4% of the particles >1.18 mm; thus only about 10% of the pef level suggested. Overall, NFFS are not very effective at contributing peNDF to dairy cattle diets. The one exception among those considered in Table 1 is cottonseed, including WCS and cottonseed hulls. The lint and hulls contribute to the fiber in WCS, and it becomes trapped in the rumen mat, increasing the residence time in the

Table 1 Nutrient composition, fiber digestibility, and particle size of selected plant-based by-product feeds[a]

Cereal grains	DM (%)	Starch (% of DM)	NDF (% of DM)	ADF (% of DM)	NDF - ADF[b]	IVNDFD48 (% of NDF)	Lignin (% of DM)	Particle size[c] (>1.18 mm) (% of DM)	Pef[c] (%)
Brewer grains	22.5 (3.6)	5.2 (3.9)	49.3 (4.9)	23.8 (2.7)	25.5	47.3 (8.6)	6.64 (1.11)	18	40
Corn gluten feed	89.2 (1.8)	15.5 (4.6)	35.7 (4.6)	11.5 (2.1)	24.2	74.1 (12.1)	2.31 (1.22)	36	40
Distiller grains with solubles and high fat	89.1 (1.7)	4.5 (1.6)	32.1 (2.7)	14.6 (1.6)	17.5	71.5 (8.2)	4.17 (0.94)	4	40
Oat hulls	91.4 (1.9)	10.6 (4.9)	73.6 (8.2)	39.6 (6.3)	34.0	Not available	6.54 (1.92)	Not available	40
Wheat bran	90.1 (1.2)	20.8 (6.2)	40.1 (6.2)	13.8 (2.5)	26.3	43.3 (9.6)	4.15 (1.02)	33	40
Wheat middlings	88.3 (1.4)	22.9 (5.6)	38.7 (4.6)	13.2 (1.8)	25.5	48.6 (8.8)	3.77 (0.94)	2	40
Oilseeds									
Cottonseed, whole linted	91.4 (1.8)	0.8 (0.9)	50.6 (4.2)	38.6 (4.3)	12.0	12.8 (8.3)	11.21 (2.58)	Not available	90
Cottonseed hulls	91.0 (1.8)	1.2 (1.2)	80.9 (6.1)	66.0 (4.8)	14.9	22.4 (7.9)	19.93 (2.71)	Not available	40, 90[d]
Soybean hulls	90.4 (1.9)	1.0 (0.7)	66.7 (3.6)	47.9 (3.1)	18.8	88.1 (7.3)	2.57 (0.75)	3	30
Other									
Beet pulp (pelleted)	92.3 (1.2)	0.6 (0.4)	46.9 (4.3)	28.2 (2.4)	18.7	79.0 (not available)	3.88 (2.13)	Not available	30
Citrus pulp (pelleted)	88.2 (1.4)	1.3 (1.3)	24.1 (2.9)	20.1 (3.3)	4.0	86.5 (not available)	2.49 (1.37)	Not available	30

[a]Values are expressed as mean (standard deviation). Data from NASEM (2021) unless noted otherwise. DM, dry matter; NDF, neutral detergent fiber; ADF, acid detergent fiber; IVNDFD48, *in vitro* NDF digestibility with 48 h incubation, pef, physically effective fiber as a factor for chewing potential.
[b]NDF - ADF = hemicellulose.
[c]Data from Mertens (2002).
[d]Based on diets without and with long-stemmed forage, respectively.

rumen in comparison to cereal grain by-products. The effectiveness of fiber from WCS is 84–90% of that of the fiber from forages (Harvatine et al., 2002 and Mertens, 1997, respectively). Cottonseed hulls are indicated as having 40–90% of the effectiveness of fiber in forages, depending on whether they are fed without or with some long-stemmed hay, respectively (Mertens, 1997).

Starch is one of the primary energy sources in diets of dairy cattle, with contributions primarily from corn silage, cereal grains, and by-product feeds. Most of the starch is digested in the rumen, with over 95% typically being digested in the total tract. Most by-product feeds contain less than 11% starch (Table 1), but corn gluten feed, wheat bran, and wheat middlings contain 15–23% starch. Yet, the amount of starch in these three by-products can vary by processor, with coefficient of variation of 25–30%. Thus, it is most important to monitor the variation in starch among by-product feeds, so diets can be formulated accurately for targeted starch concentrations. The level of starch in the diet and its rate and extent of digestibility impact ruminal pH and are reflected in the minimum levels of forage and total NDF suggested for diets. As the level of fNDF in a diet for lactating cows decreases from a minimum of 19% to 15%, the minimum total NDF is recommended to increase from 25% to 33%, and the maximum starch is recommended to decrease from 30% to 22% (Table 2). To accomplish this carbohydrate balancing act, NFFS can be used to increase total NDF which reduces fNDF. By-products with high NDF and low starch can most effectively be used to accomplish these targets. With less than 19% fNDF in diets, by-product feeds as NFFS, adequate forage particle size, a total mixed ration that minimizes sorting, feeding buffers to maintain rumen pH, and moderating rate of carbohydrate digestibility by varying starch sources must be used to provide for a stable rumen (Fig. 2). Various dietary and management practices must be monitored for rumen stability, including when fNDF is between more common levels of 19% and 25% of dietary DM. Not only does using NFFS to dilute starch in high-energy diets help to moderate the rate of carbohydrate fermentation, but in some cases (e.g. WCS and cottonseed hulls), the by-products can also contribute to the rumen mat for trapping feed

Table 2 Recommended minimum forage and total NDF and maximum starch concentrations (% of DM) for diets for lactating dairy cows (NASEM, 2021)[a]

Minimum fNDF	Minimum total NDF	Maximum starch
19	25	30
18	27	28
17	29	26
16	31	24
15	33	22

[a]NDF, neutral detergent fiber; fNDF, forage NDF.

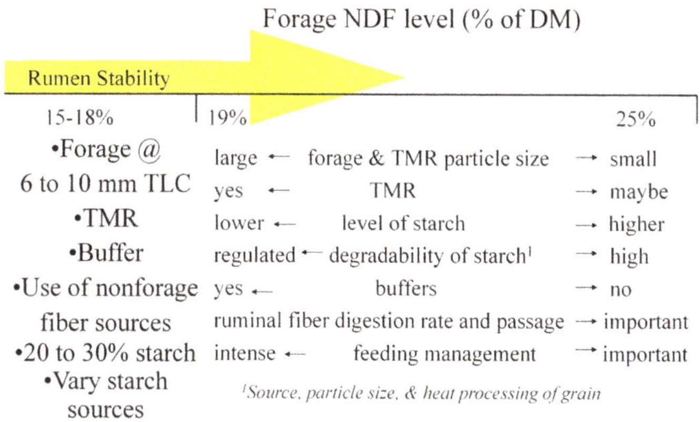

Figure 2 Factors important for consideration at different levels of dietary forage neutral detergent fiber (NDF). TLC, theoretical length of cut; TMR, total mixed ration.

particles to reduce their passage from the rumen. Although NFFS typically add limited amounts of peNDF to diets, their role in diluting starch while also being high in digestibility is the greater benefit toward maintaining rumen pH and fiber digestibility in high-energy diets. Thus, the peNDF of diets should be evaluated rather than individual feeds during ration formulation, and then animal performance monitored relative to the diets fed (NASEM, 2021).

The digestibility of NDF from NFFS varies considerably [e.g. 12.8–88% *in vitro* NDF digestibility at 48 h (IVNDFD48) among the feeds listed in Table 1]. The lowest IVNDFD48 was with WCS and cottonseed hulls, caused primarily by the high concentrations of lignin. The intermediate IVNDFD48 values were for brewer grains, wheat bran, and wheat middlings which have a lignin/NDF ratio of 0.10 to 0.13. Corn gluten feed, DGs with solubles, soybean hulls, beet pulp, and citrus pulp had the highest IVNDFD48, and either had a lignin/NDF ratio $\leq$0.10 or less than 5% of the particles were > 1.18 mm. With the exception of cottonseed and cottonseed hulls, the hemicellulose content ranged from 18% to 34% and does not appear to explain the differences in *in vitro* digestibility. Therefore, particle size, lignin/NDF ratio, and variation in carbohydrate fraction (e.g. pectin in beet and citrus pulps and lignin in cottonseed) in the by-products impact NDF digestibility of the NFFS. Water holding capacity, rate of hydration, and functional specific gravity differ among by-products (Bhatti and Firkins, 1995), and these characteristics can affect residence time in the rumen and therefore ruminal digestibility. However, actual energy available from the NFFS will be affected by the dietary levels provided and digestibility of NDF, starch, protein, and fat and just as importantly, the composition of the diet to which it is added. With the dilution of starch by the NFFS in high-grain diets, a positive associative effect may occur, resulting in a higher energy value from NFFS than estimated from the chemical composition of the feedstuff.

Market values of plant-based by-products often vary seasonally due to timing of harvest for the primary source of the by-product and the associated growing conditions and market demands for the primary source. Even with this seasonality, the market prices for most plant-based by-products in the USA follow the general trend of corn and soybeans as they are the primary grain sources for energy and protein, respectively. Many software programs are available that can be used for ration formulation, many of which provide least-cost programming features. Such computer programs, some of which have complex nutrition models, will take into account the nutritional provisions of by-product feeds and their associated costs. Over the years, other approaches have been used to determine the value of individuals feeds, outside of diet formulation, for use in making purchase decisions. Most of these have focused on energy and protein contents in the feeds. However, using a computer program named SESAME (https://dairy.osu.edu/node/23), the market value of various feeds, including by-product feeds, can be calculated based on their nutrient composition and the economical values calculated for net energy for lactation, metabolizable protein (MP), peNDF, and non-effective NDF (neNDF) from a calibration set of feed ingredients (St-Pierre, 2005). Using this software, the actual market value of a given by-product can be compared to a predicted market value, along with lower and upper confidence levels for the predicted value (White, 2022). Oftentimes, feeds with high levels of neNDF (e.g. soybean hulls) are discounted in value because of a negative value derived for neNDF. It does not mean that feeds in the over-priced category are not to be used at all. Such feeds may result in positive associate effects in diets which is not accounted for in net energy for lactation, MP, peNDF, or neNDF. They also may serve as carriers for other ingredients, be important for pellet quality, or provide some of intrinsic value to the formulation.

Besides the cost of the ingredients relative to the prices for other commodities, other limitations exist on the inclusion levels of by-products in diets for dairy cattle. Some of the limitations will occur based on the target levels of starch, NDF, and peNDF in diets. The maximum level of unsaturated fat in diets to reduce the risk for lowering ruminal fiber digestibility will limit the use of some by-product feeds (e.g. DGs with solubles and WCS). The use of some of the by-products will be limited based on the target levels of protein fractions (e.g. MP and rumen degradable and undegradable protein). Cottonseed products will be limited in diets, especially for breeding animals, to reduce the risk of gossypol toxicity (Bernard et al., 1999). During the processing of the primary grain source, one of the primary components is removed (e.g. starch) which results in an increased concentration of minerals in the by-product. Also, during processing, chemicals used in the process may alter the composition of the by-products in comparison to the primary grain product (e.g. sulfuric acid added to stop the fermentation during ethanol production). Both of these situations occur for DGs and thus can limit the levels of dietary inclusion. High levels of DGs in diets can result in overfeeding of phosphorus and thus increase phosphorus excretion which has environmental impacts. Overfeeding of DGs can cause excessive intake of sulfur which can reduce thiamine synthesis by ruminal bacteria and increase the risk for polioencephalomalacia. During years with high risk for the presence of mycotoxins, especially in cereal grains, the use of the by-products from the primary source with contamination may have to be limited or halted since the concentrations of the mycotoxins are often higher in the by-products than in the original primary source (e.g. corn grain high in mycotoxins likely will result in DGs with a higher concentration of mycotoxins than the whole grain). Any of the by-products purchased wet will have storage limitations and often must be fed within a week of delivery, depending on the type of storage and climatic conditions. In addition to rapid feedout, ensiling (Garcia, A. and Kalscheur, K., 2007) and top-dressing with a preservative, such as propionic acid (Moriel et al., 2016), are options for consideration.

3 Case study

Slater et al. (2000) investigated the effect of fNDF levels in diets for lactating dairy cows with the use of soybean hulls or cottonseed as NFFS and the grain starch sources being either corn or corn and wheat. Forty-eight cows were blocked and randomly assigned within a block to one of four diets formulated to contain 21% fNDF with corn; 16% fNDF with corn; 16% fNDF with corn and wheat (1 : 1) or 11% fNDF with cottonseed and corn. The 11% fNDF diet was formulated such that if cottonseed was assumed to have the same peNDF as for forage, then the fNDF would equal 16%. Corn and alfalfa silages were the forages used in all of the diets. Cows were fed the diets throughout weeks 10-25 of lactation. Soybean hulls were added at approximately 23.0% of DM for the 16% and 11% fNDF diets to replace forage and formulate diets with 35% non-fiber carbohydrates (NFCs). The control diet was formulated to have 42% NFCs. Actual fNDF concentrations were 17.8%, 14.0%, 13.9%, and 9.4%. Actual NFC concentrations (calculated) were 47, 42, 40, and 39, respectively, with nonstructural carbohydrates (analyzed) at 38% for control and 30% for the 16% and 11% fNDF diets. DM intake and milk yield were highest for cows fed 11% fNDF with cottonseed, resulting in the lowest numerical feed efficiency (3.5% fat-corrected milk/DM intake = 1.45). Milk fat percentage was higher for cows consuming 21% fNDF and 16% fNDF with corn than for cows fed the two other diets. Cows fed 16% fNDF with corn and wheat experienced milk fat-protein inversion, but ruminal acetate : propionate ratio was lower for cows fed 11% fNDF than cows fed 16% fNDF. Time spent chewing was similar among all diets. The researchers concluded that for cows in mid-lactation, fNDF may be reduced to 9–11% when cottonseed is at 11% of DM and dietary nonstructural carbohydrates (NSC) are at 30% of DM. fNDF may be reduced to 14–16% without cottonseed when NSC are at 30% of DM. This study illustrates the impact of starch source (corn versus wheat), levels of fNDF (9%, 14%, or 18%), and NFFS used for peNDF and as a diluter of starch (soybean hulls versus WCS) on DM intake, feed efficiency, ruminal fermentation, and chewing behavior of lactating dairy cows. It must also be noted that the lower levels of fNDF in this study are below the typical minimum recommendation of 15% fNDF (Table 2; Fig. 2).

4 Conclusion

Feed costs will remain the major cost of food production from animals, often reaching in excess of 50% of the cost of production with dairy cows. By-product feeds will continue to offer cost savings while also providing required nutrients for efficient production and healthy animals. Fiber from these by-products add value to the diet, whether as peNDF, energy, or a highly digestible fiber that slows overall carbohydrate digestibility in the rumen to improve microbial efficiency. Particle size, chemical composition, and interaction with other dietary components will affect the rumen environment and the nutritional contribution of the by-product to the animal. Under-utilization of by-products limits profitability on farms, but overfeeding certain by-products can reduce profitability, result in excessive excretion of N and P (and sometimes S; Lee et al., 2020), and put the animals at higher risk for health problems. Among the plant-based by-products discussed, DG is used in the highest volume within the US dairy industry. This occurs because of its typically competitive price, along with its energy, protein, and fiber contents that add value to the diet for high-producing animals. Yet, because of these attributes, it is often included in diets at the upper levels of recommendations. Limitations of feeding by-products need to be recognized and practiced for efficient animal production, minimizing environmental impacts and promoting animal health.

5 Future trends in research

The availability of by-product feeds will continue to increase as food processing increases to meet the growing human population. In the environmental assessment of animal production systems, the value of utilization of human-inedible plants and plant fractions by ruminants must be included. Research in the animal system area will need to provide data on the efficiency of food production relative to the inputs and the emissions of greenhouse gases under different feeding and management systems. Many different descriptions and approaches are being used to advance agricultural sustainability, including animal agriculture. For example, two systems often referred to are regenerative agriculture and circular agriculture. Regenerative agriculture often lacks consistency among users of the system, or the users may use the term without having a clear definition of practices (Newton et al., 2020). Yet, some organizations have clearly stated the definition of the system for their propagation of sustainability [e.g. Nestle has defined regenerative agriculture is an approach to farming that aims to improve soil health and soil fertility - as well as protecting water resources and biodiversity, https://www.nestle.com/sustainability/nature-environment/regenerative-agriculture (accessed 15 June 2022)]. Circular agriculture has been described as focusing on using minimal amounts of external inputs, closing nutrient loops, regenerating soils, and minimizing the impact on the environment (United Nations, 2021). Regardless of the system and how it focuses on processes, inputs, and outcomes for providing sustainability, the effective use of by-products in feeding to livestock contributes to sustainable systems.

Further development of more refined by-products with specific attributes for animals in different geographical regions and levels of production will need to occur for the most efficient use of the components within the by-products. This approach will even provide more market potential for by-products with specific fiber attributes. This area of focus could also assist in reducing the variability in composition of by-products. However, additional research is needed on how best to define fiber from by-products for its peNDF and the value of neNDF for providing energy, moderating starch level in diets, and contributing to positive associate effects in the rumen. Modification of the physical form to reduce transportation costs (e.g. limited processing versus pelleting) still needs research for some of the by-products as to how the physical form impacts ruminal digestibility. New hybrids and genetically modified plants used in the food production systems will impact the nature of the corresponding by-products and how they can be used in diets for ruminants. New by-products will arrive from new processes for food, which will need animal data for most effective utilization. Additional information is needed on how to reduce the risks associated with the anti-nutritional aspects of some of by-products, including how to best reduce the risk for mycotoxins and how to manage mycotoxin-contaminated feed on the farm. By-products will remain commonplace in the feeding of dairy cattle, whether it is a commercial operation or one focusing on regenerative agriculture, and research will need to continue to investigate on how to improve animal efficiency and improve farms' environmental sustainability.

6 Where to look for further information

- National Academies of Science, Engineering, and Medicine (NASEM) (2021). Nutrient Requirements of Dairy Cattle, 8th rev. ed. National Acad Press, Washington DC.
- INRAE-CIRAD-AFZ Feed Tables: Composition and nutritive values of feeds for cattle, sheep, goats, pigs, poultry, horses, and salmonids. Feeds are categorized into source groups and within each category, the respective feeds are listed along with composition data. https://www.feedtables.com/.
- Bernard, J. K. (2017). Considerations for using by-product feeds. University of Georgia, Athens. https://secure.caes.uga.edu/extension/publications/files/pdf/B%20862_3.PDF.
- For additional reading on peNDF for high-producing dairy cows, see Zebeli et al. (2012).

7 References

AAFCO (Association of American Feed Control Officials). (2016). *Official Publication, Association of American Feed Control Officials Incorporated*. Oxford, UK: Association of American Feed Control, Available at: https://www.aafco.org/ (accessed 13 June 2022).

Bernard, J. K., Calhoon, M. C. and Martin, S. A. (1999). Effect of coating whole cottonseed on performance of lactating dairy cows, *J. Dairy Sci.* 82(6), 1296–1304.

Bhatti, S. A. and Firkins, J. L. (1995). Kinetics of hydration and functional specific gravity of fibrous feed by-products, *J. Anim. Sci.* 73(5), 1449–1458, https://doi.org/10.2527/1995.7351449x.

Firkins, J. L. (2003). Feeding by-product high in concentration of fiber to ruminants. In: Eastridge, M. L. (Ed.), *Proceedings of the Third National Alternative Feeds Symposium Alternative Feeds for Livestock and Poultry*, 4–5 November 2003. Kansas City, MO, Columbus: The Ohio State University Press.

Garcia, A. and Kalscheur, K. (2007). Storage of wet distillers grains. In: Eastridge, M. L. (Ed.), *Proceedings of the Tri-State Dairy Nutrition Conference*, 24–25 April, Wayne, IN, Available at: https://www.tristatedairy.org/ (accessed 13 June 2022).

Harvatine, D. I., Winkler, J. E., Devant-Guile, M., Firkins, J. L., St-Pierre, N. R., Oldick, B. S. and Eastridge, M. L. (2002). Whole linted cottonseed as a forage substitute: Fiber effectiveness and digestion kinetics, *J. Dairy Sci.* 85(8), 1988–1999.

Knapp, J. R. (2015). Dairy sustainability - Using the real facts. In: Eastridge, M. L. (Ed.), *Proceedings of the Tri-State Dairy Nutrition Conference*, 20–22 April, Wayne, IN, Available at: https://www.tristatedairy.org/ (accessed 13 June 2022).

Lee, C., Morris, D. L., Lefever, K. M. and Dieter, P. A. (2020). Feeding a diet with corn distillers grain with solubles to dairy cows alters manure characteristics and ammonia and hydrogen sulfide emissions from manure, *J. Dairy Sci.* 103(3), 2363–2372, https://doi.org/10.3168/jds.2019-17524.

Mertens, D. R. (1997). Creating a system for meeting the fiber requirements of dairy cows, *J. Dairy Sci.* 80(7), 1463–1481.

Mertens, D. R. (2002). Measuring fiber and its effectiveness in ruminant diets. In: *Proceedings* of the *Plains Nutrition Conference*, San Antonio, TX, Available at: https://www.nutritionmodels.com/papers/MertensPNC2002.pdf (accessed 13 June 2022).

Moriel, P., Piccolo, M. B., Artioli, L. F. A., Santos, G. S., Poore, M. H. and Ferraretto, L. F. (2016). Method of propionic acid-based preservative addition and its effects on nutritive value and fermentation characteristics of wet brewers grains ensiled in the summertime, *Prof. Anim. Sci.* 32(5), 591–597, http://doi.org/10.15232/pas.2016-01513.

Morris, D. L., Kim, S. H., Kononoff, P. J. and Lee, C. (2018). Continuous 11-week feeding of reduced-fat distillers grains with and without monensin reduces lactation performance of dairy cows, *J. Dairy Sci.* 101(7), 5971–5983.

National Academies of Science, Engineering, and Medicine (NASEM). (2021). *Nutrient Requirements of Dairy Cattle* (8th Rev. Ed.). Washington, DC: National Acad Press.

National Alternative Feeds Symposium (NAFS). (1991). *Proceedings of the Alternative Feeds for Dairy and Beef Cattle*, Jordan, E. R. (Ed.), September 22–24, University of Missouri, Columbia, St. Louis, MO.

National Alternative Feeds Symposium (NAFS). (1995). *Proceedings of the Alternative Feeds for Dairy and Beef Cattle*, Eastridge, M. L. (Ed.), September 24–26, University of Missouri, Columbia, St. Louis, MO.

National Alternative Feeds Symposium (NAFS). (2003). *Proceedings of the Alternative Feeds for Livestock and Poultry*, Eastridge, M. L. (Ed.), November 4–5, The Ohio State University Press, Kansas City, MO, Columbus.

Newton, P., Civita, N., Frankel-Goldwater, L., Bartel, K. and Johns, C. (2020). What is regenerative agriculture? A review of scholar and practitioner definitions based on processes and outcomes, *Front. Sustain. Food Syst.* 4, October 2 issue, https://doi.org/10.3389/fsufs.2020.577723.

Reveneau, C., Ribeiro, C. V. D. M., Eastridge, M. L., St-Pierre, N. R. and Firkins, J. L. (2005). Processing whole cottonseed modulates fatty acid metabolism and improves performance by dairy cows, *J. Dairy Sci.* 88(12), 4342–4355.

Slater, A. L., Eastridge, M. L., Firkins, J. L. and Bidinger, L. J. (2000). Effects of starch source and level of forage neutral detergent fiber on performance by dairy cows, *J. Dairy Sci.* 83(2), 313–321.

St-Pierre, N. R. (2005). May 2–3, Ft. Wayne, IN. In: *Proceedings of the Tri-State Dairy Nutrition Conference (New Version) of SESAME*, The Ohio State University Press, Columbus, Available at: https://www.tristatedairy.org/_files/ugd/36a444_5dff04ac66354580af2bef5702ffe34b.pdf (accessed 13 June 2022).

Sullivan, J. L., Huber, J. T. and Harper, J. M. (1993). Performance of dairy cows fed short staple, Pima, and cracked Pima cottonseed and feed characteristics, *J. Dairy Sci.* 76(11), 3555–3561.

United Nations. (2021). Circular agriculture for sustainable rural development, *Universen/DESA Policy Brief*, Department of Economic and Social Affairs Economic Analysis (Vol. 105), Available at: https://www.un.org/development/desa/dpad/publication/un-desa-policy-brief-105-circular-agriculture-for-sustainable-rural-development/#:~:text=Circular (accessed 15 June 2022).

White, A. F. (2022). Milk prices, costs of nutrients, margins, and comparison of feedstuffs prices, *Buckeye Dairy News*, The Ohio State University Press, Columbus (Vol. 24,

Issue 2), Available at: https://dairy.osu.edu/newsletter/buckeye-dairy-news/volume-24-issue-2/milk-prices-costs-nutrients-margins-and-comparison (accessed 13 June 2022).

White, R. R. and Hall, M. B. (2017). Nutritional and greenhouse gas impacts of removing animals from US agriculture, *Proc. Natl Acad. Sci. U. S. A.* 114(48), E10301–E10308, https://doi.org/10.1073/pnas.1707322114.

Yang, K., Qing, Y., Yu, Q., Tang, X., Chen, G., Fang, R. and Liu, H. (2021). By-product feeds: Current understanding and future perspectives, *Agriculture* 11(3), 207, https://doi.org/10.3390/agriculture11030207.

Zebeli, Q., Aschenbach, J. R., Tafaj, M., Boguhn, J., Ametaj, B. N. and Drochner, W. (2012). Invited review: Role of physically effective fiber and estimation of dietary fiber adequacy in high-producing dairy cattle, *J. Dairy Sci.* 95(3), 1041–1056, http://doi.org/ 10.3168/jds.2011-4421.

Chapter 12

Assessing alternative protein sources from by-products

Zachary E. Carlson and Gregory Lardy, North Dakota State University, USA

1 Introduction

By-product feeds can be defined as feed produced in addition to the principal product, often intended for greater-value human markets. These secondary products, also called co-products, are often economical feeds with greater concentrations of protein, fiber, and minerals than that of the original crop (Grasser et al., 1995). Feeding by-products to dairy cattle is not a novel idea. Some of the first published work was performed in the late 1910s, evaluating sorghum mill refuse (McCandlish, 1920). Bath et al. (1993) identified over 350 by-product feeds utilized in livestock diets in the United States. By-product feeds typically do not meet the specifications for markets typically used for human consumption. Many are derived from the food and beverage, fiber, and bioenergy industries (NASEM, 2021). The physical and nutrient composition of by-product feeds depends on the manufacturers' processes and techniques, location, and advances in processing technology (NASEM, 2021). As manufacturing techniques and processes improve and become more efficient, greater utilization of the principal product may lead to changes in the physical and nutrient composition of the by-product feeds. Often, the use of by-products depends on the geographic location where the primary product

http://dx.doi.org/10.19103/AS.2022.0117.13

is grown, seasonal availability, and harvest schedule of the primary product (Eastbridge et al., 2006). Common grain by-products include brewers grains, corn gluten feed, distillers grains plus solubles (DGS), wheat middlings, and common oilseed by-products including canola, cottonseed, safflower, soybean, and sunflower meals. Dairy operations may also utilize by-products from citrus production, almond processing, vegetable processing, as well as rendered animal-derived by-products such as blood meal, feather meal, fish meal, and poultry meal.

There are several reasons for the dairy industry's acceptance and use of by-products. First, the increased supply of corn by-products through the establishment of the ethanol Renewable Fuel Standard from the Energy Policy Act of 2005 and expansion by the Energy Independence and Security Act of 2007 created a growing demand for corn grain production and alcohol produced by corn milling facilities in the US (USDE, 2022). Second, increased demand for principal products from food manufacturing industries for human markets (World Agriculture Outlook Board, 2020). Third, upcycling by-products, often inedible for humans, such as livestock feeds, prevent waste disposal (Grasser et al., 1995). Fourth, a reduction in the amount of human-edible food (i.e. grain) is required to produce high-quality products for the human food market (i.e. meat and milk) with the use of by-products in ruminant diets (Mottet et al., 2017). Fifth, replacing starch (i.e. grain) in ruminant diets with by-products may reduce animal health issues (i.e. ruminal acidosis) without compromising animal performance (Wickersham et al., 2004).

This chapter will update the most commonly used protein-based by-product feeds in dairy cattle diets. This chapter focuses on (1) protein quality of by-products; (2) characteristics of grain-sourced by-products; (3) characteristics of oilseed by-products; (4) rendered by-products; (5) environmental impact of by-products; and (6) future direction and trends of by-products. Review papers will be cited to direct readers interested in further detail on a specific topic when appropriate.

2 Quality and digestibility of protein

The protein quality of a feedstuff is directly related to the bioavailability and amount of essential amino acids (AAs) in a feed relative to the animal's requirement. There are nine 'essential' AA, and 'essential' is defined as if their carbon skeleton cannot be synthesized by animal cells or synthesized at a rate insufficient to meet the animal's needs (NASEM, 2021). Estimates of protein quality are based on the biological value measured by the amount of nitrogen retained in the body for physiological functions (i.e. maintenance, growth, lactation, and pregnancy). Therefore, the protein quality of feedstuffs is relative to the requirements of the animal, given its physiological state.

Dietary protein, typically referenced as crude protein (CP), was the historical standard for measuring the amount of nitrogen and essentially estimating the amount of AA in feeds. The CP estimation determined the average AA contained 16 g per 100 g of protein. In other words, 1 g of nitrogen in a feed would represent 6.25 g of CP. This meant that all protein was regarded as equal in chemical structure, physical structure, and value to the animal. To address this limitation, the Subcommittees on *Nitrogen Usage in Ruminants* (NRC, 1985) and *Nutrient Requirements of Dairy Cattle* (NRC, 1989) established that feed protein entering the rumen can have two chemical structures, either as true protein meaning the nitrogen is actually part of the AA or alternatively, the nitrogen may be in the form of non-protein nitrogen (NPN). The subcommittees acknowledged that a portion of feed protein escapes ruminal degradation. These changes lead to the development of the metabolizable protein (MP) system.

The MP system focuses on rumen degradation of dietary protein, accounting for the needs of the microorganisms and the animal. It is important for several reasons, (1) to characterize the rumen degradable protein (RDP) and digestible rumen undegradable protein (dRUP), (2) to predict the conversion of RDP into microbial crude protein (MCP), and (3) to match MP, or AA, supply and requirements to improve animal efficiency and reduce nitrogen excretion (NASEM, 2021; White et al., 2017). The RDP portion is either degraded in the rumen into ammonia and incorporated into MCP or utilized by microbes as AA or peptides. Ruminally synthesized MCP typically supplies most of the AA absorbed in the small intestine. The RUP fraction is the other source of AA absorbed in the small intestine (NRC, 2001). The AA supply from feedstuffs is derived from ruminal microbial CP and RUP content and their associated digestibility fractions. The extent of ruminal digestion of proteinaceous feedstuffs and the availability of AA for absorption in the digestive tract varies among feedstuffs (Paz et al., 2014). One of the primary factors influencing ruminal degradation is the passage rate (NRC, 2001). The most widely accepted model to describe protein degradation divides feed CP into three fractions (A, B, and C; NRC, 1985). Briefly, fraction A consists of the NPN fraction that is assumed to be solubilized immediately upon entry into the rumen. Fraction C is the undegraded portion associated with lignin, tannins, and heat-damaged proteins. Fraction B represents the potentially degradable true protein. The fraction B degraded is determined by the fractional rate of ruminal degradation potentially affected by the relative passage rate (NRC, 2001). As it relates to proteinaceous by-products, the original source of feed and the method in which the by-product was processed are important in determining the dietary contributions of MP and specific AA profiles.

A key consideration for evaluating by-products is the variation and consistency in physical and nutrient composition. For example, the NASEM (2021) reported a large coefficient of variation (CV) for CP of wet (8.8%), modified (7.5%), and dried (5.6%) DGS. Variation is likely greater across different manufacturing plants than within a manufacturing plant (Droppo et al., 1985; Spiehs et al., 2002; Kleinschmit et al., 2007; Buckner et al., 2011). For example, an analysis of 14 loads of wet corn gluten feed from one starch plant in the 1980s observed a CV for CP of 11.5% (Droppo et al., 1985). More recently, Kleinschmit et al. (2007) reported a relatively small CV of 3.6% for CP from five different dried DGS (DDGS) sources. Similar results were reported for wet (WDGS) and modified DGS (MDGS) within 1 ethanol plant over 4 months, with a CV for CP ranging from 0.92% to 3.72% (Buckner et al., 2011). Variation in protein content is dependent on the grain or raw material used in processing. Poos (1981) showed that CP from dried DGS produced from rye, wheat, milo, and corn grains were 22.1%, 33.5%, 33.3%, and 29.7%, respectively. Grain quality and variety influence the protein content of the by-product (Hatch, 1986; Olentine, 1986; Clark et al., 1987).

2.1 Rumen degradability

Several factors affect the extent of ruminal degradation of protein in feeds. Some are related to the physical and nutrient composition of the feedstuff and diet (i.e. forage:concentrate ratio, nutrient interactions, particle size, protein structure), others to the animal and the rumen environment (i.e. ruminal pH, ruminal dilution rate; Tamminga, 1979; NRC, 1985; Clark et al., 1987; Paz et al., 2014). Four types of proteins are found in feeds: albumins, globulins, prolamins, and glutelins (Sniffin, 1974; Tamminga, 1979). Albumins and globulins are light molecular weight proteins soluble in ruminal fluid. Prolamins and glutelins are heavier molecular weight proteins with disulfide bonds, making them less soluble in the rumen (Sniffin, 1974; Wohlt et al., 1976; Clark et al., 1987). Albumins and globulins typically have a better balance of AA and a greater biological value than prolamins and glutelins but are unfortunately more degradable in the rumen (Sniffin et al., 1974; Clark et al., 1987). For example, corn contains no albumins, 5-6% globulins, and the remaining protein comprises prolamins and glutelins (Sniffin, 1974). Therefore, protein in corn consists mainly of the RUP fractions (Paz et al., 2014). In contrast, soybean protein is mainly globulins, making soybean meal (SBM) more concentrated in RDP (Clark et al., 1987). In the production of DGS, brewers grains, and corn gluten meal, most of the soluble protein fractions are removed, resulting in a protein by-product concentrated in RUP (Clark et al., 1987).

Another factor influencing protein degradability is the processing techniques, especially the drying or heating processes used in producing the various by-products (Bradford and Mullins, 2012). Many by-product feeds are dried to help with transportation, storage, and spoilage before being marketed as feed. Heat denatures and coagulates proteins, making them less soluble and degradable in the rumen (Bendall, 1964; Clark et al., 1987). Cooling causes a random relinkage of chemical bonds, shrinking the protein molecules (Bendall, 1964; Clark et al., 1987). By-products that undergo drying may have heat-damaged proteins or indigestible Maillard reaction products. The Maillard reaction degrades sugar compounds to phenolic products, which destroys AA and polymerizes with AA remnants (Hodge, 1953; Van Soest and Mason, 1991). Specifically, lysine is susceptible to heat damage via the Maillard reaction due to its exposed ε-amino groups (Choi et al., 1949). For example, total-tract CP digestibility from five different sources of DDGS varied from 70.7% to 85.3%, which was less than the total-tract CP digestibility of SBM (93.9%; Kleinschmit et al., 2007). Two of the DDGS sources had significantly reduced CP digestibility estimates than that of WDGS, suggesting the degree of heat damage via the Maillard reaction, a non-enzymatic browning reaction, denatured some of the protein acid detergent insoluble nitrogen, rendering a portion of the protein indigestible. These heat-damaged proteins have been demonstrated to be less digestible (Kajikawa et al., 2012).

The extent of protein degradation in the rumen is determined by the rate of microbial digestion of feed and how feed passes from the rumen (NASEM, 2016). The rate of digestion and passage rate of feeds are highly related to each other. A faster passage rate or a slower digestion rate will decrease ruminal digestion of feed. Conversely, a slower passage rate or faster digestion rate will increase ruminal digestion (NASEM, 2016). The rate of passage of feeds through the rumen and overall digestive tract is dependent on several factors, including the composition of the diet, feed additives, feeding method, level of feed intake, and particle size of feed (Owens and Goetsch, 1988). As feed intake increases, the passage rate increases, increasing the fraction of rumen undegraded protein (Balch, 1958). However, the effect of feed intake on the passage rate of feed in the rumen and the extent of protein degradation is likely diminished as dry matter intake (DMI) increases, such as with high producing dairy cows (Clark et al., 1987).

3 Ethanol production and associated by-products

The use of corn and other grains to produce ethanol is a process that has undergone rapid growth and extensive changes since the early 2000s (Liu, 2011). Ethanol can be produced by two major processes, dry milling and wet milling. Each process yields unique by-products; therefore, it is important to understand each process's techniques to better understand the characteristics of the by-products.

3.1 Dry milling

Ethanol production from the dry-milling process yields distillers grains, an energy and protein by-product utilized in lactating dairy cow diets. The dry-milling process is used at approximately 80% of the corn ethanol facilities in the United States (NASEM, 2021). The process can use various grain sources, and the grain's quality can be poorer (Stock et al., 2000). It starts by grinding the grain (corn, barley, or sorghum) through a hammer mill to make the starch available for fermentation (see Fig. 1). The fine ground grain is converted to dextrins by liquefaction enzymes at temperatures above 50°C for 1-2 h (Wang et al., 2007). Dextrins are then hydrolyzed to simple sugars (i.e. glucose) using saccharification enzymes at the same time as fermentation using yeast to produce ethanol, carbon dioxide, and solids (Wang et al., 2007; NASEM,

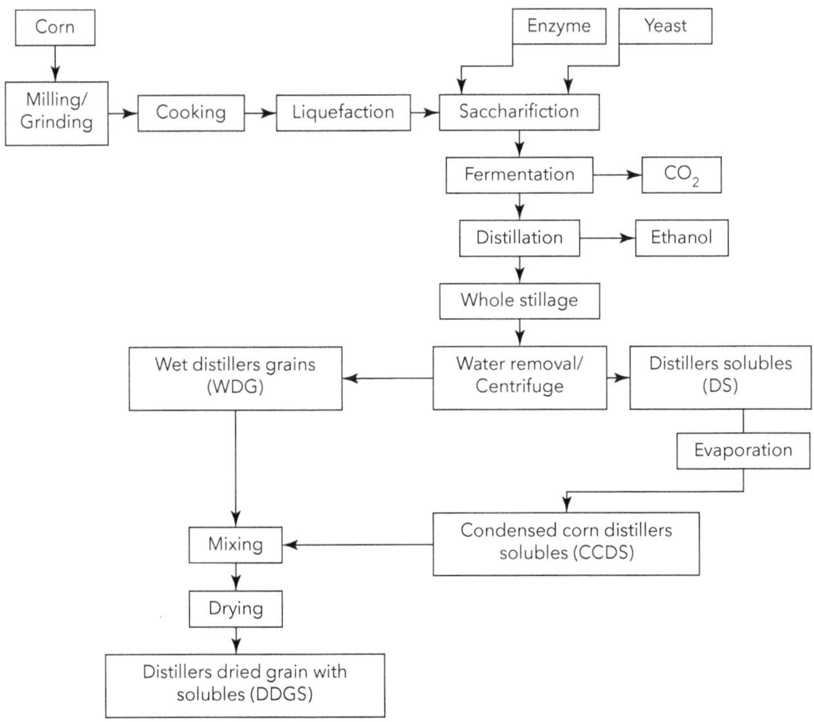

Figure 1 Overview of the corn dry-milling process. Source: Adapted from Liu and Rosentrater (2016) and Shad et al. (2021).

2016). The fermentation process takes approximately 48-72 h and yields 10-12% ethanol (Bothast and Schlicher, 2005). To separate the ethanol from the solids, the fermented mixture is distilled. After distillation, the remaining solids are called whole stillage, containing protein, fiber, oil, yeast cells, and residual starch from the grain (Bothast and Schlicher, 2005). The whole stillage can be fed as is, but this practice is not common due to the high moisture content of thin stillage (5-10% dry-matter [DM]). Instead, the whole stillage is centrifuged to separate the stillage (i.e. thin stillage) from the solids, which are then considered wet distillers grains (WDG). The nutrient composition of WDG is variable but typically averages 35.3% DM, 36.6% CP, with 9.6% crude fat (Kim et al., 2008). The thin stillage remaining is evaporated to produce condensed corn distillers solubles (CCDS), commonly referred to as syrup. The protein and fat content of CCDS also vary but averages 18.4-26.9% CP and 5.4-16.6% crude fat (NASEM, 2021; see Table 1). Condensed distillers solubles and WDG can be marketed separately but usually are combined to produce WDG plus solubles (WDGS). The fermentation process converts large proportions of the starch fraction to ethanol. Consequently, compared to corn grain, WDGS has increased concentrations of protein, fat, fiber, and nearly all minerals (approximately a threefold increase in concentration). Concentrations of Na, S, and Ca are also increased due to exogenous sources of these minerals added during the process (NRC, 2012). The nutrient composition of WDGS is dependent on the variability in grain source, grain harvest methods, variability in processing methods, the extent of starch digestion, the extent of drying and temperature, and the amount of CCDS added back to WDG (Kleinschmit et al., 2007; Liu, 2011). According to Belyea et al. (2010), the process of blending the two streams can vary and is difficult to control. If all the CCDS produced is added to WDG, the composition is approximately 80% WDG and 20% CCDS (Corrigan et al., 2009). In a review of the nutrient composition of DDGS from 1993 to 2008, the results ranged for DM (87.2-92.7%), CP (25.8-33.3%), crude fat (9.1-14.1%), starch (3.2-5.9%), neutral detergent fiber (NDF; 33.1%), and acid detergent fiber (ADF; 49.1%; Liu, 2011). The nutrient composition of dried, modified wet, and WDGS are listed in Table 1. Current ethanol industry practices generally utilize oil extraction procedures in the by-product streams, resulting in lower levels of crude fat (2-8%; Loy and Lundy, 2018).

The inclusion of DGS on lactation performance is summarized in a meta-analysis by Kalscheur (2005). Briefly, 23 studies were compiled, and overall, DMI increased when DGS was included by up to 20% (DM basis). DMI was influenced by the form of DGS (i.e. DDGS, MDGS, WDGS), with DDGS having greater DMI up to 30% inclusion than WDGS. Milk yield was greatest for cows fed 4-30% DGS, with approximately 0.4 kg/day more than diets without DGS (Kalscheur, 2005; Kalscheur et al., 2016). For more information on DGS, several reviews and book chapters have been published on the use of corn milling

Table 1 Nutrient composition and digestibility of corn milling by-products

Item	Distillers grains plus, dried, low fat[a]	DGS, dried, high protein[a]	DGS, modified wet[a]	DGS, wet[a]	Distillers solubles	Corn gluten feed, dry	Corn gluten feed, wet
DM, % as feed[b]	89.9	91.1	49.2	33.2	31.2	89.2	45.6
Ash, % DM	5.3	4.0	5.6	4.5	11.1	7.5	7.2
CP, % DM[b]	31.0	39.0	30.3	31.5	22.6	23.2	23.1
RUP, % of CP[b]	47	47	44	42	25	29	25
dRUP, % of RUP[b]	75	75	75	75	75	79	79
Crude fat, % DM	8.90	7.56	3.50	9.31	10.99	3.91	3.81
NDF, % DM[b]	30.8	37.6	27.1	31.7	4.8	35.7	36.6
ADF, % DM[b]	14.8	17.7	14.4	16.1	3.2	11.5	12.1
Lignin, % DM	3.53	5.83	3.09	3.22	0.57	2.31	1.89
Starch, % DM	6.1	6.2	4.7	6.3	4.0	15.5	15.3
Digestible energy base, Mcal/kg	3.44	3.34	3.50	3.50	3.62	3.21	3.23

Source: National Academies of Sciences, Engineering, and Medicine, (2021), *Nutrient Requirements of Dairy Cattle*: Eighth Revised Edition, Washington, DC: The National Academies Press.

[a] Distillers grains plus solubles (DGS).

[b] Dry-matter (DM); crude protein (CP); rumen degradable protein (RUP); digestibility of rumen degradable protein (dRUP); neutral detergent fiber (NDF); acid detergent fiber (ADF).

by-products and their nutrient composition in dairy cattle rations (Kalscheur, 2005; Schingoethe et al., 2009; Hollmann et al., 2011a,b; Bradford and Mullins, 2012; Paz et al., 2014; Böttger and Südekum, 2018; Liu and Rosentrater, 2016; Loy and Lundy, 2018).

3.2 Wet milling

The conventional wet-milling process uses only #1 and #2 corn, no other grains, with the primary objective to isolate starch (see Fig. 2). The process separates the individual components of corn (i.e. bran, gluten, germ) to recover starch and several by-products (i.e. steep corn liquor, wet corn gluten feed, dry corn gluten feed, corn gluten meal). The process is divided into six main processes: (1) grain handling, (2) steeping, (3) germ separation and recovery, (4) fiber separation and recovery, (5) gluten separation and recovery, and (6) starch washing and recovery (Ramirez et al., 2008). Briefly, corn is screened to remove crop residue, fines, and broken kernels. The screened corn is steeped in a dilute sulfurous dioxide solution for 40–48 h. The steepwater can be condensed to produce steep corn liquor, officially known as 'condensed fermented corn extractives' (AAFCO, 2021), which contains protein and complex sugars from the soluble portion of the corn kernel (NASEM, 2021). The corn kernel fractions are then separated by a series of grinds, differential separations, and centrifuging. At this point, the germ is isolated and removed for oil extraction. The degermed components are separated using centrifugation into gluten, starch, and bran. The starch can be dried or sold as is, converted to corn syrup and high-fructose corn sweeteners, or converted to dextrose for fermentation by yeast to produce ethanol. A by-product of ethanol production is distillers solubles, containing yeast, and residual sugars but little fat, unlike CCDS (Stock et al., 2000). The gluten is dried and sold as corn gluten meal, high in CP and RUP, and used primarily by pet food and poultry industries. The protein in corn gluten meal is primarily a mixture of glutelin and zein. Its CP is highly digestible and a valuable source of methionine and cystine but low in lysine and tryptophan (Loy and Lundy, 2018). Further enzymatic breakdown of corn gluten meal to remove residual starch produces a product termed Empyreal, containing approximately 80% CP (DuFour, 2017). The bran and bran-bound starch and gluten are ground and washed through fiber wash screens to isolate the residual starch and gluten. The filtered bran is combined with the corn steep liquor and, where available, germ meal, to produce corn gluten feed, either dried or wet (Stock et al., 2000; Ramirez et al., 2008; Loy and Lundy, 2018). Wet corn gluten feed has a nutrient composition similar to dried corn gluten feed (see Table 1). Several trials in lactating dairy cows have shown that inclusion from up to 20–38% DM of wet corn gluten feed in the diet increased DMI, milk yield, and milk protein percent but reduced milk fat percent (Gunderson et al., 1988; VanBaale et al., 2001; Kononoff et al., 2006).

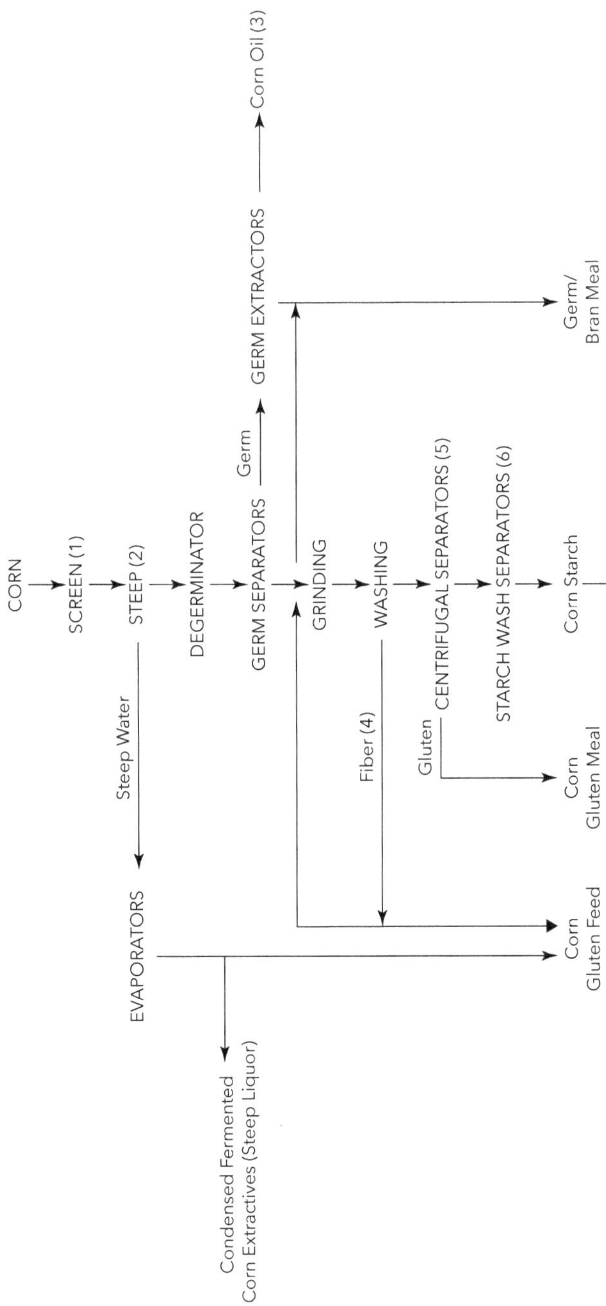

Figure 2 Overview of the corn wet-milling process (NB numbers indicate the main processes identified by Ramirez et al. (2008)). Source: Adapted from Loy and Wright (2003) and Liu and Rosentrater (2016).

3.3 Fractionation

Beginning in the early 2010s, ethanol production from the dry-milling industry took steps to improve the efficiency of ethanol production and increase the utilization of the corn kernel in an attempt to add value to the by-products, as the dry-milling process typically yields by-products with less value compared to wet-milling process (Rajagopalan et al., 2005; Kurambhatti et al., 2019; Loy and Lundy, 2018). One of these improvements has been changes in the processing methods reducing the heat damage of DDGS, demonstrated by greater lysine concentrations over time, lysine being an AA susceptible to heat damage via the Maillard reaction (Choi et al., 1949; Kim et al., 2012).

Ethanol plants have improved their corn oil separation methods by one of two processes, oil removal via centrifugation of the thin stillage before evaporation or secondary fermentation of the whole stillage (Lundy and Loy, 2014). Centrifugation of solubles produces a DGS, commonly called 'low-fat DGS,' containing approximately 5–8% fat. Solvent-extracted oil processes produce a 'deoiled' DGS containing 2–6% fat (Loy and Lundy, 2018). Removing larger amounts of oil from DGS slightly increases the CP concentration (Morris et al., 2018). Studies have reported that 'deoiled' DGS is a valuable protein feed for lactating dairy cows up to 20% of dietary DM (Mjoun et al., 2010; Castillo-Lopez et al., 2014). Ramirez-Ramirez et al. (2016) demonstrated that the inclusion of 'deoiled' DGS (6.6% crude fat) at 30% of dietary DM had no difference in milk fat percentage and yield compared to diets without DGS while diets containing DGS (12.0% crude fat) at 30% dietary DM had decreased milk fat percentage and yield than that of 'deoiled' DGS. DGS with reduced crude fat content may reduce the risk for milk fat depression when included in the diet up to 30% (Mjoun et al., 2010; Ramirez-Ramirez et al., 2016).

The ethanol industry has adopted techniques to fractionate the corn kernel to extract the bran and germ for additional fermentation and increase cellulosic fermentation capacity. These techniques are referred to as partial fractionation and can occur pre-or post-fermentation. Fractionation focuses on separating the pericarp fiber (i.e. corn bran) and germ for alternative feed markets. The corn bran can be fermented using cellulosic fermentation to produce ethanol. The germ can be marketed as germ oil (Kurambhatti et al., 2019). Nutrient composition reported by NASEM (2021) for low-fat and high-protein DGS are listed in Table 1.

3.4 Pre-fermented fractionation

Pre-fermentation fractionation techniques isolate corn components for high-value food and industrial applications because the isolation and recovery occur before fermentation for ethanol. Pre-fermentation fractionation results in a greater concentration of corn oil removed from the germ. Pre-fractionation reduces the amount of DGS produced by up to 66% but is dependent on processing methods and by-products recovered (Wang et al., 2005). The end product is a DGS containing reduced concentrations of fat (4%) and NDF (23.0%) but more CP (43.0%; Depenbusch et al., 2008) than DGS from traditional dry milling. The value in pre-fermentation fractionation originates because only the endosperm (containing the starch) is the fermentable component. The other non-fermentable components (i.e. protein, fiber, and fat) can be isolated and processed into other higher-value products (Rosentrater et al., 2016). Pre-fermentation fractionation occurs via two different techniques, dry or wet fractionation. Dry fractionation involves tempering corn kernels with hot water or steam for 15–30 min, then coarse milling and separating the individual components. Dry fractionation can have inefficient separation of components leading to poorer quality products than wet fractionation (Kurambhatti et al., 2019). Several variations of dry fractionation have been developed and implemented, including BFrac (www.poet.com; accessed 24 February 2022), Cereal Process Technologies (www.cerealprocess.com; accessed 24 February 2022), and dry degerm defiber (Murthy et al., 2006) technologies (Rosentrater et al., 2016). The reader is referred to Singh and Johnston (2009) for a more detailed discussion of dry fractionation techniques.

Wet fractionation techniques require soaking the corn kernel for 6–12 h, followed by coarse grinding to separate and recover the germ, pericarp fiber, and fine fiber before fermentation in the dry-grind process (Kurambhatti et al., 2019). Several variations of wet fractionation have been developed and implemented, including quick germ (Singh and Eckhoff, 1997), quick germ quick fiber (Wahjudi et al., 2000a; Wahjudi et al., 2000b), and enzymatic milling (Wang et al., 2005) techniques (Rosentrater et al., 2016; Kurambhatti et al., 2019). The reader is referred to Singh and Johnston (2009) for a more detailed discussion of wet fractionation techniques.

3.5 Post-fermented fractionation

Post-fermentation separation techniques focus primarily on fractionating the by-products, namely DDGS, thin stillage, or CCDS (Rosentrater et al., 2016). For DDGS, additional sieving and air separation techniques are utilized to sort fiber particles, or solvent extractants are used to separate and recover more oil. For the liquid streams (i.e. thin stillage and CCDS), additional screening and centrifugation are used to separate and recover more oil. Before evaporating the CCDS, thin stillage contains approximately 30% of the available oil from corn (Rosentrater et al., 2016). Therefore, when reduced-fat CCDS is added to WDG, the end product is considered a reduced-fat distillers grain. Greater oil recovery from the by-product streams has decreased the crude fat content of distillers grains by 40–60%, with a crude fat content of 5.5–7.5% (DM basis; DuFour, 2017). Some of the current processes developed and employed include Mean Green BioFuels (Winsness, 2006), GS CleanTech (Kirkbride McElroy, 2007; Cantrell and Winsness, 2009; Winsness and Cantrell, 2009), ICM (www.icminc.com; accessed 25 February 2022; ICM, 2009), which are physical separation techniques using centrifuges and are reported to remove 30–70% of the oil from the by-products (Rosentrater et al., 2016).

Historically, DGS was limited to 10% (DM basis) in dairy cow diets due to the relatively high crude fat content (ranging from 10% to 12%; Janicek et al., 2008). A meta-analysis of 23 studies by Kalscheur (2005) evaluated the inclusion of DGS. They concluded no significant effect on milk fat depression for DGS inclusion level or form (i.e. DDGS, WDG, or MDGS). However, dietary fat can be toxic to ruminal microbes, negatively affecting fiber digestion (Van Soest, 1994), and may contribute to milk fat depression (Pantoja et al., 1994; Bauman and Griinari, 2003). Inclusions of 'full-fat' DGS up to 30% of the diet (DM basis) have been shown to decrease milk fat percent (Janicek et al., 2008; Abdelqader et al., 2009; Benchaar et al., 2013) and milk protein concentrations (Janicek et al., 2008; Benchaar et al., 2013). Recently, the inclusion of reduced-fat DGS replacing corn, forages, SBM, and soy products has been reported not to affect milk fat percent (Castillo-Lopez et al., 2014; Foth et al., 2015) or have increased milk fat percent (Mjoun et al., 2010). To avoid milk fat depression when feeding DGS, an adequate supply of effective fiber must be provided, and the fermentability of starch and the diet must be considered (Kalscheur et al., 2016).

4 Oilseed processing by-products

Oilseed by-products include the meals after extracting the oil, which is utilized in human consumption markets (i.e. food and fuel). In most cases, oil extraction can be done via mechanical (extruded/expeller) or solvent extraction.

4.1 Soybean meal

SBM is one of the most common protein sources in dairy diets (Prestegaard-Wilson, 2021). In 2020, the top five soybean-producing countries were Brazil, the United States, Argentina, China, and India (fao.org; accessed 14 May 2022). It is popular, in part, because soybean products are characterized as a highly palatable, relatively consistent source of protein that is highly digestible, a great source of essential AA and has less fiber than most other oilseed meals

(Awawdeh et al., 2007; Isher and Varga, 2016). A comparison of 11 protein sources indicated soybean RUP is only second to microbial CP on the essential AA index (Chandler, 1989). However, it is important to ensure a balanced ratio of the first two limiting AAs, lysine and methionine (Liu et al., 2013). The current understanding is to target a lysine to methionine ratio of 3:1 to maximize the use of MP for milk production (Liu et al., 2013). The AA profile of SBM (averaged across solvent-extracted, expeller, and extruded) includes a ratio of 4.53:1, while canola meal has a ratio of 2.80:1 (NASEM, 2021).

The method of extracting oil from soybeans will influence the nutrient composition and quality of SBM (Karr-Lilienthal et al., 2006). Soybeans can be extracted for oil through solvent extraction, mechanical extrusion, or expeller methods. Table 2 lists the nutrient composition and ruminal degradability of SBM relative to the oil extraction method (i.e. solvent-extracted, expeller, and extrusion). Solvent-extracted SBM is the most common form of SBM on the market and is produced by grinding soybeans and separating the fat with organic solvents such as hexane (Isher and Varga, 2016). Solvent-extracted SBM contains more CP (50.9–54.3%) and the least amount of crude fat (0.95–2.7%) compared to extruded and expeller SBM (NASEM, 2021). Solvent-extracted SBM is marketed as either 44% CP or 48% CP (as-fed basis). The 48% meal contains about 8% NDF, whereas the 44% meal is blended with toasted ground soybean hulls to provide approximately 14% NDF (Isher and Varga, 2016). Extrusion is the process of pressure-cooking and heat-roasting at specific temperatures and pressures for a period of time, followed by forcing the material through a special opening (Jiang et al., 2011). The end product has less CP (38.3% to 42.6%; NASEM, 2021) and more crude fat (8.64%; Harper et al., 2019) relative to other SBM products. The traditional oil extraction method has used mechanical presses (hydraulic or screw presses). Expeller screw presses or expellers remove the oil by forcing seeds through a die, reducing the volume in the barrel, and compressing the seeds (Bhargavi et al., 2018). Expelled SBM protein and crude fat content is 45.5–49.8% CP and 4.8–9.5% crude fat (NASEM, 2021).

The ruminal degradability of protein from SBM is relatively high. Rumen degradability is approximately 55–67% of CP for extruded and solvent-extracted SBM, respectively (NASEM, 2021). Various methods of heat processing are used commercially to reduce the degradation of soy protein by rumen microbes. Higher temperatures during the extrusion process can reduce the rumen degradability of protein (Solanas et al., 2008; Giallongo et al., 2015). Heat treatment partially denatures protein and produces Maillard reactions among AA and sugars (Choi et al., 1949; Kim et al., 2012), making proteins more resistant to microbial enzymatic reactions, thus reducing ruminal degradation of the protein. SBM expellers alter the ruminal degradability and increase the RUP content (63% of CP; NASEM, 2021). In a survey of United States

Table 2 Nutrient composition and digestibility of oilseed meals

Item	Soybean meal, solvent-extracted[a,b]	Soybean meal, expellers[a,b]	Soybean meal, extruded[a,b,c]	Canola meal, solvent-extracted[a]	Cottonseed meal[a]	Sunflower meal[1a]	Safflower Meal[a]
DM, % as feed[d]	89.3	91.2	93.4	89.1	89.9	90.2	94.0
Ash, % DM	7.2	6.7	5.7	7.9	7.6	7.4	4.9
CP, % DM[d]	52.6	47.6	40.4	41.5	46.7	37.0	26.2
RUP, % of CP[d]	33	63	45	32	44	16	31
dRUP, % of RUP[d]	91	93	91	74	83	90	75
Crude fat, % DM	1.82	7.12	8.64	3.51	3.60	2.20	5.35
NDF, % DM[d]	11.1	19.6	18.4	29.0	28.1	40.2	55.4
ADF, % DM[d]	7.2	10.1	10.6	20.3	19.2	29.0	40.1
Lignin, % DM	1.08	2.06	2.15	8.51	7.01	9.07	13.8
Starch, % DM	1.9	1.8	1.5	1.6	1.1	1.1	1.2
Digestible energy base, Mcal/kg	3.99	3.90	4.16	3.51	3.32	2.99	2.51

[a] National Academies of Sciences, Engineering, and Medicine, (2021), *Nutrient Requirements of Dairy Cattle*: Eighth Revised Edition, Washington, DC: The National Academies Press.
[b] Soybean meal (SBM).
[c] Value for crude fat: Harper, M. T., Oh, J., Melgar, A., Nedelkov, K., Räisänen, Chen, X., Martins, C. M. M. R., Young, M., Ott, T. L., Kniffen, D. M., Fabin, R. A. and Hristov, A. N. (2019), 'Production effects of feeding extruded soybean meal to early-lactation dairy cows', *J. Dairy Sci.* 102:8999-9016.
[d] Dry-matter (DM); crude protein (CP); rumen degradable protein (RUP); digestibility of rumen degradable protein (dRUP); neutral detergent fiber (NDF); acid detergent fiber (ADF).

dairy nutritionists representing 1065 herds, heat-treated SBM was the most common plant protein source used in lactating dairy cow diets (Prestegaard-Wilson, 2021). There are two commercial expeller methods, cooker-expeller and extruder-expeller. Two commercial products utilize the cooker-expeller process, SoyPLUS® (West Central Cooperative, Ralston, IA) and Soy Best® (Grain States Soya Inc., West Point, NE). Several extruder-expeller SBM products are available, including Soy King® (Topeka, KS), InstaSoy-XP® (Soy-Co, Adams, WI), and SoyMax® (Soy-Max Proteins Inc., Cartier, MB). Another method is cooker processing of de-hulled, solvent-extracted SBM, Amino Plus® (AGP, Omaha, NE). There is also a non-enzymatically browned SBM, also referred to as lingo-sulfate-treated SBM. The process involves heating to 150°C for 30 min after mixing with xylose and adjusting the moisture content to 17% (Isher and Varga, 2016). The commercial product from this processing method is Soy Pass® (Ligno Tech USA, Overland Park, KS).

4.2 Canola meal

Canola is a variety of rapeseed trademarked and developed in Canada. The top five rapeseed-producing countries are Canada, China, India, Germany, and France (fao.org; accessed 14 May 2022). Canola is characterized by low erucic acid (less than 2% as a percent of oil) and low alkenyl glucosinolates (less than 30 μmol per gram of oil-free DM), a sulfur-bearing glucose moiety that interferes with thyroid hormone synthesis by affecting the uptake of iodine and has a bitter flavor (Bell, 1993; Woyenga et al., 2017). Canola seeds (primarily *Brassica campestris, B. napus,* or *B. juncea*) crude fat and CP composition contain approximately 36.5–44.5% crude fat and 20.7–26.1% CP (DM basis; NASEM, 2021; see Table 2). Canola meal, also called double-zero or double-low, is the by-product meal produced following oil extraction of the seed (AAFCO, 2021). In western Canada and parts of the United States, canola meal is the primary source of protein for dairy cow diets due to regional availability and one of the highest biological values of all plant protein sources (Piepenbrink and Schingoethe, 1998; Martineau et al., 2013). The AA composition of canola meal more closely resembles that of milk protein compared to blood meal, corn gluten meal, or fish meal (Piepenbrink and Schingoethe, 1998). As previously mentioned, a lysine to methionine ratio of 3:1 is ideal for maximizing the use of MP for milk production (Liu et al., 2013). Solvent-extracted canola meal has a ratio of 2.80:1, making the AA ratio of canola meal closer to that ideal ratio than other protein sources.

In addition to glucosinolates, Bell (1993) discussed other potential anti-nutritional factors (i.e. sinapine, tannins, and phytate) that may influence the nutritional value of canola meal. Sinapine, approximately 2.5–3.0% of canola meal (DM basis; Blair and Reichert, 1984), may affect feed intake, but this has yet to be tested in cattle (NASEM, 2021). Canola meal contains approximately 1.5–3.0% tannins (DM basis; NASEM, 2021) which can interfere with enzymes responsible for protein hydrolysis (Bell, 1993). Phytate binds approximately two-thirds or more phosphorus in cereal grains, oilseed meals, and grain by-products (NRC, 2001). Phosphorus-bound phytate is approximately 43% of the total phosphorus in canola meal (Bell, 1993). Ruminants have several sources of phytase, the enzyme catalyzing the hydrolysis of phytate, available: ruminal microbial phytase, endogenous mucosal phytase, large intestinal microbial phytase, plant phytase, and exogenous microbial phytase to aid in phosphorus absorption (Humer and Zebeli, 2015).

In a meta-analysis, Martineau et al. (2013) evaluated substituting a protein source with canola meal. Overall, canola meal positively affected DMI, milk yield, milk protein yield, and milk protein percentage. Lage et al. (2020) found no differences in milk composition between solvent-extracted canola meal, solvent-extracted SBM, or extruded SBM. However, canola meal increased DMI by 5.9% and 8.9% compared to extruded SBM and solvent-extracted SBM, respectively. Comparisons of SBM and canola meal report greater milk yield without affecting milk fat or protein. According to Huhtanen et al. (2011), a greater response in DMI may be due to more energy demand from improved lactational performance due to a greater supply of AA or a more balanced AA profile with canola meal than that from SBM. Moore and Kalscheur (2016) reported no difference in DMI, and greater milk yield, with no effect on milk fat, lactose, or total solids percentage for cows fed canola meal compared to SBM. Similar performance between canola meal and SBM means that pricing and availability will have a greater influence on which protein source is incorporated into dairy cow diets.

Traditional processing methods for canola oil separation are crushing and solvent extraction techniques. Briefly, the process includes seed cleaning, preconditioning, flaking, cooking at 80–105°C for 15–20 min, mechanical oil extraction through crushing, solvent extraction, and desolventizing and toasting (Heendeniya et al., 2012). Because canola seeds have greater oil content than soybeans (40.5% versus 20.7%, respectively; NASEM, 2021), an additional step to extract the oil from the seeds is required for canola (Lage et al., 2020). The seeds are first extruded to reduce the oil content to approximately 18%, followed by the solvent extraction process (Canola Council of Canada, 2022). The heat generated from mechanical extraction damages some protein structures, reducing ruminal protein degradation and increasing the RUP content of the meal (Björck and Asp, 1983; Van Soest and Mason, 1991; Giallongo et al., 2015). The mechanical extraction process of canola meal results in a RUP content greater than solvent-extracted SBM (Lage et al., 2020). Currently, the NASEM (2016) for beef cattle reports a greater RUP content for canola meal than that of solvent-extracted SBM (42.3% versus

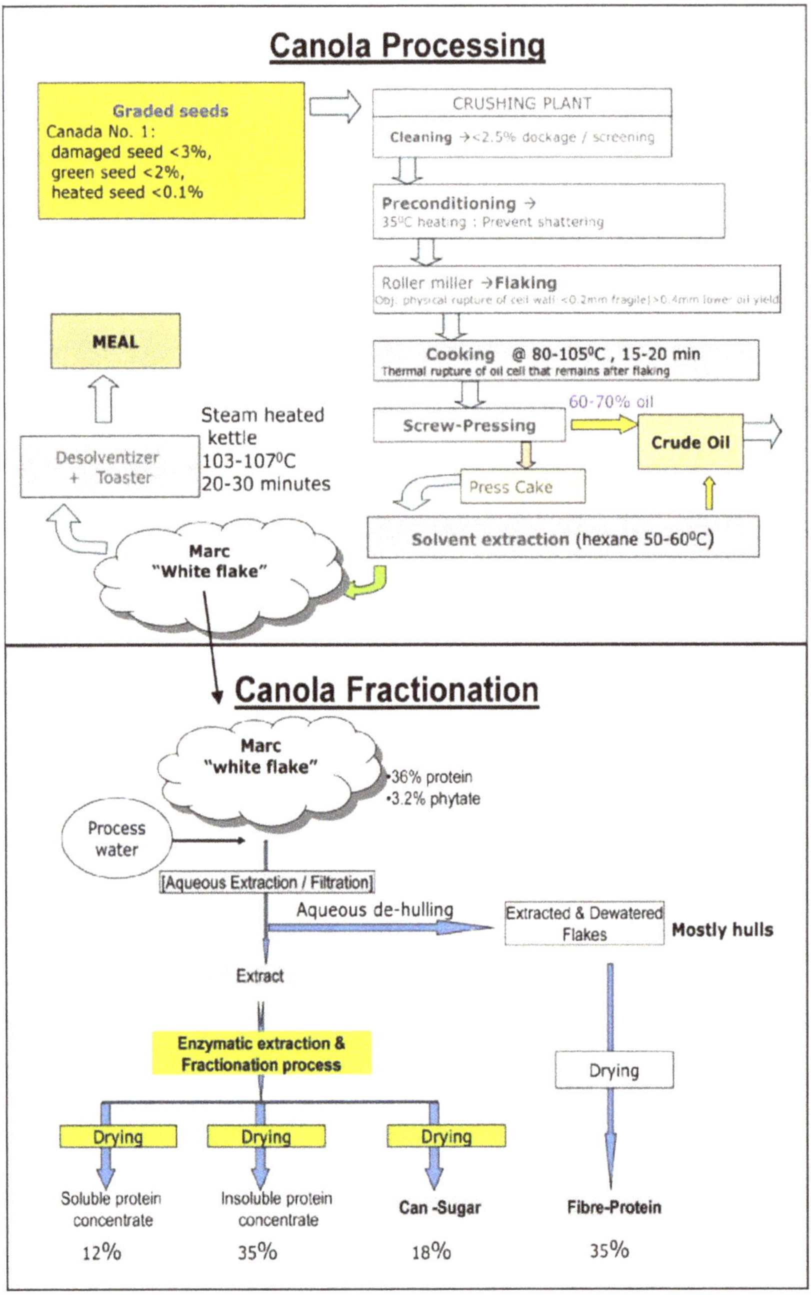

Figure 3 Processing and fractionation of canola seeds. Source: Heendeniya et al. (2012).

29.5%, respectively). However, NASEM (2021) for dairy cattle reports the RUP content of canola meal as similar to that of solvent-extracted SBM (32% versus 33%, respectively). The differences in RUP estimates between these two references are not clear. However, some recent research has reported greater RUP content for canola meal than the average RUP value reported in NASEM (2021; Broderick et al., 2016; Lage et al., 2020).

In the past two decades, processing technologies have been developed to fractionate canola meal to enhance oil extraction (see Fig. 3). The process occurs following solvent extraction but before desolventizing and toasting in the traditional process. At this stage, the processed canola seed is referred to as 'white flake' and contains approximately 36% CP. The fractionation process begins with aqueous extraction and filtration, followed by aqueous de-hulling. At this point, two product streams are created, a fiber stream and an extract stream. The fiber stream, mostly hulls, is extracted and dewatered for further drying. The end product is called Fiber-Protein with a nutrient composition of 91.8% DM, 30.9% CP, 55.6% NDF, 46.3% ADF, and 0.15% crude fat (Heendeniya et al., 2012).

For more information on canola meal, several reviews are available on canola meals' nutrient composition and performance in lactating dairy cows (Huhtanen et al., 2011; Martineau et al., 2013).

4.3 Cottonseed meal

Cottonseed meal is the by-product of oil removal from cottonseed by mechanical or solvent extraction. The protein composition of cottonseed meal must contain atleast 36% CP (AAFCO, 2021) but typically contains 42.9–50.5% CP with 44% of CP as RUP (NASEM, 2021; see Table 2). It must contain at least 36% CP (AAFCO, 2021). Milk yields are similar when replacing soybean or canola meals with cottonseed meals (Sánchez and Claypool, 1983; Brito and Broderick, 2007). Lower milk fat yield has been reported in lactating dairy cows fed 14% of dietary DM cottonseed meal compared to SBM and canola meal (Brito and Broderick, 2007). Milk fat depression has not been reported when feeding cottonseed meal (Sánchez and Claypool, 1983; Coppock et al., 1987; Brito and Broderick, 2007). However, milk fat depression has been reported in feeding whole cottonseeds at 10–15% of dietary DM (Van Horn et al., 1984). Therefore, depending on the amount of oil extracted from cottonseed meal, there could be potential for milk fat depression with cottonseed meal containing greater amounts of crude fat. The AA content, specifically lysine, of cottonseed meal may be low and partially unavailable due to gossypol, a phenolic compound found in cottonseeds, binding to the ε-amino group of lysine (Reiser and Fu, 1962; Coppock et al., 1987).

4.4 Sunflower meal

Sunflower seeds are a prominent source of vegetable oil for human consumption and biofuel production (Oliveira et al., 2018). The major sunflower-producing countries are Ukraine, Russia, Argentina, China, Turkey, Kazakhstan, Moldova, South Africa, and the United States (USDA, 2021; ipad.fas.usda.gov, accessed on 12 May 2022). Commercial varieties of sunflower seeds were initially developed in Russia and the former Soviet Union (Gupta, 2014). There are two specific types of sunflower seeds: an oil-type for oil production and a larger seed variety for direct human consumption also referred to as confectionery. Nutrient composition differences between oil-type and confectionary sunflower seeds are most notable in crude fat ranging from 29.4% to 48.6% (NASEM, 2021). Oil varieties account for 89% of total sunflower production in the United States from 2017 to 2021 (National Sunflower Association, 2022). Sunflower meal is produced when the black oilseed is mechanically crushed. The solvent extraction method relies on exerting pressure on the seed to separate the majority of oil prior to solvent extraction. Solvent extraction recovers over 40% of oil from the seed, the highest recovery rate of oil extraction methods. As in most oilseed markets, the oil has a higher economic value than the meal. Therefore with the greatest oil recovery, solvent extraction is the most common method of oil separation globally (Le Clef and Kemper, 2015). However, some plants use mechanical expeller and extrusion methods (Lardy and Anderson, 2002). Similar to other by-products, the nutrient composition of sunflower meal is affected by the seed's oil content, the extent of hull removal, and the efficiency of oil extraction. One of the largest sources of variation in nutrient composition is the degree of seed decortication or removal of the hulls prior to oil extraction (Lardy and Anderson, 2002). Sunflower hulls have low CP and high lignified fiber levels (Arija et al., 1998) and are highly resistant to rumen digestion (Van Soest, 1994). Solvent-extracted sunflower meal decorticated yields a protein content of 25–28%, partial decortication yields 34–38% CP, and complete decortication yields over 40% CP (Lardy and Anderson, 2002). Non-decorticated sunflower meal is commercialized, likely due to the additional processing costs of decortication (Oliveira et al., 2018). The nutrient composition for sunflower meal in NASEM (2021) for CP ranges from 32.7% to 41.3%, with an average of 16% of CP as RUP (see Table 2). The protein from sunflower meal is more ruminally degradable than that from SBM, cottonseed meal, and canola meal (Erasmus et al., 1994; Rodríguez et al., 2008). However, sunflower meal DM is less degradable than that of SBM, likely due to greater fiber content (Rodríguez et al., 2008; Oliveira et al., 2018). Sunflower meal has a high concentration of sulfur-containing AA, including methionine, but contains less lysine than that from SBM (NASEM, 2021). Previously, non-decorticated sunflower meal replaced SBM (Schingoethe et al., 1977) without affecting the yield or composition of milk. In more recent research, non-decorticated sunflower meal replacing SBM (Yildiz et al., 2015) or SBM and wheat middlings (Oliveira et al., 2018) reduced milk yield (Yildiz et al., 2015; Oliveira et al., 2018) and milk protein content (Oliveira et al., 2018).

4.5 Safflower meal

Safflower is grown primarily in semi-arid regions (Godfrey, 2007). The primary countries producing safflower are Kazakhstan, Russia, Mexico, the United States, and India (foa.org, accessed 13 May 2022). Safflower oil is used primarily in human consumption markets (Berglund et al., 1998). Safflower meal is also relatively high in fiber content (51.0–59.8% NDF and 36.1–44.1% ADF; NASEM, 2021; see Table 2). There is little research on feeding safflower meal in dairy cattle diets. Like canola meal, safflower meal contains two glucosinolates, Matairesinol-β-glucoside and 2-hydroxyacrtiin-β-glucoside, which can be bitter and unpalatable for cattle (Jin et al., 2010; NASEM, 2021).

5 Rendered protein meals

Humans do not consume up to 50% of animal products derived from animal food production. These animal products are not consumed because these animal by-products have unusual physical and nutrient characteristics (Ockerman and Hansen, 2000). Compared to oilseed meals, animal by-products supply more essential AA (NASEM, 2021).

5.1 Blood meal

Blood meal is a by-product concentrated in protein (92.0–98.3% CP) and RUP (75–85% of CP; NASEM, 2021). According to a survey of dairy nutritionists, blood meal ranked among the top six most common protein sources. The other protein sources were rumen-protected methionine, rumen-protected SBM, SBM, canola meal, and DDGS (Prestegaard-Wilson, 2021). It is produced from clean, fresh cattle, pig, or poultry blood, exclusive of all extraneous materials, including hair, ingesta, and urine, except for unavoidable circumstances in good manufacturing processes (AAFCO, 2021). Production of blood meal starts with a coagulation process followed by cooking to remove excess moisture, followed by drying to bring the blood meal to the desired DM and yield a granular product (Ockerman and Hansen, 2000). The process involves decantation, cooking, pressing, drying, and grinding (Almeida et al., 2013). Blood is processed and dried in several ways, including conventional cooker, flash, or spray-dried (AAFCO, 2021). Limited amounts of blood meal are produced using conventional cooker drying methods, which involve cooking, steaming, or hydrolyzation. For the flash-dried method, moisture is removed through a mechanical dewatering process or by condensing by cooking to produce a semi-solid product (AAFCO, 2021). Then, the semi-solid blood meal is transferred to a rapid drying process using either drum, roller, or ring driers (Almeida et al., 2013; AAFCO, 2021). The spray-dried process removes moisture at low temperatures until the product reaches approximately 30% solids. The semi-solid material is sprayed into a draft of warm, dry air to finish the drying process (AAFCO, 2021). Blood meal may be deficient in sulfur-containing AAs (Klemesrud et al., 2000) but is a valuable source of lysine and histidine (Ockerman and Hansen, 2000), containing approximately 8.8% and 6.0% of CP as lysine and histidine, respectively (NASEM, 2021). The drying process may affect protein and AA availability (Almeida et al., 2013), specifically lysine, due to its ε-amino group being susceptible to heat damage via the formation of Maillard reaction products (Choi et al., 1949; Kim et al., 2012).

5.2 Feather meal

Feather meal is a concentrated RUP by-product (86.9–94.4% CP; 72% of CP as RUP) from the poultry industry (Goedeken et al., 1990). Research on several methods of processing feather meal showed various methods might cause differences in the nutrient composition and availability. The process generally utilizes steam, pressure, and agitation to hydrolyze and break down the keratinous feathers. This process lasts from several minutes up to an hour. Then the feathers are dried and ground to produce feather meal. Feather meal is concentrated in CP, cystine, threonine, and arginine but contains less methionine, lysine, histidine, and tryptophan (El Boushy et al., 1990). Therefore, feeding feather meal as a single source of RUP results in poorer protein efficiency than that of blood meal (Goedeken et al., 1990). The blending of feather meal and blood meal improves the AA composition of the blend, mainly supplying additional lysine (Goedeken et al., 1990).

6 Environmental impact of feeding by-products

Utilizing by-products in ruminant diets is an efficient use of secondary product streams. Regarding environmental impact, there are two essential advantages to feeding by-products: (1) to reduce the use of human-edible grains used in livestock diets and (2) to prevent additional waste in landfills (Grasser et al., 1995). Livestock are an important segment of the lifecycle of human consumption markets. Ruminants contribute significantly to sustainable food systems due to their ability to convert human inedible by-products into high-quality proteins (i.e. meat and milk) that would otherwise need to be disposed of (Oltjen and Beckett, 1996). All by-products from the agriculture and food industries account for approximately 30% of global agriculture production (Ajila et al., 2012). Steinfeld et al. (2006) described the environmental impacts of livestock production are related to emissions of greenhouse gasses, nitrogen and phosphorus, land degradation, water pollution, and biodiversity losses. Production, transport, and utilization of livestock feed significantly influence these environmental impacts (Gerber et al., 2013). Russomanno et al. (2010) modeled by-products, namely corn gluten feed, distillers grains, SBM, whole cottonseed, and canola meal, as feed for dairy cattle compared to if the same by-products were combusted. The authors reported that by recycling carbon in by-products by feeding dairy cattle when accounting for animal tissue requirements, milk yield, and fecal excretion, the amount of carbon dioxide excreted was more than fivefold less than levels emitted if the same carbon was combusted. Similarly, total greenhouse gas emissions, including methane expressed as carbon dioxide equivalence, were less when by-products were fed to dairy cattle than combusted (Russomanno et al., 2010). Lee et al. (2003) estimated oilseed meals to have less potential methane production than bran, hulls, and grains. The methane potential was reported in ascending order from canola meal < SBM < sunflower meal < cottonseed meal (Lee et al., 2003). Reduced methane production has been observed with feeding distiller grains (Salami et al., 2019). However, this may be due to the fat content of distillers grains, as recent research has not reported reductions in methane production with distillers grains with less fat content (Salami et al., 2019).

Understanding the impacts of feeding protein and the consequences of over-feeding protein on nutrient runoff is important for the dairy industry and manufacturing industries producing by-products. Arriola Apelo et al. (2014) estimated that in the United States, 1.3 million tonnes of reactive nitrogen in the form of nitrous oxide, ammonia, and nitrates are excreted by dairy cattle every year. Many states have mandated manure nutrient testing as part of their nutrient management plan in many concentrated animal feeding operations (Hellerstein et al., 2019). The use of ration formulation models (i.e. Cornell Net Carbohydrate and Protein System, NASEM, 2021) utilizing the MP system, which was previously discussed in the section of this chapter on protein quality and digestibility, rather than the CP system has improved the conversion of feed protein into milk protein. This concept is commonly referred to as nitrogen use efficiency (i.e. Feed into Milk) and has led to decreased nitrogen emissions into the environment (Thomas, 2004; Haque et al., 2012; Sinclair et al., 2014).

Circular food systems, or circular agriculture, focus on converting by-products from the global food system and grass resources into valuable food and manure. Producing livestock or milk in this circular system would shift from the current linear food production systems in which crops are grown as feedstuffs for livestock. The circular system would potentially close the production loop and reduce resource consumption and environmental emissions. In the circular system, methods to accomplish this paradigm shift include minimizing material losses by recovering and recycling nutrients, searching for practices and technologies that minimize the use of finite resources (i.e. phosphate), making more use of regenerative energy (i.e. wind and solar), preventing leakage of natural resources from food systems (i.e. nitrogen and phosphorus), and improving the recycling of inevitable resource losses (i.e. feces) to add value to the food system (Van Zantan et al., 2019; see Fig. 4).

In circular systems, the priority for by-product use is first for human food, if possible (i.e. wheat middlings for use in muesli, an oatmeal dish). Then the value of by-products as animal feed is considered. As discussed in this chapter, many products have value as animal feed. However, the general philosophy of circular food systems suggests by-products with low nutritional value should be considered for use as fertilizer and soil health rather than livestock feed (Van Zanten et al., 2019).

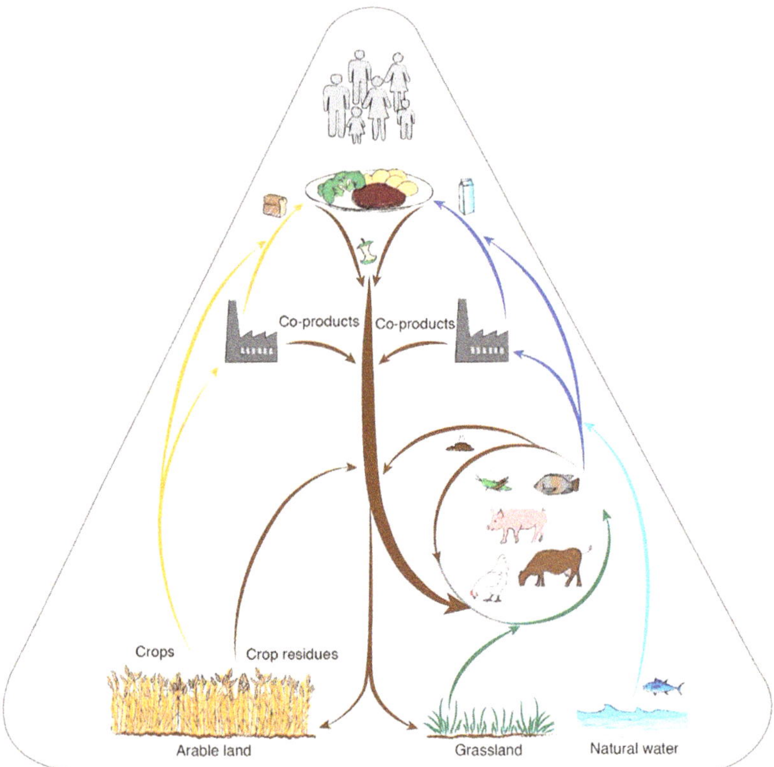

Figure 4 The concept of circular food systems: arable land is primarily used for food production; remaining inarable land unsuited for human consumption is used as animal feed; and by-products or co-production are used to maintain animals and soil fertility. Source: Van Zantan et al. (2019).

Circular food systems will require new metrics for assessing environmental footprints to address the feed vs. food vs. fuel debate. For example, using animal manure as fertilizer is not always more resource use efficient than using chemical fertilizers (Bhatt et al., 2019). However, relative to circular food systems, it is favorable to use manure rather than chemical fertilizers. Therefore, metrics are needed to address these economic and environmental impacts (Van Zanten et al., 2019).

Currently, in The Netherlands, circular food systems are being developed to address sustainability issues within livestock production by developing alternative feeds from food waste and production of alternative plant-based feed sources (i.e. seaweed, microalgae, and locally produced soybeans). Their goals are that circular agriculture will stop and prevent environmental impacts from food waste, use natural resources efficiently, and have fair and accessible options for livestock producers while maintaining animal welfare and wellbeing (Puente-Rodriguez et al., 2022). According to De Boer and van Ittersum (2018), 'plant biomass is the basic building block for food and should be used by humans first. By-products from food production, processing, and consumption should be recycled back into the food system. Use animals for what they are good at, transforming human inedible plants and organic matter into high quality human edible proteins.' The Netherlands are focused on developing and expanding seaweed cultivation to reduce the arable land use for livestock feed production, increase biomass production from marginal lands (i.e. riparian areas, ditches, and roadsides) to expand local grazing resources, decrease their dependency on imported protein feeds by expanding local production of soybeans and other proteins feedstuffs used in dairy cow diets (Puente-Rodriguez et al., 2022).

7 Future trends in research

Makkar and Ankers (2014) outlined the framework of sustainable animal diets must be balanced in all nutrients, free from harmful components, meet the operation's production objectives, and generate animal products safe for human consumption. Ruminants have an important role in recycling and converting non-edible food and by-products into nutritional food (Van Soest, 1994). Greater awareness of the sustainability of production practices in animal agriculture and feed manufacturing industries drives the need to develop 'new' and alternative protein feeds.

7.1 Microalgae

Growing phototrophic algae and feeding algae to livestock is not a novel idea (Hintz et al., 1966; Lum et al., 2013). Recently, microalgae production has experienced rapid growth and continues to receive renewed interest as a sustainable protein feed resource due to its yield potential per area of land use, estimated less energy use and demand leading to reduced greenhouse gas emissions, and ability to be cultivated on marginal or non-arable land compared to other crops (Walsh et al., 2015; Lamminen et al., 2019a). Several industrial applications have found uses for growing microalgae, including food formulation, livestock feed, cosmetics, health products, fertilizers, wastewater treatment, and biofuel production (Camacho et al., 2019). However, current production costs inhibit the rapid expansion of this feed resource but could change in the future with advancements in techniques and governmental policy (Lamminen et al., 2019a). For example, the production and commercialization of heterotrophic algae grown in dark fermenters for biofuels and oil have increased in recent years, leading to a demand for their large-scale production (Van Emon et al., 2015; Norman, 2021). The nutrient composition of microalgae varies greatly with variation in AA content (Lamminen et al., 2019a) and protein degradability (Costa et al., 2016). Microalgae species (*Spirulina platensis* and *Chlorella vulgaris*) have been recently examined as alternative protein sources in dairy diets. The protein content for *spirulina* and *chlorella* is around 70% and 50–60%, respectively, with crude fat contents below 1% (Lamminen et al., 2019a). *Spirulina* and *chlorella* microalgae are valuable sources of methionine but poor in histidine. However, the remaining AA profile is comparable to SBM (Lamminen et al., 2019a).

The inclusion of microalgae in ruminant diets can reduce DMI due to relatively poor palatability (Hintz et al., 1966; Van Emon et al., 2015; Lamminen et al., 2019b). Replacing rapeseed meal with microalgae (*spirulina*) affected the composition of DMI with reduced intake of the microalgae but did not affect overall DMI (Lamminen et al., 2017). In the same study, feeding microalgae reduced milk yield, protein yield, and nitrogen utilization efficiency, indicating reduced protein value than feeding rapeseed meal. Similarly, replacing SBM with *spirulina* and *chlorella* did not reduce DMI or milk yield and increased milk fat percentage (Lamminen et al., 2019a). While certain microalgae species may be useful as a protein source in ruminant diets, issues with palatability and DMI must be addressed before it becomes a widely adopted feedstuff in the dairy industry.

7.2 Hemp

In the United States, feeding hemp by-products still requires approval from the U.S. Food and Drug Administration and will be done on an individual by-product request (FDA, 2020). This is primarily due to limited research on the bioavailability of specific compounds in cannabis, like tetrahydrocannabinol (THC) and cannabidiol (CBD) in ruminants (EFSA-FEEDAP, 2011). Hemp

(*Cannabis sativa* L.) by-products include seed, oil, oilseed cake, hulls, and leaves (Semwogerere et al., 2020). Currently, in the United States, the hemp industry has grown exponentially and is currently an 824 million-dollar (USD) industry with more than 21 000 hectares planted in 2021 (NASS, 2022). The physical and nutrient composition of hemp by-products largely depends on seed variety, pressing, and seed treatment methods (Semwogerere et al., 2020). Hemp oilseed cake by-product, referred to as hempseed cake, is relatively concentrated in CP (29.1–39.1%) with an AA profile comparable to SBM with the exception of poorer lysine and tryptophan content but similar in sulfur-containing AAs (Callaway et al., 2004; Semwogerere et al., 2020; Bailoni et al., 2021). The ruminal degradation of hempseed cake appears to be relatively low (77.4% of CP as RUP; Mustafa et al., 1999). Like many oilseed meals, mechanical extraction results in more oil content in the hempseed cake than that of solvent extraction methods (Bailoni et al., 2021). The cannabinoids, THC and CBD, are condensed into hempseed cake during the oil extraction process. Therefore, cleaning and de-hulling of seed are required to minimize the transfer of THC and CBD to the oilseed cake (Semwogerere et al., 2020). Some anti-nutritional compounds that reduce protein absorption and other nutrients are found in hempseed (Bailoni et al., 2021), with phytate (22.5 mg/g) and glucosinolates (3.8 µmol/g) being most abundant (Pojić et al., 2014).

There is limited data on feeding hempseed cake to dairy cows (Bailoni et al., 2021). Karlsson et al. (2010) is the only published data feeding hempseed cake to lactating dairy cows. The authors reported that the inclusion of hempseed cake at levels up to 31.8% of the diet (DM) had no effect on DMI but did quadratically increase milk yield, protein yield, and fat yield. The curvilinear response indicated the maximum production was achieved at 14% inclusion of hempseed cake (Karlsson et al., 2010). Further research is needed to establish appropriate inclusion levels of hempseed cake and the potential effects on milk quality.

8 Where to look for further information

8.1 Further reading

- A brief description of by-products commonly used in dairy cattle can be found in Chapter 15 of 'Nutrient Requirements of Dairy Cattle' National Academies of Sciences, Engineering, and Medicine (2021), Eighth Revised Edition, Washington, DC: The National Academies Press.
- A thorough description and update on ethanol production by-products is Chapter 23: Nutritional properties and feeding value of corn and its co-products by Loy and Lundy from 'Corn: Chemistry and Technology,' 3rd ed, by Serna-Saldivar.

- An in-depth review of the sustainability of by-products is Salami et al. (2019).
- An update on all subjects related to distillers grains is 'Distillers Grains: Production, Properties, and Utilization' by Liu and Rosentrater, 1st ed., Boca Raton, US: CRC Press.
- A recent survey on various aspects of the dairy industry from a nutritionist's perspective is Prestegaard-Wilson et al. (2021).
- A journal article explaining circular food systems by Van Zanten et al.

8.2 Key journals and conferences

- The Journal of Dairy Science is the premier journal for dairy research.
- American Dairy Science Association (ADSA) Annual Meeting provides national and international research updates.
- The European Federation of Animal Science (EAAP) hosts a bi-annual meeting called the 'International Symposium on Energy and Protein Metabolism and Nutrition (ISEP)' that can provide great insight into current dairy nutrition trends.

9 References

AAFCO. (2021). *Official Publication*. The Association of American Feed Control Officials. Available at: https://www.aafco.org.

Abdelqader, M. M., Hippen, A. R., Kalscheur, K. F., Schingoethe, D. J. and Garcia, A. D. (2009). Isolipidic additions of fat from corn germ, corn distillers grains, or corn oil in dairy cow diets, *J. Dairy Sci.* 92(11):5523–5533. doi:10.3168/jds.2008-1867.

Ajila, C. M., Brar, S. K., Verma, M., Tyagi, R. D., Godbout, S. and Valéro, J. R. (2012). Bio-processing of agro-byproducts to animal feed, *Crit. Rev. Biotechnol.* 32(4):382–400. doi:10.3109/07388551.2012.659172.

Almeida, F. N., Htoo, J. K., Thomson, J. and Stein, H. H. (2013). Comparative amino acid digestibility in US blood products fed to weanling pigs, *Anim. Feed Sci. Technol.* 181(1–4):80–86. doi:10.1016/j.anifeedsci.2013.03.002.

Arija, I., Brenes, A., Viveros, A. and Elites, R. (1998). Effects of inclusion of full-fat sunflower kernels and hulls in diets for growing broiler chickens, *Anim. Feed Sci. Technol.* 70(1–2):137–149. doi:10.1016/S0377-8401(97)00077-1.

Arriola Apelo, S. I., Knapp, J. R. and Hanigan, M. D. (2014). Invited review: Current representation and future trends of predicting amino acid utilization in the lactating dairy cow, *J. Dairy Sci.* 97(7):4000–4017. doi:10.3168/jds.2013-7392.

Awawdeh, M. S., Titgemeyer, E. C., Drouillard, J. S., Beyer, R. S. and Shirley, J. E. (2007). Ruminal degradability and lysine bioavailability of soybean meals and effects on performance of dairy cows, *J. Dairy Sci.* 90(10):4740–4753. doi:10.3168/jds.2007-0210.

Bailoni, L., Bacchin, E., Trocino, A. and Arango, S. (2021). Hemp (*Cannabis sativa l.*) seed and coproducts inclusion in diets for dairy ruminants: A review, *Animals (Basel)* 11(3):856–867. doi:10.3390/ani11030856.

Balch, C. C. (1958). Observations on the act of eating in cattle, *Br. J. Nutr.* 12(3):330–345. doi:10.1079/BJN19580044.

Bath, D. L., Dunbar, J. R., King, J. M., Berry, S. C. and Olbrich, S. L. (1993). By-products and unusual feedstuffs, *Feedstuffs* 65:32–37.

Bauman, D. E. and Griinari, J. M. (2003). Nutritional regulation of milk fat synthesis, *Annu. Rev. Nutr.* 23(1):203–227. doi:10.1146/annurev.nutr.23.011702.073408.

Bell, J. M. (1993). Factors affecting the nutritional value of canola meal: A review, *Can. J. Anim. Sci.* 73(4):689–697. doi:10.4141/cjas93-075.

Belyea, R. L., Rausch, K. D., Clevenger, T. E., Singh, V., Johnston, D. B. and Tumbleson, M. E. (2010). Sources of variation in composition of DDGS, *Anim. Feed Sci. Technol.* 159(3–4):122–130. doi:10.1016/j.anifeedsci.2010.06.005.

Benchaar, C., Hassanat, F., Gervais, R., Chouinard, P. Y., Julien, C., Petit, H. V. and Massé, D. I. (2013). Effects of increasing amounts of corn dried distillers grains with solubles in dairy cow diets on methane production, ruminal fermentation, digestion, N balance, and milk production, *J. Dairy Sci.* 96(4):2413–2427. doi:10.3168/jds.2012-6037.

Bendall, J. R. (1964). Meat proteins. In: Schultz, H. W. (Ed.), *Symposium on Foods: Proteins and Their Reactions*. Westport, CT: AVI Publ. Co., Inc., p. 255.

Berglund, R. D., Riveland, N. and Bergman, J. (1998). Safflower production, *North Dakota State Univ. Extension Serv.* A-870.

Bhargavi, G., Nageswara Rao, P. and Renganathan, S. (2018). Review of the extraction methods of crude oil from all generation biofuels in last few decades. *IOP Conf. Ser.: Mater. Sci. Eng.* 330 012024. doi:10.1088/1757-899X/330/1/012024.

Bhatt, M. K., Labanya, R. and Joshi, H. C. (2019). Influence of long-term chemical fertilizers and organic manures on soil fertility - A review, *Univ. J. Agric. Res.* 7(5):177–188. doi:10.13189/ujar.2019.070502.

Björck, I. and Asp, N. G. (1983). The effects of extrusion cooking on nutritional value - A literature review. *Journal of Food Engineering*, 2(4): pp. 281-308. doi:10.1016/0260-8774(83)90016-X.

Blair, R. and Reichert, R. D. (1984). Carbohydrate and phenolic constituents in a comprehensive range of rapeseed and canola fractions: Nutritional significance for animals, *J. Sci. Food Agric.* 35(1):29–35. doi:10.1002/jsfa.2740350106.

Bothast, R. J. and Schlicher, M. A. (2005). Biotechnical processes for conversion of corn into ethanol, *App. Microbiol.* 67:19–25. doi:10.1007/s00253-004-1819-8.

Böttger, C. and Südekum, K.-H. (2018). Review: Protein value of distillers dried grains with solubles (DDGS) in animal nutrition as affected by the ethanol production process, *Anim. Feed Sci. Technol.* 244:11–17. doi:10.1016/j.anifeedsci.2018.07.018.

Bradford, B. J. and Mullins, C. R. (2012). Invited review: Strategies for promoting productivity and health of dairy cattle by feeding nonforage fiber sources, *J. Dairy Sci.* 95(9):4735–4746. doi:10.3168/jds.2012-5393.

Brito, A. F. and Broderick, G. A. (2007). Effects of different protein supplements on milk production and nutrient utilization in lactating dairy cows, *J. Dairy Sci.* 90(4):1816–1827. doi:10.3168/jds.2006-558.

Broderick, G. A., Colombini, S., Costa, S., Karsli, M. A. and Faciola, A. P. (2016). Chemical and ruminal in vitro evaluation of Canadian canola meals produced over 4 years, *J. Dairy Sci.* 99(10):7956–7970. doi:10.3168/jds.2016-11000.

Buckner, C. D., Wilken, M. F., Benton, J. R., Vanness, S. J., Bremer, V. R., Klopfenstein, T. J., Kononoff, P. J. and Erickson, G. E. (2011). Nutrient variability for distillers grains plus solubles and dry matter determination of ethanol by-products, *Prof. Anim. Sci.* 27(1):57–64. doi:10.15232/S1080-7446(15)30445-9.

Callaway, J. C. (2004). Hempseed as a nutritional resource: An overview, *Euphytica* 140(1-2):65–72. doi:10.1007/s10681-004-4811-6.

Camacho, F., Macedo, A. and Malcata, F. (2019). Potential industrial applications and commercialization of microalgae in the functional food and feed industries: a short review. *Mar Drugs*. 17:312–337. doi:10.3390/md17060312.

Canola Council of Canada. (2022). Steps in oil and meal processing. Accessed on May 14, 2022. Available at: https://www.canolacouncil.org/oil-and-meal/what-is-canola/how-canola-is-processed/steps-in-oil-and-meal-processing.

Cantrell, D. F. and Winsness, D. J. (2009). *Method Process: Ethanol Byproducts Relat. Subsystems*, U.S. Patent No. 7,601,858.

Castillo-Lopez, E., Ramirez Ramirez, H. A., Klopfenstein, T. J., Hostetler, D., Karges, K., Fernando, S. C. and Kononoff, P. J. (2014). Ration formulations containing reduced-fat dried distillers grains with solubles and their effect on lactation performance, rumen fermentation, and intestinal flow of microbial nitrogen in Holstein cows, *J. Dairy Sci.* 97(3):1578–1593. doi:10.3168/jds.2013-6865.

Chandler, P. T. (1989). Achievement of optimum amino acid balance possible, *Feedstuffs* 61(26):14–25.

Choi, R. P., Koncus, A. F., O'Malley, C. M. and Fairbanks, B. W. (1949). A proposed method for the determination of color of dry products of milk, *J. Dairy Sci.* 32(6):580–586. doi:10.3168/jds.S0022-0302(49)92084-6.

Clark, J. H., Murphy, M. R. and Crooker, B. A. (1987). Supplying the protein needs of dairy cattle from by-product feeds, *J. Dairy Sci.* 70(5):1092–1109. doi:10.3168/jds.S0022-0302(87)80116-9.

Coppock, C. E., Lanham, J. K. and Horner, J. L. (1987). A review of the nutritive value and utilization of whole cottonseed, cottonseed meal and associated by-products by dairy cattle, *Anim. Feed Sci. Technol.* 18(2):89–129. doi:10.1016/0377-8401(87)90041-1.

Corrigan, M. E., Klopfenstein, T. J., Erickson, G. E., Meyer, N. F., vander Pol, K. J., Greenquist, M. A., Luebbe, M. K., Karges, K. K. and Gibson, M. L. (2009). Effects of level of condensed distillers solubles in corn dried distillers grains on intake, daily body weight gain, and digestibility in growing steers fed forage diets1, *J. Anim. Sci.* 87(12):4073–4081. doi:10.2527/jas.2009-1969.

Costa, D. F. A., Quigley, S. P., Isherwood, P., McLennan, S. R. and Poppi, D. P. (2016). Supplementation of cattle fed tropical grasses with microalgae increases microbial protein production and average daily gain, *J. Anim. Sci.* 94(5):2047–2058. doi:10.2527/jas.2016-0292.

De Boer, I. J. M. and van Ittersum, M. K. (2018). *Circularity in Agricultural Production*. Wageningen University & Research, Wageningen, The Netherlands.

Depenbusch, B. E., Loe, E. R., Quinn, M. J., Corrigan, M. E., Gibson, M. L., Karges, K. K. and Drouillard, J. S. (2008). Corn distillers grains with solubles derived from a traditional or partial fractionation process: Growth performance and carcass characteristics of finishing feedlot heifers, *J. Anim. Sci.* 86(9):2338–2343. doi:10.2527/jas.2007-0501.

Droppo, T. E., Macleod, G. K. and Grieve, D. G. (1985). Composition and storage characteristics of wet corn gluten feed. *Can. J. Anim. Sci.* 65(1):265–268. doi:10.4141/cjas85-032.

DuFour, E. I. (2017). *Advancing chemical characterization of feedstuffs commonly included in dairy cow rations*. Master's Thesis, University of Nebraska, Lincoln.

Eastbridge, M. L. (2006). Major advances in applied dairy cattle nutrition, *J. Dairy Sci.* 89(4):1311–1323. doi:10.3168/jds.S0022-0302(06)72199-3.

EFSA-FEEDAP (2011). Scientific Opinion on the safety of hemp (Cannabis genus), *Europ. Food Safety Author. J.* 9(3). doi:10.2903/j.efsa.2011.2011.

El Boushy, A. R., van der Poel, A. F. B. and Walraven, O. E. D. (1990). Feather meal - A biological waste: Its processing and utilization as a feedstuff for poultry, *Biol. Wastes* 32(1):39–74. doi:10.1016/0269-7483(90)90071-Y.

Erasmus, L. J., Botha, P. M., Cruywagen, C. W. and Meissner, H. H. (1994). Amino acid profile and intestinal digestibility in dairy cows of rumen-undegradable protein from various feedstuffs, *J. Dairy Sci.* 77(2):541–551. doi:10.3168/jds.S0022-0302(94)76982-4.

Foth, A. J., Brown-Brandl, T., Hanford, K. J., Miller, P. S., Garcia Gomez, G. and Kononoff, P. J. (2015). Energy content of reduced-fat dried distillers grains with solubles for lactating dairy cows, *J. Dairy Sci.* 98(10):7142–7152. doi:10.3168/jds.2014-9226.

Gerber, P., Steinfeld, H., Henderson, B., Mottet, A., Opio, C., Dijkman, J., Falcucci, A. and Tempio, G. (2013). *Tackling Climate Change through Livestock – A Global Assessment of Emissions and Mitigation Opportunities.* Food and Agriculture Organization of the United Nations, Rome, Italy.

Giallongo, F., Oh, J., Frederick, T., Isenberg, B., Kniffen, D. M., Fabin, R. A. and Hristov, A. N. (2015). Extruded soybean meal increased feed intake and milk production in dairy cows, *J. Dairy Sci.* 98(9):6471–6485. doi:10.3168/jds.2015-9786.

Godfrey, L. R. (2007). *Influence of method of processing and feeding level of safflower seeds on the performance of dairy cows.* Master's Thesis, Utah State University, Logan.

Goedeken, F. K., Klopfenstein, T. J., Stock, R. A., Britton, R. A. and Sindt, M. H. (1990). Protein value of feather meal for ruminants as affected by blood additions, *J. Anim. Sci.* 68(9):2936–2944. doi:10.2527/1990.6892936x.

Grasser, L. A., Fadel, J. G., Garnett, I. and DePeters, E. J. (1995). Quantity and economic importance of nine selected by-products used in California dairy rations, *J. Dairy Sci.* 78(4):962–971. doi:10.3168/jds.S0022-0302(95)76711-X.

Gunderson, S. L., Aguilar, A. A., Johnson, D. E. and Olson, J. D. (1988). Nutritional value of wet com gluten feed for sheep and lactating dairy cows, *J. Dairy Sci.* 71(5):1204–1210. doi:10.3168/jds.S0022-0302(88)79675-7.

Gupta, M. K. (2014). Sunflower oil: History, applications and trends, *Lipid Technol.* 26(11–12):260–263. doi:10.1002/lite.201400068.

Haque, M. N., Rulquin, H., Andrade, A., Faverdin, P., Peyraud, J. L. and Lemosquet, S. (2012). Milk protein synthesis in response to the provision of an "ideal" amino acid profile at 2 levels of metabolizable protein supply in dairy cows, *J. Dairy Sci.* 95(10):5876–5887. doi:10.3168/jds.2011-5230.

Harper, M. T., Oh, J., Melgar, A., Nedelkov, K., Räisänen, S., Chen, X., Martins, C. M. M. R., Young, M., Ott, T. L., Kniffen, D. M., Fabin, R. A. and Hristov, A. N. (2019). Production effects of feeding extruded soybean meal to early-lactation dairy cows, *J. Dairy Sci.* 102(10):8999–9016. doi:10.3168/jds.2019-16551.

Hatch, R. H. (1986). The nutrient bioavailability of distillers feeds: Bioavailability of nutrients in feedstuffs. In: *Proceedings of the National Feed Ingredients Association*. West, Des Moines, IA.

Heendeniya, R. G., Christensen, D. A., Maenz, D. D., McKinnon, J. J. and Yu, P. (2012). Protein fractionation by-product from canola meal for dairy cattle, *J. Dairy Sci.* 95(8):4488–4500. doi:10.3168/jds.2011-5029.

Hellerstein, D., Vilorio, D. and Ribaudo, M. (Eds.). (2019). *Agricultural Resources and Environmental Indicators*. United States Department of Agriculture Economic Research Service. Accessed on March 12, 2022.

Hintz, H. F., Heitman, H., Weir, W. C., Torell, D. T. and Meyer, J. H. (1966). Nutritive value of algae grown on sewage2, *J. Anim. Sci.* 25(3):675–681. doi:10.2527/jas1966.253675x.

Hodge, J. E. (1953). Dehydrated foods, chemistry of browning reactions in model systems, *J. Agric. Food Chem.* 1(15):928–943.

Hollmann, M., Allen, M. S. and Beede, D. K. (2011a). Diet fermentability influences lactational performance responses to corn distillers grains: A meta-analysis, *J. Dairy Sci.* 94(4):2007-2021. doi:10.3168/jds.2010-3711.

Hollmann, M., Allen, M. S. and Beede, D. K. (2011b). Dietary protein quality and quantity affect lactational responses to corn distillers grains: A meta-analysis, *J. Dairy Sci.* 94(4):2022–2030. doi:10.3168/jds.2010-3712.

Huhtanen, P., Hetta, M. and Swensson, C. (2011). Evaluation of canola meal as a protein supplement for dairy cows: A review and a meta-analysis, *Can. J. Anim. Sci.* 91(4):529–543. doi:10.4141/cjas2011-029.

Humer, E. and Zebeli, Q. (2015). Phytate in feed ingredients and potentials for improving the utilization of phosphorus in ruminant nutrition, *Anim. Feed Sci. Technol.* 209:1-15. doi:10.1016/j.anifeedsci.2015.07.028.

ICM. (2009). Oil recovery. ICM, Inc. Accessed March 7, 2022. Available at: www.icminc .com.

Isher, V. and Varga, G. (2016). Soybeans and soybean by-products for dairy cattle, *The Pennsylvania State University Cooperative Extension Fact Sheet DAS 00-3*.

Janicek, B. N., Kononoff, P. J., Gehman, A. M. and Doane, P. H. (2008). The effect of feeding dried distillers grains plus solubles on milk production and excretion of urinary purine derivatives, *J. Dairy Sci.* 91(9):3544–3553. doi:10.3168/jds.2007-0777.

Jiang, N., Zhang, A. Z., Yang, R. Q. and Zhang, Y. C. (2011). An experimental approach to optimize several processing conditions when extruding soybeans, *Anim. Feed Sci. Technol.* 170(3-4):277–283. doi:10.1016/j.anifeedsci.2011.09.005.

Jin, Q. Z., Zou, X. Q., Shan, L., Wang, X. G. and Qiu, A. Y. (2010). β-D-glucosidase-catalyzed deglucosidation of phenylpropanoid amides of 5-hydroxytryptamine glucoside in safflower seed extracts optimized by response surface methodology, *J. Agric. Food Chem.* 58(1):155–160. doi:10.1021/jf902623v.

Kajikawa, H., Miyazawa, K., Yanase, A., Tanabe, Y., Tsuchida, Y., Mitsumoto, Y., Kozato, Y. and Mitsumori, M. (2012). Variation in chemical composition of corn dried distillers grains with solubles in relation to in situ protein degradation profiles in the rumen, *Anim. Sci. J.* 83(4):299–304. doi:10.1111/j.1740-0929.2011.00956.x.

Kalscheur, K. F. (2005). Impact of feeding distillers grains on milk fat, protein, and yield. In: *Proceedings of the Distillers Grains Technology Council 9th Annual Symposium*, Louisville, KY.

Kalscheur, K. F., Hippen, A. R. and Garcia, A. D. (2016). *Distillers Grains: Production, Properties, and Utilization* (1st edn.). Edited by Liu, K. and Rosentrater, K. A. CRC Press, Boca Raton, pp. 265–296.

Karlsson, L., Finell, M. and Martinsson, K. (2010). Effects of increasing amounts of hempseed cake in the diet of dairy cows on the production and composition of milk, *Animal* 4(11):1854-1860. doi:10.1017/S1751731110001254.

Karr-Lilienthal, L. K., Bauer, L. L., Utterback, P. L., Zinn, K. E., Frazier, R. L., Parsons, C. M. and Fahey, G. C. (2006). Chemical composition and nutritional quality of soybean meals prepared by extruder/expeller processing for use in poultry diets, *J. Agric. Food Chem.* 54(21):8108–8114. doi:10.1021/jf061425t.

Kim, B. G., Kil, D. Y., Zhang, Y. and Stein, H. H. (2012). Concentrations of analyzed or reactive lysine, but not crude protein, may predict the concentration of digestible lysine in distillers dried grains with solubles fed to pigs, *J. Anim. Sci.* 90(11):3798-3808. doi:10.2527/jas.2011-4692.

Kim, Y., Mosier, N. S., Hendrickson, R., Ezeji, T., Blaschek, H., Dien, B., Cotta, M., Dale, B. and Ladisch, M. R. (2008). Composition of corn dry-grind ethanol by-products: DDGS, wet cake, and thin stillage, *Bioresour. Technol.* 99(12):5165–5176. doi:10.1016/j.biortech.2007.09.028.

Kirkbride McElroy, A. (2007). Corn oil extraction opens new markets, *Distillers Grains Q.* 2(1):7–8.

Kleinschmit, D. H., Anderson, J. L., Schingoethe, D. J., Kalscheur, K. F. and Hippen, A. R. (2007). Ruminal and intestinal degradability of distillers grains plus solubles varies by source, *J. Dairy Sci.* 90(6):2909–2918. doi:10.3168/jds.2006-613.

Klemesrud, M. J., Klopfenstein, T. J. and Lewis, A. J. (2000). Evaluation of feather meal as a source of sulfur amino acids for growing steers, *J. Anim. Sci.* 78(1):207–215. doi:10.2527/2000.781207x.

Kononoff, P. J., Ivan, S. K., Matzke, W., Grant, R. J., Stock, R. A. and Klopfenstein, T. J. (2006). Milk production of dairy cows fed wet corn gluten feed during the dry period and lactation, *J. Dairy Sci.* 89(7):2608–2617. doi:10.3168/jds.S0022-0302(06)72338-4.

Kurambhatti, C., Kumar, D. and Singh, V. (2019). Impact of fractionation process on the technical and economic viability of corn dry grind ethanol process, *Processes* 7(9):578. doi:10.3390/pr7090578.

Lage, C. F. A., Räisänen, S. E., Stefenoni, H., Melgar, A., Chen, X., Oh, J., Kniffen, D. M., Fabin, R. A. and Hristov, A. N. (2020). Lactational performance, enteric gas emissions, and plasma amino acid profile of dairy cows fed diets with soybean or canola meals included on an equal protein basis, *J. Dairy Sci.* 104(3):3052–3066. doi:10.3168/jds.2020-18851.

Lamminen, M., Halmemies-Beauchet-Filleau, A., Kokkonen, T., Jaakkola, S. and Vanhatalo, A. (2019a). Different microalgae species as a substitutive protein feed for soya bean meal in grass silage based dairy cow diets, *Anim. Feed Sci. Technol.* 247:112-126. doi:10.1016/j.anifeedsci.2018.11.005.

Lamminen, M., Halmemies-Beauchet-Filleau, A., Kokkonen, T., Vanhatalo, A. and Jaakkola, S. (2019b). The effect of partial substitution of rapeseed meal and faba beans by *Spirulina platensis* microalgae on milk production, nitrogen utilization, and amino acid metabolism of lactating dairy cows, *J. Dairy Sci.* 102(8):7102–7117. doi:10.3168/jds.2018-16213.

Lamminen, M., Halmemies-Beauchet-Filleau, A., Kokkonen, T., Simpura, I., Jaakkola, S. and Vanhatalo, A. (2017). Comparison of microalgae and rapeseed meal as supplementary protein in the grass silage based nutrition of dairy cows, *Anim. Feed Sci. Technol.* 234:295–311. doi:10.1016/j.anifeedsci.2017.10.002.

Lardy, G. P. and Anderson, V. (2002). Canola and sunflower meal in beef cattle diets. *Vet Clin North Am Food Anim Pract.* 2002 July; 18(2):327–338. doi:10.1016/s0749-0720(02)00018-x. PMID: 12235664.

Le Clef, E. and Kemper, T. (2015). Sunflower seed preparation and oil extraction. In: Martinez-Force, E., Dunford, N. T. and Salas, J. J. (Eds.), *Sunflower: Chemistry, Production, Processing, and Utilization.* Elsevier Inc., pp. 187–226.

Lee, H. J., Lee, S. C., Kim, J. D., Oh, Y. G., Kim, B. K., Kim, C. W. and Kim, K. J. (2003). Methane production potential of feed ingredients as measured by in vitro gas test, *Asian. Australas. J. Anim. Sci.* 16(8):1143–1150.

Liu, K. (2011). Chemical composition of distillers grains, a review, *J. Agric. Food Chem.* 59(5):1508–1526. doi:10.1021/jf103512z.

Liu, K. and Rosentrater, K. A. (2016). *Distillers Grains: Production, Properties, and Utilization* (1st edn.). CRC Press, Boca Raton.

Liu, Y. G., Peng, H. H. and Schwab, C. G. (2013). Enhancing the productivity of dairy cows using amino acids, *Anim. Prod. Sci.* 53(11):1156–1159. doi:10.1071/AN13203.

Loy, D. D. and Lundy, E. L. (2018). *Nutritional Properties and Feeding Value of Corn and Its Co-products* (3rd edn.). Elsevier Inc.

Loy, D. D. and Wright, K. N. (2003). Nutritional properties and feeding value of corn and its by-products. In: White, P. J. and Johnson, L. A. (Eds.), *Innovation.* Corn Chemistry and Technology, American Association of Cereal Chemists, St Paul, MN.

Lum, K. K., Kim, J. and Lei, X. G. (2013). Dual potential of microalgae as a sustainable biofuel feedstock and animal feed, *J. Anim. Sci. Biotechnol.* 4(1):53. doi:10.1186/2049-1891-4-53.

Lundy, E. L. and Loy, D. (2014). Ethanol coproducts for beef cattle: Changing distillers grains, Iowa beef center, *Iowa State University Extension and Outreach Fact Sheet IBCR 200B.*

Makkar, H. P. S. and Ankers, P. (2014). Towards sustainable animal diets: A survey-based study, *Anim. Feed Sci. Tech.* 198: 309–322. doi:10.1016/j.anifeedsci.2014.09.018.

Martineau, R., Ouellet, D. R. and Lapierre, H. (2013). Feeding canola meal to dairy cows: A meta-analysis on lactational responses, *J. Dairy Sci.* 96(3):1701–1714. doi:10.3168/jds.2012-5740.

McCandlish, A. C. (1920). The digestibility of sorghum mill refuse, *J. Dairy Sci.* 3(5):367–369. doi:10.3168/jds.S0022-0302(20)94281-9.

Mjoun, K., Kalscheur, K. F., Hippen, A. R., Schingoethe, D. J. and Little, D. E. (2010). Lactation performance and amino acid utilization of cows fed increasing amounts of reduced-fat dried distillers grains with solubles, *J. Dairy Sci.* 93(1):288–303. doi:10.3168/jds.2009-2377.

Moore, S. A. E. and Kalscheur, K. F. (2016). Canola meal in dairy cow diets during early lactation increases production compared to soybean meal, *J. Dairy Sci.* 99:718–732. doi:10.2527/jam2016-1506.

Mottet, A., de Haan, C., Falcucci, A., Tempio, G., Opio, C. and Gerber, P. (2017). Livestock: On our plates or eating at our table? A new analysis of the feed/food debate, *Glob. Food Sec.* 14:1–8. doi:10.1016/j.gfs.2017.01.001.

Morris, D. L., Kim, S. H., Kononoff, P. J. and Lee, C. (2018). Continuous 11-week feeding of reduced-fat distillers grains with and without monensin reduces lactation performance of dairy cows, *J. Dairy Sci.* 101(7):5971–5983. doi:10.3168/jds.2017-14170.

Murthy, G. S., Singh, V., Johnston, D. B., Rausch, K. K. and Tumbleson, M. E. (2006). Evaluation and strategies to improve fermentation characteristics of modified dry-grind corn processes, *Cereal Chem.* 83(5):455–459. doi:10.1094/CC-83-0455.

Mustafa, A. F., McKinnon, J. J. and Christensen, D. A. (1999). The nutritive value of hemp meal for ruminants, *Can. J. Anim. Sci.* 79(1):91–95. doi:10.4141/A98-031.

National Academies of Sciences, Engineering, and Medicine. (2016). *Nutrient Requirements of Beef Cattle* (8th rev. edn.). The National Academies Press, Washington, DC.

National Academies of Sciences, Engineering, and Medicine. (2021). *Nutrient Requirements of Dairy Cattle* (8th rev. edn.). The National Academies Press, Washington, DC.

National Research Council. (1985). *Ruminant Nitrogen Usage*. The National Academies Press, Washington, DC.

National Research Council. (1989). *Nutrient Requirements of Dairy Cattle* (6th rev. edn.). The National Academies Press, Washington, DC.

National Research Council. (2001). *Nutrient Requirements of Dairy Cattle* (7th rev. edn.). The National Academies Press, Washington, DC.

National Research Council. (2012). *Nutrient Requirements of Swine* (11th rev. edn.). The National Academies Press, Washington, DC.

National Sunflower Association. (2022). Sunflower statistics. *USDA Report*. Accessed March 10, 2022.

Norman, M. M. (2021). *The role of fatty acids in ruminant diets and novel feed ingredients high in omega-3 fatty acids fed in feedlot diets*. Master's Thesis, University of Nebraska, Lincoln.

Ockerman, H. W. and Hansen, C. L. (2000). *Animal By-Product Processing and Utilization*. CRC Press, Boca Raton, FL.

Olentine, C. (1986). Ingredient profile: Distillers feeds. In: *Proceedings of the Distillers* Feed *Conference* 41:13–24.

Oliveira, A. S., Campos, J. M. S., Ogunade, I. M., Caixeta, D. S., Viana, E. P. and Alessi, K. C. (2018). Performance and utilization of nutrients in dairy cows fed with sunflower meal, *J. Agric. Sci.* 156(10):1233–1240.

Oltjen, J. W. and Beckett, J. L. (1996). Role of ruminant livestock in sustainable agricultural systems, *J. Anim. Sci.* 74(6):1406–1409. doi:10.2527/1996.7461406x.

Owens, F. N. and Goetsch, A. L. (1988). Ruminal fermentation. In: Church, D. C. (Ed.), *The Ruminant Animal: Digestive Physiology and Nutrition*. Prentice-Hall Inc., Englewood Cliffs, NJ, pp. 145–171.

Pantoja, J., Firkins, J. L., Eastridge, M. L. and Hull, B. L. (1994). Effects of fat saturation and source fiber on site of nutrient digestion and milk production by lactating dairy cows, *J. Dairy Sci.* 77(8):2341–2356.

Paz, H. A., Klopfenstein, T. J., Hostetler, D., Fernando, S. C., Castillo-Lopez, E. and Kononoff, P. J. (2014). Ruminal degradation and intestinal digestibility of protein and amino acids in high-protein feedstuffs commonly used in dairy diets, *J. Dairy Sci.* 97(10):6485–6498. doi:10.3168/jds.2014-8108.

Piepenbrink, M. S. and Schingoethe, D. J. (1998). Ruminal degradation, amino acid composition, and estimated intestinal digestibilities of four protein supplements, *J. Dairy Sci.* 81(2):454–461. doi:10.3168/jds.S0022-0302(98)75597-3.

Pojić, M., Mišan, A., Sakač, M., Dapčević Hadnađev, T., Šarić, B., Milovanović, I. and Hadnađev, M. (2014). Characterization of byproducts originating from hemp oil processing, *J. Agric. Food Chem.* 62(51):12436–12442. doi:10.1021/jf5044426.

Poos, M. I. (1981). Ruminant nutritional utilization of alcohol production by-products, *Feedstuffs* 53(26):19–21.

Prestegaard-Wilson, J. M., Daley, V. L., Drape, T. A. and Hanigan, M. D. (2021). A survey of United States dairy cattle nutritionists' practices and perceptions of reducing crude protein in lactating dairy cow diets, *Appl. Anim. Sci.* 37(6):697–709. doi:10.15232/aas.2021-02179.

Puente-Rodriguez, D., van Laar, H. and Veraart, M. (2022). A circularity evaluation of new feed categories in the Netherlands-Squaring the circle: A review, *Sustainability* 14(4):2352–2380. doi:10.3390/su14042352.

Rajagopalan, S., Ponnampalam, E., McCalla, D. and Stowers, M. (2005). Enhancing profitability of dry mill ethanol plants: Process modeling and economics of conversion of degermed defibered corn to ethanol, *Appl. Biochem. Biotechnol.* 120(1):37–50. doi:10.1385/ABAB:120:1:37.

Ramirez, E. C., Johnston, D. B., McAloon, A. J., Yee, W. and Singh, V. (2008). Engineering process and cost model for a conventional corn wet milling facility, *Ind. Crops Prod.* 27(1):91–97. doi:10.1016/j.indcrop.2007.08.002.

Ramirez-Ramirez, H. A., Castillo Lopez, E., Jenkins, C. J. R., Aluthge, N. D., Anderson, C., Fernando, S. C., Harvatine, K. J. and Kononoff, P. J. (2016). Reduced-fat dried distillers grains with solubles reduces the risk for milk fat depression and supports milk production and ruminal fermentation in dairy cows, *J. Dairy Sci.* 99(3):1912–1928. journaldoi:10.3168/jds.2015-9712.

Reiser, R. and Fu, H. C. (1962). The mechanism of gossypol detoxification by ruminant animals, *J. Nutr.* 76(2):215–218. doi:10.1093/jn/76.2.215.

Rodríguez, C. A., González, J., Alvir, M. R. and Caballero, R. (2008). Effects of feed intake on in situ rumen microbial contamination and degradation of feeds, *Livest. Sci.* 116(1–3):108–117. doi:10.1016/j.livsci.2007.09.013.

Rosentrater, K. A., Ileleji, K. and Johnston, D. B. (2016). Manufacturing of fuel ethanol and distillers grains-current and evolving processes. In: Liu, K. and Rosentrater, K. A. (Eds.), *Distillers Grains: Production, Properties, and Utilization* (1st edn.). CRC Press, Boca Raton, pp. 73–102.

Russomanno, K. L., Van Amburgh, M. E. and Higgs, R. J. (2012). Utilization of byproducts from human food production as feedstuffs for dairy cattle and relationships to greenhouse emissions and environmental efficiency. *Proc. Cornell Nutr. Conf. Feed Manufac.* pp. 130–145.

Salami, S. A., Luciano, G., Grady, M. N. O., Biondi, L., Newbold, C. J., Kerry, J. P. and Priolo, A. (2019). Sustainability of feeding plant by-products: A review of the implications for ruminant meat production, *Anim. Feed Sci. Technol.* 251:37–55. doi:10.1016/j.anifeedsci.2019.02.006.

Sánchez, J. M. and Claypool, D. W. (1983). Canola meal as a protein supplement in dairy rations, *J. Dairy Sci.* 66(1):80–85. doi:10.3168/jds.S0022-0302(83)81756-1.

Schingoethe, D. J., Kalscheur, K. F., Hippen, A. R. and Garcia, A. D. (2009). Invited review: The use of distillers products in dairy cattle diets, *J. Dairy Sci.* 92(12):5802–5813. doi:10.3168/jds.2009-2549.

Schingoethe, D. J., Rook, J. A. and Ludens, F. (1977). Evaluation of sunflower meal as a protein supplement for lactating cows, *J. Dairy Sci.* 60(4):591–595. doi:10.3168/jds.S0022-0302(77)83906-4.

Semwogerere, F., Katiyatiya, C. L. F., Chikwanha, O. C., Marufu, M. C. and Mapiye, C. (2020). Bioavailability and bioefficacy of hemp by-products in ruminant meat production and preservation: A review, *Front. Vet. Sci.* 7:572906. doi:10.3389/fvets.2020.572906.

Shad, Z. N., Venkitasamy, C. and Wen, Z. (2021). Corn distillers dried grains with solubles: Production, properties and potential uses, *Cereal Chem.* 98:1–21.

Sinclair, K. D., Garnsworthy, P. C., Mann, G. E. and Sinclair, L. A. (2014). Reducing dietary protein in dairy cow diets: Implications for nitrogen utilization, milk production, welfare and fertility, *Animal* 8(2):262–274. doi:10.1017/S1751731113002139.

Singh, V. and Eckhoff, S. R. (1997). Economics of germ pre-separation for dry-grind ethanol facilities, *Cereal Chem.* 74(4):462–466. doi:10.1094/CCHEM.1997.74.4.462.

Singh, V. and Johnston, D. B. (2009). Fractionation technologies for dry-grind corn processing. In: Ingledew, M. W., Kelsall, D. R., Austin, G. D. and Kluhspies, C. (Eds.), *The Alcohol Textbook* (5th edn.). Nottingham Univesity Press, Nottingham, UK, pp. 193–207.

Sniffin, C. J. (1974). Nitrogen utilization as related to solubility of NPN and protein in feeds. In: *Proceedings of the Cornell Nutrition Conference*, p. 12.

Solanas, E. M., Castrillo, C., Jover, M. and de Vega, A. (2008). Effect of extrusion on in situ ruminal protein degradability and in vitro digestibility of undegraded protein from different feedstuffs, *J. Sci. Food Agric.* 88(15):2589–2597. doi:10.1002/jsfa.3345.

Spiehs, M. J., Whitney, M. H. and Shurson, G. C. (2002). Nutrient database for distiller's grains with solubles produced from new ethanol plants in Minnesota and South Dakota, *J. Anim. Sci.* 80(10):2639–2645. doi:10.1093/ansci/80.10.2639.

Steinfeld, H., Gerber, P., Wassenaar, T., Castel, V., Rosales, M. and De Haan, C. (2006). *Livestock's Long Shadow*. Food and Agriculture Organization of the United Nations (FAO), Rome, Italy.

Stock, R. A., Lewis, J. M., Klopfenstein, T. J. and Milton, C. T. (2000). Review of new information on the use of wet and dry milling feed by-products in feedlot diets, *J. Anim. Sci.* 77(E-Suppl):1-12. doi:10.2527/jas2000.77E-Suppl1w.

Tamminga, S. (1979). Protein degradation in the forestomachs of ruminants, *J. Anim. Sci.* 49(6):1615–1630. doi:10.2527/jas1979.4961615x.

Thomas, C. (2004). *Feed into Milk: A New Applied Feeding System for Dairy Cows*. Nottingham University Press, Nottingham.

United States Department of Agriculture and Foreign Agricultural Service (FAO). Sunflower world production. Accessed May 12, 2022. Available at: https://ipad.fas.usda.gov/cropexplorer/cropview/commodityView.

United States Department of Agriculture and National Agricultural Statistics Service (NASS). (Released February 17, 2022). *National Hemp Report*. Accessed March 14, 2022.

United States Department of Energy. (2022). *Alternative Fuels Data Center*. Accessed January 17, 2022.

United States Food and Drug Administration (FDA). (2020). FDA regulation of cannabis and cannabis-derived products, including cannabidiol (CBD). US Food & Drug Administration, p. 18. Accessed on March 14, 2022. Available at: https://www.fda.gov/news-events/public-health-focus/fda-regulation-cannabis-and-cannabis-derived-products-including-cannabidiol-cbd#dietarysupplements%0A.

VanBaale, M. J., Shirley, J. E., Titgemeyer, E. C., Park, A. F., Meyer, M. J., Lindquist, R. U. and Ethington, R. T. (2001). Evaluation of wet corn gluten feed in diets for lactating dairy cows, *J. Dairy Sci.* 84(11):2478–2485. doi:10.3168/jds.S0022-0302(01)74698-X.

Van Emon, M. L., Loy, D. D. and Hansen, S. L. (2015). Determining the preference, in vitro digestibility, in situ disappearance, and grower period performance of steers fed a novel algae meal derived from heterotrophic microalgae, *J. Anim. Sci.* 93(6):3121-3129. doi:10.2527/jas.2014-8654.

Van Horn, H. H., Harris, B., Taylor, M. J., Bachman, K. C. and Wilcox, C. J. (1984). By-product feeds for lactating diary cows: Effects of cottonseed hulls, sunflower hulls, corrugated paper, peanut hulls, sugarcane bagasse, and whole cottonseed with additives of fat, sodium bicarbonate, and spergillus oryzae product on production. *J. Dairy Sci.* 67(12): 2922–2938. doi: 10.3168/jds.S0022-0302(84)81656-2.

Van Soest, P. J. (1994). *Nutritional Ecology of the Ruminant* (2nd edn.). Cornell University Press, Ithaca, NY.

Van Soest, P. J. and Mason, V. C. (1991). The influence of the Maillard reaction upon the nutritive value of fibrous feeds, *Anim. Feed Sci. Technol.* 32(1–3):45–53. doi:10.1016/0377-8401(91)90008-G.

Van Zantan, H. H. E., Van Ittersum, M. K. and Da Boer, I. J. M. (2019). The role of farm animals in a circular food system, *Glob. Food Sec.* 21(1):18–22. doi:10.1016/j.gfs.2019.06.003.

Wahjudi, J., Xu, L., Wang, P., Singh, V., Buriak, P., Rausch, K. D., McAloo, A. J., Tumbleson, M. E. and Eckho, S. R. (2000a). Quick fiber process: Effect of mash temperature, dry solids, and residual germ on fiber yield and purity, *Cereal Chem.* 77(5):640–644. doi:10.1094/CCHEM.2000.77.5.640.

Wahjudi, J., Xu, L., Wang, P., Buriak, P., Singh, V., Tumbleson, M., Rausch, R. D. and Eckhoff, S. (2000b). The "quick fiber" process: Effect of temperature, specific gravity, and percentage of residual germ, *Cereal Chem.* 77(5):640–644. doi:10.1094/CCHEM.2000.77.5.640.

Walsh, B. J., Rydzak, F., Palazzo, A., Kraxner, F., Herrero, M., Schenk, P. M., Ciais, P., Janssens, I. A., Peñuelas, J., Niederl-Schmidinger, A. and Obersteiner, M. (2015). New feed resources key to ambitious climate targets, *Carbon Balance Manag.* 10(1):26. doi:10.1186/s13021-015-0040-7.

Wang, P., Singh, V., Xu, L., Johnston, D. B., Rausch, K. D., McAloon, A. J. and Tumbleson, M. E. (2005). Comparison if enzymatic (e-mill) and conventional dry-grind corn processes using a granular starch hydrolyzing enzyme, *Cereal Chem.* 82(6):734–738. doi:10.1094/CC-82-0734.

Wang, P., Singh, V., Xue, H., Johnston, D. B., Rausch, K. D. and Tumbleson, M. E. (2007). Comparison of raw starch hydrolyzing enzyme with conventional liquefaction and saccharification enzymes in dry-grind corn processing, *Cereal Chem.* 84(1):10–14. doi:10.1094/CCHEM-84-1-0010.

White, R. R., Roman-Garcia, Y., Firkins, J. L., Kononoff, P. J., VandeHaar, M. J., Tran, H., McGill, T., Garnett, R. and Hanigan, M. D. (2017). Evaluation of the Nationa Research Council (2001) dairy model and derivation of new prediction equations. 2. Rumen degradable and undegradable protein, *J. Dairy Sci.* 100(5):3611–3627. doi:10.3168/jds.2015-10801.

Wickersham, E. E., Shirley, J. E., Titgemeyer, E. C., Brouk, M. J., DeFrain, J. M., Park, A. F., Johnson, D. E. and Ethington, R. T. (2004). Response of lactating dairy cows to diets containing wet corn gluten feed or a raw soybean hull-corn liquor pellet, *J. Dairy Sci.* 87(11):3899–3911. doi:10.3168/jds.S0022-0302(04)73529-8.

Winsness, D. J. (2006). Increase ethanol industry profits with corn oil extraction technology. *Ethanol Producer Magazine*, February, pp. 108–110.

Winsness, D. J. and Cantrell, D. F. (2009). Method of freeing the bound oil present in whole stillage and thin stillage. U.S. Patent No. 7,608,729.

Wohlt, J. E., Sniffen, C. J., Hoover, W. H., Johnson, L. L. and Walker, C. K. (1976). Nitrogen metabolism in wethers as affected by dietary protein solubility and amino acid profile, *J. Anim. Sci.* 42(5):1280–1289. doi:10.2527/jas1976.4251280x.
World Agriculture Outlook Board. (2020). *USDA Agricultural Projections to 2029, January 17, 2022*. United States Department of Agriculture, Washington, DC.
Woyenga, T. A., Beltranena, E. and Zijlstra, R. T. (2017). Effect of anti-nutritional factors of oilseed co-products on feed intake of pigs and poultry, *Anim. Feed Sci. Technol.* 233:76–86. doi:10.1016/j.anifeedsci.2016.05.006.
Yildiz, E., Todorov, N. and Nedelkov, K. (2015). Comparison of different dietary protein sources for dairy cows, *Bulg. J. Agric. Sci.* 21:199–208.

Index

www.ingramcontent.com/pod-product-compliance
Ingram Content Group UK Ltd.
Pitfield, Milton Keynes, MK11 3LW, UK
UKHW021825100926
8556IPUK00003B/161

9 781801 462051